Annual Meeting of the International Society of Exposure Science (ISES 2023)

Connecting Communities & Science through Addressing Environmental Exposures

Book of Abstracts

Chicago, Illinois, USA
27 – 31 August 2023

ISBN: 978-1-7138-9644-9

Printed from e-media with permission by:

Curran Associates, Inc.
57 Morehouse Lane
Red Hook, NY 12571

Some format issues inherent in the e-media version may also appear in this print version.

International Society of Exposure Science (ISES)
c/o Infinity Conference Group
1035 Sterling Road, Suite 202
Herndon VA 20170
USA

Phone: (703) 925-9455
Fax: (703) 925-9453

info@intlexposurescience.org

Additional copies of this publication are available from:

Curran Associates, Inc.
57 Morehouse Lane
Red Hook, NY 12571 USA
Phone: 845-758-0400
Fax: 845-758-2633
Email: curran@proceedings.com
Web: www.proceedings.com

ISES 2023 ANNUAL MEETING CHICAGO, IL

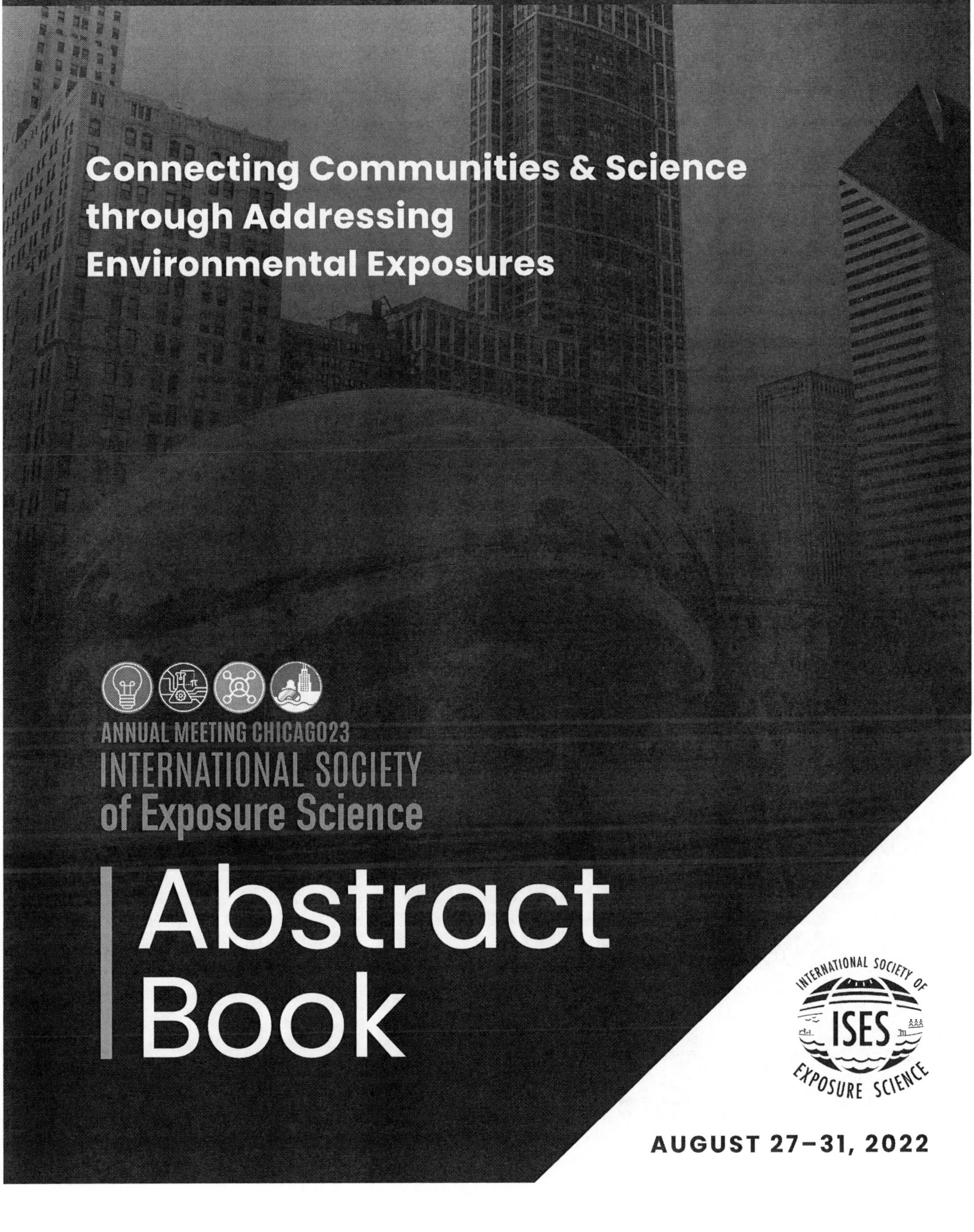

Connecting Communities & Science through Addressing Environmental Exposures

ANNUAL MEETING CHICAGO23
INTERNATIONAL SOCIETY of Exposure Science

Abstract Book

INTERNATIONAL SOCIETY OF EXPOSURE SCIENCE
ISES

AUGUST 27–31, 2022

ISES 2023
Annual Meeting

Chicago, IL USA
August 27–31, 2023

Connecting Communities & Sciences
through Addressing
Environmental Exposures

INTERNATIONAL SOCIETY OF EXPOSURE SCIENCE

2023 MEETING CO-CHAIRS

Amina Salamova

Dr. Salamova is an Associate Scientist in the O'Neill School of Public and Environmental Affairs at Indiana University Bloomington where she has been a member of the Environmental Sciences faculty since 2014. She received her Ph.D. in Environmental Science from Indiana University in 2011 and was a post-doctoral research associate from 2011-2014. She was the recipient of the Joan M. Daisey Outstanding Young Scientist Award in 2021. Dr. Salamova uses analytical chemistry and novel exposure assessment tools to investigate environmental exposures to harmful chemical contaminants and their effects on human health. She specifically focuses on exposures to a large group of organic contaminants, semi-volatile organic compounds (SVOCs), in vulnerable populations, such as children, older adults, indigenous and rural communities. Dr. Salamova's research interests focus on determining exposure sources of well-known and emerging SVOCs and on developing effective interventions to prevent or reduce these exposures. Her research is funded by the United States National Institutes of Health, Environmental Protection Agency, and Department of Agriculture.

Dana Barr

Dr. Dana Boyd Barr is a Professor of Environmental Health at Emory University's Rollins School of Public Health, Gangarosa Department of Environmental Health. Dr. Barr's research portfolio is broad and ranges from emerging issues such as COVID and health disparities to more traditional environmental studies. Her main research focuses on maternal-child health particularly involving prenatal exposures to environmental chemicals. She is actively studying birth cohorts in Atlanta, Thailand, Peru, Guatemala, India and Rwanda evaluating environmental exposures to chemicals, mitigating exposure pathways, and their relation to child development and disease. More recently, she has begun evaluating SARS-2-CoV seroconversion in these cohorts to determine the onset of COVID infection in these populations. She co-directs Emory's Laboratory for Exposure Assessment and Development in Environmental Research. She has devoted much of her 33-year career to the development of methods for assessing human exposure to a variety of environmental toxicants including current-use pesticides, phthalates, organochlorine chemicals (pesticides and PCBs), phytoestrogens, diethylene glycol, methyl eugenol, vinyl chloride and others. Dr. Barr has authored or coauthored over 350 peer-reviewed publications and book chapters. She is past President of the International Society of Exposure Science (ISES) and the former Editor- in-Chief of its official journal, Journal of Exposure Science and Environmental Epidemiology. She is currently Deputy Editor of the flagship journal Environmental Health Perspectives. As a result of her research, Dr. Barr has received many awards including ISES's Award for Outstanding Investigator, two HHS Secretary's awards for exposure-health investigations involving diethylene glycol and methyl parathion poisoning, the 2004 Federal Scientific Employee of the Year, CDC's Mackel Award for outstanding collaboration among epidemiology and laboratory, and EPA's Silver Medal for outstanding service to environmental health. Dr. Barr received her BS in Biology from Brenau University in 1987 and her Ph.D. in Analytical Chemistry from Georgia State University in 1994.

Mary Turyk

Dr. Salamova is an Associate Scientist in the O'Neill School of Public and Environmental Affairs at Indiana University Bloomington where she has been a member of the Environmental Sciences faculty since 2014. She received her Ph.D. in Environmental Science from Indiana University in 2011 and was a post-doctoral research associate from 2011-2014. She was the recipient of the Joan M. Daisey Outstanding Young Scientist Award in 2021. Dr. Salamova uses analytical chemistry and novel exposure assessment tools to investigate environmental exposures to harmful chemical contaminants and their effects on human health. She specifically focuses on exposures to a large group of organic contaminants, semi-volatile organic compounds (SVOCs), in vulnerable populations, such as children, older adults, indigenous and rural communities. Dr. Salamova's research interests focus on determining exposure sources of well-known and emerging SVOCs and on developing effective interventions to prevent or reduce these exposures. Her research is funded by the United States National Institutes of Health, Environmental Protection Agency, and Department of Agriculture.

TECHNICAL ORGANIZING COMMITTEE

Elizabeth Marder | CalEPA Office of Environmental Health Hazard Assessment

Kai Zhang | SUNY Albany

Amanda Wilson | University of Arizona

Inkyu Han | Temple University College of Public Health

Courtney Carignan | Michigan State University

Sa Liu | Purdue University

Peng Gao | University of Pittsburgh

Anindita Dutta | University of Chicago

Robin Dodson | Silent Spring Institute

Nicole Deziel | Yale University School of Public Health

Donghai Liang | Emory University

Esther Erdei | University of New Mexico Health Sciences Center

Chibuzor Abasilim | University of Illinois-Chicago

Timothy J. Buckley | U.S. EPA / ORD / CCTE

Kristen Malecki | University of Illinois-Chicago

Claudiu T. Lungu | University of Alabama at Birmingham (UAB)

Mohammed Abbas Virji | CDC/NIOSH

Seung-Hyun Cho | RTI International

Francesca Borghi | University of Insubria

Ray Smith | U.S. EPA, Office of Research and Development

Kristin Isaacs | U.S. Environmental Protection Agency

Xueying Zhang | Icahn School of Medicine at Mount Sinai

Brad Goodwin | ATSDR

Eve Pinsker | University of Illinois-Chicago

Daisy Magana | University of Illinois-Chicago

Nadia Pack | University of Illinois-Chicago

2023 SCHEDULE AT-A-GLANCE

27 SUNDAY

- 1:00-3:00 PM — **Welcome and Opening Plenary Speaker** Rob Bilott
- 2:00-4:00 PM — **Movie Screening: Dark Waters**
- 4:00-5:00 PM — **Community Panel Discussion**
- 5:00-7:00 PM — **Opening Reception and Networking**

28 MONDAY

- 8:00-9:00 AM — **Morning Poster Sessions, New Attendee Gathering**
- 9:00-10:15 AM — **Plenary Speakers:** Pam Miller
  - 10:15–10:30 AM | BREAK
- 10:30 AM-12:30 PM — **Scientific Sessions** (8 concurrent, 6 presentations each)
  - 12:30–1:45 PM | LUNCH
- 1:45-3:45 PM — **Scientific Sessions** (8 concurrent, 6 presentations each)
- 3:45–4:30 PM — **Afternoon Poster Sessions**
- 4:30-6:30 PM — **Community Panel Discussion**
- 6:30-8:00 PM — **Community Leader Reception**

International Society of Exposure Science ISES — Annual Meeting Chicago 2023 — INTERNATIONAL SOCIETY of Exposure Science

29 TUESDAY

- 8:00-9:00 AM — **Morning Poster Sessions**
- 9:00-10:15 AM — **Plenary Speakers:** Chris Frey, Ayana Perkins
  - 10:15–10:30 AM | BREAK
- 10:30 AM-12:30 PM — **Scientific Sessions** (8 concurrent, 6 presentations each)
  - 12:30–1:45 PM | LUNCH
- 1:45-3:45 PM — **Scientific Sessions** (8 concurrent, 6 presentations each)
- 3:45-4:30 PM — **Afternoon Poster Sessions**
- 4:30-5:30 PM — **Exposure Science Technology Fair, Committee Fair**
- 5:45-7:45 PM — **Women's Networking Event** Sponsored In-Part By: Bayer
- 7:45 PM — **Student, New Researcher Networking Event**

30 WEDNESDAY

- 9:00-10:15 AM — **Morning Poster Sessions**
- 10:15-10:30 AM — **Plenary Speakers:** Heather Stapleton, Miriam Diamond
  - 10:30 AM–12:30 PM | BREAK
- 12:30-1:45 PM — **Scientific Sessions** (8 concurrent, 6 presentations each)
  - 1:45–3:45 PM | LUNCH
- 3:45-4:30 PM — **Scientific Sessions** (8 concurrent, 6 presentations each)
- 6:00 PM — **Entertainment Evening – Architectural Lake Tour**
  - *ADDITIONAL FEE REQUIRED

31 THURSDAY

- 8:00-9:00 AM — **Morning Poster Sessions**
- 9:00-10:15 AM — **Plenary Speakers:** Serap Erdal, Marie Lynn Miranda
  - 10:30 AM–12:30 PM | BREAK
- 10:30 AM-12:30 PM — **Scientific Sessions** (8 concurrent, 6 presentations each)
  - 12:30–1:45 PM | LUNCH
- 1:45-3:45 PM — **Scientific Sessions** (8 concurrent, 6 presentations each)
- 3:45-4:45 PM — **ISES Awards Ceremony and Celebration**

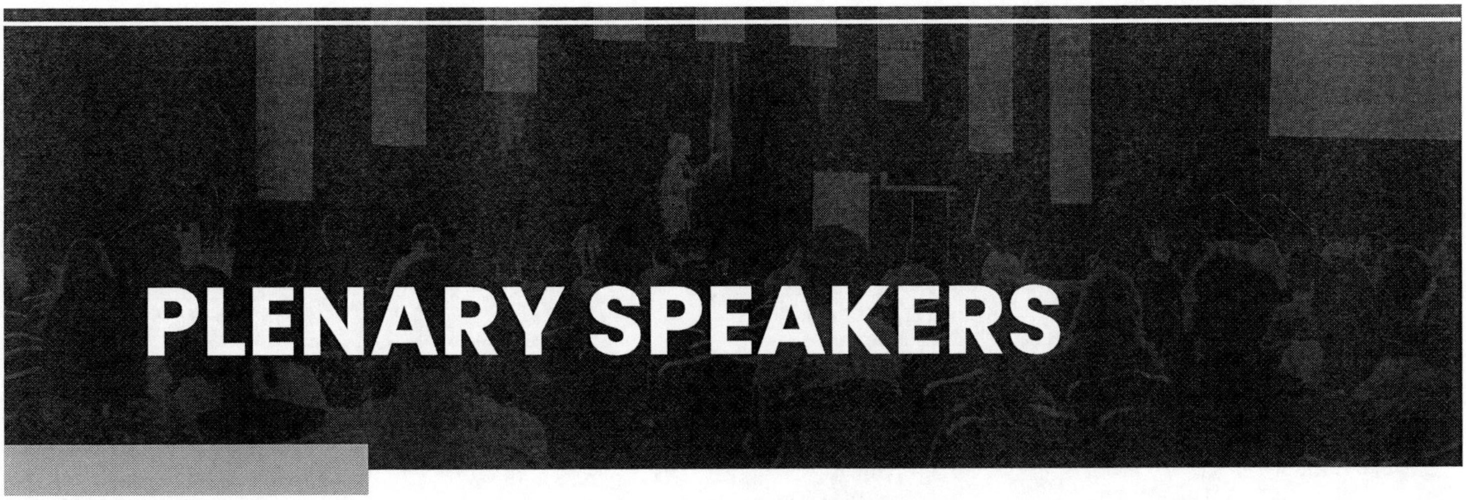

PLENARY SPEAKERS

Robert Bilott
Partner, Taft Stettinius & Hollister LLP

Presented: Sunday, August 27th
Topic: Exposing Worldwide PFAS "Forever Chemical" Contamination: Why Did It Take This Long?

Miriam Diamond
Professor, University of Toronto

Presented: Wednesday, August 30th
Topic: Community-based Environmental Health Research in the Alaskan Arctic and Protecting the Health of Future Generations

Pam Miller
Executive Director and Principal Investigator, Alaska Community Action on Toxics

Presented: Monday, August 28th
Topic: Community-based Environmental Health Research in the Alaskan Arctic and Protecting the Health of Future Generations

Heather Stapleton
Ronie-Richelle Garcia-Johnson Distinguished Professor, Duke University

Presented: Wednesday, August 30
Topic: Using Silicone Wristbands to Advance Research on Exposure and the Exposome (Characterizing personal exposures to organic chemicals using silicone wristbands and examples of their applications in health studies

Ayana Perkins
Chief Health Equity Officer, Centers for Disease Control and Prevention (CDC)

Presented: Tuesday, August 29th
Topic: Why Humane Housing is a Vital Pathway to Achieving Health Equity

Marie Lynn-Miranda
Chancellor, University of Illinois-Chicago (UIC)

Presented: Thursday, August 31st
Topic: Leveraging Geospatial Approaches to Address Enduring Health Injustices

Chris Frey
Assistant Administrator for Research and Development, United States Environmental Protection Agency (EPA)

Presented: Tuesday, August 29th
Topic: Linking Exposure Science to Community Health and Well-Being through Partnerships

Serap Erdal
Associated Professor, University of Illinois-Chicago School of Public Health, Environmental and Occupational Health Sciences

Presented: Wednesday, August 30th
Topic: Community-driven Air Quality Research and Community Empowerment for Environmental Justice

2024 COMMUNITY PANEL PARTICIPANTS

9 – Characterization of physicochemical properties affecting the applied dose of Pb from airborne particles: A systematic literature review of existing data and methods

Natalia Neal-Walthall[1], Peter Byrley[2], James Brown[3],
[1]Oak Ridge Institute for Science and Education, Postdoctoral Fellow, RTP, NC, USA. [2]US EPA, Physical Scientist, RTP, NC, USA. [3]US EPA, Health Scientist, RTP, NC, USA

Abstract
U.S. ambient air Pb concentrations have decreased significantly due to implementation of Pb emissions regulations. However, concentrations may still be elevated near remaining point and non-point sources, potentially resulting in increased Pb exposure and body burdens in populations residing near sources and/or workers in high Pb-use occupations. In addition to airborne Pb concentration, accurately modeling the Pb dose an individual receives from exposure to airborne Pb-bearing particulate matter (Pb-PM) requires information on physicochemical characteristics of individual particles that affect deposition and dissolution in the respiratory tract (e.g., size, density, shape, and chemical speciation). These characteristics vary by source, potentially resulting in different absorbed Pb doses for individuals exposed to similar air Pb concentrations. While many studies have characterized Pb-PM, the resulting data have not been collated in a manner allowing for easy incorporation into dose models. To that end, we have conducted a systematic review of peer-reviewed literature on characterization of Pb-PM for current U.S. sources including mining and mine waste, leaded aviation fuel, industrial activities, wildfires, and resuspended particles. We will present a summary of these data by source, discuss potential implications for exposure-related risk, and identify critical research gaps and limitations likely to impede accurate dosimetric modeling for Pb-PM. The results of this systematic review will improve future efforts to model Pb dose received from airborne exposure originating from various sources, allowing for better assessment of associated risk. Views expressed here are the authors' and do not necessarily reflect the views or policies of the U.S. EPA.

10 – POSTER: The joint effect of temperature and humidity on airborne mold and bacteria concentration in indoor environment: a machine learning approach for cost-effective intervention

SungChul Seo[1], Dohyeong Kim[2], Hyosun Lee[3], Boyeon Kwon[3]
[1]Seokyeong University, Professor, Seoul, Korea, Republic of. [2]University of Texas at Dallas, Professor, Richardson, USA. [3]Seokyeong University, Researcher, Seoul, Korea, Republic of

Abstract
Researchers have attempted to predict indoor bacteria and mold levels, known to cause respiratory allergies, but practical applications have been limited due to difficulties in measuring concentrations and complex interactions among factors. To address this, this study collected data on temperature, humidity, and airborne mold and bacteria concentrations from 3,555 indoor locations in Korea from August 2, 2021, to December 31, 2022 and used the data to develop a deep learning model to predict risky ranges in mold and bacteria concentration (over 500 CFU/m3 for mold and over 800 CFU/m3 for bacteria) and cautionary ranges (250-500 CFU/m3 for mold and 500-800 CFU/m3 for bacteria) by facility and season based on temperature and humidity combinations. Results showed that bacteria prediction performance was better than that for mold, with the best performance seen in the spring and summer and in underground subway stations compared to childcare centers. The algorithm developed in this study could create the most cost-effective mold and bacteria reduction scenarios by facility and season, provide empirical evidence for establishing practical indoor temperature and humidity control guidelines, and develop big data-based automatic temperature and humidity control devices.

12 – Paired liver: plasma PFAS concentration ratios from adolescents in the Teen-LABS study and derivation of empirical and mass balance models to predict and explain liver PFAS accumulation

Brittney Baumert[1], Fabian Fischer[2], Flemming Nielsen[3], Philippe Grandjean[4,5], Scott Bartell[6], Nikos Stratakis[7], Douglas Walker[8], Damaskini Valvi[9], Rohit Kohli[10,11], Thomas Inge[12,13], Justin Ryder[14,13], Todd Jenkins[15], Stephanie Sisley[16], Stavra Xanthakos[17], Sarah Rock[18], Michele La Merrill[19], David Conti[20], Rob McConnell[20], Lida Chatzi[20]
[1]University of Southern California, Post doctoral fellow, Los Angeles, USA. [2]Harvard University, Post doctoral fellow, Boston, USA. [3]University of Southern Denmark, Senior Scientist, Odense, Denmark. [4]University of Southern Denmark, Professor, Odense, Denmark. [5]Harvard University, Adjunct Professor, Boston, USA. [6]University of California, Irvine, Professor, Irvine, USA. [7]ISGlobal, Postdoctoral Researcher, Barcelona, Spain. [8]Emory University, Associate Professor, Atlanta, USA. [9]Mount Sinai, Assistant Professor, NYC, USA. [10]CHLA, Chief, Division of Gastroenterology, Hepatology and Nutrition, Los Angeles, USA. [11]University of Southern California, Professor of Pediatrics, Los Angeles, USA. [12]Lurie Children's Hospital of Chicago, Surgeon-in-Chief, Chicago, USA. [13]Northwestern Feinberg School of Medicine, Professor, Chicago, USA. [14]Lurie Children's Hospital of Chicago, Vice Chair of Research, Chicago, USA. [15]University of Cincinnati College of Medicine, Assistant Professor,

Cincinnati, USA. [16]Baylor College of Medicine, Assistant Professor, Houston, USA. [17]University of Cincinnati College of Medicine, Associate Professor, UC Department of Pediatrics, Cincinnati, USA. [18]University of Southern California, PhD Student, Los Angeles, USA. [19]University of California, Davis, Assisstant Professor, Davis, USA. [20]University of Southern California, Professor, Los Angeles, USA

Abstract
Animal studies have shown the liver is a hotspot for per- and polyfluoroalkyl substances (PFAS) accumulation and toxicity, however, these findings have not been replicated in human populations. We measured concentrations of 7 PFAS in matched liver and plasma samples collected at the time of bariatric surgery from 64 adolescents from the Teen-Longitudinal Assessment of Bariatric Surgery (Teen-LABS) study. Liver:plasma concentrations ratios were perfectly explained ($r^2 > 0.99$) in a multi-linear regression (MLR) model based on toxicokinetic (TK) descriptors consisting of binding to tissue constituents and membrane permeabilities. Of the 7 matched plasma and liver PFAS concentrations compared in this study, the liver:plasma concentration ratio for perfluoroheptanoic acid (PFHpA) was considerably higher than the liver:plasma concentration ratio for other PFAS congeners. Comparing the MLR model with an equilibrium mass balance model (MBM) suggested that complex kinetic transport processes are driving the unexpectedly high liver:plasma concentration ratio of PFHpA. Intra-tissue MBM modeling pointed to membrane lipids as the tissue constituents that drive the liver accumulation of long-chain, hydrophobic PFAS, whereas measured albumin binding of hydrophobic PFAS dominated PFAS distribution in plasma. The liver:plasma concentration dataset, empirical MLR model, and mechanistic MBM modeling allow the prediction of liver from plasma concentrations measured in human cohort studies. Our study demonstrates that combining biomonitoring data with mechanistic modeling can identify mechanisms underlying internal distribution and specific target organ toxicity of PFAS in humans.

13 – Exposure assessment of Slovenian children and adolescents to phthalates and DINCH

Agneta Runkel[1], Tina Kosjek[2], Žan Rekar[3], Darja Mazej[4], Janja Snoj Tratnik[1], Milena Horvat[5]
[1]Jožef Stefan Institute, Research assistent with PhD, Ljubljana, Slovenia. [2]Jožef Stefan Institute, Associate Professor, Ljubljana, Slovenia. [3]Jožef Stefan Institute, PhD student, Ljubljana, Slovenia. [4]Jožef Stefan Institute, Assistent with PhD, Ljubljana, Slovenia. [5]Jožef Stefan Institute, Professor, Ljubljana, Slovenia

Abstract
Phthalates are synthetic chemicals with numerous applications. As they are not chemically bound to the matrix, they can leach into the surrounding media. Phthalates are concerning, due to their endocrine activity and restrictions have been introduced to reduce exposure. Simultaneously, phthalate alternatives, such as Hexamoll® DINCH have been on the rise, however, information on the exposure to these compounds is scarce.

We assessed the exposure of 449 Slovenian children (age 6 - 9) and 356 adolescents (age 11 - 15) to nine phthalates and DINCH via the analysis of 16 metabolites in urine.

Following deconjugation, 0.5mL sample was extracted on an Oasis-HLB well plate prior to analysis by reverse-phase UHPLC coupled to MS/MS (Sciex AB QTRAP 4500). Most compounds were detected in 97 - 100% of samples, except for MCHP and MINCH (detected in 56% and 58% of samples, respectively).

The range of density-adjusted concentrations was high (<LOQ - 2879 ng/mL). We observed significant age and sex-dependent differences in exposure. Using detailed questionnaire data covering the 7 days prior to sampling, we identified potential determinants of exposure: consumption of food stored/heated in plastic containers, consumption of fast food, meat, chocolate, soft drinks, and cooking oils, use of cosmetics, the age of plastic bottles and family cars, use of air freshener, playing with plastic toys, and parental education.

This study is part of the ongoing second national human biomonitoring program in Slovenia and the observed trends will be re-evaluated in the future on a larger dataset including all study areas.

14 – POSTER: New statistical approach for generating historical $PM_{2.5}$ exposure estimates in the US for 1980-2019 for epidemiological research

Jianzhao Bi[1], Sun-Young Kim[2], Adam Szpiro[3], Joel Kaufman[4]
[1]University of Washington, Postdoc, Seattle, USA. [2]National Cancer Center Korea, Associate Professor, Goyang-Si, Korea, Republic of. [3]University of Washington, Associate Professor, Seattle, USA. [4]University of Washington, Professor, Seattle, USA

Abstract
Large-scale cohort studies for $PM_{2.5}$ are best conducted using highly resolved exposures estimated by spatiotemporal statistical models. This approach, however, makes it challenging to estimate historical exposures prior to the widespread deployment of ground $PM_{2.5}$ monitors (e.g., pre-2000 in the US). We developed a novel approach generating spatially resolved annual $PM_{2.5}$ estimates for 1980-2019 in the US for epidemiological research. We developed a spatiotemporal model with geographical variables and annual $PM_{2.5}$ measurements from the Inhalable Particulate Network (IPN, available for 1980-1983) and Air Quality System (AQS) networks (sparse before 1999). Our model first fitted a non-linear time basis characterizing the general temporal trend of $PM_{2.5}$ for 1980-2019. The time basis was innovatively fitted based on spatially matched IPN and nearby AQS monitors. The model then regressed spatially varying scaling factors of time basis to recover $PM_{2.5}$ time-series at individual locations with a spatiotemporal extension of universal kriging. We evaluated $PM_{2.5}$ estimates based on leave-monitors-out cross-validation (CV) with the IPN/AQS data and external validation with Harvard Six Cities PM2.5 data for 1980-1988 (not included in model development). Our model achieved a good CV performance with spatial R^2 (root-mean-square error, RMSE) of 0.87 (1.08 μg/m^3) and spatiotemporal R^2 (RMSE) of 0.87 (1.26 μg/m^3). Our model accurately estimated $PM_{2.5}$ at the Six Cities monitoring locations, with spatial R^2 (RMSE) of 0.83 (2.37 μg/m^3) and spatiotemporal R^2 (RMSE) of 0.78 (2.95 μg/m^3). We demonstrated that using a non-linear time basis substantially improved spatial CV R^2 by 0.35 compared to a linear time basis.

15 – Evaluating low-cost monitoring designs for $PM_{2.5}$ exposure assessment with a spatiotemporal modeling approach

Jianzhao Bi[1], Amanda Gassett[2], Lianne Sheppard[3]
[1]University of Washington, Postdoc, Seattle, USA. [2]University of Washington, Research Scientist, Seattle, USA. [3]University of Washington, Professor, Seattle, USA

Abstract
Determining feasible and cost-effective approaches to improve $PM_{2.5}$ exposure assessment with low-cost monitors (LCMs) will enhance the quality of epidemiological inferences. This analysis investigated features of fixed-site LCM designs that most impact $PM_{2.5}$ exposure estimates to be used in epidemiological inference for the Adult Changes in Thought Air Pollution (ACT-AP) study in Washington. ACT-AP collected and calibrated LCM-based $PM_{2.5}$ measurements at the fortnight level from April 2017 to September 2020 (N of monitors [measurements] = 82 [502]). We also acquired regulatory $PM_{2.5}$ measurements for January 2010 to September 2020 (N of monitors [measurements] = 78 [6186]) from US EPA and Washington's local agencies. We utilized a spatiotemporal modeling approach to predict $PM_{2.5}$ exposures using either all measurements or varying subsets, including designs with reduced temporal or spatial coverage. We evaluated the models based on a combination of cross-validation and external validation at all LCM locations (to increase comparability among models). Temporally, using the first and last measurements from each LCM resulted in the closest model's performance (R^2 [RMSE] = 0.79 [0.99 μg/m^3]) to the baseline (0.84 [0.88 μg/m^3]). The models with the first or last measurement had a decreased performance (0.77 [1.05 μg/m^3]), while still outperforming the model without LCM measurements (0.69 [1.20 μg/m^3]). Spatially, the performance peaked at using randomly selected 90% of LCMs (0.84 [0.88 μg/m^3]), which linearly decreased to 0.74 (1.11 μg/m^3) when 10% of LCMs were included. LCMs in densely populated, road-proximate areas improved the model more than those in moderately populated, road-distant areas.

16 – A Prospective Multi-city Study of Environmental Contaminant Levels in Green vs Non-green Low-income Housing

Felicia Rabito[1], Derek Werthmann[2], Rebecca Straubing[3], Gary Adamkiewicz[4], Tiina Reponen[5], Peter Ashley[6], Ginger Chew[7]
[1]Tulane School of Public Health and Tropical Medicine, Professor, New Orleans, USA. [2]Tulane School of Public Health and Tropical Medicine, PhD, New Orleans, USA. [3]Tulane School of Public Health and Tropical Medicine, MPH, New Orleans, USA. [4]Harvard T.H. Chan School of Public Health, Associate Professor, Boston, USA. [5]University of Cincinnati, Professor (emeritus), Cincinnati, USA. [6]Department of Housing and Urban Development, Office of Lead Hazard Control and Healthy Homes, ScD, Washington, DC, USA. [7]CDC, Senior Scientific Advisor, Atlanta, USA

Abstract
Background and Aims: The importance of housing on health is well established. The condition of the home is a strong predictor of exposure to environmental contaminants with low-income households particularly vulnerable. Therefore, improving housing standards is a priority. Housing built to "green" standards has been suggested to reduce contaminants due to improved building methods and materials. However, there is limited data supporting this. The Green Housing Study was conducted to test whether housing built to green standards has lower concentrations of environmental contaminants compared to conventional housing.

Methods: A repeated measures cohort study was performed in three U.S. cities. Data were collected at three home visits over 12 months. Exposures included: PM2.5, black carbon; dust allergens, pesticides, mold.

Results: In adjusted GEE models black carbon (β= -0.22, 95% CI=-0.38- -0.06, p = 0.01); permethrin (OR=0.28, 95% CI = 0.15-0.49, p < 0.0001), and mold (OR=0.29, 95% CI= 0.13-0.68, p= 0.003) were significantly lower in green homes versus non-green homes. Cockroach antigen was also lower in green homes (OR=0.59, 95% CI = 0.33-1.08, p = 0.09) although not statistically significant. We found that 68% of PM2.5 was explained by dwelling type and smoking and 42% of black carbon was explained by venting while cooking and use of a gas stove.

Conclusions: This study provides quantitative evidence of the benefits of incorporating green building practices on the level of numerous environmental contaminants known to be associated with health. Occupant behavior, particularly smoking is an important contributor to indoor air pollution.

18 – In-Silico Prediction of Dislodgeable Foliar Residues and Regulatory Implications for Agrochemicals

Edward Chikwana[1], Yi Shi[2]
[1]Corteva Agriscience, Human Health Risk Assessor, Indiana, USA. [2]IUPUI, Grad Student, Indianapolis, USA

Abstract
In instances where experimentally determined dislodgeable foliar residue (DFR) values are not available, agrochemical regulatory agencies use conservative default DFR values as a first-tier approach to assessing post-application dermal exposures to plant protection products (PPPs). Random forest, a machine learning algorithm, was used to develop an in-silico prediction model for DFR values after the last application (DFR0) once the spray has dried. The model was trained using more than 100 in-house DFR studies from Corteva Agriscience. Several variables related to what PPP was applied, what crop it was applied on, and location-specific climatic conditions at the time of last application were considered as model parameters. The proposed random forest regression model serves two purposes; (1) it predicts DFR values for use in risk assessment taking into consideration a large range of PPPs, agronomic practices, crop groupings, and variability in geographical locations; (2) it enables decision making for conducting a field study by demonstrating 98% prediction accuracy that if a predicted DFR0 is less than the EU DFR0 default value (3 μg/cm2/kg a.i./ha), it will highly likely be less than the default if the study is conducted.

20 – Occupational Exposures to Glyphosate and Insecticides in El Salvador and Nicaragua

Kathryn Rodgers[1], Jocelyn Fimbres[1], Juan Jose Amador Velázquez[2], Ramon Garcia-Trabanino[3,4,5], Emmanuel Jarquin[3], Morgan Robinson[6], Kurunthachalam Kannan[7], Jessica Leibler[8], Daniel Brooks[9], Madeleine Scammell[10]
[1]Department of Environmental Health, Boston University School of Public Health, PhD Student, Boston, USA. [2]Centro Médico del Pacífico (CENMED), Physician, Masaya, Nicaragua. [3]Agencia para el Desarrollo y la Salud Agropecuaria (AGDYSA), Physician, San Salvador, El Salvador. [4]Centro de Hemodialisis, Physician, San Salvador, El Salvador. [5]Emergency Social Fund for Health, Physician, Tierra Blanca, El Salvador. [6]Department of Pediatrics and Department of Environmental Medicine, New York University School of Medicine, Research Scientist, New York, USA. [7]Department of Pediatrics and Department of Environmental Medicine, New York University School of Medicine, Reserach Professor, New York, USA. [8]Department of Environmental Health, Boston University School of Public Health, Associate Professor, Boston, USA. [9]Department of Epidemiology, Boston University School of Public Health, Associate Professor, Boston, USA. [10]Department of Public Health, Boston University School of Public Health, Associate Professor, Boston, USA

Abstract
Background: Globally, pesticide use in agriculture is among the highest in Central America, where there is also an epidemic of kidney disease referred to Mesoamerican Nephropathy. The Mesoamerican Nephropathy Occupational Study (MANOS) includes 569 males aged 18-45 across five industries (sugar, corn, plantain, bricks and construction) in El Salvador and Nicaragua.

Objective: Evaluate differences in urinary pesticide concentrations between country, industry, and job task. We will also explore other potential sources of pesticide exposures and compare our cohort's levels to National Health and Nutrition Examination Survey (NHANES).

Methods: Post work-shift spot urine samples were analyzed for glyphosate and its metabolite AMPA in 285 participants, and 12 pesticides (organohalogens, organophosphates, pyrethroids) were analyzed in 50 participants at the same time-point. Participant questionnaires collected information on job tasks, pesticide use outside of work, and diet. We used Ordered Quantile transformation to produce normally distributed urinary pesticide concentrations and estimated predictors using linear regression.

Results: Workers in sugarcane had significantly higher concentrations of glyphosate and AMPA, one organohalogen (2,4-dichlorophenoxyacetic acid (2,4-D)), and two non-specific pyrethroid metabolites after controlling for country and age. When we used urine creatinine-standardized concentrations, results remained the same, though glyphosate had borderline significance. Overall, MANOS participants had lower glyphosate and AMPA concentrations and higher pyrethroid, organohalogen, and organophosphate concentrations compared with NHANES males aged 18-45.

Conclusion: This analysis will help determine occupational and non-occupational sources of glyphosate and insecticides in MANOS. Future analyses will evaluate potential relationships between pesticide exposures and kidney injury and function.

21 – Untargeted analysis and complex mixture dosing strategies to inform human biomonitoring beyond legacy PFAS

Carrie McDonough[1], David Dukes[2], Dilani Perera[3]
[1]Carnegie Mellon University, Assistant Professor, Pittsburgh, PA, USA. [2]Stony Brook University, Ph.D. Student, Stony Brook, NY, USA. [3]Carnegie Mellon University, Postdoctoral Scientist, Pittsburgh, PA, USA

Abstract
Establishing a comprehensive understanding of the toxicity and toxicokinetics of per/polyfluoroalkyl substances (PFASs) relevant for human exposure is a challenge due to the large number of existing PFASs and the absence of neat standards for many of these compounds. For this reason, in vivo and in vitro dosing with complex, exposure-relevant mixtures followed by high-resolution mass spectrometry (HRMS) characterization is a promising approach for prioritization of bioaccumulative novel PFASs that may otherwise be overlooked. Here, I will present results from high resolution mass spectrometry (HRMS) analysis of tissues and fluids from C57BL/6 mice dosed with an historical aqueous film-forming foam (AFFF) containing hundreds of known and novel PFASs. This AFFF contained significant contributions from sulfonamide precursors, which could theoretically be metabolized to perfluoroalkyl acids. We identified sulfonamides as well as their N-glucuronidated phase II metabolic conjugates in urine collected at multiple timepoints from mice in this study, though these conjugates were not identified in other tissues. Identification was further confirmed by generating these compounds in vitro. Relative abundance of conjugates was dose-dependent and increased over time. Analysis of liver, kidney, and brain tissue from AFFF-dosed mice suggests chain-length dependent tissue-serum ratios for targeted perfluoroalkyl acids as well as novel substituted perfluoroalkyl acids. Observations regarding tissue distribution, metabolism, and excretion will be discussed here with a focus on how this type of information can inform human biomonitoring campaigns, and examples where human results have echoed our findings.

22 – The Association Between Short-term Exposure to Particulate Matter and All-cause Mortality by Demographic Characteristics

Jiyoung Shin[1], Hee-Chung Kang[2], Hyeyun Kim[3], Kyeezu Kim[4], Jongmin Oh[5,6,7]
[1]Department of Health Care Policy Research, Korea Institute for Health and Social Affairs, Associate Research Fellow, Sejong, Korea, Republic of. [2]Department of Health Care Policy Research, Korea Institute for Health and Social Affairs, Senior Research Fellow, Sejong, Korea, Republic of. [3]Department of Health Care Policy Research, Korea Institute for Health and Social Affairs, Senior Researcher, Sejong, Korea, Republic of. [4]Department of Preventive Medicine, Northwestern University Feinberg School of Medicine, Postdoctoral Scholar, Chicago, IL, USA. [5]Department of Environmental Medicine, College of Medicine, Ewha Womans University, Researcher, Seoul, Korea, Republic of. [6]Institute of Ewha-SCL for Environmental Health (IESEH), College of Medicine, Ewha Womans University, Researcher, Seoul, Korea, Republic of. [7]Department of Human System Medicine, College of Medicine, Seoul National University, Ph.D student, Seoul, Korea, Republic of

Abstract
Background/Aim: Although exposure to particulate matter (PM) is known to have an adverse effect on various health risks, studies about exposure to PM and mortality according to demographic characteristics are limited. This study aimed to evaluate the association between short-term PM and all-cause mortality and estimate the stratified association according to demographic characteristics.

Methods: This study used the PM (PM2.5, PM10) data of Seoul, Korea extracted from the AirKorea website. Meteorological information was obtained from the Korea Meteorological Administration (KMA). Participants' mortality information and demographic characteristics were from the Cause of Death Statistics database (Microdata integrated Service). A generalized additive model with quasi-Poisson distribution was applied to estimate the relationship. We stratified the model by sex, age, education level, occupation, and marital status.

Results: A total of 221,042 death counts were used in the analysis. We observed 2% increase of all-cause mortality per interquartile increase of PM2.5 (lag0) among women (RR=1.015, 95% CI=1.006-1.024). The age group over 65 was more susceptible to PM2.5 (lag0) compared to other age groups (RR=1.010, 95% CI=1.001-1.015). Finally, manual workers (i.e., construction, transportation, and manufacturing industry) were more susceptible to PM compared to other occupation groups. The association between PM exposure and mortality was not significantly associated in the total group.

Conclusions: Our findings suggest that certain demographic and socioeconomic subgroups might be more susceptible to adverse health outcomes related to air pollution. Future research is needed to provide a scientific basis for appropriate health interventions targeting vulnerable populations.

23 – Determination of terpene levels after the use of essential oil diffusers in vehicles and studio apartments using passive sampling

Pil-Gon Kim[1], Jung-Hwan Kwon[2]
[1]Korea university, Research professor, Seoul, Korea, Republic of. [2]Korea university, Professor, Seoul, Korea, Republic of

Abstract
The exposure levels of selected terpenes (limonene, a- and b-pinenes, and g-terpinene) emitted by essential oil diffusers in vehicles and studio apartments were assessed using a passive sampling method. A previously developed passive sampler composed of an expanded polytetrafluoroethylene membrane and adsorbent (ePTFE PS) was enlarged and made wearable. Before field deployment, the sampling performance of the modified ePTFE PS for selected terpenes was compared with that of active sampling in a lab-scale 5 m^3 test chamber under constant exposure conditions, supporting that passive sampling provides reasonable estimates of the time-weighted exposure concentration. Fifty volunteers were recruited and asked to wear the ePTFE PS while using an essential oil diffuser inside their own vehicle while commuting and in their studio apartment while sleeping. Terpene levels without an essential oil diffuser in this study, extremely low concentrations in the vehicles and 47, 3.6, 1.6, and 0.62 $\mu g\ m^{-3}$ for average concentrations of limonene, a- and b-pinenes, and g-terpinene in studio apartments respectively were concurred with those reported in previous studies. The indoor concentrations of all selected terpenes in vehicles and studio apartments were elevated by the use of essential oil diffusers, especially in vehicles. The average concentration of limonene in vehicles after the use of essential oil diffusers was 11 $\mu g\ m^{-3}$, which was greater than that before use by a factor of 30. Therefore, cautious usage of essential oil diffusers indoors where the volume is limited, such as a vehicle, is needed to reduce exposure to terpenes.

24 – School Uniform: a potential source of per- and Polyfluoroalkyl Substances (PFAS) exposure to children

Chunjie Xia[1], Miriam Diamond[2], Graham Peaslee[3], Arlene Blum[4], Zhanyun Wang[5,6], Anna Shalin[7], Heather Whitehead[8], Megan Green[8], Heather Schwartz-Narbonne[9], Diwen Yang[9], Marta Venier[10]
[1]Indiana University Bloomington, Visiting Research Associate, Bloomington, USA. [2]University of Toronto, Professor, Toronto, Canada. [3]University of Notre Dame, Professor, Notre Dame, USA. [4]Green Science Policy Institute, Director, Berkeley, USA. [5]Empa - Swiss Federal Laboratories for Materials Science and Technology, Researcher, St. Gallen, Switzerland. [6]Institute of Environmental Engineering, Researcher, Zürich, Switzerland. [7]ETH Zurich, Research assistant, Zürich, Switzerland. [8]University of Notre Dame, Research Assistant, Notre Dame, USA. [9]University of Toronto, Research Assistant, Toronto, Canada. [10]Indiana University Bloomington, Assistant Professor, Bloomington, USA

Abstract
Public concern over the health risk posed by per- and polyfluoroalkyl substances (PFAS) has triggered increasing number of studies conducted to address their environmental occurrence and associated toxic effects. However, knowledge remains limited in children exposure to PFAS. We analyzed 72 children's textile products marketed as stain-resistant from US and Canadian stores, particularly school uniforms, to assess if clothing represents a significant route of exposure to per- and polyfluoroalkyl substances (PFAS). Products were first screened for total fluorine (total F) using particle-induced gamma ray emission (PIGE) spectroscopy (N=72), followed by targeted analysis of 49 neutral and ionic PFAS (N=57). PFAS were detected in all products from both markets, with the most abundant compound being 6:2 fluorotelomer alcohol (6:2 FTOH). Total targeted PFAS concentrations for all products collected from both countries ranged from 0.0281–38100 $\mu g/m^2$ with a median of 24.0 $\mu g/m^2$. Total targeted PFAS levels in school uniforms were significantly higher than in other textile items, but comparable to outdoor wear. Higher total targeted PFAS concentrations were found in school uniforms made of 100% cotton than synthetic blends. Perfluoroalkyl acids (PFAAs) precursors were abundant in school uniforms based on the results of hydrolysis and TOP assay. The estimated median potential children's exposure to PFAS via dermal exposure through school uniforms was 1.03 ng/kg bw/day. This study raises concerns over the use of PFAS in children's textile products, especially those that are worn close to the body and thus have a higher potential for dermal absorption.

25 – The Impact of Wildfires on Air Pollution and Health Across Land Use Categories in Brazil: Fires Originating From Sugarcane and Soybean Crops are one the Most Impactful

Igor Cobelo[1], Francisco Castelhano[2], Rafael Borge[3], Heresh Amini[4], Matthew Adams[5], Petros Koutrakis[6], Weeberb Requia[7]
[1]Getúlio Vargas Foundation, Reseacher, Brasília, Brazil. [2]University Federal do Rio Grande do Norte, Assistant Professor, Natal, Brazil. [3]Universidad Politécnica de Madrid, Professor, Madrid, Spain. [4]Icahn School of Medicine at Mount Sinai, Associate Professor, New York, USA. [5]University of Toronto Mississauga, Assistant Professor, Toronto, Canada. [6]Harvard University, Professor, Boston, USA. [7]Getulio Vargas Foundation, Assistant Professor, Brasilia, Brazil

Abstract:
We proposed to test two hypotheses in this study: i) the wildfires in Brazil have increased the levels of air pollution and posed a health hazard in 2003-2018, and ii) the magnitude of this phenomenon depends on the type of land use and land cover. We used a framework that derives "wildfire penalty" by accounting for the differences of the linear pollutant annual trend (β) values between two models to test these hypotheses. The first model was adjusted for Wildfire-related Land Use activities (WLU), considered as an adjusted model. In the second model, defined as an unadjusted model, we removed the wildfire variable (WLU). Our findings suggest that wildfire events between 2003 and 2018 have increased the levels of air pollution and posed a significant health hazard in Brazil, supporting our first hypothesis. Our results also confirm the second hypothesis. We observed that the greatest impact of wildfires on $PM_{2.5}$ concentrations occurred in soybean areas in the Amazon biome. During the 16 years of the study period, wildfires originating from soybean areas in the Amazon biome were associated with a total penalty of $0.72\mu g/m^3$ on $PM_{2.5}$, causing an estimated 3,872 excess deaths. Sugarcane crops were also a driver of deforestation-related wildfires in Brazil, mainly in Cerrado and Atlantic Forest biomes. Our findings suggest that between 2003 and 2018, fires originating from sugarcane crops in these biomes were associated with a total penalty of $0.11\mu g/m^3$ on $PM_{2.5}$, resulting in an estimated 1,632 excess deaths during the study period.

26 – Disparities of the Weather Impact on Air Quality in Brazil by Race/Ethnicity and Income

Weeberb Requia[1], Francisco Castelhano[2]
[1]Getulio Vargas Foundation, Assistant Professor, Brasilia, Brazil. [2]University of Rio Grande do Norte, Assistant Professor, Natal, Brazil

Abstract
Several studies have quantified the air pollution exposure disparities across racial and income groups. However, there is still a lack of investigations assessing disparities related to the impacts of weather on air pollution, which could indicate target air pollution reduction strategies under different climate scenarios. Our study aims to address this gap by estimating the economic and racial disparities of the weather impact on air quality in Brazil between 2003 and 2018. First, we used a generalized additive approach to estimate the weather-related changes in PM2.5. This framework derived "weather penalty", which a positive penalty suggests that an increase in PM2.5 was associated with long-term weather changes in the study period. Then, we estimated the population-weighted weather penalty for racial and income groups. Average penalty for the White population (the most-exposed group) was 31% higher than that of the Pardo population (the least-exposed group, mainly people of light brown skin color) in Brazil. In the stratification analysis by region, the Midwest and South were the regions where the black population was the most-exposed group. For the income group, our results indicate that the high-income population group was the most-exposed group in all analyses, including the national and the regional analyses. These findings are somewhat surprising, as previous studies have shown that minority and low-income populations tend to be more exposed to air pollution, than white and higher-income populations. However, our study suggests that disparities in exposure to air pollution may be more complex and nuanced than previously thought.

27 – Perfluoroalkyl Ether Acids in Well Water and Blood Serum from Private Well Users Residing Near a Fluorochemical Facility, Fayetteville, North Carolina

Nadine Kotlarz[1], Claire Critchley[2], David Collier[3], C. Suzanne Lea[3], Zachary Hopkins[4], Detlef R. U. Knappe[5], Jane A. Hoppin[5]
[1]North Carolina State University, Research Scholar, Raleigh, USA. [2]North Carolina State University, Study Manager, Raleigh, USA. [3]East Carolina University, Professor, Greenville, USA. [4]North Carolina State University, PhD Student, Raleigh, USA. [5]North Carolina State University, Professor, Raleigh, USA

Abstract
For ~40 years, a fluorochemical manufacturing facility in North Carolina discharged perfluoroalkyl ether acids (PFEAs), a subgroup of fluorochemicals, to air and water. Those discharges contaminated ~6,000 private wells near the facility.

In 2019, we enrolled 153 participants ages 6 and older (median age = 56 years) using 84 private wells into the GenX Exposure Study. We measured for 24 PFAS, including nine PFEAs, in blood serum and 26 PFAS (11 PFEAs) in well water using liquid chromatography-high resolution mass spectrometry. We assessed relationships between PFAS levels in serum and water using regression modeling.

Most wells (74%) had >5 detectable PFEAs; median ΣPFEAs was 842 ng/L (interquartile range [IQR]=197-1,760 ng/L). Low molecular weight PFEAs (PMPA, GenX, PEPA, PFO2HxA) were frequently detected in water and had the highest median concentrations (PMPA=342 ng/L, GenX=107 ng/L, PEPA=81 ng/L, PFO2HxA=107 ng/L) of the PFAS measured but these compounds were not detectable in participants' serum. However, Nafion byproduct 2, a seven-carbon diether, was detected in 73% of wells (median=14 ng/L, IQR=not detected [ND]-30 ng/L) and 56% of serum samples (median=0.2 ng/mL, IQR= ND-0.6 ng/mL). In an adjusted generalized estimating equations model, cumulative drinking water intake (well concentration × years lived at address) was associated with higher Nafion byproduct 2 serum levels, while more time spent outside the home was associated with lower levels. These results demonstrate the importance of measuring environmental and biological media to characterize human exposure to PFAS.

28 – POSTER: Association between thallium exposure and decreased renal function in Chinese adults: A cross-sectional study from the China National Human Biomonitoring

zheng zhang[1], Yuebin Lv[2], Yingli Qu[2], Xiaoming Shi[3]
[1]National Institute of Environmental Health, Chinese Center for Disease Control and Prevention, STUDENT, beijing, China. [2]National Institute of Environmental Health, Chinese Center for Disease Control and Prevention, associate professor, beijing, China. [3]National Institute of Environmental Health, Chinese Center for Disease Control and Prevention, professor, beijing, China

Abstract
BACKGROUND AND AIM: Thallium exposure is related to several adverse health consequences. However, little is known about the effects of exposure to thallium on the decreased renal function in adults.To elucidate the effects of long-term Tl exposure on renal function, we investigated the association between urinary Tl levels and renal function indicators based on data from China National Human Biomonitoring (CNHBM).
METHODS: A total of 9 815 subjects aged ≥ 18 years old with urinary Tl and serum creatinine were recruited from the CNHBM, from 2017 to 2018.

RESULTS: The weighted geometric mean (95%CI) of urinary thallium was 0.25 μg/L. After adjusting for potential confounding factors, the urinary thallium levels in the highest tertile had a significantly higher odds of CKD (odds ratio (OR: 1.68, 95% CI: 1.19, 2.38)) when compared with the lowest tertile. The urinary thallium levels (ln-transformed) may contributed to decreased eGFR (β=-5.51, 95% CI: -9.82, -1.19) and increased serum creatinine (β=1.67, 95% CI: 0.02, 3.32).Ln-transformed Tl was nonlinear associated with the risk of CKD in the logistic regression with restricted cubic spline (P for non-linearity = 0.0175).

CONCLUSIONS: Our study showed that Tl exposure was significantly associated with decreased glomerular filtration rate, elevated serum creatinine and increased risk of CKD, to some extent demonstrating nephrotoxicity. In view of the severe situation of metal pollution, it is of great significance to control or eliminate the exposure of renal toxic metals to reduce the burden of renal function decline.

29 – Antimony may impair cognitive function in older adults: Evidence from a population-based, prospective cohort study in China

Yidan Qiu[1,2], Saisai Ji[3], Yuebin Lv[4], Xiaoming Shi[5]
[1]China CDC Key Laboratory of Environment and Population Health, National Institute of Environmental Health, Chinese Center for Disease Control and Prevention, student, Beijing, China. [2]Department of Big Data in Health Science, School of Public Health, Zhejiang University, student, Hangzhou, Zhejiang,, China. [3]China CDC Key Laboratory of Environment and Population Health, National Institute of Environmental Health, Chinese Center for Disease Control and Prevention, none, Beijing, China. [4]China CDC Key Laboratory of Environment and Population Health, National Institute of Environmental Health, Chinese Center for Disease Control and Prevention, Deputy department director, Beijing, China. [5]China CDC Key Laboratory of Environment and Population Health, National Institute of Environmental Health, Chinese Center for Disease Control and Prevention, director, Beijing, China

Abstract
Background: Antimony (Sb) has been identified as a novel neurotoxicant that affects neurological functions in animal studies, but it is unknown in the population. Therefore, we aim to study the relationship of Sb and cognitive impairment in older adults.

Methods: We conducted a prospective cohort study based on Healthy Ageing and Biomarkers Cohort Study (HABCS). Participants were recruited in 2017-2018 and followed up in 2020-2021. Blood Sb (B-Sb) and urine Sb (U-Sb) were determined by inductively coupled plasma mass spectrometry. Mini-Mental State Examination was used to assess cognitive function. Cox proportional hazards models were used to explore the association of Sb exposure and cognitive impairment. Restricted cubic splines (RCS) were used to explore the presence of nonlinear relationships.

Results: After adjusting for potential covariates, the risk of cognitive impairment increased by 56.5% for each e-fold of U-Sb [hazard ratio (HR)=1.565, 95% confidence interval (95%CI): 1.230-1.991]. The results remained robust in B-Sb (HR=1.523, 95%CI: 1.100-2.109).

Conclusion: Our study suggests that Sb exposure is considered to be a risk factor for cognitive function.

30 – POSTER: The adverse effect of per- and polyfluoroalkyl substances (PFAS) may be reduced on prediabetes and diabetes via seafood intake in Chinese adults.

Yingli Qu
China CDC Key Laboratory of Environment and Population Health, National Institute of Environmental Health, Chinese Center for Disease Control and Prevention, professor, Beijing, China

Abstract
Although seafood is known to contain nutrients which has beneficial effects on glucose metabolism, some people may reduce their intake of seafood due to anxiety regarding high PFAS exposure from seafood. It is imperative to clarify the association among seafood consumption, PFAS exposure, and the risk of diabetes. This study was performed to evaluate the association of 8 PFAS (PFOA, PFNA, PFDA, PFUnDA, PFHxS, PFHpS, PFOS, and 6:2 Cl-PFESA) with prediabetes and diabetes and also the modifying effect of the seafood ingestion. We conducted a cross-sectional study of China National Human Biomonitoring, which is a nationwide and representative survey including questionnaire, physical and laboratory examinations. We found that single exposure of PFAS were significantly positive associated with the odds of prediabetes and/or diabetes, however, showing non-monotonic inverted-U dose-response patterns. Compared with the lowest exposure of PFAS, the multivariate adjusted OR (95%CI) of prediabetes and/or diabetes in subjects with intermediate levels are higher than with the highest exposure levels. The WQS index of PFAS mixtures was positively associated with prediabetes and combined diabetes, with the largest contributor being PFOA. Stratification analysis showed monotonically increasing relationships between PFAS and prediabetes/diabetes among the subjects with no seafood consumption, whereas no such associations were found among those having seafood (P for interaction<0.05). Our results suggest that PFAS exposures were significantly associated with prediabetes and/or diabetes and showing non-monotonic inverted U-shaped patterns. The seafood ingestion may reduce the harmful effects of PFAS on the odds of prediabetes and diabetes.

32 – POSTER: 1,3-butadiene occupational exposure in French industries: results of a sampling campaign conducted 3-year sampling campaign

Gautier Mater[1], Jean-François Sauvé[2]
[1]INRS, Laboratory manager, Nancy, France. [2]INRS, Researcher, Nancy, France

Abstract
Background: A survey conducted by the INRS in 2017 in workplaces associated with 1,3-butadiene use showed a lack of knowledge on the risks and exposure levels to this carcinogenic and mutagenic substance.

Methods: Between 2019-2022, the INRS and the eight chemical laboratories of the French regional health insurance funds conducted a sampling campaign across multiple industries in France to collect exposure data to 1,3-butadiene. Both long-term (LT) samples (≥ 60 minutes) and short term (ST) samples (≤ 15 minutes) were collected using thermal desorption tubes.

Results: Sampling was conducted in 51 companies from 18 different activity sectors. This effort yielded 541 exposure results (88% LT samples and 12% ST samples). The median LT concentration was 0.003 mg/m3 and 0.11 mg/m3 for ST samples. Half of all samples were under the limit of quantification. Production and primary use of 1,3-butadiene showed similar exposure levels with few concentrations above the binding European occupational exposure limit (2.2mg/m3) but ≥25% of samples were above the concentration of 0.08 mg/m3 recommended by the National Agency for Food, Environmental and Occupational Health Safety (Anses). Moreover, this national campaign identified elevated 1,3-butadiene concentrations in less traditional sectors due to its presence as residues or impurities in the products fabricated or handled, such as liquefied propane gas.

Discussion: This campaign provided increased knowledge on the sources and levels of 1,3-butadiene exposure in French companies, which will be used to improve prevention efforts in reducing exposures to this non-threshold carcinogen.

33 – A tool to query exposure measurements collected in French workplaces since 1987

Gautier Mater[1], Andrea Emili[2], Barbara Savary[2], Audrey Humbert[3], Jean-François Sauvé[2]
[1]INRS, Laboratory manager, Nancy, France. [2]INRS, Researcher, Nancy, France. [3]INRS, Engineer, Nancy, France

Abstract
Background: Occupational exposure databases represent important sources of information for exposure surveillance. Unfortunately, the information contained in these databases is rarely publically accessible outside of crude descriptive syntheses by industrial sector and agent. We aimed to develop an online tool to allow external users to access detailed statistical exposure data for specific exposure scenarios by leveraging data collected in French workplaces since 1987 and stored in the Colchic database.

Methods: The tool rests on a database containing over 800,000 exposure results to 614 chemical agents. For each agent, queries can be made for a combination of ten exposure variables: time period, industrial sector, occupation, task, company size, process type (e.g. open, closed), exposure frequency, product associated with the exposure (e.g. "paints and varnishes"), general ventilation, and local exhaust ventilation. Ring charts dynamically display the distribution of data across categories for each exposure variable to guide users in their queries. Descriptive statistics (e.g. number of samples, quantiles, proportion of samples above the exposure limit) are reported when at least fifty number of measurements are available. A detailed printable report providing enhanced graphics and information on coexposures and time trends can also be generated.

Results: The online tool is freely available in French and in English on the INRS website at https://outil-expo-rch-rb.inrs.fr/Outil110.

Discussion: The tool provides information historical and recent exposure levels across a wide range of chemical substances and work situations that can be useful to exposure scientists and hygienists for a priori occupational risk assessment and prevention.

34 – POSTER: Assessing occupational mixtures of chemical substances using a maximum cumulative ratio approach applied to a large administrative database

Jean-Francois Sauve[1], Andrea Emili[1], Sarazin Philippe[2], Gautier Mater[3]
[1]INRS, Researcher, Vandoeuvre-les-Nancy, France. [2]IRSST, Researcher, Montreal, Canada. [3]INRS, Laboratory manager, Vandoeuvre-les-Nancy, France

Abstract
Background: The maximum cumulative ratio (MCR) approach has been applied in environmental sciences to assess the relative contribution of substances to the overall risk posed by exposure to a given mixture. We applied this method to a dataset of occupational exposures to evaluate whether health risks of mixtures are driven by one or several agents.

Methods: We extracted exposure data to 123 chemical agents from the French Colchic database for the period 2010-2019. We grouped measurements by work situation (WS, combination of sector, occupation, task, and year). We calculated hazard quotients (HQ) by dividing the agents' concentrations by their occupational exposure limit. We calculated mixtures' hazard indices (HI) for 21 toxicological classes using the MiXie France tool. We calculated MCRs as the ratio between the HI and the maximum HQ in the mixture.

Results: We identified 14,170 combinations of WS and toxicological class with exposure to at least two chemical agents. 2501 had an HI>1, indicating overexposure. 86% of those were due to ≥1 agent with HQ>1. Of the remaining mixtures with no single substance with a HQ >1 , 12% had their HI dominated by one substance, and 2% had their HI driven by multiple substances. Trends were relatively similar between toxicological classes.

Conclusions: Overexposures of mixtures in Colchic were predominantly driven by a single substance, suggesting that prevention efforts in most of the cases observed should focus on the substance with the highest HQ. Future work should assess whether this trend occurs in other occupational exposure databases.

35 – Indoor Air Quality Assessment in Grocery Stores

Marina Almeida-Silva[1,2], Nuno Canha[3], Teresa Baptista[4], Dário Silva[5], Carlos Diogo[5]
[1]H&TRC-Health & Technology Research Center, ESTeSL-Escola Superior de Tecnologia da Saúde, Instituto Politécnico de Lisboa, 1990-096 Lisboa, Portugal, Professor, Lisbon, Portugal. [2]OSEAN—Outermost Regions Sustainable Ecosystem for Entrepreneurship and Innovation, 9000-039 Funchal, Portugal, Researcher, Funchal, Portugal. [3]Centro de Ciências e Tecnologias Nucleares, Instituto Superior Técnico, Universidade de Lisboa, E.N. 10 ao Km 139,7, 2695-066 Bobadela, Portugal, Researcher, Bobadela, Portugal. [4]H&TRC-Health & Technology Research Center, ESTeSL-Escola Superior de Tecnologia da Saúde, Instituto Politécnico de Lisboa, 1990-096 Lisboa, Portugal, Master Student, Lisbon, Portugal. [5]IDMEC, Instituto Superior Técnico, Universidade de Lisboa, Av. Rovisco Pais, 1049-001 Lisboa, Portugal, Researcher, Lisbon, Portugal

Abstract
Indoor Air Quality (IAQ) is a public and occupational health concern, as we are exposed to air pollutants daily given that we spend a great amount of our time in indoor environments. IAQ can be affected by numerous factors, from outdoor pollutants that get indoors through ventilation to building materials, furnishings, and activities carried by the building occupants. Exposure to air pollutants has been linked to a panoply of adverse effects on our health, well-being, and performance. The aim of this study was to assess the IAQ in grocery stores (GSs) in the municipality of Cascais in the Lisbon metropolitan area (Portugal) to characterize the workers' occupational exposure to air pollutants. The study was conducted in 13 small "family" grocery stores. The IAQ monitoring campaign was conducted using low-cost sensor technologies and focused on several parameters, namely: carbon dioxide (CO2), volatile organic compounds (VOCs), particulate matter (PM10 and PM2.5), temperature (T), and relative humidity (RH). Overall, the IAQ of the studied GSs complied with Portuguese legislation, except for PM2.5, where 23% of GSs presented levels above the established limit value. The mean inhaled dose of workers during a workday was estimated to be 157.7 ± 57.2 µg for PM2.5 and 165.8 ± 56.0 µg for PM10. The IAQ assessment allowed the identification of the best strategies to improve worker experience in indoor workplace environments.

36 – POSTER: Assessment of potential occupational multiple-exposures in the workplace using company-reported product inventory data

Jean-Francois Sauve[1], Frederic Clerc[1], Emili Andrea[1], Gautier Mater[2]
[1]INRS, Researcher, Vandoeuvre-les-Nancy, France. [2]INRS, Laboratory manager, Vandoeuvre-les-Nancy, France

Abstract
Background: Identifying multiple exposures of concern in workplaces is challenging because of the variety of substances that may be present. Data stored in chemical inventories as part of risk assessment activities could represent a surrogate information source on multiple exposures.
Objective: To describe multiple exposure patterns in a sample of work situations in France using company-provided chemical inventory data.
Methods: Following a call for data, we received anonymized standardized product inventories recorded in the Seirich software from 63 companies across 48 sectors. Each inventory contained information on work areas, tasks performed, and chemical products used. We analyzed 23 inventories which reported specific chemical substances in the products. We analyzed mixtures of substances across work situations (WS; unique combinations of work area and task) using pattern mining methods, and used the MiXie software to link substances sharing similar health effects.

Results: We collected exposure data to 1,918 agents across 456 WS (median 4 agents per WS, maximum 415). Frequent combinations of substances included {1,4-naphthoquinone, cumene hydroperoxide} (12% of all WS, 4 inventories), {butane, propane} (10% of WS, 10 inventories), and {ethanol, isopropanol} (7% of all WS, 9 inventories). The most prevalent health effects identified in the mixtures involved ocular, respiratory, skin, and central nervous system damage.

Discussion: Collecting company-reported information on products used in workplaces can be informative to highlight prevalent multiple exposure situations, especially those that have combined health effects. However, the small number of product inventories we received limits the generalizability of our findings.

37 – A critical assessment of the TVOC value

Tunga Salthammer
Fraunhofer WKI, Deputy Director, Braunschweig, Germany

Abstract
The primary goal of indoor hygiene studies is to create a pleasant and healthy environment for people. This requires the setting of climatic parameters that lead to thermal comfort. However, the concentration of air pollutants must also be kept as low as possible. A complete measurement program to study indoor air quality can be both time-consuming and expensive. For this reason, parameters that can be measured easily and quickly are often used for an initial assessment of the situation. One of these parameters is TVOC (Total Volatile Organic Compounds). There are several methods to measure and calculate TVOC, the most common being thermal desorption GC/MS. Within the framework of surveys TVOC can make contributions to identifying statistical trends and reference values can be extracted from the data collected in this way. The online measured TVOC values are at best suitable for investigating time courses. However, it must also be made clear that they are limited to this. Consequently, the application of TVOC in relation to health and exposure related questions is inappropriate. The use of TVOC in connection with health-related and/or odor-related questions is neither recommended nor possible. TVOC does not take into account the toxicological properties of the individual substances for inhalation exposure. This would not even be possible, since different substances usually have different toxicological endpoints that cannot simply be summed up for a reliable risk assessment. TVOC is not a toxicologically based parameter and is therefore only suitable for a limited number of screening purposes.

39 – Exposure of children to glyphosate in Morocco: Urinary levels andpredictors of exposure

Imane Berni[1,2], Aziza Menouni[3,4], Matteo Creta[5], Lode Godderis[6], Radu Duca[5], Samir El jaafari[7]
[1]Higher institute of nursing professions and health techniques, Postdoc, Laayoune, Morocco. [2]Cluster of Competency "Health and Environment", Moulay Ismail University, Postdoc, Meknes, Morocco. [3]Cluster of Competency "Health and Environment", Moulay Ismail University, Phd student, Meknes, Morocco. [4]Environment and Health Unit, Department of Public Health and Primary Care, Katholieke Universiteit Leuven, PhD student, Leuven, Morocco. [5]Unit Environmental Hygiene and Human Biological Monitoring, Department of Health Protection, National Health Laboratory (LNS), Researcher, Dudelange, Luxembourg. [6]Environment and Health Unit, Department of Public Health and Primary Care, Katholieke Universiteit Leuven, Professor, Leuven, Belgium. [7]Cluster of Competency "Health and Environment", Moulay Ismail University, Professor, Meknes, Morocco

Abstract
Glyphosate has recently been classified as probably carcinogenic to humans by the International Agency for Research and Cancer (IARC). We aimed to evaluate the urinary levels of glyphosate and its metabolite aminomethylphosphonic acid (AMPA) in Moroccan children, to identify the main predictors and to perform a risk assessment. Data was collected during a cross sectional study of 48 children from an intensive agricultural area. Measurements included a questionnaire on life-style, socio-demographic and herbicide exposures. Overall, glyphosate and AMPA were detected in 73% and 75% of urine samples, respectively. The mean concentrations were 0.97 μg L- 1 (2.06 μg/gcreatinine) for glyphosate and 0.79 μg L- 1 (1.52 μg/gcreatinine) for AMPA. Children younger than 5 years had a higher AMPA and glyphosate urine concentration than children aged 6–12. Children living near the pesticide spraying fields (<50 m) had 14.91 μg L- 1 and 2.35 μg L- 1 more glyphosate and AMPA, respectively, than children living in urban counties. AMPA concentration varied significantly with the source of drinking water, AMPA was higher among children that used water from open water sources compared to those using water from closed water sources. With the regard to the health risk assessment, estimated daily intake (EDIs), hazard quotient (HQs), and a hazard index (HI) were calculated. The GMs of EDI were 4.38 and 2.26 μg/kg of body weight BW/day for glyphosate and AMPA, respectively. This study provides further evidence on factors associated with glyphosate exposure, especially in developing countries.

40 – A Hybrid Air Dispersion Exposure System (HADES) for rapid local exposure assessments: Traffic interventions through an environmental justice lens.

Calvin Jephcote[1], John Gulliver[2]
[1]University of Leicester, Lecturer in GIS for Environmental Health, Leicester, United Kingdom. [2]University of Leicester, Professor of Environmental and Exposure Sciences, Leicester, United Kingdom

Abstract
Urban air pollution is traditionally estimated by land use regression (LUR) or dispersion modelling approaches. LUR indirectly quantifies the relationship between measured pollutant concentrations and proximity to polluting activities and may incorporate spatially detailed surrogates, including descriptions of the local geometry (i.e., street canyons) and satellite derived transboundary concentrations in the troposphere. Dispersion models quantify the relationship between emission inventories and airflows, informed by temporally detailed meteorological parameters, however, these benefits come with increased data, computational and financial cost.

As such, we have developed a GIS-dispersion model from open access datasets that are globally available, which applies the benefits of LUR to a streamlined Gaussian-dispersion approach. This Hybrid Air Dispersion Exposure System (HADES) creates hourly-annual and monthly multipollutant estimates at a 10m resolution to address complex spatiotemporal health policy questions.

HADES has been used to examine the local impact of the 2020 traffic interventions, resulting from extended COVID-19 lockdowns in Great Britain. The national monitoring network recorded 23%, 22%, and 25% reductions in annual-average nitrogen dioxide levels at rural (n=14), suburban (n=4), and urban (n=64) background sites, respectively. This is reflective of a broad behavioural shift, restricting movements to essential local travel. In some instances, these increased local movements may have offset the expected environmental gains of some communities. Furthermore, the flight of affluent urban communities to second-homes in rural locations was widely observed. We believe this to be the first hyperlocal examination of lockdown traffic interventions and air quality changes by social class at a neighbourhood level.

41 – Association of PM2.5 Concentration and Variation with Mortality: Evidence from a Taiwanese Cohort Study

JInjian Chen[1], Tsung Yu[2], Xiang Qian Lao[3]
[1]the Chinese University of Hong Kong, PhD candidate, Hong Kong, Hong Kong. [2]National Cheng Kung University, Associate Professor, Tainan, Taiwan. [3]City University of Hong Kong, Associate Professor, Hong Kong, Hong Kong

Abstract
This study aimed to investigate the relationship between long-term exposure to ambient PM2.5 and its variation with mortality in Taiwan. While epidemiological studies typically rely on the mean concentration of PM2.5 to estimate its health effects, there is evidence that its variations may also be responsible for excess mortality. Therefore, the study analyzed the associations of long-term exposure to PM2.5 and its variation with mortality among 434,443 Taiwanese individuals with 976,913 medical records from a large ongoing cohort between 2000 and 2016.

The results of the study showed that there is a significant association between the average concentrations, standard deviation (SD), and range of annual and summertime PM2.5 with all-cause mortality. Specifically, every 1 µg/m3 increase in annual, summer, and winter mean concentration corresponded to a 0.5%, 0.4%, and 0.2% increase in all-cause mortality, respectively. Each 1 µg/m3 increase in the SDs of annual, summer, and winter concentrations was associated with 0.8%, 0.8%, and 0.2% increases in all-cause mortality. Meanwhile, every 1 µg/m3 increase in the ranges of annual, summer, and winter concentrations was associated with 0.2%, 0.3%, and 0.1% increases in all-cause mortality. However, the study did not find any significant interaction between the means and variations of PM2.5 with mortality.

In conclusion, the study suggests that long-term exposure to ambient PM2.5 and its variations were positively associated with increased mortality. This finding provides valuable insights for policymakers and public health professionals to develop more effective strategies to reduce the adverse health effects of air pollution in Taiwan.

42 – POSTER: Measuring cotinine and trans-3'-hydroxycotinine in Bronchoalveolar Lavage Fluid

TIFFANY SEYLER[1], BENJAMIN BLOUNT[2], LANQING WANG[3]
[1]CDC, TEAM LEAD, ATLANTA, USA. [2]CDC, Branch Chief, ATLANTA, USA. [3]CDC, Lab Chief, ATLANTA, USA

Abstract
In 2019 an outbreak of severe lung injury was identified primarily in young people from around the United States. The Centers for Disease Control and Prevention (CDC) responded to the outbreak, which was formally defined as e-cigarette, or vaping, product use–associated lung injury (EVALI). CDC laboratories rapidly developed assays to analyze potentially harmful substances in bronchoalveolar lavage (BAL) fluid collected from EVALI case patients. This report describes the development and validation of a high-throughput isotope dilution HPLC-MS/MS method for measuring two tobacco specific biomarkers, cotinine (COT) and trans-3'-hydroxycotinine (HCT), in BAL fluid samples. COT and HCT are major metabolites of nicotine, the addictive alkaloid presents in tobacco products. The method has good sensitivity: the limit of detection is 0.033 ng/mL for COT and 0.0165 ng/mL for HCT, despite using only 200 µL of sample. Excellent selectivity was confirmed by the absence of detectable interferences. The within-day and between-day precision was 2-10%. The overall accuracy, calculated from recovery in three different sample matrices spiked at three concentrations, was 92.5 to 94%. We applied this novel HPLC-MS/MS method to characterize recent tobacco exposure in EVALI patients. This method is useful for characterizing recent tobacco exposure based on COT and HCT in BAL fluid.

43 – Discovering and Re-discovering Global PFAS Contamination at 3M: "Strategic Science" in Action

Kris Hansen
Savanna Science, Consultant, Afton, USA

Abstract
As an analytical chemist on the front lines of 3M's "discovery/rediscovery" of PFAS contamination in humans and in the environment during the late 1990s, it's interesting to reflect on the various strategies employed by 3M to control the public conversation about PFAS. Evaluating my experience in the 3M Environmental Lab between 1996-2000 against the plethora of documents released in the 2018 lawsuit brought against 3M by the MN Attorney General, provides the opportunity to evaluate if and how 3M protected critical information as "confidential", shared partial data to support an inaccurate narrative, and delayed action pursuing "impact-oriented science" relevant to affected communities. 3M boasts of maintaining an inclusive and collaborative culture; it's interesting to consider the internal barriers erected in the late 1990s to protect information and limit data sharing even amongst 3M's own front-line scientists working to unravel the biggest environmental issue of 3M's 120 year history. An examination of the PFAS crisis from inside 3M is useful in considering what kinds of expectations and requirements we should place on industry in introducing new chemicals and new products.

44 – Marginal Quantile Regression for Longitudinal Data with Non-Detects and Time-Dependent Covariates – Application to Modeling Carbon Nanotube and Nanofiber Exposures.

I-Chen Chen[1], Stephen Bertke[1], Matthew Dahm[2]
[1]Division of Field Studies and Engineering, National Institute for Occupational Safety and Health, Centers for Disease Control and Prevention, Statistician, Cincinnati, Ohio, USA. [2]Division of Field Studies and Engineering, National Institute for Occupational Safety and Health, Centers for Disease Control and Prevention, Supervisory Environmental Health Specialist, Cincinnati, Ohio, USA

Abstract
Longitudinal exposure and biomonitoring data from occupational studies are frequently right-skewed and left-censored. To address right-skewed data, data are generally log-transformed and analyzed modeling the geometric mean under the assumption that the data are log-normally distributed. However, modeling the mean of exposure may lead to bias and loss of efficiency if the transformed data do not follow a known distribution. Additionally, left censoring occurs when measurements are below the limit of detection (LOD). Often, a value is substituted (e.g., LOD/2 or LOD/square root of 2), however, there is no unique substitution value and results have been shown to be biased when the censoring proportion is large. Maximum likelihood estimation has been advocated to reduce this bias, but this method also requires an assumption about the distribution of the data. In a longitudinal setting some of the covariates may change their values over time, and the topic of time-dependent covariates has not been explored in the literature. As a result, we propose the use of quantile regression that does not assume any specified distribution, which in the presence of time-dependent covariates includes a strategy to select a working type of time-dependency. We demonstrate the proposed methods in application to the carbon nanotube and nanofiber (CNT/F) exposure study. Left-censored mass of elemental carbon at both the respirable and inhalable aerosol size fractions are the dependent variables. Worker-level determinants measured over time and based upon previously published CNT/F exposure studies are treated as independent variables or potential time-dependent covariates.

45 – Indoor Air Pollution in Multigenerational Households and COVID-19 Infection in the Cheyenne River Sioux Tribe, South Dakota

Anna Tillery[1], Rae O'Leary[2], Angela Aherrera[3], Jaylynn Stocklin[4], Erena Hovhannisyan-Romero[5], Megan Wilks[6], Chris Heaney[7], Ana M. Rule[1], Esther O Erdei[8]
[1]Johns Hopkins University, PhD Candidate, Baltimore, USA. [2]Missouri Breaks Industries Research Inc, Site PI, Eagle Butte, USA. [3]Johns Hopkins University, Post Doctoral Fellow, Baltimore, USA. [4]Missouri Breaks Industries Research Inc, Coordinator, Eagle Butte, USA. [5]University of New Mexico Health Sciences Center College of Pharmacy, PhD Study, Albuquerque, USA. [6]Johns Hopkins University, Research Associate, Baltimore, USA. [7]Johns Hopkins University, Associate Professor, Baltimore, USA. [8]University of New Mexico Health Sciences Center College of Pharmacy, Assistant Professor, Albuquerque, USA

Abstract
Introduction: The Cheyenne River Sioux Tribe (CRST), as other Indigenous communities in the US, have disproportionately high tobacco use compared to other demographic groups. Of concern is exposure to environmental tobacco smoke (ETS) and NO2 in indoor environments and their potential impact on risk of infection from COVID-19. The purpose of our study was to assess indoor air pollution and COVID-19 infection in multi-generation households within the CRST.

Methods: The CRST COVID-19–Wayakta He study ("Are you on guard against COVID-19?") recruited multigenerational households with 2 adults at least 9 years apart. All participants provided responses to a Tribally-validated questionnaire on COVID-19 infection, risk factors, housing conditions, and environmental health factors. Participants deployed passive ETS and NO2 air monitors in their homes for one week.

Results: In this community-based study, 239 households have been recruited. The median ETS concentration was 0.017 μg/m3 (range: 0.017-5.35). Of the 24.2% samples with results above LOD (0.034), 12.4% of households had exposures qualified as high. The median NO2 concentration was 12.6 ppb (range: 0.65-138). Twenty-six percent of participants have been infected with COVID-19 at least once based on self-report.

Discussion: ETS and NO2 exposure are known to be associated with more severe upper respiratory infections. The combination of health disparities and systemic racism causes Indigenous communities to be at higher risk of COVID-19 infection. High exposures to household ETS, NO2, and COVID-19 in the CRST highlight the importance of exposure assessment to design targeted interventions in Tribal communities.

46 – Environmental injustice in estimated cancer risk from airborne toxics exposure across United States communities

Patrick Hurbain[1], Matthew Strickland[2], Yan Liu[3], Dingsheng Li[3]
[1]University of Nevada, Reno, Graduate Student, Reno, USA. [2]University of Nevada, Reno, Professor, Reno, USA. [3]University of Nevada, Reno, Assistant Professor, Reno, USA

Abstract
Negative environmentally linked health outcomes have been associated with socio-demographic status (SDS) factors in different regions of the United States. US Census Bureau SDS factors were matched to US EPA National Air Toxics Assessment and Air Toxics Screening Assessment estimated cancer risk by census tract. Spearman correlations showed modest associations with various SDS factors and estimated cancer risk, both nationally and by state. Multiple linear regressions were performed to relate Z-scored estimated cancer risk with SDS factors. Multivariate regression analyses show increases in the numbers of minority populations relative to the number of White households were positively associated with increased estimated cancer risk in urban areas ($p < 0.05$), but only for high proportions of Black populations in suburban and rural areas. While in general environmental justice seems to have improved across the examined years, equity is far from achieved and future work should incorporate targeted exploration of extreme values in estimated risks to reveal the magnitude in disparity in estimated risk.

47 – In vivo neutron activation analysis (IVNAA) system improvement for metal exposure assessment

Song Yue[1], Linda Nie[2]
[1]Purdue Univerisity, PhD student, West Lafayette, USA. [2]Purdue Unveirsity, Professor, West Lafayette, USA

Abstract
Human exposure to toxic metals and excessive exposure to essential metals can lead to many health problems, such as neurodegeneration. Our lab has been developing a DD neutron generator-based neutron activation analysis system to quantify heavy metals and elements in human body in vivo. The technology provides a useful way to assess metals exposures and to estimate elemental deposition, storage and biokinetics. It has great potential to be applied in the occupational and environmental health field to study the association of metal exposure and various health outcomes, as well as in the nutrition field to study the intake of essential elements and human health. However, the relatively low sensitivity of the IVNAA system has greatly limited its applications. We developed a novel method to improve the sensitivity of the system for elemental measurement. The improvement of the new design predicted by Monte Carlo simulation for the quantification of one of the elements, Mn, was verified by experimental irradiation of Mn-dopped bone equivalent phantoms. For the same radiation dose, a 67.77% thermal neutron flux enhancement is reached. With only 4.32% increase of radiation dose, the simulated thermal neutron flux and activation can be further enhanced to 84.87%. A 100% thermal neutron enhancement is also achievable with acceptable dose change. The experimental results clearly show higher Mn activation gamma counts for each specific phantom. This greatly decreased the detection limit of the system for elemental quantification, which will broadly expand its application in human environmental exposure study.

49 – Predictive analysis for respiratory illnesses among household women exposed to indoor cooking fumes and biomass burning

Faiza Naseem[1], Mian Nazish Adnan[2], Afshan Naseem[3]
[1]Rawalpindi Women University, Assistant Professor, Rawalpindi, Pakistan. [2]Human Geography Department, F.G. Postgraduate College for Men, Sector H-9, Associate Professor, Islamabad, Pakistan. [3]National University of Sciences and Technology (NUST), Assistant Professor, Islamabad, Pakistan

Abstract
The use of solid fuels is predominant for energy needs in impoverished households in Pakistan where utilization of biomass for heating and cooking is widespread. To estimate the risk of respiratory illnesses in women exposed to cooking fumes, we conducted a case-control study among the indigent urban population in Rawalpindi. Working women and housewives aged between 21 and 45 years were examined (n=129) comprised of 70 cases (with respiratory illnesses) and 59 control. Active monthly monitoring of respiratory illnesses was performed, and the population-attributable risk (PAR) for the type of fuel used was computed. Results of Multinomial Logistic Model showed use of solid fuels or biomass stoves (OR=3.16;$p<0.01$), kerosene stoves (OR=2.37;$p<0.05$), was associated with increase in respiratory illnesses compared to use of clean fuels for cooking. Additionally, indoor heating with coal, wood, or kerosene was significantly linked to a high incidence of respiratory illness (OR=3.12; $p<0.01$). In comparison to working women who use electricity or gas for heating, housewives had a population-attributable risk of 20.3% and 12.4% for respiratory illnesses due to poor ventilation and the use of biomass as a fuel, respectively. Our study shows that using biomass as a fuel in households is a substantial risk factor for respiratory illness, particularly for housewives who spend the majority of their time at home. Furthermore, we present a novel finding that poor ventilation in small dwellings and the use of biomass for cooking may pose a greater risk for housewives to develop respiratory illnesses than for working women.

50 – Understanding the Potential for Low Cost Sensors in Occupational Hygiene

Miranda Loh[1], Rebecca Canham[2], Liesbeth Gilissen[3], Le Feber Maaike[4], Katie Clabon[5], Karen Galea[6], Eelco Kuijpers[4]
[1]Institute of Occupational Medicine (IOM), Director of Science and Engagement, Edinburgh, United Kingdom. [2]Institute of Occupational Medicine (IOM), Principal Scientist, Edinburgh, United Kingdom. [3]IDEWE, Onderzoeker, Leuven, Belgium. [4]TNO, Principal Scientist, Utrecht, Netherlands. [5]Institute of Occupational Medicine (IOM), Research Scientist, Edinburgh, United Kingdom. [6]Institute of Occupational Medicine (IOM), Team Lead, Edinburgh, United Kingdom

Abstract
Low cost sensors (LCS) have become more commonly used to monitor community air pollution and other exposure related information (e.g. physical activity, location). LCS could also have applications in the occupational environment for managing exposures and risks, however, such settings pose some differences in considerations compared to non-occupational use. These may include higher concentrations of substances or hazards, different hazards of interest, mixed exposures, motivations for use, privacy and employer-employee relationships. As part of the EPHOR project, we are using a co-design approach to develop an online tool to facilitate the use of LCS. We conducted three focus groups, in the UK, Netherlands, and Belgium, with occupational health professionals. The focus groups examined needs, limitations, and benefits of sensing technology in occupational health and hygiene applications. Results of the focus groups have been used to develop a validation survey for a wider group of participants. A number of common themes arose, including those around the need for guidance on LCS use, the robustness and appropriateness of use in occupational environments, the role of LCS in monitoring and compliance, the need for contextual information, and the need for buy-in by stakeholders. Next steps will be to incorporate results from the focus groups and surveys into a prototype tool which will be tested by end-users.

51 – POSTER: Exposure inequalities in poor urban neighbourhoods causing rise in diarhhoeal incidence

Masood Mahmood[1], Yasir Rehman[2], Aniqa Batool[3], Ikhtiar Uddin[1]
[1]Arid Agriculture University, Research Fellow, Rawalpindi, Pakistan. [2]Quaid-i-Azam University, PhD scholar, Islamabad, Pakistan. [3]Arid Agriculture University, Lecturer, Rawalpindi, Pakistan

Abstract
The rising urban inequalities is causing rise in unneccessary public exposures various risk factors resulting in high diarrheal incidence. In urban areas of Pakistan, diarrhea is a significant public health problem, and is frequently connected to community exposure to poor environmental sanitation and hygiene. For this purpose, we investigate poor urban neighborhoods. Population sampling through questionnaires (n=411), and samples of solid waste were analyzed to identify their characteristics, such as organic matter, papers/

cardboards, metals, plastics, and others. The study found that food waste accounted for the highest percentage (39%). The prevalence ratio (PR) of diarrhea related to poor hand hygiene was estimated to be 1.02 for adults and 1.04 for children, with females having a lower prevalence of diarrhea than males. The results of the binary logistic regression analysis identified several factors associated with diarrheal occurrences, including exposure of children to garbage (OR=3.31; 95% CI=1.67–6.3), unsafe waste disposal within the household (OR=4.85; 95% CI=1.85–16.9), and poor house hygiene conditions (OR=2.36; 95% CI=1.37–6.4). Our study revealed a significant correlation between the presence of garbage both inside and outside of households and the incidence of diarrhea. Therefore, we strongly advocate for the enhancement of hygiene conditions such as improved sanitation in the poor urban neighbourhoods as a means of reducing diarrheal diseases. In conclusion, improving waste disposal methods to minimize garbage exposure in the urbanized areas of Rawalpindi city could prove to be an effective intervention for controlling diarrheal illnesses.

52 – Investigating Mitigation Pathways for Urban Heat and Air Pollution

Laura Minet[1], Forood Azargoshasbi[2]
[1]University of Victoria, Assistant Professor, Victoria, Canada. [2]University of Victoria, PhD Student, Victoria, Canada

Abstract
Urban centers house more than half of the global population, and yet, they can be locations of extreme heat and air pollution, burdens that are respectively causing 490,000 and 4.2 million premature deaths per year globally. With climate change leading to an increase in the frequency, duration and magnitude of extreme temperature and air pollution events, we need to actively take action against these threats to human health.

Using Victoria and Vancouver (British Columbia) as a testbed, this research aims to assess the combined effectiveness of heat and air pollution mitigation strategies, focusing specifically on the implementation of cool roofs, and the reduction of heat release and air pollutant emissions from the transportation sector.

To achieve this, the Weather and Research Forecast model coupled with the Urban Canopy Model (WRF-UCM), and the Community Multiscale Air Quality (CMAQ) modeling system were set up using updated land-use data and local climate zone datasets. The heat and air pollution generated by transportation was estimated using traffic information processed with the MOtor Vehicle Emission Simulator (MOVES).

Scenarios of cools roofs were examined by altering the surface albedo. Scenarios of vehicle electrification were also explored, since electric vehicles do not release heat and do not have exhaust emissions.

Preliminary results show a 0.5°C increase in temperature due to on-road transportation, with higher impacts in high-density urban areas. Cool roofs could help decrease the surface temperature by up to 1.5°C. Impacts on ambient air quality are mitigated, given that lower temperatures could trigger higher ozone concentrations.

53 – Arsenic in soil-plant-human continuum: exposure and risk assessment

Debasis Golui[1], Md Basit Raza[2], Jajati Mandal[3]
[1]ICAR-Indian Agricultural Research Institute, Scientist, New Delhi, India. [2]ICAR-Indian Institute of Soil and Water Conservation, Senior Technical Officer, Odisha, India. [3]University of Salford, PhD Scholar, Salford, United Kingdom

Abstract
This article presents the regional-level health risks due to the consumption of arsenic contaminated rice in the three regions of Asia. Such macro-level study has not been reported so far, while there are micro-level reports for smaller geographic areas. The study also suggests a possible safe limit of bioavailable arsenic in soil for a smaller geographic area based on the solubility-free ion activity model. A discussion on risk assessment analyses for better appraisal of arsenic risks in soil-plant-human system is also included. It was found that adults in Asian countries are prone to a high risk of cancer due to the consumption of arsenic contaminated rice. South Asia (SA), South East Asia (SEA), and East Asia (EA) regions exceeded the USEPA-prescribed safe limit for cancer risk with ~100 times higher probability of cancer due to rice consumption. The hazard quotient for the ingestion of arsenic containing rice was found to be 4.526±5.118 for SA, 2.599±0.801 for SEA, and 2.954±2.088 for EA, which is much above the safe limit of HQ of 1. This study presents rice consumption related carcinogenic and non-carcinogenic risks to adults. A model was tested to calculate the safe limit of bioavailable arsenic in paddy soils. The methods and findings of this study are expected to be useful for regional level policy making and resource mobilization to alleviate the public health issues related to arsenic and the work can be expanded to arsenic present in drinking water.

54 – Evaluating the Environmental Impacts of Cruise Ship Activity-Induced Traffic in Victoria, British Columbia

Laura Minet[1], Muhammad Qasim[2]
[1]University of Victoria, Assistant Professor, Victoria, Canada. [2]University of Victoria, MASc Student, Victoria, Canada

Abstract
Victoria, British Columbia, is a popular cruise ship destination, and attracts more than 600,000 tourists every year, a number that is constantly growing. For many years, the inhabitants of the neighbourhood of James Bay, where the cruise ship terminal is located, have reported increases in local traffic and air pollution from April to October, but the actual contribution of the cruise ship tourism to local traffic emissions and air pollution has never been quantified. This research aims to evaluate the environmental impacts of cruise ship-induced traffic in James Bay, Victoria.

In collaboration with the James Bay Neighbourhood Association, we are simultaneously conducting traffic counts and air pollutant (fine particulate matter - $PM_{2.5}$, nitrogen dioxide - NO_2, ozone - O_3, and black carbon - BC) concentration measurements at key locations in the James Bay neighbourhood before and during the 2023 cruise ship season.

Traffic counts are used to estimate the increase in traffic volume associated with the cruise ship activity and determine traffic flows on the major links of the neighborhood, to estimate traffic-related air pollutant emissions using the Motor Vehicle Emission Simulator (MOVES) model. Those results are analyzed along with air pollutant concentration measurements to identify the relationship between cruise ship-induced traffic and ambient air quality and provide valuable insights on the impact of cruise ship tourism on the population's exposure to air pollution. By evaluating the environmental impacts of cruise ship activity-induced traffic congestion, we can develop effective measures to reduce emissions and improve air quality in the area.

55 – How to calculate the airborne concentration of an organic compound from vapor pressure data for acute inhalation studies

Tunga Salthammer[1], Jeffrey R. Kelsey[2], Emma Rogers[3], Rainer Otter[4]
[1]Fraunhofer WKI, Scientist, Braunschweig, Germany. [2]Chemsage Ltd., Scientist, Fleet, United Kingdom. [3]Labcorp Early Development Laboratories Ltd., Scientist, Alconbury, United Kingdom. [4]BASF SE, Scientist, Ludwigshafen, Germany

Abstract
Acute inhalation toxicity studies with small mammals often require exposure experiments close to the maximum attainable saturation vapor concentration of the target compound in air. However, application of the ideal gas law and assuming equilibrium conditions does not necessarily mean that this theoretical concentration can be achieved under realistic dynamic testing conditions. Molecules interact with surfaces (often reversible), diffuse into irreversible sinks and are sorbed by the test organism, particularly fur. In addition, for many compounds the vapor pressure and its temperature dependence is insufficiently known. In our study, we describe and discuss, using the solvent 2-butoxyethanol as an example, the methods to ensure the highest saturated vapor concentration that can be experimentally achieved while avoiding aerosol formation. We consider the practical difficulties and implications of this target compound at or close to saturation conditions and the criteria that should be used to critically assess such studies. The measured, extrapolated and calculated vapor pressures of 2-butoxyethanol are between 84 Pa and 115 Pa in the temperature range of 293 – 295 K. Error estimation depends on numerous factors, especially the purity of the chemical, temperature stability and the accuracy of measuring devices. Even when choosing the most reliable thermodynamic data our results show that the expected saturation concentration of 2-butoxyethanol at 295 K in an exposure chamber can vary between approximately 2 mg/L and 4.5 mg/L depending on the experimental error of vapor pressure, the enthalpy of evaporation, as well as sink and sorption effects.

56 – The Proximity of Children's Public Schools to Nearby Pollution Sources in the United States

Marsha Morgan[1], Wei-Lun Tsai[2], Anne Neale[3], Megan Mehaffey[4]
[1]US EPA, Research Environmental Health Scientist, Research Triangle Park, USA. [2]US EPA, Postdoctoral Research Associate, Research Triangle Park, USA. [3]US EPA, Physical Scientist, Research Triangle Park, USA. [4]US EPA, Branch Chief, Research Triangle Park, USA

Abstract
In the United States (US), no federal or consistent state regulations currently exist for the safe siting distance of children's public schools to nearby environmental hazards. Our study objective was to determine the number of K-12 public schools (regular, special education,

career/technical, and alternative education; n=93,736) located within 0.5 miles of pollution sources reporting annually to the US Environmental Protection Agency (i.e., Brownfield sites, Superfund sites, Resource Conservation & Recovery Act [RCRA] and Toxic Release Inventory facilities). We found that 41.3% (n=38,700) of all public schools were located within 0.5 miles of greater than 10 pollution sources. A total of 7.6% (n=7,079) of all public schools had more than 50 nearby pollutant sources within 0.5 miles. Among all states, New York had the highest number of schools (n=1,819) within 0.5 miles of more than 50 pollution sources. The most common source within 0.5 mile of schools was RCRA hazardous, non-hazardous, and municipal waste sites (91.5%). In addition, 57.4% of the special education schools in New York were within 0.5 miles of over 50 pollution sources. Our assessment of K-12 schools siting indicated that over 40% of US public schools are currently located within 0.5 miles of more than 10 pollution sources. These significant findings suggest the need for a closer evaluation of the public-school environment to determine children's potential exposures and health risks to chemicals originating from nearby pollution sources. This document has been reviewed by the U.S. EPA, Office of Research and Development, and approved for publication.

58 – Investigating the Movement of Parent PAHs and Alkylated PAHs Between Air and Soil Before, During, and After a Wildfire to Understand Potential Human Exposure

Kelly O'Malley[1], Christine Ghetu[1], Diana Rohlman[2], Brian Smith[3], Richard Scott[4], Kaley Adams[3], Kim Anderson[5]
[1]Oregon State University, Graduate Student, Corvallis, USA. [2]Oregon State University, Assistant Professor, Corvallis, USA. [3]Oregon State University, Faculty Research Assistant, Corvallis, USA. [4]Oregon State University, Senior Chemist, Corvallis, USA. [5]Oregon State University, Professor, Corvallis, USA

Abstract
There is an increasing number of large wildfires (>1000 acres) in the Western United States (U.S.), resulting in a need for rapid responses to measure human exposure to wildfire smoke. This can be facilitated by working with trained community members to deploy low-density polyethylene (LDPE) passive samplers before, during, and after wildfires to measure chemical contaminants. Wildfire smoke is a complex mixture and recent research has shown vapor-phase polycyclic aromatic hydrocarbons (PAHs) increased during a wildfire, concurrent with increased Air Quality Index (AQI), yet little is known about PAH movement, a valuable parameter for exposure risk assessment. Paired soil-poreair and air LDPE passive samplers were deployed by community members in the Western U.S. to measure diffusive flux (direction and magnitude) of PAHs and alkyl-PAHs between soil and air using a gas chromatography mass spectrometry (GC-MS/MS) method. We hypothesized that PAHs and alkyl-PAHs will deposit to soil during high (>200) AQI wildfires and volatilize after. Generally, we found 2 ringed PAHs diffused into air before, during, and after a wildfire, while 3 ringed PAHs deposited to soil, and 3-4 ringed alkyl-PAHs were in equilibrium or deposited at lower magnitudes compared to parent PAHs. During high AQI, there was a greater magnitude of 3-7 ringed PAHs and 3 ringed alkyl-PAHs depositing during a wildfire compared to pre-wildfire, and some post-wildfire locations showed volatilization of 4 ringed PAHs and 3 ringed alkyl-PAHs This study examines both the direction and magnitude of PAHs and alkyl-PAHs movement during all stages of a wildfire event.

59 – POSTER: Atmospheric pollutants variation and distribution over South Asia

Muhammad Zeeshaan Shahid
College of Earth and Environmental Sciences, University of the Punjab, Lahore Pakistan, Assistant Professor, Lahore, Pakistan

Abstract
Air pollution is the biggest environmental threat to public health, emissions of atmospheric pollutants are rapidly increasing in South Asia. Due to a lack of air pollution monitoring and modeling, a scientific challenge is to better understand the seasonal variations in air pollution concentrations over South Asia. Therefore, this study investigates aerosol patterns and trends over some major cities in the Indo-Gangetic Plain of South Asia, i.e., Islamabad, Lahore, Delhi, and Dhaka, by using simulations from the Modern-Era Retrospective Analysis for Research and Applications, version 2 (MERRA-2) model and satellite measurements (Moderate Resolution Imaging Spectroradiometer, (MODIS)) from 2000 to 2020. The results show that seasonal MODIS–aerosol optical depth (AOD) during 2000-2020 in Lahore is 0.5, 0.52, 0.92, and 0.71, while in Islamabad 0.25, 0.32, 0.45, and 0.38, in Delhi 0.68, 0.6, 1.0, and 0.77, and in Dhaka 0.79, 0.75, 0.78 and 0.55 values are observed during different seasons, i.e., winter, spring, summer, and autumn, respectively. The analysis reveals a significant increase in aerosol concentrations by 25%, 24%, 19%, and 14%, and maximum AOD increased by 15%, 14%, 19%, and 22% during the winter of the last decade (2011–2020) over Islamabad, Lahore, Delhi, and Dhaka, respectively. Thus, this study provides the aerosol spatial and temporal variations over the South Asian region and would help policymakers to strategize suitable mitigation measurements.

61 – The effect of relative humidity modified by air pollution on lung function in adolescents

Ashtyn Areal[1,2], Nidhi Singh[3], Qi Zhao[3,4], Dietrich Berdel[5], Sibylle Koletzko[6,7], Andrea von Berg[8], Monika Gappa[9], Joachim Heinrich[10,11,12], Marie Standl[13,11], Tamara Schikowski[14]
[1]IUF-Leibniz Research Institute for Environmental Medicine, PhD student, Düsseldorf, Germany. [2]Heinrich-Heine University Düsseldorf, PhD Student, Düsseldorf, Germany. [3]IUF-Leibniz Research Institute for Environmental Medicine, Post-Doc, Düsseldorf, Germany. [4]Cheeloo College of Medicine, Shandong University, Professor, Jinan, China. [5]Marien-Hospital Wesel, Professor, Wesel, Germany. [6]Department of Pediatrics, Dr. von Hauner Children's Hospital Munich, University Hospital, LMU Munich, Professor, Munich, Germany. [7]Department of Pediatrics, Gastroenterology and Nutrition, School of Medicine Collegium Medicum University of Warmia and Mazury, Professor, Olsztyn, Poland. [8]Marien-Hospital Wesel, Doctor, Wesel, Germany. [9]Evangelisches Krankenhaus, Professor, Düsseldorf, Germany. [10]Institute and Clinic for Occupational, Social and Environmental Medicine, University Hospital, LMU Munich, Doctor, Munich, Germany. [11]German Center for Lung Research (DZL), Doctor, Munich, Germany. [12]Melbourne School of Population and Global Health, The University of Melbourne, Doctor, Melbourne, Australia. [13]Institute of Epidemiology, Helmholtz Zentrum München - German Research Center for Environmental Health, Doctor, Neuherberg, Germany. [14]IUF-Leibniz Research Institute for Environmental Medicine, Workgroup Head, Düsseldorf, Germany

Abstract
Background: Weather and air pollution both affect respiratory morbidity and mortality. Yet, how air pollution and weather variables interact and how this interaction impacts lung function is unknown. This study assesses the interactive effect of relative humidity (RH) and air pollution on lung function in German adolescents.
Methods: A total of 2116 participants from the GINIplus and LISA birth cohorts with z-scores of forced vital capacity (FVC) and forced expiratory volume in one second (FEV1), were included. Daily meteorological data were obtained from the German Weather Service's COSMO-REA6. Air pollution exposure was assessed as average concentrations of 24-hour Ozone (O3), Nitrogen dioxide (NO2) and particulate matter <2.5µm (PM2.5). Multivariable linear models were fitted using an interaction term between the two-day moving average of RH and the categorical variable for air pollution.
Results: A 5% increase in RH was associated with a 4.46% to 4.58% increase in FEV1 (p=0.004). A 5% increase in RH was consistently associated with a decrease in FEV1 on days with high O3 levels (p=0.011). A similar result was observed for FVC (p=0.002). We also found that a 5% increase in RH was associated with a decrease in FEV1 and FVC on days with high PM2.5 and NO2 concentrations, however, these results were not statistically significant.
Conclusions: RH is associated with an increase in FEV1. A 5% increase in RH on high O3 days was consistently associated with a decrease in lung function in this cohort of German adolescents.

62 – Human Health Exposure Analysis Resource (HHEAR) Laboratory Quality Assurance Efforts to Increase the Quality of Exposome Measurement

Susan Teitelbaum[1], Matthew Mazzella[2], Chris Gennings[1], HHEAR Lab Network[3]
[1]Icahn School of Medicine at Mount Sinai, Professor, New York, USA. [2]Icahn School of Medicine at Mount Sinai, Biostatistician, New York, USA. [3]Human Health Exposure Analysis Resource, Lab Network, Multiple Cities, USA

Abstract
Exposomic health research requires accuracy and validity in environmental exposure measurement. The HHEAR laboratory network have been collaborating with researchers to provide state-of-the-art exposome biomarkers measured in a wide variety of biological matrices. NIEHS/NIH funded, the laboratories work toward "interchangeability" in the exposure measurements provided to investigators and deposited in the HHEAR public data repository where they can be harmonized across studies. HHEAR labs undertake a comprehensive quality assurance (QA) program to promote measurement harmonization for a wide range of biomarkers. All HHEAR labs routinely participate in proficiency testing (PT)/external quality assurance (EQA) programs. All labs also include and analyze aliquots of urine- or blood-based HHEAR designed common quality control (QC) pools across all analytic batches. The labs validate their analytic methods against NIST Standard Reference Materials® (SRMs). Additionally, blinded duplicates are analyzed as well as other QC samples included by the HHEAR project investigator.

We will present the extensive quality control data that has been generated by HHEAR over the past six years including HHEAR QC pools that have been included in over 30 different studies. We will show how the HHEAR Data Center has employed multivariate control charts to evaluate inter-study biomarker variability using these common HHEAR QC pools. We will demonstrate how continued lab participation in PT/EQA programs has enhanced confidence in data quality. Ultimately, we will highlight the laboratories' unique and extensive quality control work as part of the HHEAR QA program and provide support for combining exposomic datasets across studies and laboratories.

63 – Pesticide Exposure and Asthma Morbidity in Children Residing in Urban, Multi-family Housing

Derek Werthmann[1], Felicia Rabito[2], Gary Adamkiewicz[3], Tiina Reponen[4], Antonia Calafat[5], Maria Ospina[6], Ginger Chew[7]
[1]Tulane University SPHTM, Post Doctoral Fellow, New Orleans, USA. [2]Tulane University SPHTM, Professor, New Orleans, USA. [3]Harvard T.H. Chan School of Public Health, Associate Professor, New Orleans, USA. [4]University of Cincinnati, Professor-Emiritus, Cincinnati, USA. [5]Centers for Disease Control and Prevention, Chief of Organix Analytical Toxicology, Atlanta, USA. [6]Centers for Disease Control and PRevention, Research Chemist, Atlanta, USA. [7]Centers for Disease Control and Prevention, Deputy Associate Director for Science, Atlanta, USA

Abstract
Background: Children are potentially more susceptible to the effects of pesticides due to more sensitive organ systems and lower capacity to eliminate chemicals compared to adults. The health risks are particularly concerning for children with asthma, living in low-income, multi-family housing because of their impaired respiratory health, and factors associated with low-income, multi-family environments.

Objective: To assess the association between pesticide exposure and asthma morbidity among children 7–12 years residing in low-income, multi-family housing.

Methods: The concentrations of seven urinary pesticide biomarkers: 3,5,6-trichloro-2-pyridinol (TCPy), 2-isopropyl-4-methyl-6-hydroxypyrimidine, para-nitrophenol (PNP), 3-phenoxybenzoic acid (3-PBA), 4-fluoro-3-phenoxybenzoic acid, trans-3-(2,2-dichlorovinyl)-2,2-dimethyl-cyclopropane-1-carboxylic acid, and 2,4-dichlorophenoxyacetic acid (2,4-D) were measured in children (n = 162) over one year with three measures of pesticides. Adjusted associations between individual biomarkers and asthma attack (AA), asthma related health care utilization (HCU), and fraction of exhaled nitric oxide (FeNO), were examined with Generalized Estimating Equations (GEE). Weighted Quantile Sum regression was used to examine the effect of pesticide mixture on AA and HCU.

Results: In adjusted GEE models, positive non-significant associations were found between PNP and HCU (aOR:2.05 95% CI:0.76–5.52) and null associations for 3-PBA and HCU (aOR:1.07 95% CI: 0.88–1.29). Higher concentrations of PNP and 2,4-D were associated with lower FeNO levels (PNP: -17.4%; 2,4-D:-19.74%). The biomarker mixture was positively associated with HCU in unadjusted models (OR: 1.56 97.5% CI: 1.08–2.27) but not significant in adjusted models (aOR: 1.40 97.5% CI: .86–2.29). The pyrethroid biomarker 3-PBA at baseline contributed the greatest weight to the index (45%).

65 – Critical issues in exposure to PFAS: Introduction

Tom Webster[1], Kathleen Attfield[2], Elizabeth Marder[3]
[1]Boston University, Professor, Boston, USA. [2]California Department of Public Health, Research Scientist Supervisor, Richmond, USA. [3]University of California Davis, Adjunct Professor, Davis, USA

Abstract
Exposure to PFAS via consumption of drinking water is well established as important, particularly in contaminated communities. Consequently, USEPA has proposed water quality standards for a small number of PFAS. In this paper and symposium we will review key issues in PFAS exposure that need further research.

1. What is the contribution of drinking water to PFAS exposure in general populations, e.g., non-occupationally exposed people who do not live in contaminated communities. Is the USEPA's default 20% relative source contribution appropriate?
2. Mass balance studies of human serum indicate substantial gaps between extractable organic fluorine and the dozens of PFAS typically measured. What is the other organic fluorine, conventional PFAS that we don't typically measure? A recent US study suggests it may not be fluorinated pharmaceuticals.
3. Diet is often proposed as the main source of PFAS exposure for general populations. But while older studies examining the association between serum PFAS and diet (measured by food frequency questionnaires) show substantial associations, recent measurements by the US FDA show relatively few foods had detectable levels of PFAS. Has the situation changed?
4. While large amounts of EOF have been found in cosmetics, little is known about dermal exposure to PFAS with only a few studies measuring skin permeability.
5. Several studies have found associations between serum PFAS levels in general populations and indoor air concentrations of volatile precursors. How important is indoor exposure, including dust, and what are the major sources?
6. What are the most important occupational exposures?

67 – POSTER: The potential health risks posed by the cumulative exposure of children to antibiotics in rural environment

Abid Ali[1], Habib Ullah[2], Asfandyar Shahab[3], Qaisar Mahmood[4], Audil Rashid[5]
[1]PMAS Arid Agriculture University, PhD student, Rawalpindi, Pakistan. [2]Zhejiang University, Post-Doc Fellow, Hangzhou, China. [3]Guilin University of Technology, Assistant Professor, Guilin, China. [4]COMSATS University, Professor, Abbottabad, Pakistan. [5]PMAS Arid Agriculture University, Professor, Rawalpindi, Pakistan

Abstract
The indiscriminate use of medicines by humans and veterinary purposes has led to severe environmental contamination by antibiotics, causing harmful health effects to people exposed through various pathways. Therefore, it is crucial to conduct a comprehensive health risk assessment of the population in rural areas. There have been limited studies on the cumulative health risks of antibiotics through aggregative pathways, especially among children, who are more vulnerable to the harmful effects. We aimed to correlate concentrations of antibiotic residues in soil and water with respect to exposure frequency of children in rural areas of Punjab to determine the potential health risks. The antibiotics detected in soil were predominantly comprised of human use such as erythromycin (31.2 μg kg^{-1}), doxycyclines (21.56 μg kg^{-1}) and ciprofloxacin (14.32 μg kg^{-1}). However, multiple pathways contributed to water contamination of antibiotics which include poultry, veterinary and human use. Dust ingestion accounted for over 70% of the total daily exposure dose for most antibiotics. The risks of allergic symptoms (OR 3.14, 95% confidence interval (CI) 1.43-6.75) and atopic dermatitis (OR 2.21, 95% CI 1.13-5.77) were higher among exposed children. Allergic rhinoconjunctivitis risks were mainly attributed to frequent water contact, which was 30 times higher than children living away from stagnant water bodies. Our study emphasizes the importance of conducting a comprehensive exposure assessment to evaluate potential health risks and highlights the need to curtail overwhelming use to antibiotics to reduce the harmful health effects on children living in rural areas.

68 – Changes in Knowledge, Attitudes, and Beliefs Regarding Air Pollution and Research Among Study Participants After Receiving Personal Air Sampling Results

Patrick Ryan[1,2], Sherrill Ingram[3], Chris Wolfe[4], Erin Haynes[5]
[1]Department of Pediatrics, University of Cincinnati, Professor, Cincinnati, USA. [2]Division of Biostatistics and Epidemiology, Professor, Cincinnati, USA. [3]Division of Biostatistics and Epidemiology, Clinical Research Coordinator III, Cincinnati, USA. [4]RTI International, Exposure Scientist, Research Triangle Park, USA. [5]Department of Epidemiology and Environmental Health, Professor, Lexington, USA

Abstract
Background: We previously developed, in collaboration with research participants, a novel, interactive, and electronic report-back document to provide study participants their personal air sampling results. The objective of this study was to assess changes in knowledge, attitudes, and beliefs regarding air pollution and scientific research among study participants after receiving their monitoring results.

Methods: Adolescents enrolled in a personal air monitoring study and their caregivers were invited to receive their individual sampling results, including times and locations where exposures were recorded, information about air pollution, and potential sources of exposure. A questionnaire to ascertain participants' and caregivers' knowledge of air pollution and attitudes / beliefs regarding research was completed before and after receiving their study results. Changes in knowledge, attitudes, and beliefs were analyzed using paired t-tests.

Results: Adolescent (n = 37) participants and caregivers (n = 68) completed the questionnaire before and after receiving their individual study results. Both adolescent and caregivers reported significant increases in their knowledge of health effects related to air pollution and awareness of specific times, locations, and activities where their or their child's exposure may be increased. Adolescents and caregivers also indicated increased enthusiasm for research and the use of personal air monitors after receiving their study results.

Conclusion: Providing the results of personal air monitoring to research participants increases awareness of times, locations, and activities where exposures are elevated and interest in scientific research. Future studies should consider individual report back as a potential intervention to decrease personal exposure to air pollution.

69 – Enhancing Food Safety in Germany - The Interaction of Food Monitoring of the Federal States with the National Total Diet Study

Anna Elena Kolbaum[1], Oliver Lindtner[2]
[1]German Federal Institute for Risk Assessment (BfR), Research Associate, Berlin, Germany. [2]German Federal Institute for Risk Assessment (BfR), Head of Unit Dietary Exposure and Aggregated Exposure, Berlin, Germany

Abstract
The German Food Monitoring Programme coordinates food control activities of independent administrations and laboratories of the German Federal States. The aim is to have a subset of all samples investigated by the State laboratories that follows a random sampling which allows for long-term, short-term and consumer-loyal dietary exposure scenarios. However, the orientation on regulated foods and substances causes data gaps. These are missing data on not regulated items, but also on systematic stratification (e.g. regions, production type). Furthermore, foods are measured at producer or retail level, but for many foods or substances the 'table-ready' condition is of interest. The implementation of Germanys' first Total Diet Study (BfR MEAL Study) fills these gaps by covering >90% of the diet, considering potential influencing factors (e.g. regions, shopping habits), and analysing foods 'as consumed' capturing realistic substance intakes. With its food and substance choice, the BfR MEAL Study improves dietary exposure assessment by supplementing >100 substances to the monitoring dataset and for those substances already included in the monitoring, the BfR MEAL Study broadens the food spectrum. With this approach the BfR MEAL Study provides background contamination data for most realistic long-term total dietary exposure assessments. Thus, both programmes together enable a case-dependent selection and combination of the datasets for dietary exposure assessment and can benefit from each other to act more focused. This new combination approach opens new possibilities for exposure and risks assessors as well as more reliable information for risk managers to conclude on management measures or risk communication strategies.

70 – Guided Optimization of ToxPi Model Weights Using a Semi-Automated Approach

Jonathon Fleming[1,2], Jessie Chappel[1], John House[3], Alison Motsinger-Reif[4], David Reif[4]
[1]NCSU, PhD Candidate, Raleigh, USA. [2]NIEHS, Trainee, Durham, USA. [3]NIEHS, Staff Scientist, Durham, USA. [4]NIEHS, Chief, Durham, USA

Abstract
The Toxicological Prioritization Index (ToxPi) is a versatile framework for analysis and visualization of multi-dimensional feature data, which involves developing prioritization rank scores for samples based on a set of user-specified feature weights. Initially used for ranking of chemical toxicity, ToxPi has since been extended to several applications, including assessment of county-level Covid-19 risk and GIS-based exposure risks. Defining appropriate feature weights for ToxPi can be challenging, and existing prediction methods are often ineffective. To address this issue, we propose a new semi-supervised method that predicts feature weights using user-defined response level categories, such as low, moderate, or high toxicity. Ordinal regression is used to predict a probable set of feature weights using a small subset of samples with known response levels, and a custom genetic algorithm is then employed to fine tune the feature weights by incorporating any unknown samples. We assessed error with a simulation study and show the genetic algorithm can improve the results of ordinal regression. We utilize both methods with existing published polycyclic aromatic compound measurements to build models that rank bioactivity levels of petroleum substances. We confirmed the viability of these methods to predict feature weights using existing cell-assay data models. Our method provides a practical approach for determining feature weights in scenarios with minimal response data, thus enhancing the utility of ToxPi for a wide range of applications including exposure assessments. The new functionality will be publicly-available in the next release of the open-source toxpiR package.

72 – POSTER: Applying a mass-balance model to predict human in vitro dermal absorption of data poor chemicals

Daniel Vallero[1], Hunter Fisher[2,3], Marian Evans[4], Elaine Cohen Hubal[5]
[1]U.S. Environmental Protection Agency, Center for Computational Toxicology & Exposure, Senior Research Physical Scientist, Research Triangle Park, NC, USA. [2]Oak Ridge Associated Universities, Student Services Contractor, Oak Ridge, TN, USA. [3]U.S. Environmental Protection Agency, Center for Computational Toxicology & Exposure, SSC, RTP, NC, USA. [4]U.S. Environmental Protection Agency, Center for Computational Toxicology & Exposure, Biomedical Engineer, RTP, NC, USA. [5]Center for Public Health and Environmental Assessment, Supervisory Physical Scientist, RTP, NC, USA

Abstract
The U.S. Environmental Protection Agency recently developed a computational dermal model for human exposure consisting of a series of four well-mixed compartments. These compartments represent the source solution (vehicle), stratum corneum, viable tissue, and

receptor fluid. In addition, steps are taken to account for chemical evaporation out of the source compartment and, once that compartment has evaporated, evaporation continues from the stratum corneum. This model was evaluated by comparing model outputs to experimental measurements for 26 data-rich, cosmetically relevant chemicals. In this study, we plan to apply the model to a set of data poor PFAS compounds that require dermal exposure predictions for accurate aggregate exposure estimates. In the absence of experimental data to validate these model predictions with, we will focus our results on bounding the potential dermal uptake and identifying data gaps. Since experimental values for partition coefficients and permeabilities may not be available for many of these compounds, we will determine methods to obtain these inputs, including documentation of uncertainties. This study's aims are to evaluate the feasibility of using this dermal model to predict dermal uptake for PFAS and other data poor chemicals, and to explore methods for obtaining required input values for such chemicals.
Disclaimer: This abstract does not necessarily reflect Agency policy.

73 – POSTER: A Biomonitoring Surveillance System to Measure Concentrations of Per- and Polyfluoroalkyl Substances (PFAS) in Serum of Michigan Firefighters.

Priyashi Manani[1], Eva Kool[2], Carissa Hueter[2], Elizabeth Wasilevich[3], Jennifer Gray[4]
[1]MDHHS, LEAD EPIDEMIOLOGIST, MCKINNEY, USA. [2]MDHHS, EPIDEMIOLOGIST, LANSING, USA. [3]MDHHS, SENIOR EPIDEMIOLOGIST, LANSING, USA. [4]MDHHS, Senior Toxicologist, Lansing, USA

Abstract
INTRODUCTION: The Centers for Disease Control and Prevention (CDC) awarded a five-year grant to the Michigan Department of Health and Human Services (MDHHS) to expand the monitoring of state residents' exposure to toxic environmental chemicals. Firefighters are at increased risk of exposure to per- and polyfluoroalkyl substances (PFAS) as they are commonly found in Class B firefighting foams and protective equipment. In 2021 MDHHS launched a surveillance system to assess exposures to PFAS among Michigan firefighters.

OBJECTIVE: To determine PFAS serum concentrations in Michigan firefighters and identify subpopulations of highest risk of exposure.

METHODS: This is a multi-year, cross-sectional project. The project population is recruited from municipal and airport fire departments in Michigan utilizing a stratified sampling approach. Current career and volunteer firefighters from the selected departments are invited to participate. Project participants have their blood tested for more than 40 PFAS and complete an online survey that aims to identify sources of PFAS exposure. The drinking water at participating fire stations is also being tested for PFAS.

RESULTS: Since 2021, 177 fire departments have been invited to participate and 63 (35.6%) of selected departments enrolled in the project. To date, 955 firefighters have completed the online survey and provided a blood sample. Analysis is underway to characterize PFAS exposures.
CONCLUSION: The results from this project will establish PFAS serum concentrations among Michigan firefighters and will help local, state, and federal agencies understand more about PFAS exposure and inform decisions to minimize these exposures.

74 – Leveraging Exposomics and Multi-omics to Investigate the Pathobiology of Asthma

Peng Gao[1], Carla Ng[2], Aaron Barchowsky[3], Patricia Opresko[3], Sally Wenzel[4]
[1]University of Pittsburgh, Assistant Professor, Pittsburgh, USA. [2]University of Pittsburgh, Associate Professor, Pittsburgh, USA. [3]University of Pittsburgh, Professor, Pittsburgh, USA. [4]University of Pittsburgh, Professor and Chair, Pittsburgh, USA

Abstract
Asthma develops through repeated cycles of inflammation often triggered by environmental factors, which can lead to progressive obstruction of the airways. Currently, guidance is urgently needed from a preventative environmental health perspective. As such, further studies to explore the unknown environmental exposures related to asthma severity and control, particularly in monitoring exposures at molecular and individual levels, and understanding the underlying biological mechanisms are urgently needed. This innovative study utilizes biomonitoring and bioinformatics approaches to identify specific toxic-biologic pathways and validate them in human cellular systems. Pittsburgh and Allegheny County both have high asthma prevalence, as well as indications of more severe disease. The County also has some of the worst air quality in the nation. Here, as a pilot study, external and internal exposome data are collected from fifty asthmatics from the existing Asthma and Environmental Lung Health Institute Asthma Registry at UPMC with differing exposures to potential airborne pollutants to determine whether these pollutants link to the differences in the personal external exposome, the internal fingerprint, and the biologic response signatures. Specifically, the following three specific aims are achieved: Aim 1 - Characterize asthma-related organic contaminants and xenobiotics from the external and internal exposome profiles; Aim 2 -

Determine asthma-related internal metabolome and proteome in both cross-sectional blood samples; and Aim 3 - Perform multivariate mediation analysis based on the results of the exposome profiles, internal metabolome and proteome, and asthma biomarkers, then validate the hypothesized xenobiotic-biomolecule-asthma pathways by human cell lines. Altogether, these studies form a novel exposomics approach by integrating these three aims on environmental monitoring, biomonitoring, molecular epidemiological, statistical inferential, and in vitro experimental approaches. This project generates a better understanding of environmental influences on asthma phenotypes, severity, and control by connecting external exposures with internal biomolecules at both molecular and clinical levels.

75 – Community-centric management of waterborne diseases and evidence based exposure assessment to predict health risks for policy interventions

Shahzad Azeem[1], Faiza Naseem[2], Muhammad Irfan Khawar[3,4], Malik Faisal Azeem[5,6], Malik Tahir Hayat[7]
[1]University of Gujrat, MS student, Gujrat, Pakistan. [2]Rawalpindi Women University, Assistant Professor, Rawalpindi, Pakistan. [3]Holy Family Hospital, Lab Analyst, Rawalpinid, Pakistan. [4]National University of Sciences and Technology (NUST), PhD student, Islamabad, Pakistan. [5]TÉLUQ University of Quebec, Research Fellow, Quebec, Canada. [6]COMSATS University Islamabad, Assistant Professor, Islamabad, Pakistan. [7]COMSATS University Islamabad, Associate Professor, Abbottabad, Pakistan

Abstract
Due to substandard water quality and insufficient sanitation facilities, Pakistan bears a heavy load of waterborne illnesses, notably diarrhea followed by malaria, gastroenteritis, typhoid fever, and hepatitis A. The WHO approximating 16,500 fatalities annually among children aged below five from diarrhea alone. We conducted a community assessment study to determine its empowering status to take ownership of their water and sanitation needs (n=594) from suburban areas of Rawalpindi and Abbottabad districts. Concern over inadequate health budget was expressed by 45% of local communities. Multinomial logistic model revealed high Diarrheal incidence owing to drinking unpurified water (odds ratio= 2.4, 95% CI=1.28-4.68), using public tap water (odds ratio=3.16, 95% CI=1.82-5.48). The Principal Component Analysis predicted areas with poor sanitation are likely to have high incidence of waterborne diseases especially for family with large size (>5 individuals) and poor income group (<$4 a day). Areas where local communities actively collaborate with district government, high awareness was observed for awareness good hygiene practices, improved sanitation systems, and waste management practices. Although we noticed differences in estimated Escherichia coli exposure ranging from 10^3 to 10^8 CFU/day across various neighborhoods and age groups however, due to community-centric management, exposure to fecal contamination was significantly reduced resulting in decline in child mortality rates (70 to 56 deaths per 1,000 live births) during last three years. These findings emphasize the necessity of policy intervention to increase public awareness and evaluate water quality using multiple sources, to monitor access to safe drinking water.

76 – Prenatal residential air quality and childhood neurobehavioral diseases - A nationwide birth cohort follow-up approach

Shu-Li (Julie) Wang[1], Chih-Da Wu[2]
[1]National Health Research Institutes, P.I., Zhunan town, Miaoli county, Taiwan. [2]National Cheng-Kung University, Associate Professor, Tainan, Taiwan

Abstract
Air pollution was associated with decreased brain functions. However, representative large prospective follow-up studies with prenatal air quality data are scarce. We aimed to investigate the incidence of children's prevalent complex diseases and their association with pre-pregnancy and perinatal exposure to air pollution in a population-based birth cohort across Taiwan.

There were 478,735 singletons born in 2007~2011 and follow-up till 8 years of age to study the neurobehavioral disorder of attention deficit hyperactivity disorder (ADHD) or autism spectrum disorder (ASD) without major pregnancy complications nor congenital diseases. We utilized hybrid land-use regression models and machine learning methods to assess daily air concentrations of NO2 by integrating the data from 73 stationary air monitors and considered air quality for each child from 8 weeks before conception till birth.

The children experienced increased neurobehavioral diseases from the 2nd compared to the 1st Quartile of NO2 level (>13 ppb) before birth. Further, the Hazard Ratio (HR) increased in a dose-dependent manner. For example, the HRs were 1.10 (confidence interval 0.98~1.24), 1.16 (1.01~1.32), and 1.27 (1.10~1.47) for Q2~Q4 for ADHD in girls, adjusted for sex, maternal smoking status, insurance premium, and parental history of ADHD, PM2.5 and urbanization in the Cox's model. The results are similar for ASD to ADHD.

The result corresponded to the new WHO guideline of optimized 20 ppb or even lower for NO2 as a surrogate exposure index. Early-life traffic-related air pollution control might be essential for preventing children's ADHD and ASD.

78 – Reevaluation of Minamata disease provides insights into the protective role of selenium against methylmercury toxicity

Mineshi Sakamoto[1], Masumi Marumoto[2], Koichi Haraguchi[3], Takashi Toyama[4], Yoshiro Saito[5], Steven Balogh[6], Chiharu Tohyama[7], Masaaki Nakamura[8]
[1]National Institute for Minamata Disease, Special Adviser to Director-General, Minamata, Japan. [2]National Institute for Minamata Disease, Chief Researcher, Minamata, Japan. [3]National Institute for Minamata Disease, Section Chief, Minamata, Japan. [4]Tohoku University, Lecturer, Sendai, Japan. [5]Tohoku University, Professor, Sendai, Japan. [6]Metropolitan Council Environmental Service, Principal Research Scientist, Saint Paul, USA. [7]University of Tokyo, Professor Emeritus, Tokyo, Japan. [8]National Institute for Minamata Disease, Director of Clinical Medicine, Minamata, Japan

Abstract
Selenium (Se) protects organisms against methylmercury (MeHg) toxicity. However, the functional role of Se in human MeHg intoxication remains unclear. Preserved samples collected in Minamata during the methylmercury poisoning epidemic can provide important information on Hg and Se interactions in human toxicity. We reanalyzed samples related to Minamata disease for Se, including patients' organs and sediments, fish, and shellfish from Minamata Bay. Increased Se levels were observed in all specimens compared to controls. Additionally, residual wastes in the factory that discharged MeHg to the bay contained high Se concentrations, suggesting that Se was also discharged from the factory. This is the first study to report that patients with Minamata disease were probably exposed to elevated levels of MeHg and Se simultaneously. The Hg/Se molar ratio in preserved shellfish samples was approximately 3. The Hg/Se molar ratio in patients' cerebrum and cerebellum samples ranged from 2–3, suggesting that MeHg exposure and uptake exceeded that of Se. Based on these findings, we further performed animal experiments to estimate the role of increased Se levels on MeHg exposures among patients. In rat experiments with simultaneous exposure to MeHg and selenomethionine (SeMet) at a Hg/Se molar ratio of 3, the protective effects of Se against MeHg toxicity were limited. Sufficient protection against MeHg toxicity was demonstrated at a 1:1 molar ratio of SeMet supplementation. Our results provide insight into the importance of the Hg/Se molar ratio in seafood for protecting humans against MeHg toxicity.

79 – POSTER: Limitations in Current Methods to Measure Heat Stress in Outdoor Workplaces and Proposed Alternatives

Rod Harvey
RHP Risk Management, Director IH and Field Services, Chicago, USA

Abstract
The most common metrics used to monitor the level of heat stress in the workplace include, in order of sophistication and efficacy, Heat Index, Wet Bulb Globe Temperature (WBGT), and calculation of effective WBGT for comparison to the American Conference of Governmental Industrial Hygienists® (ACGIH®) thermal stress Threshold Limits Values® (TLVs®). There are deficiencies with each of these methods. When combined with the inter- and intra-personal differences in the physiological responses to heat stress, their ability to determine when heat stress presents a hazard in a given workplace is limited.

Because the Occupational Safety and Health Administration (OSHA) currently does not have a federal Heat Stress standard, all citations made by OSHA are under Section 5(a) of the Occupational Safety and Health Act of 1970, which is generally referred to as the General Duty Clause. This clause states that all employers "shall furnish to each of his employees employment and a place of employment which are free from recognized hazards that are causing or likely to cause death or serious physical harm to his employees."

This presentation will discuss the deficiencies in measuring heat stress and discuss how they have been a factor in recent decisions published by the Occupational Safety and Health Review Commission (OSHRC). The OSHRC is an agency created by Congress to decide contests of citations issued by OSHA to employers. The presentation will also discuss potential alternatives to measuring environmental heat to determine if and when heat presents a hazard, such as physiological monitoring.

80 – The HBM4EU E-Waste Study: Exploratory Survey of Occupational Exposures in Waste Processing

Paul T.J. Scheepers[1], Susana Viegas[2], Radu-Corneliu Duca[3], Jelle Verdonck[4], Emilie Hardy[5], Radu-Corneliu Duca[3], Adrian Covaci[6], Paulien Cleys[7], Adam Cseresznye[8], Thomas Goën[9], Karen S. Galea[10], Jelle Verdonck[4], Katrien Poels[4], Lode Godderis[11,12], Elizabeth Leese[13], Henriqueta Louro[14,15], Maria Silva[14,15], Sopie Ndaw[16], Simo Porras[17], Selma Mahiout[17], Tiina Santonen[18]

[1]Radboud Institute for Biological and Environmental Sciences, Radboud University, Associate Professor, Nijmegen, Netherlands. [2]NOVA National School of Public Health, Universidade NOVA de Lisboa, Assistant professor, Lisboa, Portugal. [3]Laboratoire National de Santé (LNS), Department of Health Protection, Unit head, Luxembourg, Luxembourg. [4]Centre for Environment and Health, Department of Public Health and Primary Care, KU Leuven, Post-doc, Leuven, Belgium. [5]Laboratoire National de Santé (LNS), Department of Health Protection, Researcher, Luxembourg, Luxembourg. [6]Dept. Pharmaceutical Sciences, University of Antwerpen, Professor, Antwerp, Belgium. [7]Dept. Pharmaceutical Sciences, University of Antwerpen, Post-doc, Antwerp, Belgium. [8]Dept. Pharmaceutical Sciences, University of Antwerpen, PhD candidate, Antwerp, Belgium. [9]Friedrich-Alexander-Universität Erlangen-Nürnberg, Professor, Erlangen-Nürnberg, Germany. [10]Institute of Occupational Medicine (IOM), Senior researcher, Edinburgh, United Kingdom. [11]Centre for Environment and Health, Department of Public Health and Primary Care, KU Leuven, Professor, Leuven, Belgium. [12]IDEWE, External service for prevention and protection at work, CEO, Heverlee, Belgium. [13]Health & Safety Executive, Buxton SK17 9JN, Post-doc, Buxton, United Kingdom. [14]INSA, National Institute of Health Dr. Ricardo Jorge, Department of Human Genetics, Av. Lisbon, Portugal, Research scientist, Lisbon, Portugal. [15]Centre for Toxicogenomics and Human Health, Universidade NOVA de Lisboa, Research Scientist, Lisbon, Portugal. [16]INRS, French National Research and Safety Institute, Senior researcher, Vandœuvre-lès-Nancy, France. [17]Finnish Institute of Occupational Health, Researcher, Helsinki, Finland. [18]Finnish Institute of Occupational Health, Research Professor, Helsinki, Finland

Abstract
So far, human biomonitoring (HBM) has not been much used to study exposure of workers involved in the processing of e-waste in the EU. In this study we aimed to explore exposures of workers to chemical contaminants, contribute to raise awareness of potential hazards and to further improve work practices. The study was conducted in eight European countries in a target population of 195 exposed and 73 controls. Biomarkers of exposure were used for selected metals and organic contaminants. Occupational hygiene sampling methods and contextual information were collected to facilitate the interpretation of the biomarker results. We found somewhat elevated exposures in workers for cadmium and mercury in blood and urine compared to controls. Blood analysis indicated high lead levels in post-shift compared to pre-shift in battery workers. Some urinary phthalate metabolite levels indicated a contribution from work-related exposures and were more pronounced in battery workers. Only small differences were observed in urinary excretion before and after the shift for organophosphorus flame retardants. Brominated flame retardant and PCB serum levels were in the range of general population background. From this exploratory study we conclude that more studies are needed to better understand chemical exposure in the processing of e-waste.

81 – A Review of PFAS Exposure in Occupational Settings

Brian Christensen, Miriam Calkins
National Institute for Occupational Safety and Health, Research Industrial Hygienist, Cincinnati, USA

Abstract
Industry began synthesizing long-alkyl chain per- and polyfluoroalkyl substances (PFAS) in the 1950s to enhance the water repellant, oil repellant, thermal stability, wetting, foaming, or dispersing properties of various industrial and consumer products. However, some of the same molecular properties that made these PFAS desirable in industry applications also contributed towards their persistence in both environmental and biological matrices. Over the last 20 years, voluntary phaseout efforts by industry paired with increased state- and federal-level regulations led to major reductions in US production and usage of long-alkyl chain PFAS. These reductions have led to a steep decline in long-alkyl chain PFAS serum levels in the general US population, but the relationship between reduction efforts and long-alkyl chain PFAS serum levels in workers has been less pronounced. Many workers still exhibit long-alkyl chain PFAS serum levels that exceed levels reported for the general US population and, with frequency, health-based advisory thresholds recommended by organizations such as NASEM. Furthermore, as increasingly robust analytical methods have become available over time, researchers have detected a wide variety of other PFAS with increased regularity in workplace settings, including short-alkyl chain PFAS, 'precursor PFAS', and other long-alkyl chain alternatives. This presentation is based on a recently published review article in the Journal of Exposure Science and Environmental Epidemiology and will highlight 1) temporal trends in methods used to analyze PFAS, 2) key environmental and biological exposure monitoring findings for different occupational groups, and 3) major research gaps in the current occupational exposure to PFAS literature.

83 – Understanding the role of environmental metals exposures on COVID-19 disease severity in New Mexico

Daniel Beene[1], Curtis Miller[2], Melissa Gonzales[3], Esther Erdei[4]
[1]University of New Mexico Health Sciences Center, Data Manager, Albuquerque, USA. [2]University of New Mexico Health Sciences Center, Statistician, Albuquerque, USA. [3]Tulane University, Professor and Chair, New Orleans, USA. [4]University of New Mexico Health Sciences Center, Associate Professor, Albuquerque, USA

Abstract
The impact of the COVID-19 pandemic on New Mexico is influenced by several environmental, sociodemographic, and political factors. Risk factors include an inability to mount a protective immune response against the virus, as well as sociodemographic and physiological covariates. We argue that chronic exposure to environmental metals renders communities at greater risk of mortality due to COVID-19. Environmental concentrations of metals are elevated through anthropogenic activities related to resource extraction. Large portions of the state are host to such activities, which are often situated in rural areas and in close proximity to underserved communities. In collaboration with the NM Department of Health, we paired geospatial models of environmental transport of mine waste with datasets reflecting measured concentrations of metals in environmental media to predict ZIP code-level counts of COVID-19-related deaths. Because a large proportion of ZIP codes report no deaths during the study timeframe, we leveraged zero-inflated models to predict a) ZIP codes with at least one death using a binary logistic regression, and b) the log ratio of deaths/overall COVID-19 tests recorded per ZIP code. Geographically weighted regression models demonstrate locally heterogeneous relationships between environmental pathways of metal exposures and COVID-19-related deaths while global models counterintuitively suggest that ZIP codes downwind from contamination sources are at a decreased risk. Results indicate that environmental and sociodemographic stressors coproduce increased risks of death, but that these relationships are obfuscated in non-local models. The current study contributes to environmental health policy evaluation, pandemic preparedness, and spatial non-stationarity in geographic health research.

84 – Emerging Chemical Exposures in Tribal Environments and Potential Immunological Health Effects

Esther Erdei[1], Elena O'Donald[2], Rae O'Leary[3], Lesliam Quiros-Alcala[4], Ana Rule[4]
[1]University of New Mexico Health Sciences Center, Associate Professor, Albuquerque, USA. [2]University of New Mexico Health Sciences Center, Research Scientist, Albuquerque, USA. [3]Missouri Breaks Industries Research Inc., Managing Director, Eagle Butte, USA. [4]Johns Hopkins University, Assistant Professor, Baltimore, USA

Abstract
The Cheyenne River is the only public drinking water source for the Cheyenne River Sioux Tribe (CRST) in South Dakota, home of 13,000 Tribal members. Thus, emerging chemicals such as Per- and Polyfluoroalkyl Substances (PFAS) that enter the river add to the cumulative exposure to legacy chemicals (e.g. metals) due to extensive mining in the area and pose a public health concern to the community. Our goal was to assess exposure to a panel of 13 PFAS in sera and 23 phthalates in urine among 50 participants from the CRST and examine immunological effects. Given the potential immunotoxicity of many of the studied compounds, we assessed antinuclear and tissue-specific autoantibody prevalence as a proxy of dysfunction in immune tolerance. Furthermore, we examined associations between autoimmunity markers and exposure to metals and emerging contaminants. Positive correlations were detected between serum silver and serum PFHxS concentrations while serum PFOA were correlated with serum zinc concentrations. Univariate quantile regression analyses revealed significant associations between serum PFHxS concentrations and the detection of serum anti-thyroglobulin autoantibodies. The β estimate was found as 4.2 for every quantile increase of concentration in age and gender-adjusted model. Similarly, serum PFOA concentrations significantly predicted increased serum anti-thyroid peroxidase autoantibodies. Several serum PFAS levels were also predictors of anti-native DNA and anti-chromatin autoantibodies. These results indicate changes to several physiological systems and alterations to genetic material that warrant further research and planned community action setting Tribal water protection policies.

85 – State-Level Biomonitoring Surveillance in Michigan: the Michigan Chemical Exposure Monitoring (MiChEM) Project

Rachel Long, Matthew Geiger, Kory Groetsch, Jennifer Gray, Elizabeth Wasilevich
Michigan Department of Health and Human Services, Lansing, USA

Abstract
The Michigan Department of Health and Human Services (MDHHS) launched the Michigan Chemical Exposure Monitoring (MiChEM) biomonitoring surveillance project in 2022. Funded in part by the Centers for Disease Control and Prevention (CDC) cooperative agreement to build states' capacity to conduct biomonitoring, the MiChEM project aims to establish reference concentrations for certain chemicals of concern among adult, non-institutionalized Michigan residents. This presentation will describe the design, methods, and expected outcomes of the MiChEM project. Lessons learned for building state biomonitoring surveillance programs and considerations for integrating biomonitoring data into state environmental health tracking portals will be discussed. MiChEM utilizes a three-stage stratified cluster sample. Recruitment involves mailings to selected addresses and promotion via traditional and social media and community outreach. Specimens are collected using a mobile laboratory at locations throughout the state. The MDHHS Bureau of Labs is measuring certain PFAS, polychlorinated biphenyls (PCBs), organochlorine pesticides (OCPs), polybrominated diphenyl ethers (PBDEs), polybrominated biphenyls (PBBs), and metals (including arsenic, cadmium, lead, and mercury) in collected specimens. Participants complete a survey to collect information about factors related to contaminant exposure and demographic data to identify differences in chemical concentrations among subpopulations. To date, 373 participants from 15 communities across Michigan have provided biospecimens and survey data. In addition to traditional results dissemination, results from the MiChEM project will be featured in the MiTracking data portal. The results will allow MDHHS to identify subpopulations with elevated exposures, track trends in exposure over time, and inform ways to reduce exposures among Michigan residents.

86 – POSTER: Identifying gut microbial cliques in association with prenatal Pb exposure using Microbial Co-occurrence Analysis (MiCA)

Vishal Midya[1], Jamil Lane[2], Libni A. Torres-Olascoaga[3], Robert O. Wright[4], Manish Arora[4], Martha María Téllez-Rojo[5], Chris Gennings[4], Shoshannah Eggers[6]
[1]Icahn School of Medicine at Mount Sinai, Assistant Professor, New York, USA. [2]Icahn School of Medicine at Mount Sinai, Postdoctoral Fellow, New York, USA. [3]National Institute of Public Health, Research Scientist, Cuernavaca, Mexico. [4]Icahn School of Medicine at Mount Sinai, Professor, New York, USA. [5]National Institute of Public Health, Professor, Cuernavaca, Mexico. [6]University of Iowa College of Public Health, Assistant Professor, Iowa City, USA

Abstract
Background and Method: We present a novel analytical approach to identify multiple bacterial taxa (microbial cliques) within the gut microbiome of children at 9-11 years associated with prenatal Pb exposure. Data came from a subset (n=123) of the Programming Research in Obesity, Growth, Environment, and Social Stressors cohort. Pb concentrations were measured in maternal whole blood from the second and third trimesters of pregnancy using ICPMS. Stool samples collected at 9-11 years old underwent metagenomic sequencing. In a novel analytical approach called Microbial Co-occurrence Analysis (MiCA), we first paired a machine-learning algorithm with a causal-inference framework to identify microbial cliques predictive of prenatal Pb exposure and then estimate causal association between prenatal Pb exposure and microbial clique abundance.

Results: Mean prenatal blood Pb was 33.6(SE=2.1)ug/L and 34.9(SE=2.1)ug/L at the second and third trimesters, respectively. With second-trimester Pb exposure, we identified a microbial clique that included Bifidobacterium adolescentis and Ruminococcus callidus, two common probiotic bacteria. Increasing second-trimester Pb exposure showed increased odds of having the microbial clique below the 50th percentile relative abundance (OR=1.02,95%CI[1.00-1.04]). We identified a dose-response effect, such that increasing Pb exposure showed higher odds of below-median relative abundance of the microbial clique. For people with second-trimester Pb exposure in the 50th-75th percentile, the mean odds of having the microbial clique below median relative abundance were 2.98(95%CI[2.79-3.19]).

Discussion: Using a novel combination of machine-learning and causal-inference, MiCA identified an association between second-trimester Pb exposure and reduced abundance of a microbial-clique.

87 – Examining the Geographical Distribution of Air Pollution Disparities Across Different Racial and Ethnic Groups: Incorporating Workplace Addresses

eunhye yoo[1], Abigail Cooke[1], Youngseob Eum[2]
[1]university at buffalo, associate professor, buffalo, USA. [2]Department of Geography and Earth Sciences, University of North Carolina at

Charlotte, Assistant professor, Charlotte, USA

Abstract
Air pollution exposure disparities among racial and ethnic groups in the United States have been extensively documented; however, most of the previous studies have only taken into account residential address. Nonetheless, relying solely on residence-based exposures and disparities could lead to bias, as individuals spend a significant portion of their time outside their homes, including their workplace. Furthermore, air pollution exposure disparities are rooted in complex and multifaceted social and geographical issues, such as underlying patterns of residential segregation, clustering of employment sites, the localized spatial patterns of air pollutant concentration.

To address these issues, we investigated a set of racial and ethnic disparities in multiple air pollution exposures, estimated using census data based on residence, as well as information on homes and workplaces of workers. Annual-averaged air pollution concentrations for three pollutants were obtained from spatio-temporal models. Air pollution exposure disparities were quantified at the census tract level across 8 core-based statistical areas by incorporating home and workplace addresses of workers.

Our findings indicated that taking into account workplace addresses had a significant impact on the exposure disparities observed among the race/ethnic groups, especially in the case of NO2. Furthermore, our results revealed regional differences in these disparities were substantial. We also found that Non-White were consistently the highest levels of exposure to air pollution, regardless of their residential workplace addresses and the overall population composition.
We concluded that consideration of both region-specific contexts and activity patterns is important to understand population level air pollution exposure disparities.

88 – Gene-environment Interaction in Long-term Effects of Polychlorinated Biphenyls Exposure on Glucose Homeostasis and Type 2 Diabetes Among Chinese Community Residents: The Modifying Effects of Genetic Risk and Lifestyle

Qiyou Tan[1], Shijie Yang[1], Bin Wang[2], Mengyi Wang[1], Linling Yu[1], Ruyi Liang[1], Wei Liu[1], Yanjun Guo[3], Jiahao Song[1], Min Zhou[2], Weihong Chen[4]
[1]School of Public Health, Tongji Medical College, Huazhong University of Science and Technology, Student, Wuhan, China. [2]School of Public Health, Tongji Medical College, Huazhong University of Science and Technology, Lecturer, Wuhan, China. [3]School of Public Health, Tongji Medical College, Huazhong University of Science and Technology, Associate Professor, Wuhan, China. [4]School of Public Health, Tongji Medical College, Huazhong University of Science and Technology, Professor, Wuhan, China

Abstract
The longitudinal relationships of polychlorinated biphenyls (PCBs) exposure with glucose homeostasis and type 2 diabetes (T2D) risk among Chinese population have not been assessed, and interactions of PCB exposure with genetic susceptibility and lifestyle are unclear. A total of 2,676 Chinese adults were enrolled at baseline from the Wuhan-Zhuhai cohort and 1,705 participants were followed up over 6 years. Fasting plasma glucose (FPG) and insulin (FPI) and seven serum indicator-PCBs were measured for each participant. Besides, we constructed the polygenic risk score (PRS) of T2D and healthy lifestyle score. Each 1-unit increase of ln-transformed PCB-118 was associated with a 0.142 mmol/L, 11.924 pmol/L, 0.662, and 76.7% increase in FPG, FPI, homeostasis model assessment of insulin resistance, and incident T2D risk over 6 years, respectively. Each 1-unit increment in T2D-PRS was related with a 0.169 mmol/L elevation of FPG and a 78.2% elevation of incident T2D risk over 6 years. Compared with participants who had low T2D-PRS and low PCB-118 exposure, participants with high T2D-PRS and high PCB-118 exposure showed a significant increase in FPG level (0.193 mmol/L; P interaction < 0.001) and incident T2D risk [ratio risk (RR) = 2.752]. Participants with low PCB-118 exposure, low genetic risk, and healthy lifestyle had the lowest risk of incident T2D, with RR (95% CI) of 0.207 (0.089, 0.480). Our findings highlighted the importance of reducing PCB exposure and improvement in lifestyle for T2D prevention and management, especially for individuals with higher genetic risk of T2D.

89 – Pesticides Biomonitoring: Feces as New Matrix and Comparison with Urine Data

Hans Mol[1], Jonatan Dias[1], Ling Yen[1], Paul Scheepers[2], Nelson Abrantes[3], Francisco Alcon[4], Abdallah Alaoui[5], Virginia Aparicio[6], Isabelle Baldi[7], Paula Harkes[8], Jakub Hofman[9], Matjaz Glavan[10], Trine Norgaard[11], Igor Paskovic[12], Daria Sgargi[13], Esperanza Huetra Lwanga[8], Coen Ritsema[14], Violette Geissen[14], Vera Silva[8], Anne Vested[15], Vivi Schlünssen[16]
[1]Wageningen Food Safety Research, researcher, Wageningen, Netherlands. [2]Radboud Institute for Biological and Environmental Sciences, researcher, Nijmegen, Netherlands. [3]Centre for Environmental and Marine Studies and Department of Biology, University of Aveiro, researcher, Aveiro, Portugal. [4]Universidad Politécnica de Cartagena, researcher, Cartagena, Spain. [5]Institute of Geography, University of Bern, researcher, Bern, Switzerland. [6]Instituto Nacional de Tecnología Agropecuaria - INTA, researcher, Buenos Aires, Argentina. [7]INSERM

U1219, EPICENE Team, Bordeaux University, researcher, Nouvelle-Aquitaine, France. [8]Soil Physics and Land Management Group, Wageningen University & Research, researcher, Wageningen, Netherlands. [9]Research Centre for Toxic Compounds in the Environment (RECETOX), Masaryk University, researcher, Brno, Czech Republic. [10]Biotechnical Faculty, University of Ljubljana, researcher, Ljubljana, Slovenia. [11]Department of Agroecology, Aarhus University, researcher, Aarhus, Denmark. [12]Institute of Agriculture and Tourism, Department of Agriculture and Nutrition, researcher, Porec, Croatia. [13]Cesare Maltoni Cancer Research Center, Ramazzini Institute, researcher, Bologna, Italy. [14]Soil Physics and Land Management Group, Wageningen University & Research, Professor, Wageningen, Netherlands. [15]Department of Public Health, Research Unit for Environment, Occupation and Health, Danish Ramazzini centre, researcher, Aarhus, Denmark. [16]Department of Public Health, Research Unit for Environment, Occupation and Health, Danish Ramazzini centre, Professor, Aarhus, Netherlands

Abstract
Humans are exposed to pesticides through their diet and the environment. Simultaneous exposure to multiple pesticides is a reality. For biomonitoring of pesticides, urine is the most common matrix but analysis is usually restricted to a limited number of pesticide biomarkers (metabolites of pesticides for which standards are available). In the EU project 'SPRINT' (https://sprint-h2020.eu/), for the first time, feces was included as HBM matrix for a wide scope of pesticides in a study covering 10 European countries and Argentina. One reason was to cover a higher number of pesticides (parent compounds, unabsorbed fraction and/or biliary excretion). A second reason was the increased interest in possible effects of pesticides on intestinal microbiota, calling for real-life occurrence data. Three methods were developed and validated for feces analysis, two multi-residue methods (LC-MS/MS, GC-MS/MS) and an additional method for glyphosate/AMPA, together covering 188 pesticides. In total 725 samples of feces were analyzed. Overall, 94 pesticides were detected, typically in the 1-10 µg/kg range, but also 10-100 µg/kg or even higher. Glyphosate, fipronil-sulfone, pyrethroids, and organochlorine pesticides were most frequently detected. The results were compared with results from analysis of urine samples concurrently taken from the same subjects and analyzed for 50 urinary biomarkers, including glyphosate/AMPA. Due to the larger scope, the number of pesticides found in feces was substantially higher than the number that could be measured in urine (94 versus 31). The fecal data were considered of added value for human exposure assessment of pesticides.

90 – Health risk assessment of arsenic-contaminated surface soils in West Bengal, India

Rahul Mishra[1], S. P. Datta[2], M. C. Meena[3], D. Golui[4]
[1]ICAR-Indian Institute of Soil Science, Scientist, Bhopal, India. [2]ICAR-Indian Institute of Soil Science, Director, Bhopal, India. [3]ICAR-Indian Agricultural Research Institute, Senior Scientist, New Delhi, India. [4]ICAR-Indian Agricultural Research Institute, Scientist, New Delhi, India

Abstract
Arsenic (As) contamination in soil and water directly or indirectly affect sustainable development goals and human health. This study aims to assess the extent of As contamination in soil and its effects on human health. For this purpose, 201 soil and plant samples were collected and analysed. Hazard quotient (HQ) for As intake by humans through the rice grain varied between 0.27 and 0.83. Total As in soil ranges from 3.47 to 53.1 mg kg^{-1} with a mean value of 19.3 mg kg^{-1}. The HQ and HI (hazard index) as the index of noncarcinogenic risk of As in soil through ingestion, dermal and inhalation pathways for adults and children were computed. The average HQ for ingestion, dermal and inhalation were 9.19E-02, 6.82E-03 and 2.11E-05, respectively, for adults; the corresponding value for children were 8.58E-01, 3.35E-03, 5.87E-05, respectively. The mean HI value for adults and children were 9.87E-02 and 8.61E-01, respectively. The trend of HQ was HQingestion > HQdermal > HQinhalation for both adults and children. Cancer risk (CR) observed that mean CRingestion, CRdermal, and CRinhalation for As for adults were 4.13E-05, 1.26E-06 and 3.89E-09, respectively, while the corresponding value for children was 3.86E-04, 6.17E-07 and 1.08E-08, respectively. Lifetime cancer risk (LCR) for adults and children was 4.26E-05 and 3.87E-04, respectively. The threshold limit for LCR is 1.0E-04, indicating significant carcinogenic risk due to As in the soils of the study region, and children are more vulnerable to the carcinogenic effect of As than adults.

91 – Is Ambient Temperature an Emerging Exposomal Factor to Induce Skin Aging? Evidence from German and Indian Women

Nidhi Singh[1], Prince Vijay[2], Harish Phuleria[3], Jean Krutmann[4], Tamara Schikowski[5]
[1]IUF-Leibniz Research Institute for Environmental Medicine, Postdoc, Düsseldorf, Germany. [2]Indian Institute of Technology Bombay, PhD student, Mumbai, India. [3]Indian Institute of Technology Bombay, Associate Professor, Mumbai, India. [4]IUF-Leibniz Research Institute for Environmental Medicine, University Professor and Scxientific Director, Düsseldorf, Germany. [5]IUF-Leibniz Research Institute for Environmental Medicine, Working group leader, Düsseldorf, Germany

Abstract
Extrinsic skin aging is significantly associated with chronic exposure to ultraviolet radiation (UVR) and air pollution (PM2.5). Rising temperature has been established as a potential threat to human health. Nevertheless, its effects on skin aging remain largely unknown. We analysed the association between environmentally-induced skin aging phenotypes and long-term exposure to mean temperature (Tmean). We initially used the data of 806 women (aged 66–79 years) from the German SALIA cohort and 1510 women (aged 20-91 years) from the Indian CASAI cohort. Five-year mean exposure to Tmean and other exposures were assigned at the residential addresses. Extrinsic skin aging phenotypes were assessed using validated SCINEXA™ visual scores. We employed an ordinal multivariate logistic regression model and adjusted for the potential confounding effect of UVR, PM2.5, and relative humidity (RH). We found a positive and significant association of Tmean with facial pigment spots (e.g. on the forehead for SALIA, OR:1.74, 95%CI: 1.01,2.98; for CASAI, OR: 6.85, 95% CI; 5.65, 8.31) and with coarse wrinkles (e.g. under the eyes for SALIA, OR:5.29, 95%CI: 3.15,8.87; in crow's feet for CASAI, OR:2.75, 95%CI: 2.34,3.22). In general, the mean extrinsic aging scores for pigment spots and coarse wrinkle signs on the face were higher in German women than Indian women, despite, the association was found to be higher in Indian women, which could be due to the relatively stronger effect of UV in German women. In conclusion, the higher ambient temperature might induce premature skin aging in combination with other potential environmental factors.

92 – Preliminary Results of Blood Analysis from the Sustainable Plant Protection Transition (SPRINT) Study

Paul T.J. Scheepers[1], Martien Graumans[2], Nina Wieland[2], Maurice van Dael[3], Franck Saint-Marcoux[4], Souleiman El Balkhi[5], Hans Mol[6], Nelson Abrantes[7], Francisco Alcon[8], Abdallah Alaoui[9], Virginia Aparicio[10], Isabelle Baldi[11], Paula Harkes[12], Jakub Hofman[13], Daniele Mandrioli[14], Matjaz Glavan[15], Trine Norgaard[16], Igor Pasković[17], Daria Sgargi[18], Esperanza Huerta Lwanga[19], Coen J. Ritsema[20], Violette Geissen[20], Vera Silva[12], Anne Vested[21], Vivi Schlünssen[22]
[1]Radboud Institute for Biological and Environmental Sciences, Radboud University, Associate Professor, Nijmegen, Netherlands. [2]Radboud Institute for Biological and Environmental Sciences, Radboud University, PhD candidate, Nijmegen, Netherlands. [3]Radboudumc, Senior technician, Nijmegen, Netherlands. [4]Toxicologie analytique environnementale et santé au travail Service de Pharmacologie, Toxicologie et Pharmacovigilance CHU de Limoges, Prof, Limoges, France. [5]Toxicologie analytique environnementale et santé au travail Service de Pharmacologie, Toxicologie et Pharmacovigilance CHU de Limoges, Senior technician, Limoges, France. [6]Wageningen Food Safety Research, Senior researcher, Wageningen, Netherlands. [7]Centre for Environmental and Marine Studies and Department of Environment and Planning, University of Aveiro, Research Assistant, Aveiro, Portugal. [8]Universidad Politécnica de Cartagena, Senior researcher, Cartagena, Spain. [9]Institute of Geography, University of Bern, Senior Research Scientist, Bern, Switzerland. [10]Instituto Nacional de Tecnología Agropecuaria, Senior researcher, Buenos Aires, Argentina. [11]INSERM, Bordeaux University, Professor, Nouvelle-Aquitaine, France. [12]Soil Physics and Land Management Group, Wageningen University & Research, Wageningen, Post-doc, Wageningen, Netherlands. [13]Research Centre for Toxic Compounds in the Environment (RECETOX), Faculty of Science, Masaryk University, Professor, Brno, Czech Republic. [14]Cesare Maltoni Cancer Research Center, Ramazzini Institute, Director, Bologna, Italy. [15]Biotechnical Faculty, University of Ljubljana, Senior researcher, Ljubljana, Slovenia. [16]Department of Agroecology, Aarhus University, Post-doc, Aarhus, Denmark. [17]Institute of Agriculture and Tourism, Department of Agriculture and Nutrition, Senior researcher, Poreč, Croatia. [18]Cesare Maltoni Cancer Research Center, Ramazzini Institute, Post-doc, Bologna, Italy. [19]Soil Physics and Land Management Group, Wageningen University & Research, Wageningen, Researcher, Wageningen, Netherlands. [20]Soil Physics and Land Management Group, Wageningen University & Research, Wageningen, Professor, Wageningen, Netherlands. [21]Department of Public Health, Aarhus University, Post-doc, Aarhus, Denmark. [22]Department of Public Health, Aarhus University, Professor, Aarhus, Denmark

Abstract
The current use of pesticides has an impact on health and the environment. Collection of real-life data on pesticide exposure is vital to inform changes in current agricultural practices. A study was conducted by human biomonitoring in 10 European countries and Argentina. Blood was collected from 736 subjects including farmers, and rural and urban consumers during the growth season of 2021 and analyzed for 198 pesticides (incl. metabolites) by LC-MS/MS and GC-MS/MS.

The median number of pesticides detected per blood sample was 2 (range: 0-10) for all participants from European countries and 5.5 (range: 2-10) in Argentina. Median concentrations ranged from 0.1 to 1 µg/L for 92 pesticides and >1 µg/L for chlorotoluron (n=4) and MCPA (n=2). p,p'DDE was the most abundant compound with 338 detects (45%). Glyphosate and AMPA were not found at an LOQ of 0.1 µg/L. In blood samples from Argentina, bromoxynil, fipronil-sulfone and chlorpyrifos-methyl were detected at relatively high frequencies (69, 67 and 53 out of 73 samples, respectively). These results were presented on group level and individually in a letter with a list of pesticides observed >LOQ.

This is the first study to report on such a large number of pesticides in human blood. Most pesticides were observed at levels <1 µg/L. Detection frequencies were low except for p,p'DDE and three pesticides in Argentina.

Acknowledgments: The SPRINT project receives funding from the European Union's Horizon 2020 Programme for research & innovation under grant no. 862568.

93 – Shifting for the Better in Microbial Risk Assessments

Caroline Rose[1], Mackenzie Lowe[2], Valar Anoop[3]
[1]Health Canada, Senior Evaluator, Ottawa, Canada. [2]Health Canada, Scientific Evaluator, Ottawa, Canada. [3]Health Canada, Section Head - Biotechnology 1, Ottawa, Canada

Abstract
The New Substances program (the program) of Health Canada and Environment and Climate Change Canada is responsible for conducting microbial risk assessments (MRA) to determine potential risks posed by micro-organisms to the environment, its biodiversity, or human health under the New Substances Notification Regulations (Organisms) [NSNR (Organisms)] of the Canadian Environmental Protection Act, 1999 (CEPA 1999). Until recently, those MRAs have been predominantly qualitative, and their outcomes supported by descriptive narratives. To reduce the reliance on solely qualitative assessments, a shift to a semi-quantitative approach is being implemented by the program. This poster will highlight the new approach which incorporates a modernized microbial risk assessment framework (MRAF) and a validated supporting computational tool (CT). An overview of exposure parameters and considerations for vulnerable populations, as well as the challenges and uncertainties associated with this new approach will be discussed. The modernized MRAF and the CT will strengthen the risk assessment process by improving efficiency and consistency. In addition, the CT will effectively use the quantitative data specific to the micro-organism under assessment, adding to the confidence of the risk assessment conclusion. In the future, it will enable the development of predictive modelling for environmental and human health risk assessments. The MRAF will be published for regulated parties and interested stakeholders to better understand the approach taken for MRAs conducted under CEPA 1999. Future work to improve the CT as new information and data become available will be discussed.

94 – New Tools to Support Understanding and Action During Personal Report-Back of Chemical Exposures

Katherine Boronow[1], Amy Padula[2], Sarah Geiger[3], Erin DeMicco[2], Aaron Maruzzo[1], Allison Landowski[2], Phil Brown[4], Rachel Morello-Frosch[5], Julia Brody[1]
[1]Silent Spring Institute, Newton, USA. [2]University of California San Francisco, San Francisco, USA. [3]University of Illinois Urbana-Champaign, Urbana-Champaign, USA. [4]Northeastern University, Boston, USA. [5]University of California Berkeley, Berkeley, USA

Abstract
Report-back is a valuable opportunity for participants in biomonitoring studies to build environmental health literacy because it creates a sensitive window for communications about chemical exposures. To support participants' ability to understand and act on their personal results, we developed and fielded two novel interventions: a tutorial designed to build confidence in reading individual exposure graphs and stimulate curiosity about personal exposure sources, and follow-up emails that reminded participants about taking exposure reduction actions. We returned results for chemical exposures measured during pregnancy to participants in the Chemicals in Our Bodies (CIOB) and Illinois Kids Development Study (IKIDS) cohorts using digital reports produced using DERBI. Participants were randomly assigned to the interventions. We assessed environmental health knowledge and exposure-related behaviors before and after report-back, and we recorded participants' responses during the graph reading tutorial. The tutorial was effective at building graph reading skills, especially among participants without a bachelor's degree. Prior to report-back, participants understood core concepts related to chemical exposures and health but held important misconceptions about chemicals regulation in the U.S. Report-back strengthened participants' confidence in their pre-existing knowledge and improved understanding about chemicals regulation, although future reports should continue to elevate messaging around this topic. Participants reported using the information in the report to take concrete actions in their lives, including checking product labels for information and making choices about products to stop using or replace.

96 – Overcoming Challenges in Conducting National-scale Chemical-specific Groundwater Exposure Assessments Under the Toxic Substance Control Act

Shawn Shifflett[1], Susanna Wegner[2], Nerija Orentas[3]
[1]USEPA, Physical Scientist, Washington DC, USA. [2]USEPA, Toxicologist, Washington DC, USA. [3]USEPA, Team Lead, Washington DC, USA

Abstract
The Office of Pollution Prevention and Toxics (OPPT) evaluates the risk of industrial chemical substance exposures to human health and the environment under the Toxic Substance Control Act (TSCA). As part of this effort, assessors consider the potential for chemical contamination of sourced drinking water including groundwater. Conducting national groundwater exposure assessments is challenging when considering sentinel exposures to susceptible subpopulations relying on vulnerable private drinking water wells. Groundwater contamination and the extent to which humans may be exposed are affected by aquifer characteristics; local climate; drinking water well location; type, frequency, and duration of chemical release; and drinking water consumption rates. An approach that accounts for these variables and provides estimated exposure including sentinel exposures is desired to overcome these challenges. This presentation will provide options for accounting for these variables in the development of exposure estimates at a national scale for use in TSCA Risk Evaluations. Specifically, it will explore models developed by EPA for groundwater-sourced drinking water exposure assessments and TSCA-specific considerations that are necessary to consider (e.g., conditions of use, environmental chemical releases, etc.).

Disclaimer: The views expressed in this abstract are those of the authors and do not necessarily represent the views or the policies of the U.S. Environmental Protection Agency.

97 – Emissions of Fluorotelomer Alcohols from Consumer Products

Xiaoyu Liu[1], Zachary Robbins[2], Peter Egeghy[3]
[1]US Environmental Protection Agency, Physical Scientist, Durham, USA. [2]Oak Ridge Institute for Science and Education, Postdoctoral Fellow, Durham, USA. [3]US Environmental Protection Agency, Research Environmental Health Scientist, Durham, USA

Abstract
Although airborne fluorotelomer alcohols (FTOHs) have been characterized in occupational settings, the residential indoor air burden of FTOHs from consumer products, building materials, and household dust is not well understood. Data for key parameters controlling emissions of FTOHs from consumer products (i.e., initial concentrations, emission rates, etc.) are still limited in the literature. This work reports FTOHs emissions from consumer products, mattress pad 1 (USA), mattress pad 2 (China), and stone cleaner/resealer 1 (USA). The FTOHs studied are: 4:2, 6:2, 8:2, and 10:2 FTOHs. The emission tests were conducted using a Markes Micro-Scale Chamber/Thermal Extractor™ (μ-CTE™) system that was set at 25 °C, 34% relative humidity, and approximately 100 mL/min of clean air. Air samples were collected using thermal desorption tubes at different time interval during tests. 4:2 FTOH was not detected in any of the products. Emissions of 8:2 and 10:2 FTOHs were determined from the mattress pad products with a relatively slow decay rate. Emissions of 6:2, 8:2, and 10:2 FTOHs were measured from the liquid stone cleaner/sealer product. The emission data collected was fitted by the least-square method to the first order decay model using the SCIENTIST® program (Micromath Scientific Software) to obtain the initial emission factor (EF0, 4.43E-04 to 0.25 mg/m2/h) and the first order decay rate constant (k, 0.03 to 1.03 /h) based on ASTM D5116. The emission pattern and the emission parameters obtained from this work will provide information and support per- and polyfluoroalkyl substances inhalation exposure assessment by modelers.

98 – Per- and Polyfluoroalkyl Substances (PFAS) in House Dust and Blood Serum of Pregnant People and their Relationships with Sociodemographic and Lifestyle Factors

Dylan Wallis[1], Nicole DeLuca[2], James McCord[3], Kent Thomas[4], Elaine Cohen Hubal[5], Jeffrey Minucci[6]
[1]Environmental Protection Agency, ORISE Post Doctoral Candidate, Research Triangle Park, NC, USA. [2]Environmental Protection Agency, Post Doctoral Researcher, Research Triangle Park, NC, NC, USA. [3]Environmental Protection Agency, Chemist, Research Triangle Park, NC, USA. [4]Environmental Protection Agency, Research Physical Scientist, Research Triangle Park, NC, USA. [5]Environmental Protection Agency, Senior Science Advisor, Research Triangle Park, NC, USA. [6]Environmental Protection Agency, Physical Scientist, Research Triangle Park, NC, USA

Abstract
Per- and polyfluoroalkyl substances (PFAS) are linked to substantial negative health effects and have become ubiquitous in the environment. Understanding the contribution of house dust to PFAS exposure and what factors influence PFAS in dust is important to our understanding of indoor PFAS exposure.

This investigation measured sixteen PFAS in 104 paired blood and dust specimens archived from the National Children's Study's (NCS) Vanguard Study and characterized patterns of exposure for these participants. The relationships between serum and dust PFAS levels were characterized for the sixteen PFAS and the summed concentration of PFAS. Survey data was used to elucidate associations between lifestyle factors, sociodemographic factors, and PFAS levels in house dust.

Levels of the sixteen PFAS were detected in at least 50% of samples for four PFAS in serum and nine in dust. The geometric mean concentrations in serum ranged from 0.0062 ng/mL (PFPeS) to 4.1 ng/mL (PFOS) and 0.077 ng/g (PFNS) to 17 ng/g (PFOS) in dust. Perfluorooctanoic acid (PFOA) levels and total PFAS levels in dust were associated with serum levels. Higher levels of one or more PFAS in dust were linked to several sociodemographic and lifestyle characteristics.

This research draws important connections between PFAS exposure through dust and biomonitoring data, and connects exposure to characteristics of a vulnerable population—pregnant people and children.

Disclaimer: The views expressed in this presentation are those of the author(s) and do not necessarily represent the views or the policies of the U.S. Environmental Protection Agency.

100 – A Novel Urinary Biomarker for Benzene Exposure Assessment

Deepak Bhandari[1], Brett Bowman[2], Benjamin Blount[3]
[1]Centers for Disease Control and Prevention, Lead Chemist, Atlanta, USA. [2]Centers for Disease Control and Prevention, Chemist, Atlanta, USA. [3]Centers for Disease Control and Prevention, Supervisory Chemist, Atlanta, USA

Abstract
Benzene is a common environmental pollutant and a known human carcinogen. Urinary phenylmercapturic acid (PhMA) is a commonly used biomarker of benzene exposure, but a recent interlaboratory comparative study identified significant quantitative limitations in currently used PhMA methods. In this study, we present prePhMA [N-acetyl-S-(6-hydroxy-2,4-cyclohexadien-1-yl)-L-cysteine], a precursor to PhMA, as a novel and specific biomarker of benzene exposure and compare it with the existing PhMA biomarker.

We developed and validated the first quantitative method for measuring prePhMA in human urine using LC-ESI-MS/MS. Pre-PhMA stability was tested at various urine pHs. The quantitative agreement of urinary pre-PhMA concentrations and the ruggedness of the method were compared using LC columns with different phase selectivity. In addition, urinary pre-PhMA and PhMA concentrations were measured in 100 anonymized human urine samples and the impact of tobacco smoke exposure modeled via linear regression analysis.

The pre-PhMA biomarker was stable across the normal urine pH range, and the method exhibited sub-two nanogram per milliliter sensitivity, linearity over three orders of magnitude, and precision and accuracy within 10%. Smoking was associated with significantly higher urinary pre-PhMA concentrations compared to non-smoking individuals. Regression analysis of pre-PhMA vs. PhMA indicated a strong positive correlation ($r=0.951$) and a slope of 5.34, indicating the quantitative disagreement in the molar concentration of two biomarkers in urine samples.

The pre-PhMA biomarker reduces analytical variation between laboratories compared with PhMA, and therefore may have important applications in benzene exposure assessment in occupational and environmental settings.

101 – Pediatric Psychosocial Problems and Exposure to Consumer Product Chemicals among Children in an Urban Environment

Grant Tore[1], Rachelle Koehl[2], Elizabeth C. Matsui[3,4], Nadia N. Hansel[5], Meredith C. McCormack[6,7], Lesliam Quirós-Alcalá[8]
[1]Department of Environmental Health and Engineering, Johns Hopkins University, PhD Student, Baltimore, USA. [2]Bridging Research, Lung Health, and the Environment (BREATHE) Center, Johns Hopkins University, Research Program Manager, Baltimore, USA. [3]Department of Population Health, Dell Medical School at UT Austin, Professor, Austin, USA. [4]Department of Pediatrics , Dell Medical School at UT Austin, Professor, Austin, USA. [5]School of Medicine, Johns Hopkins University, Professor, Baltimore, USA. [6]Bridging Research, Lung Health, and the Environment (BREATHE) Center, Johns Hopkins University, Director, Baltimore, USA. [7]School of Medicine, Johns Hopkins University, Associate Professor, Baltimore, USA. [8]Department of Environmental Health and Engineering, Johns Hopkins University, Assistant Professor, Baltimore, USA

Abstract
Environmental phenols and parabens commonly found in consumer products may interfere with biological pathways involved in brain development. While some epidemiologic studies suggest that prenatal exposure to some of these chemicals impacts behavior, studies on postnatal exposures are scarce. We examined associations between exposure to phenols and parabens and behavior among 110

children (8-17 years) with asthma, participating in an indoor air pollution study in Baltimore City.

We measured 12 consumer product chemicals (2,4-dichlorophenol, 24DCP; 2,5-dichlorophenol, 25DCP; bisphenol A, BPA; bisphenol S, BPS; bisphenol F, BPF; methylparaben, MPB; ethylparaben, EPB; butylparaben, BPB; propylparaben, PBP; benzophenone-3, BP3; triclocarban, TCC; and triclosan, TCS) in urine samples. Child behaviors were assessed using the Child Behavior Checklist (CBCL/6-18). We conducted covariate-adjusted linear regression to examine associations between biomarker concentrations and internal, external, or total problems(T-scores).

Participants were an average age of 11 years (± 2), African American (86%), male (57%), and receiving public health insurance (83%). 24DCP, 25DCP, BP3, BPA, BPS, MPB, and PPB were detected in >90% of children. BPS, MPB and PPB geometric mean concentrations were higher than among U.S. children. The mean±SD CBCL T-scores were: internalizing(54.9±0.6), externalizing(55.3±10.9), and total(56.2±10.6). BPA exposure was associated with increased externalizing (β=5.9,p = 0.02) and total(β=5.2,p = 0.04) problems and 24-DCP exposure was associated with increased internalizing problems(β=4.8,p = 0.02).
Findings suggest that select phenols may increase risk of psychosocial impairments. Further investigation is warranted to elucidate pathways between exposure to consumer product chemicals and behavioral development.

102 – Source Attribution of PFAS measured in Serum from California Residents in 2018-2020

Emily H. Pennoyer[1], Toki Fillman[2], Wendy Heiger-Bernays[3], Kathleen Attfield[2], Thomas F. Webster[4]
[1]Boston University School of Public Health, Doctoral Candidate, Boston, USA. [2]Biomonitoring California, Environmental Health Investigations Branch, California Department of Public Health, Research Scientist, Richmond, USA. [3]Boston University School of Public Health, Clinical Professor, Boston, USA. [4]Boston University School of Public Health, Professor, Boston, USA

Abstract
Exposure to per- and polyfluoroalkyl substances (PFAS) occurs through multiple pathways, the most studied being ingestion via drinking water. Dietary contributions have not been well characterized and previous studies on dietary PFAS exposure do not account for exposure via water, which may confound the diet-PFAS relationship. We aim to determine the relative contributions of diet and drinking water to the body burden of PFAS.

This study uses biomonitoring data for 12 PFAS measured in serum from southern California residents (n=872) participating in the California Regional Exposure Study (2018-2020). Participants' geocoded addresses were linked by the corresponding public water system (PWS) to occurrence of PFAS in drinking water from EPA's third Unregulated Contaminant Monitoring Rule (UCMR3) database (2013-2015). Dietary contributions were assessed using information collected on food consumption frequency.

Seven PFAS were detected in ≥65% of participants. Geometric mean concentrations for these compounds ranged from 0.05-2.19 ng/mL, the highest being PFOS (2.19 ng/mL; 95% CI: 1.72, 2.66). Of the 729 participants served by a PWS in UCMR3, 53 (7%) had at least one detection of PFOA, PFOS, or PFHxS in water. Preliminary unadjusted analyses show serum concentrations of PFOA and PFHxS significantly differed between participants with and without detectable measures of these compounds in water (p-values <0.05).

These observed associations suggest contributions of PFOA and PFHxS from water for residents of southern California. Levels of PFOS in serum were not initially associated with detections of PFOS in water, supporting evaluation of other known pathways, including dietary fish consumption.

103 – Estimating Releases from Chemical Processes for Occupational and Fenceline Community Exposure Assessments

Raymond Smith[1], David Meyer[1], Gerardo Ruiz-Mercado[1], William Barrett[1], Michael Gonzalez[2], Sudhakar Takkellapati[3], John Chea[4]
[1]U.S. Environmental Protection Agency, Research Chemical Engineer, Cincinnati, USA. [2]U.S. Environmental Protection Agency, Associate Division Director, Cincinnati, USA. [3]U.S. Environmental Protection Agency, Chemist, Cincinnati, USA. [4]Oak Ridge Institute for Science and Education, ORISE Participant, Cincinnati, USA

Abstract
New chemicals are introduced into commerce regularly, and existing chemicals are sometimes repurposed for new uses. In both cases the chemicals may present health risks to occupational workers that can be evaluated via exposure and risk assessments. Additionally, chemical releases from facilities can represent increased exposures and risks to communities near the facility fenceline. Quantification

of releases is an important step in conducting these analyses, along with exposure and hazard assessments. This contribution describes research at the U.S. Environmental Protection Agency's Office of Research and Development to support risk evaluations, using mechanistic, database, and machine learning models to estimate releases. A type of mechanistic model, known as a Generic Scenario, relates chemical usage, activities, and properties to describe releases. Other mechanistic models, known as bottom-up models, have been built from process simulation and models of unit operations that combine to form a chemical process. Database, or top-down, models are derived directly from inventories available through EPA and others who collect and curate the information. Often collected at a facility level, context-based filter rules are applied to the release data to more accurately represent chemical processes. Finally, machine learning models will be described for work in progress using the predictive power of training and test data sets to estimate releases. These models will be presented for a case study with resulting estimates for releases.
The views expressed in this presentation are those of the authors and do not necessarily represent the views or policies of the U.S. Environmental Protection Agency.

104 – POSTER: Occupational exposure to PAHs in life cycle assessment: a novel characterization approach based on chemical body burden and working hours

Ikhtiar Uddin[1], Iffat Ms[2], Wahaj Uddin[3], Atif Kamal[4], Bing Wang[5], Aizaz Ullah Shah[6], Audil Rashid[7]
[1]EcoHealth Lab, PMAS Arid Agriculture University, Senior Research Fellow, Rawalpindi, Pakistan. [2]Shaheed Zulfiqar Ali Bhutto Medical University, M.Phil scholar, Islamabad, Pakistan. [3]Islamic International University, Research Assistant, Islamabad, Pakistan. [4]Quaid-i-Azam University, Assistant Professor, Islamabad, Pakistan. [5]Department of Epidemiology and Health Statistics, School of Public Health, Peking University, Associate Professor, Beijing, China. [6]2. Molecular Biology & Biochemistry, Shaheed Zulfiqar Ali Bhutto Medical University, M.Phil Schlar, Islamabad, Pakistan. [7]EcoHealth Lab, PMAS Arid Agriculture University, Professor, Rawalpindi, Pakistan

Abstract
There are various methods available for assessing occupational health, including those that focus on morbidity and symptomatic cases only, those that assess the number of work-days lost. In this context, life cycle impact assessment (LCIA) is key attribute to depict exposure impacts to worker health. The study proposes describes LCIA of occupational exposure to PAHs using a novel approach based on skin contact and inhalation. We have combined workers statistics and PAH body burden from petroleum related occupations (petrol pump workers, auto-mechanics, car-spray painters). The characterization factors shown in DALY per hour worked and are provided with confidence intervals for 6 PAHs estimated in serum and 3 metabolites in urine. Findings of job-specific questionnaires for exposed groups revealed their association with potential carcinogens 1-hydroxypyrene in urine and benzo[a]pyrene in serum. Lung caner risk found strongly associated with PAH exposure after adjustment for smoking and poor health (OR=4.6, 95%CI=1.7–12.8) and has shown apparent dose-response (linear trend $p<0.01$) for exposure duration. This approach makes it possible to consistently calculate the potential human health impacts of occupational exposure to PAHs by combining body burden estimates and working hour data. Additionally, low economic status, poor ventilation at workplace and absence of safety protocols are also found significant confounding factors for health vulnerability with noticeable impact on life cycle assessment of occupational PAH exposure. Our findings broadens the LCIA scope to analyze population exposure at workplace environment across various occupations.

105 – Mortality burden due to ambient nitrogen dioxide pollution in China: Application of high-resolution models

Xinyue Li[1], Peng Wang[2], Weidong Wang[1], Hongliang Zhang[3], Su Shi[1], Tao Xue[4], Jintai Lin[4], Yuhang Zhang[5], Mengyao Liu[5], Renjie Chen[3], Haidong Kan[3], Xia Meng[2]
[1]Fudan University, Student, Shanghai, China. [2]Fudan University, Associate Professor, Shanghai, China. [3]Fudan University, Professor, Shanghai, China. [4]Peking University, Professor, Beijing, China. [5]Peking University, Student, Beijing, China

Abstract
Background: A large gap exists between the latest Global Air Quality Guidelines (AQG 2021) and Chinese air quality standards for NO2. Assessing whether and to what extent Chinese standards for NO2 should be tightened requires a comprehensive understanding of the spatiotemporal characteristics of NO2 exposure and related health risks.

Objective: We predicted ground NO2 concentrations with high resolution in mainland China, explored exposure characteristics to NO2 pollution, and assessed the related mortality burden.

Methods: Daily NO2 concentrations in 2019 were predicted at 1km resolution in mainland China using forest models incorporating multiple predictors. From high-resolution predictions, we explored the spatiotemporal distribution and exceedance rates of NO2 exposure, and premature deaths related to long- and short-term NO_2 exposure in China.

Results: The cross-validation R^2 and root mean squared error of the NO_2 model were 0.80 and 7.78 μg/m3 at the daily level in 2019. The percentage of people with annual NO_2 exposure over AQG guideline of 10 μg/m3 reached 99.68%. NO_2 levels and population exposure risk were elevated in urban areas. Long- and short-term exposures to NO_2 were associated with 285,036 and 121,263 non-accidental deaths, respectively, in China in 2019. Tightening standards in steps gradually would increase the potential health benefit.

Conclusion: In China, NO_2 pollution is associated with significant mortality burden. Spatial disparities exist in NO_2 pollution and exposure risks. China's current air quality standards may no longer objectively reflect the severity of NO_2 pollution and exposure risk. Tightening the standards for NO_2 will lead to significant health benefits.

106 – POSTER: Excessive Herbicide Application: The Cases Study in Nan and Saraburi Provinces in Thailand

Densak Yogyorn[1], Sara Arphorn[2]
[1]Mahidol University, Assistant Professor, BANGKOK, Thailand. [2]Mahidol University, Associate Professor, Bangkok, Thailand

Abstract
Contaminated water in most rivers in Thailand linked to agrochemical run-off from farms directly related to excessive application of insecticides and herbicides, particularly by crops and vegetables growers. Since agrochemical application increases yield and makes crops look good appearance, satisfy the customers in the cities. Spraying crops daily, without proper knowledge, seems to be the norm for the crop's growers. On the other hand, the effectiveness of crop protection from soil-incorporated herbicide depends on the selection of spray nozzles, droplets size and soil permeable characteristic.

Objective: This study was aimed at the crop-grower's knowledge on the selection of appropriate spray nozzles. Method: This was a cross-sectional study, questionnaires developed, and 418 respondents recruited in two agricultural crops growing areas, Nan and Saraburi provinces. Result: This study found that 31.5% of the crop farmers used the flood-type nozzles which is the most effective type, produce large droplets, less prone to drift, reducing off-target contamination. While most crop farmers (63.5%) used the less effective nozzles, flat fan nozzles (43.5.0%) and cone type nozzles (25.0%). Conclusions: Excessive spraying of herbicide harm to health and safety of the crop's growers themselves and pollutes rivers. Thus, this study recommends that training on the use of spraying equipment must be provided for crop growers around the country.

Keywords: Herbicide, Spray, Nozzles, Health, Pollute

107 – The Exposome Atlas: Everything, Everywhere, All the time, All at Once

Marie Lynn Miranda[1], Rosalind Wright[2], Robert Wright[3]
[1]University of Illinois-Chicago, Chancellor, Chicago, USA. [2]Icahn School of Medicine at Mount Sinai, Dean, Translation Medicine, New York, USA. [3]Icahn School of Medicine at Mount Sinai, Chair, Environmental Medcine and Public Health, New York, USA

Abstract
Exposomics is directed to understanding all variation in environmental exposure across the lifespan. To move the field forward, innovative tools and infrastructure are needed. As an example, the human genome project mapped a reference sequence and the genome's variability and correlation structure in the HapMap project to facilitate genome wide association studies. Exposomics needs analogous reference databases and new statistical approaches to leverage the deep variability and complex correlational structure in exposome wide association studies. Because exposomics involves multiple data types (assays, sensor data, questionnaires, satellite measures, images, etc), it requires a multi-dimensional atlas. The exposome varies over space and time and by life stage and socioeconomic status. Its correlational structure will vary along these gradients, as well as with structural societal factors such as racism. An international exposome atlas will enable the reconstruction of life stage specific environmental exposures through modeling of key "exposomic tag measures"- analogous to haplotype tag SNPs. Geospatial modelers, demographers, social scientists, chemists, exposure scientists, toxicologists, epidemiologists, clinicians, infectious disease experts, and physicists are needed. The atlas should be annotated with evidence-based studies of biological effects including their contextual nature (e.g., the differential impact of lead exposure in poor urban areas vs wealthy suburban communities). This talk will highlight the potential of an exposome atlas to characterize past and present individual and community-level exposomes and the challenges involved in constructing and analyzing spatiotemporal data architectures. Even in early stages, an Exposome Atlas will be transformational and will fuel an exponential growth in exposomic research.

108 – Longitudinal Effects of Single and Multiple Phthalate Exposures on Inflammatory Biomarkers in Pregnant African Americans

Kaitlin R. Taibl[1], Anne L. Dunlop[2], Dana Boyd Barr[2], P. Barry Ryan[2], Parinya Panuwet[3], Jasmin A. Eatman[1], Youran Tan[1], Donghai Liang[3], Stephanie M. Eick[3]
[1]Emory University, PhD Student, Atlanta, Georgia. [2]Emory University, Professor, Atlanta, Georgia. [3]Emory University, Assistant Professor, Atlanta, Georgia

Abstract
Background/Aim: During pregnancy, the immune system undergoes unique adaptations and is particularly sensitive to immunogenic chemicals, including phthalates. Current knowledge of how phthalate exposures impact inflammation in pregnant people is limited by cross-sectional study designs and single pollutant models. Here, we estimate the longitudinal effect of single and multiple phthalates on inflammatory biomarkers between early and late pregnancy.

Methods: In the Atlanta African American Maternal-Child Cohort, samples were collected up to two times during pregnancy (visit 1 mean=11 weeks gestation; visit 2 mean=26 weeks gestation; N=216). We measured urinary concentrations of eight phthalate metabolites and serum concentrations of five inflammatory biomarkers, including C-reactive protein (CRP), interferon-γ (IFNγ), interleukin-6 (IL-6), IL-10, and tumor necrosis factor-α (TNFα). Linear mixed effects regression and quantile g-computation estimated the associations for single and multiple phthalates, respectively. All models included gestational age at sample collection, urinary creatinine, maternal age, education, BMI, and use of addictive substances as potential confounders and a random intercept for participant ID.

Results: Mono-2-ethyl-5-carboxypentyl phthalate (MECPP) and mono-2-ethyl-5-hydroxyhexyl) phthalate (MEHHP) were positively associated with IFNγ (MECPP %Δ=12.75, 95% CI=0.24, 26.82; MEHHP %Δ=8.33, 95% CI=0.16, 17.16). TNFα increased in response to mono-benzyl phthalate (MBzP) (%Δ=5.13, 95% CI=1.09, 9.33) and the phthalate exposure mixture (%Δ=0.10, 95% CI=-0.01, 0.21). We did not observe consistent patterns between the other biomarkers.

Conclusion: Our findings suggest phthalate exposures increase inflammation in pregnant African Americans. The inflammatory biomarkers, IFNγ and TNFα, may play a role in the relationship between prenatal phthalate exposure and adverse pregnancy outcomes.

109 – Novel US nationwide estimates of regulated public water contaminants at various spatial and temporal resolutions for epidemiologic study

Anne E. Nigra[1], Tessa R. Bloomquist[2], Adina Cazacu-De Luca[3], Ishika Kaushal[3], Irene Martinez-Morata[2], Ana Navas-Acien[4], Kevin Patterson[2], Filippo Ravalli[3], Maya Spaur[2], Raenita Spriggs[2]
[1]Columbia University Mailman School of Public Health, Assistant Professor, New York, USA. [2]Columbia University Mailman School of Public Health, Doctoral student, New York, USA. [3]Columbia University Mailman School of Public Health, Student, New York, USA. [4]Columbia University Mailman School of Public Health, Professor, New York, USA

Abstract
Previously, there were no nationwide estimates of contaminants regulated by the US Environmental Protection Agency (USEPA) in community water systems (CWSs) that could be readily leveraged for epidemiologic research. We generated contaminant estimates nationwide at the CWS-, Census block-, zip code tabulation area (ZCTA)-, and county-level for epidemiologic research.

We evaluated >13 million USEPA-compiled compliance monitoring records (2000-2019, >92% of systems) for N=76 contaminants (inorganics, organics, radionuclides, and disinfection byproducts, including arsenic, fluoride, total trihalomethanes, total haloacetic acids, nitrate, and others). We aggregated contaminant concentrations to USEPA's Standardized Monitoring Framework periods (e.g. quarterly for disinfection byproducts, triennially for inorganics), accounting for treatment. We developed population-weighted averages by joining city-served and system name to Census Place-Relationship Files (ZCTA-level nationwide), and by spatially joining CWS service area and administrative boundaries (ZCTA- and Census block-level, N=14 states).

Estimates are available nationwide at the CWS (N>45,000), ZCTA (N>14,300), and county (N>2,800) levels; and for N>1.9 million Census blocks. We linked these estimates to several NIH-funded cohorts, including the Strong Heart Study, the Multi-Ethnic Study of Atherosclerosis, the National Health and Nutrition Examination Survey, and dozens of birth cohorts in the Environmental influences on Child Health Outcomes consortium. We will present epidemiologic associations with urinary biomarkers of internal dose, cardiometabolic and cardiovascular disease, and birth outcomes.

Our novel approach enables the epidemiologic study of regulated water contaminant concentrations and diverse adverse health outcomes nationwide. Individual-level, time-weighted cumulative and/or average contaminant exposure estimates can be generated for other cohorts by leveraging residential addresses/histories.

110 – Measured Air Concentrations of Pesticides for the Estimation of Exposure to Vapour

Edgars Felkers[1], Felix Kluxen[2], Sarah Adham[3], Anne-Kim Vinck[4], Neil Morgan[5]
[1]Bayer AG, Occupational and Residential Safety Expert, Monheim, Germany. [2]ADAMA Deutschland GmbH, Global Regulatory Toxicologist, Cologne, Germany. [3]Corteva Agriscience, Global Human Health Risk Leader, Abingdon, United Kingdom. [4]Bayer SAS, Occupational and Residential Safety Expert, Sophia Antipolis, France. [5]Syngenta, Jealott's Hill International Research Centre, Principal Technical Expert, Bracknell, United Kingdom

Abstract
In order to address the identified need for a revision of default air concentration values and assumptions applied in assessing vapour exposure in the risk assessment of bystanders and residents to plant protection products in Europe, we evaluated inhalation exposure via vapour using previously unpublished data from 29 field and wind tunnel studies. The database comprises 35 trials with 11 active ingredients covering a wide range of scenarios with respect to vapour pressure, crops, application rates and European regions. Of the 961 individual measurements, 634 were below the Limit of Detection (LOD), 282 were between the LOD and Limit of Quantification (LOQ) and only 45 (4.7%) were quantifiable. Ten individual non-normalized samples exceeded 0.1 µg/m3. Of the 81 first day measurements after the application, 36 were <LOD, and quantifiable mean, 75th and 95th percentiles values were 0.114, 0.083, 0.552 µg/m3/kg AI applied/ha, respectively. No robust correlations between air concentration and temperature, leaf coverage, humidity, wind speed, and field size were identified; there is very limited correlation between air concentration and vapour pressure and Henry's constant in a subset of the data. These data indicate that potentially inhalable pesticide vapour within or near fields occurs only at very low concentrations in real conditions.

111 – Protection by Ordinary Light Clothing Against Pesticide Spray Drift for Bystanders and Residents

Edgars Felkers[1], Felix Kluxen[2], Christian Kuester[3], Sarah Adham[4]
[1]Bayer AG, Occupational and Residential Safety Expert, Monheim, Germany. [2]ADAMA Deutschland GmbH, Global Regulatory Toxicologist, Cologne, Germany. [3]Bayer AG, Regulatory Scientific Affairs Manager, Monheim, Germany. [4]Corteva Agriscience, Global Human Health Risk Leader, Abingdon, United Kingdom

Abstract
There are stringent EU regulatory requirements to assess pesticide exposure to bystanders and residents to direct spray drift. A "light clothing" adjustment factor (AF) of 0.82 is applied in the exposure assessment, based on simple assumptions for covered body surface and penetration through clothing. To assess the appropriateness of the AF, we collated data from 32 field studies. The mean and 25th percentile % reduction from ordinary light clothing ("reduction %") in children and adults for all crops and standard and drift-reducing nozzles were 42.7% and 36.2%, resulting in AF of 0.573 and 0.638, respectively. Sources of variation were investigated, e.g., crop type, leaf coverage, buffer, spray pressure, and nozzle type, which indicated that reduction % could be impacted by several conditions. The reduction % is similar between crops; therefore, a single AF value covering all crops can be derived. One exception was for early-stage vineyard scenarios (the reduction % is lower (27%) than late stage (42–47%)) and could be considered individually to avoid unnecessary conservatism for the other scenarios. This evaluation demonstrates the current AF to be overly precautionary, and a more realistic, exposure scenario-relevant value could be applied for bystander/resident risk assessments.

112 – Using Multiple Data Streams to Evaluate a Landmark Environmental Regulation: California's Proposition 65 and Chemical Exposures

Megan Schwarzman[1], Robin Dodson[2], Kristin Knox[3], Jennifer Ohayon[2], Claudia Polsky[4], Ruthann Rudel[5]
[1]University of California, Berkeley, Associate Project Scientist and Continuing Lecturer, Berkeley, USA. [2]Silent Spring Institute, Research Scientist, Newton, USA. [3]Silent Spring Institute, Staff Scientist, Newton, USA. [4]University of California, Berkeley, Clinical Professor of Law, Berkeley, USA. [5]Silent Spring Institute, Director of Research, Newton, USA

Abstract
The Safe Drinking Water and Toxics Enforcement Act, commonly known as Prop 65, is a landmark right-to-know law passed by California voters in 1986. In separate studies, we examined how Prop 65 has affected exposures to toxics by examining trends in national biomonitoring data, analyzing an inventory of VOCs in consumers products sold in CA, interviewing businesses, and investigating the law's non-enforcement-related impacts on the use and governance of Prop 65 chemicals. We found that (1) Californians generally had lower levels of biomonitored chemicals compared to the rest of the US population, although most Prop 65-listed chemicals are not biomonitored; (2) Volatile Prop 65-listed chemicals are ubiquitous in consumer products and workplaces, with over 5,000 pounds released indoors in California in one year; (3) the law has prompted industry to avoid some Prop 65-listed chemicals, sometimes substituting related but un-listed chemicals, and manufacturers reformulate products nationally, rather than manufacturing products specifically for California; and (4) beyond legal enforcement actions, Prop 65 has had hidden impacts by spurring direct chemical regulation and other policy actions and by disseminating information on toxics in otherwise information-poor marketplaces. Taken together these findings illustrate the diverse and wide-ranging effects of the law and suggest it has been an important tool for reducing exposure to carcinogens and reproductive/developmental toxicants.

113 – POSTER: Advancing the Sociome For Social and Health Equity: Pediatric Asthma on the South Side of ChicagoAdvancing the Sociome For Social and Health Equity

Sandra Tilmon[1], Sanjay Krishnan[2], Ellen Cohen[3], Doriane Miller[4], Jonathan Ozik[5], Julian Solway[6], Samuel Volchenboum[4], Ankur Bhargava[7], Suzi Birz[8], Brian Furner[9], Anthony Solomonides[10], Carlos Santos[11], Sharmilee Nyenhuis[4], Bruno Barbarioli[12], Bradley Carlson[13], Emily Dillon[14], Zhong Huang[13], Julie Johnson[15], Kevin Lazenby[13], Kaitlyn Li[16], Doriane Miller[6], Marc Sleiman[13]
[1]University of Chicago, Healthcare Data Scientist, Chicago, USA. [2]University of Chicago, Assistant Professor, Chicago, USA. [3]University of Chicago, Executive Director, Chicago, USA. [4]University of Chicago, Associate Professor, Chicago, USA. [5]Argonne National Laboratory, Principal Computational Scientist, Chicago, USA. [6]University of Chicago, Professor, Chicago, USA. [7]University of Chicago, Informatics Fellow, Chicago, USA. [8]University of Chicago, Consultant, Chicago, USA. [9]University of Chicago, Senior Director of Data and Technology, Chicago, USA. [10]North Shore University Health System, Program Director, Evanston, USA. [11]Rush University, Associate Professor of Medicine, Chicago, USA. [12]University of Chicago, PhD, Chicago, USA. [13]University of Chicago, Medical student, Chicago, USA. [14]Rush University, Postdoctoral Fellow, Chicago, USA. [15]University of Chicago, Executive Director, CRI, Chicago, USA. [16]University of Chicago, Data Science Fellow, Chicago, USA

Abstract
Beyond the biological mechanisms of disease, social, environmental, behavioral, and psychological factors play a critical role in human health. These factors – that we collectively call the "sociome" – interact with human biology to promote wellness or to cause or exacerbate disease and injuries. But in contrast to genomic and clinical data, sociome factors have not been comprehensively collected, codified, and quantified for the big data and machine learning techniques transforming modern healthcare.

The University of Chicago is building a Sociome Data Commons. Led by the Institute for Translational Medicine (comprised of six Chicago-area medical centers and universities), the initiative partners with divisional and departmental entities within UChicago, including Biological Sciences, Data Science Institute, Spatial Data Science, Public Policy, and Sociology. The Sociome group is building the scalable and extensible infrastructure and architecture that is assembling, quantifying, and organizing as much of the entirety of the social context of disease as possible. Once cataloged, researchers can link this information to biologic and clinical data at individual and population scale to understand, predict, and treat health and health-related outcomes. Available data will be reviewed with community-based organizations.

A proof-of-concept project is using this approach to build predictive models for children with asthma on the South Side of Chicago. The project intends to generate hypotheses for future research as well as to identify novel opportunities for immediate and long-term interventions to improve asthma outcomes. Interim results from the infrastructure and datasets to date, including predictive modeling and spatial analyses, will be shown.

114 – Effects of Glutathione S-Transferase Polymorphism on Blood Ethylene Oxide Hemoglobin Adduct Levels in Adults

Tara Serio[1], Peter Kowalski[2], Patricia Ruiz[3], Hubert Vesper[4], Enada Archibold[5], Kara Fluharty[6], Zheng Li[7]
[1]Agency for Toxic Substances and Disease Registry (ATSDR), Associate Service Fellow, Atlanta, USA. [2]Agency for Toxic Substances and Disease Registry (ATSDR), Team Lead, Exposure Investigations Section, Atlanta, USA. [3]Agency for Toxic Substances and Disease Registry (ATSDR), Chief, Simulation Science Section, Atlanta, USA. [4]Centers for Disease Control and Prevention (CDC), Supervisory Research

Chemist, Atlanta, USA. [5]Centers for Disease Control and Prevention (CDC), Lead Research Chemist, Atlanta, USA. [6]CDC/NIOSH, BIO SCI LAB TECH, Morgantown, USA. [7]Agency for Toxic Substances and Disease Registry (ATSDR), Director, Office of Capacity Development and Applied Prevention Science, Atlanta, USA

Abstract
Hemoglobin adducts of ethylene oxide (HbEO) have been historically used as a biomarker to assess occupational exposure to ethylene oxide (EtO), a human carcinogen. Beyond smoking, presence of a Glutathione S-Transferase Theta 1 (GSTT1) polymorphism may significantly influence HbEO levels in humans. This study aims to investigate the effect of the GSTT1 polymorphism on blood HbEO levels in adults with no known occupational exposure to EtO. Adult blood samples (N=501) representing diverse racial/ethnic groups from a commercial biorepository were measured for presence of GSTT1, HbEO, and hemoglobin adducts of acrylonitrile as a smoking biomarker. Descriptive statistics for HbEO by demographics and genotyping results and multivariate regression were used to investigate the relationship between the GST polymorphism and HbEO levels. The sample population was 53% male, 30% non-white, and aged 16-88 years (mean: 43.1 years). Among 103 participants (21%) with the null genotype for GSTT1, Asians had the highest prevalence of the null genotype (36%). HbEO levels were over 1.5 times higher in participants with the null genotype (geometric mean: 31.2 pmol/g Hb) than in participants with the normal genotype (19.0 pmol/g Hb). GSTT1 genotype was significantly associated with blood HbEO levels ($p<0.001$) in multivariate models adjusting for presence of the polymorphism, age, gender, acrylonitrile adducts, and race/ethnicity. This study found significant impact of genotype on blood HbEO levels in adults after adjusting for covariates and contributes to the understanding of key factors that could affect the exposure-biomarker relationship when using HbEO for measuring non-occupational EtO exposures.

115 – Health effects of Subdaily Exposure to Fine Particular Matter: Insight into Real-time Data

Andrew Nguyen[1], Keita Ebisu[2], Rupa Basu[3], Xiangmei Wu[2]
[1]Office of Environmental Health Hazard Assessment, Student Assistant, Oakland, USA. [2]Office of Environmental Health Hazard Assessment, Research Scientist, Oakland, USA. [3]Office of Environmental Health Hazard Assessment, Section Chief, Oakland, USA

Abstract
Introduction: Real-time monitoring of fine particulate matter (PM2.5) is widespread nowadays. To interpret real-time data accurately, we need to understand health impacts of subdaily exposure to PM2.5.

Methods: Modeled hourly PM2.5 concentrations with 6*6 km grids covering Southern California during May 2017 – March 2020 and daily emergency department visits (EDVs) for cardiovascular and respiratory conditions at ZIP Code Tabulation Area (ZCTA) levels were obtained. ZCTAs were aggregated into 35 clusters based on geography and sociodemographic similarities to ensure sufficient daily EDV counts. Daily exceedance concentration hours (DECH) above 9, 12, and 15 ug/m3, and daily maximum and average PM2.5 concentrations were calculated for each cluster. We conducted two-stage time-series analyses to estimate the risk of daily EDVs associated with various exposure metrics and lags.

Results: DECH metrics exhibited same direction but smaller effects of cardiovascular and respiratory EDVs compared to daily average metric, e.g., excess risks for cardiovascular EDVs at lag 7 were 1.77% (95% CI: 1.20, 2.34), 1.04% (0.61, 1.47), and 2.67% (1.98, 3.37) per each interquartile range increase of DECH-9, DECH-12, and daily average PM2.5. For respiratory EDVs, risks associated with DECH metrics were stabilized at lag day 7 to 14, while risks estimated from daily metric decreased after lag 7, suggesting that effects of elevated subdaily exposure retained longer on respiratory outcomes.

Conclusion: Subdaily PM2.5 exposure above standards/guidelines did not pose higher health risks than daily average exposure. Referencing current AQI system in the U.S. to interpret hourly PM2.5 concentrations would confer sufficient public health protection.

116 – Advanced Methodologies for Collection and Estimation of Soil/Dust Ingestion for Young Children

Alesia Ferguson[1], Foluke Adelabu[2], Jenna Honan[3], Emmanuel Obeng-Gyasi[4], Christina Fayad Martinez[5], Maribeth Gidley[6], Olusola Ogunseye[7], Elizabeth Gabriele[8], Mathew Roca[5], Paloma Beamer[9]
[1]North Carolina Agricultural and Technical State University, Professor and Chair, Greensboro, USA. [2]North Carolina Agricultural and Technical State University, Student, Greensboro, USA. [3]University of Arizona, Student, Tucson, USA. [4]North Carolina Agricultural and Technical State University, Associate Professor, Greensboro, USA. [5]University of Miami, Student, Miami, USA. [6]University of Miami, Research Associate, Miami, USA. [7]Unversity of Arizona, Student, Tucson, USA. [8]University of Miami, Student, Tucson, USA. [9]University of Arizona, Professor, Tucson, USA

Abstract
Metals, pesticides and a host of contaminants found in soil and dust pose a health risk to young children through the ingestion route. Soil/dust ingestion rates for young children can be estimated using micro-level activity time series (MLATS) as model inputs. MLATS allows for the generation of frequency and duration of children's contact activities, along with sequential contact patterns. Models using MLATS consider contact types, and transfer dynamics to assign mechanisms of contact and appropriate exposure factors for cumulative estimates of ingestion rates. The objective of this presentation is to describe field implementation, data needs, advanced field collection and laboratory methodologies, and challenges for integrating into and updating a previously validated physical-stochastic MLATS-based model called the Cumulative and Aggregate Simulation of Exposure (CASE) model. Our multidisciplinary research implemented this study (Dust Ingestion childRen sTudy (DIRT)) across three US regions: the metropolitan areas of Greensboro, North Carolina, Tucson, Arizona, and Miami, Florida. We recruited 450 participants to complete a 4-part household survey, where 100 participated in a field study. The field study focused on videotaping children's natural play using advanced 3600 unattended and mounted cameras for participant tracking and ultimately conversion to MLATS. Additionally, children's hand rinses were collected before and after recording, along with indoor dust and outdoor soil sampling, with advanced mass analysis. Soil/dust ingestion rates will be quantified by region, sociodemographic variables, age groups (from 6 months to 6 years), and other variables related to indoor/outdoor settings.

117 – Occurrence of 1,3-Diphenylguanidine, 1,3-Di-o-tolylguanidine, and 1,2,3-Triphenylguanidine in Indoor Dust and Human Urine

Zhongmin Li[1,2], Kurunthachalam Kannan[3,4]
[1]Wadsworth Center, New York State Department of Health, Research Affiliate, Albany, USA. [2]New York University Grossman School of Medicine, Postdoctoral Fellow, New York, USA. [3]Wadsworth Center, New York State Department of Health, Research Scientist, Albany, USA. [4]New York University Grossman School of Medicine, Professor`, New York, USA

Abstract
1,3-Diphenylguanidine (DPG), 1,3-di-o-tolylguanidine (DTG) and 1,2,3-triphenylguanidine (TPG) are widely used as rubber additives. Little is known about their occurrence in indoor dust and human urine. We measured these chemicals in 332 indoor dust samples collected from 11 countries. DPG, DTG and TPG were found in 100%, 62%, and 76% of the house dust samples, at median concentrations of 140, 2.3 and 0.9 ng/g, respectively. The sum concentrations of DPG, DTG and TPG varied among the countries in the following decreasing order: Japan (median: 1300 ng/g) > Greece (940) > South Korea (560) > Saudi Arabia (440) > USA (250) > Kuwait (160) > Romania (140) > Vietnam (120) > Colombia (100) > Pakistan (33) > India (26). DPG accounted for ≥ 87% of the sum concentrations of the three compounds in all countries. Estimated human exposure to DPG through dust ingestion were in the range of 0.07–4.40, 0.09–5.20, 0.03–1.70, 0.02–1.04, and 0.01–0.87 ng/kg body weight (BW)/day for infants, toddlers, children, teenagers, and adults, respectively.[1] Subsequently, we examined the occurrence of these chemicals in human urine samples from the United States. DPG was found in 73% of the children's urine samples (n = 15) at median concentration of 0.05 ng/mL (range: < LOD–2.94 ng/mL). DPG was also found in adults' urine (n = 20) at a lower detection frequency (20%).[2] The safety of DPG warrants more attention.

References:
[1] Li Z-M, Kannan K. ES&T 2023
[2] Li Z-M, Kannan K. ES&T (submitted) 2023

118 – More Than Just Data Points: Evaluating the experiences of participants in the GenX Exposure Study

Katy May[1,2], Andrew Binder[1,3], Lacey Brown[4], Jane Hoppin[5]
[1]Center for Human Health and the Environment, North Carolina State University, Co-Director, Community Engagement Core, Raleigh, USA. [2]Center for Environmental and Health Effects of PFAS, Director, Community Engagement Core, Raleigh, USA. [3]North Carolina State University, Associate Professor, Department of Communication, Raleigh, USA. [4]Center for Human Health and the Environment, North Carolina State University, Community Engagement Specialist, Raleigh, USA. [5]North Carolina State University, Professor, Department of Biological Sciences, Raleigh, USA

Abstract
Reporting data back to research participants is ethically responsible, can increase study retention, and can lead to gains in knowledge, science literacy, and potentially Environmental Health Literacy. We evaluated the experiences of people who participated in and received results from the GenX Exposure Study, a cohort study looking at per- and polyfluoroalkyl substances (PFAS) exposure and potential health effects in residents of the Cape Fear River Basin in North Carolina. Our evaluation systematically explored the

motivations, understanding, experiences, and perceptions of the people who have made the GenX Exposure Study possible. We designed and administered a questionnaire that collected qualitative and quantitative data from households currently enrolled in the Study (N=345).

Our data indicate that, overall, participants had a positive experience, understanding the importance of their contributions, learning via the report back process, and are willing to participate in future studies. Our qualitative findings provide descriptive insight into what motivates members of these highly-exposed communities to participate in research studies, what participants identify as helpful, and the opportunities they see for improvement. When stratifying by sociodemographic characteristics (e.g., geographic region, education level), our data have important implications for future engagement and data reporting efforts for this and other exposure studies.

119 – POSTER: Children's Behaviors and Housing Dynamics that Influence Soil/Dust Ingestion Rates: Result of a Survey

Foluke Adelabu[1], Neema Olagbemiro[1], Helena Solo-Gabriele[2], Cristina Fayad Martinez[3], Maribeth Gidley[2], Paloma Beamer[4], Jenna Honan[5], Olusola Ogunseye[5], Emmanuel Obeng-Gyasi[6], Alesia Ferguson[6]
[1]North Carolina Agricultural and Technical State University, Graduate Student, Greensboro, USA. [2]University of Miami, Faculty, Coral Gables, USA. [3]University of Miami, Graduate Student, Coral Gables, USA. [4]University of Arizona, Faculty, Tucson, USA. [5]University of Arizona, Graduate Student, Tucson, USA. [6]North Carolina Agricultural and Technical State University, Faculty, Greensboro, USA

Abstract
Young children are vulnerable to environmental exposures because they are in the critical developmental stages and exhibit activities that increase exposure (e.g., object-to-mouth contacts, and play close to the ground). Contaminants found in soil and dust in the home can pose health risks (e.g., lead in the home affects development growth). To help improve the estimate of soil/dust ingestion rates, quantification of children's behaviors that contribute to ingestion exposure is needed. Our multidisciplinary research team implemented a multi-faceted study (Dust Ingestion childRen sTudy (DIRT)) across three US regions: the metropolitan areas of Greensboro, North Carolina, Tucson, Arizona, and Miami, Florida to quantify children's soil/dust ingestion rates. One component of the study is an activity and household survey delivered to 326 participants with children aged 6 months to 6 years. Presented here are results for the parents' responses to their child's observed behaviors and cleaning practices in the home related to soil/dust ingestion activities. The survey includes socio-demographic and housing layout sections allowing the presentation of findings across age, income, race, regions, and housing dynamics. Findings revealed that more parents of non-Hispanic children reported washing their hands before meals or snacks (37%), compared to parents of Hispanic children (18%). Additionally, children both male and female between 6-12 months and 1-2 years have similar mouthing behavior of "putting items in the mouth". Differences in behaviors across groups may indicate differences in exposure levels and targeted prevention outreach strategies.

120 – The Green Housing Study: A Novel Volatile Organic Compound (VOC) Sensor to Predict Airway Inflammation in Children

Ginger Chew[1], Fred Sieling[2], Gary Adamkiewicz[3], Erica Forzani[4], Kanta Sircar[1]
[1]CDC, Epidemiologist, Atlanta, USA. [2]Guidehouse, Inc., Data Scientist, McLean, USA. [3]Harvard T.H. Chan School of Public Health, Professor, Boston, USA. [4]Arizona State University, Professor, Phoenix, USA

Abstract
Introduction: Improved methods to measure volatile organic compounds (VOCs) in real-time can enhance environmental epidemiology studies.

Methods: As part of the Green Housing Study (GHS), 26 children (ages 7-12 with asthma) were visited once over the course of one year. Exhaled nitric oxide (eNO) was assessed for the children, and VOC sampling devices were placed in their homes. Traditional VOC passive dosimeters were left in the home on Day 1 and retrieved on Day 5 of each home visit. Concurrently, the VOC sensor collected measurements every three minutes for approximately one hour on Days 1 and 5. Home characteristics were reported by the field staff on the days of sampling.

Results: The VOC sensor measurements were associated with those of the traditional integrated passive dosimeters (Pearson's correlation r= 0.27, p=0.15). However, the sensor's area under the curve (AUC) was a stronger predictor of eNO ($r2$= 0.24, p=0.01) than the passive dosimeters ($r2$= 0.03, p=0.32). In adjusted models (for potential confounders such as race/ethnicity, season, and air freshener/ smoke/ cleaning agent odors in the home), the sensor lost significance, but was still superior in predictive ability compared with the traditional VOC dosimeters.

Conclusions: Real-time VOC sensors may provide better predictive ability for airway inflammation than traditional passive dosimeters alone.

122 – Association between styrene and ethylbenzene exposure and risk of type 2 diabetes: A gene-environment interaction study

Linling Yu[1], Wei Liu[1], Jiahao Song[1], Min Zhou[2], Qiyou Tan[1], Lieyang Fan[1], Bin Wang[2], Weihong Chen[3]
[1]Huazhong University of Science and Technology, none, Wuhan, China. [2]Huazhong University of Science and Technology, lecturer, Wuhan, China. [3]Huazhong University of Science and Technology, professor, Wuhan, China

Abstract
Objective: Styrene and ethylbenzene (S/EB) are hazardous air pollutants that have received worldwide attention. The present study aimed to evaluate the relationship between S/EB exposure and risk of type 2 diabetes (T2D) and assess the interaction effect between S/EB exposure and genetic variation on T2D.

Methods: Among 2,312 Chinese adults, associations of urinary S/EB metabolites (mandelic acid, MA; phenylglyoxylic acid, PGA) as credible biomarkers of S/EB exposure with risk of T2D were analyzed by logistic regression models. Polygenic risk score (PRS) for T2D was constructed based on 137 genetic loci. The effect of PRS as well as the joint effect of PRS with S/EB exposure on risk of T2D were assessed. And the potential interaction between S/EB exposure and PRS on T2D risk was further investigated on multiplicative and additive scales.

Results: Our results showed a remarkable increase in the risk of T2D with a unit increase in MA [odds ratios (OR) = 1.51, 95% confidence interval (95% CI): 1.19-1.91], PGA (1.28, 1.12-1.48), or MA+PGA (1.58, 1.26-1.98) and a SD increase in PRS (1.47, 1.26-1.71). Compared with participants with low S/EB exposure and low PRS, those with high S/EB exposure and high PRS had the highest risk of T2D. S/EB exposure and PRS had a significant additive interaction effect on T2D.

Conclusions: Our study provided the first evidence that S/EB exposure was significantly related to the increased risk of T2D, and the association was aggravated by genetic predisposition.

123 – POSTER: Improving Assessment of Long-Term Exposure to Air Pollution

Allison Patton[1], Dan Crouse[1], Hanna Boogaard[2]
[1]Health Effects Institute, Senior Scientist, Boston, USA. [2]Health Effects Institute, Consulting Principal Scientist, Boston, USA

Abstract
Air pollution exposure assessment can be challenging because concentrations of many air pollutants vary rapidly over scales of hundreds of meters, and hours to days. In recent years, there has been increased application of new exposure assessment methods and approaches to support policy and epidemiological studies, though the limits of those new approaches still have not been fully explored. Investigators funded as part of the Health Effects Institute's (HEI's) ongoing research program on improving long-term exposure assessment of outdoor air pollution for health studies are developing and evaluating exposure assessment methods including mobile monitoring, low-cost sensors, human activity tracking, and statistical approaches to model development. They are combining novel methods for measuring air pollution and diverse exposure assessment approaches to improve exposure assignment in well-established cohorts. They are also testing the added value of incrementally more complex statistical modeling approaches to improving exposure assessment. The presentation will describe the various innovations in approaches and present preliminary findings with their implications for design of air pollution exposure assessment. Taken as a whole, these five studies will provide guidance related to the design of exposure studies, how limitations in the input data affect results of exposure assessment and associated health studies, and generalizability of the results.

124 – The potentials of participatory science in revolutionising air quality monitoring and influencing behavioural changes targeted at air pollution management and control in Ghana: Methodist Girls' High School Case Study

Collins Gameli Hodoli
UESD, Lecturer, Somanya, Ghana. CAOA, Founder and DOO, Accra, Ghana

Abstract
$PM_{2.5}$ is a major contributor to air pollution-induced deaths in Africa. Our ability to determine $PM_{2.5}$ pollution in a specific setting in Africa is a major step to developing site-specific evidence-based mitigation strategies to control and manage $PM_{2.5}$. Anthropogenic sources of $PM_{2.5}$ pollution can largely be controlled in Africa by influencing behavioural changes but this is challenging due to patchy data, limited public knowledge and expertise. In this case study, critical consideration has been given to the usefulness of participatory science to deploy, manage and supervise PurpleAir PA-II monitor and the subsequent meaningful communication, visualisation, analysis and interpretation of the reported data at the Methodist Girls' High School (MEGHIS), Mamfe in the Eastern Region of Ghana. It was found that participatory science using air sensors has the potential to influence behavioural changes targeted at air pollution mitigation if students and teachers are engaged, educated and empowered to take ownership of air quality monitoring on campus and the reported data meaningfully communicated tied to background activities. Also, the reported data presented in time series and daily variation plots led to behavioural changes at MEGHIS such as waste segregation, handpicking on bare soils, tree and grass planting and the introduction of a science garden to support climate science education, ban on waste burning and community outreach on air pollution. This case study has the potential to shift the paradigm regarding air quality monitoring, meaningful data communication, and air pollution mitigation in Ghana and wider Africa using participatory science involving students and teachers.

125 – Environmental Release and Occupation Exposure Models Used in Chemical Risk Evaluation

William Barrett[1], Raymond Smith[1], David Meyer[1], Sudhakar Takkellapati[2], Michael Gonzalez[2]
[1]US Environmental Protection Agency, Chemical Engineer, Cincinnati, USA. [2]US Environmental Protection Agency, Chemist, Cincinnati, USA

Abstract
Under the Toxic Substances Control Act (TSCA), the US Environmental Protection Agency (US EPA) is required to determine whether a chemical substance poses an unreasonable risk before the chemical is manufactured in or imported into the United States. This presentation provides an overview of the environmental release and occupational exposure models incorporated into Generic Scenario (GS) and Emissions Scenario Documents (ESD) used by the USEPA to conduct these chemical risk evaluations. Additionally, this presentation provides an overview of methods used to model occupational exposures and environmental releases as part of the chemical risk evaluation process used in other jurisdictions, along with work being performed to improve these models. Finally, the alternative methods to evaluate occupational exposures and environmental releases that may be used as part of the decision-making process regarding a chemical are identified. The contribution provides a path forward for reducing the time required and improving the data availability for use in chemical risk evaluations.

126 – POSTER: Associations of Persistent Organic Pollutants (POPs) with Incident Prediabetes and Diabetes in the Hispanic Community Health Study/ Study of Latinos (HCHS/SOL)

Konstantina Tsintsifas[1], Robert Sargis[2,3], Chibuzor Abasilim[4], Tessa Day[5], Sally Freels[6], Martha Daviglus[7], Robert Kaplan[8], Gregory Talavera[9], Jianwen Cai[10], Kyeezu Kim[11], Andreas Sjodin[12], Mark Davis[12], Richard Jones[12], Victoria Persky[13], Mary Turyk[13]
[1]University of Illinois at Chicago, PhD Candidate, Chicago, IL, USA. [2]University of Illinois at Chicago, Associate Professor of Medicine, Chicago, IL, USA. [3]Medical Service, Jesse Brown VA Medical Center, Section Chief, Endocrinology, Chicago, IL, USA. [4]University of Illinois at Chicago, Postdoctoral Research Associate, Chicago, IL, USA. [5]University of Illinois at Chicago, Chicago, Graduate Student, Chicago, IL, USA. [6]University of Illinois at Chicago, Associate Professor Emerita, Chicago,IL, USA. [7]University of Illinois at Chicago, Associate Vice Chancellor for Research, Chicago,IL, USA. [8]Albert Einstein College of Medicine, Professor, New York, USA. [9]San Diego State University, Professor, San Diego, USA. [10]University of North Carolina at Chapel Hill, Professor, Chapel Hill, NC, USA. [11]Northwestern University, Postdoctoral Scholar, Chicago,IL, USA. [12]Centers for Disease Control and Prevention, Chemist, Atlanta, GA, USA. [13]University of Illinois at Chicago, Professor, Chicago,IL, USA

Abstract
Background: Global diabetes prevalence has risen rapidly in recent decades. Exposure to POPs has been identified as a possible risk factor for diabetes, but the impact on diabetes progression is unclear. We explored associations of serum POPs with incident prediabetes and diabetes in post-menopausal Hispanic/Latino women.

Methods: HCHS/SOL is the largest multi-ethnic cohort of Hispanics/Latinos in the US. Our sample included 310 normoglycemic and 399 prediabetic women not taking medications related to diabetes or hormones, with an average 6 years of follow-up. The effect of baseline

POPs levels on progression from 1) normoglycemia at baseline to prediabetes at follow-up (162/310) and 2) prediabetes at baseline to diabetes at follow-up (195/399) was evaluated with multivariable Poisson regression with robust variance analysis, adjusting for sampling design, demographic, clinical, and acculturation characteristics.

Results: p',p'-dichlorodiphenyldichloroethylene (IRR:1.39;95%CI:(1.07,1.82) and β-hexachlorocyclohexane (IRR:1.61;95%CI:(1.16,2.23) had positive linear relationships with progression from prediabetes to diabetes. Sum polychlorinated biphenyls(PCBs), sum dioxin-like PCBs, sum non-dioxin-like PCBs, sum 3-methylcholanthrene inducer PCBs, sum phenobarbital-type inducer PCBs, hexachlorobenzene, and trans-nonachlor showed inverted U-shaped relationships with progression from prediabetes to diabetes. Mirex (IRR:0.75;95%CI:(0.59,0.94) and o',p'-dichlorodiphenyltrichloroethane(IRR:1.89;95%CI:(1.19,3.01) had significant linear associations for progression to prediabetes from normoglycemia.

Conclusion: Our findings confirm positive associations between POPs exposure and risk of diabetes. The non-linear associations suggest a threshold effect of certain POPs and diabetes risk. The positive associations are mainly driven by conversion from prediabetes to diabetes, indicating effects of POPs may occur later in disease development.

127 – Quantifying The Infectious Flu Virus (Influenza A) Present In The Air Before And After A Resuspension Episode

Mahender Singh Rawat[1], Alan Roberts[2], Deborah M. Brown[3], Andrea R. Ferro[4]
[1]Department of Civil and Environmental Engineering, Clarkson University, PhD student, Potsdam, USA. [2]Trudeau Institute, 154 Algonquin Avenue, Senior Research Assistant, Saranac Lake, USA. [3]Trudeau Institute, 154 Algonquin Avenue, Manager, Joint Educational Programs, Saranac Lake, USA. [4]Department of Civil and Environmental Engineering, Clarkson University, Professor and Assoc. Director for Research, Potsdam, USA

Abstract
Resuspended particles from the surfaces can be a major source of airborne microbiological contaminants in indoor environments. Many studies have investigated the exposure and risk of infection from direct transmission of infectious viruses, but the infection risk from resuspended virus particles settled on indoor surfaces is still unknown. For our study, we nebulized H3N2 strain x31 of influenza A virus (stock count of 10^9 in 7 ml of phosphate buffered saline) for 30 minutes in a BSL-2 laboratory. The virus aerosol was allowed to settle for several hours, after which an investigator entered the room and performed normal walking activity for 30 minutes. Impingers were used to collect 5.5 h air samples during both the initial emission and resuspension periods, petri dishes were utilized for collecting deposition samples, and particle concentrations were monitored continuously. The quantification of viral load obtained from impinger and deposition samples was conducted using RT-PCR. The concentration of the virus in the air during and after the resuspension activity (3.7×10^3 viral RNA copies m^{-3}) was about two orders of magnitude lower than the concentration during and after the initial viral emission (1.1×10^5 viral RNA copies m^{-3}). Therefore, the potential for transmission through resuspension of respiratory viruses appears to be much less than that for the initial emission. The risk would be dependent on the infectious dose and viability of the specific virus. To our knowledge, this is the first experimental study designed to quantify the degree of viral resuspension from human activity.

128 – Are More Diverse Indoor Microbes Beneficial to Respiratory Health in School Workers?

Ju-Hyeong Park[1], Angela Lemons[2], Tara Croston[2], Brett Green[2], Jean Cox-Ganser[3]
[1]CDC/National Institute for Occupational Safety and Health (NIOSH), Research Industrial Hygienist, Morgantown, USA. [2]CDC/National Institute for Occupational Safety and Health (NIOSH), Research Biologist, Morgantown, USA. [3]CDC/National Institute for Occupational Safety and Health (NIOSH), Epidemiologist, Morgantown, USA

Abstract
The effects of indoor microbial diversity on health in adults are understudied. We examined associations between microbial diversity and respiratory health among school workers. We vacuumed floor dust from 500 classrooms in 50 elementary schools. Genomic DNA was extracted from dust, and fungal internal transcribed spacer 1 and bacterial 16S rRNA were sequenced by Illumina MiSeq. The sequences were clustered into operational taxonomic units (OTUs). Health information was collected from a web-based questionnaire survey. The average number of OTUs per school/building floor were used to assign individual participants' exposures. Multiple logistic regression adjusted for age, gender, race, ethnicity, smoking status, being a teacher, physician-diagnosed hay fever, home mold odor, and classroom dampness was used to examine the associations with wheeze, chest tightness, shortness of breath, attacks of cough, stuffy-itchy nose, sinus symptoms, dry throat, and watery-itchy eyes in the last 12 months. Of 1,529 participants (50.9% participation), 82.1% were teachers, 85% female, 6.3% current smoker, 70% white, and 4.3% Hispanic. We found that more species richness in the

Actinobacteria was negatively associated with wheeze (adjusted odds ratio=0.994, 95% confidence interval=0.988–1.0) and cough (0.991, 0.985–0.997). More species richness in the phylum Firmicutes was negatively associated with cough (0.993, 0.987–1.0). The number of total fungal or yeast OTUs were associated with stuffy-itchy nose symptoms (0.993, 0.987–0.999; 0.970, 0.941–1.0, respectively). In other non-significant associations, negative relationships were generally observed. Our study findings indicated that more diverse indoor microbiota may be beneficial to respiratory health in school workers.

129 – TAME 2.0: Expanding the inTelligence And Machine lEarning (TAME) Toolkit for Introductory Data Science, Chemical-Biological Analyses, Predictive Modeling, and Database Mining for Environmental Health Research

Lauren Koval[1,2], Kyle Roell[3], Alexis Payton[4], Elise Hickman[5,6], Oyemwenosa Avenbuan[2,7], Rebecca Boyles[8], Grace Patlewicz[9], Caroline Ring[9], Cynthia Rider[10], Cavin Ward-Caviness[11], David Reif[10], Ilona Jaspers[12,13,14,15,16], Rebecca Fry[17,13,18], Julia Rager[19,20,21,22]
[1]The Institute for Environmental Health Solutions, Gillings School of Global Public Health, The University of North Carolina at Chapel Hill, Chapel Hill, North Carolina, USA, Graduate Research Assistant, Chapel Hill, USA. [2]Department of Environmental Sciences and Engineering, Gillings School of Global Public Health, The University of North Carolina at Chapel Hill, Chapel Hill, North Carolina, USA, Graduate Research Assistant, Chapel Hill, USA. [3]The Institute for Environmental Health Solutions, Gillings School of Global Public Health, The University of North Carolina at Chapel Hill, Chapel Hill, North Carolina, USA, Data Analyst Project Manager, Chapel Hill, USA. [4]Center for Environmental Medicine, Asthma and Lung Biology, School of Medicine, University of North Carolina, Chapel Hill, North Carolina, USA, Data Analyst, Chapel Hill, USA. [5]Department of Environmental Sciences and Engineering, Gillings School of Global Public Health, The University of North Carolina at Chapel Hill, Chapel Hill, North Carolina, USA, Postdoctoral Researcher, Chapel Hill, USA. [6]Curriculum in Toxicology and Environmental Medicine, School of Medicine, University of North Carolina, Chapel Hill, North Carolina, USA, Postdoctoral Researcher, Chapel Hill, USA. [7]Curriculum in Toxicology and Environmental Medicine, School of Medicine, University of North Carolina, Chapel Hill, North Carolina, USA, Graduate Research Assistant, Chapel Hill, USA. [8]Research Computing, RTI International, Durham North Carolina, USA, Director of the Center for Data Modernization Solutions, Durham, USA. [9]Center for Computational Toxicology and Exposure, US Environmental Protection Agency, Research Triangle Park, North Carolina, USA, Research Scientist, Research Triangle Park, USA. [10]Division of Translational Toxicology, National Institute of Environmental Health Sciences, Research Triangle Park, North Carolina, USA, Research Scientist, Research Triangle Park, USA. [11]Center for Public Health and Environmental Assessment, US Environmental Protection Agency, Chapel Hill, North Carolina, USA, Research Scientist, Research Triangle Park, USA. [12]The Institute for Environmental Health Solutions, Gillings School of Global Public Health, The University of North Carolina at Chapel Hill, Chapel Hill, North Carolina, USA, Professor, Chapel Hill, USA. [13]Department of Environmental Sciences and Engineering, Gillings School of Global Public Health, The University of North Carolina at Chapel Hill, Chapel Hill, North Carolina, USA, Professor, Chapel Hill, USA. [14]Center for Environmental Medicine, Asthma and Lung Biology, School of Medicine, University of North Carolina, Chapel Hill, North Carolina, USA, Deputy Director, Chapel Hill, USA. [15]Curriculum in Toxicology and Environmental Medicine, School of Medicine, University of North Carolina, Chapel Hill, North Carolina, USA, Director, Chapel Hill, USA. [16]Department of Pediatrics, Microbiology and Immunology, School of Medicine, University of North Carolina, Chapel Hill, North Carolina, USA, Professor, Chapel Hill, USA. [17]The Institute for Environmental Health Solutions, Gillings School of Global Public Health, The University of North Carolina at Chapel Hill, Chapel Hill, North Carolina, USA, Director, Chapel Hill, USA. [18]Curriculum in Toxicology and Environmental Medicine, School of Medicine, University of North Carolina, Chapel Hill, North Carolina, USA, Director of Graduate Studies, Chapel Hill, USA. [19]The Institute for Environmental Health Solutions, Gillings School of Global Public Health, The University of North Carolina at Chapel Hill, Chapel Hill, North Carolina, USA, Assistant Professor, Chapel Hill, USA. [20]Department of Environmental Sciences and Engineering, Gillings School of Global Public Health, The University of North Carolina at Chapel Hill, Chapel Hill, North Carolina, USA, Assistant Professor, Chapel Hill, USA. [21]Center for Environmental Medicine, Asthma and Lung Biology, School of Medicine, University of North Carolina, Chapel Hill, North Carolina, USA, Assistant Professor, Chapel Hill, USA. [22]Curriculum in Toxicology and Environmental Medicine, School of Medicine, University of North Carolina, Chapel Hill, North Carolina, USA, Assistant Professor, Chapel Hill, USA

Abstract
Research in environmental health is becoming increasingly reliant upon computational methods to better identify relationships between environmental exposures and disease outcomes. Still, there remains a critical gap surrounding the training of researchers on these in silico methods. We originally aimed to address this gap by developing the inTelligence And Machine lEarning (TAME) Toolkit, promoting trainee-driven data generation, management, and analysis methods to "TAME" data in environmental health studies. This resource has recently been expanded through the organization of "TAME 2.0", which features additional training resources on topics identified as being of the greatest importance to trainees. Training modules were developed to provide applications-driven examples of data organization and analysis methods that can be used to address environmental health questions relevant to exposure science, epidemiology, toxicology, and bioinformatics/cheminformatics. Modules were developed by coauthors using annotated script and were organized into six chapters within a GitHub Bookdown site. The chapters focus on (1) an introduction to data science, (2) an introduction to R, (3) basics of data analysis and visualizations, (4) machine learning, (5) applications in toxicology and exposure science, and (6) environmental health database mining. Example analyses include the following: machine learning and predictive modeling; mixtures analyses; exposomics analyses; toxicokinetic modeling; and read-across toxicity predictions. Databases span those covering chemical

exposures, health outcomes, and environmental justice indicators. Training modules and associated data are publicly available. Together, this continually expanding resource provides unique opportunities to obtain introductory-level training on current data analysis methods applicable to 21st century science and environmental health.

130 – Development of an integrated model framework for multi-air-pollutant exposure assessments in high-density cities and the implications for epidemiological research

Zhiyuan Li[1,2], Kin-Fai Ho[3], Harry Fung Lee[4], Steve Hung Lam Yim[5]
[1]Sun Yat-sen University, Associate Professor, Guangzhou, China. [2]The Chinese University of Hong Kong, Postdoctoral Fellow, Hong Kong, Hong Kong. [3]The Chinese University of Hong Kong, Professor, Hong Kong, Hong Kong. [4]The Chinese University of Hong Kong, Associate Professor, Hong Kong, Hong Kong. [5]Nanyang Technological University, Associate Professor, Singapore, Singapore

Abstract
Exposure models for some criteria air pollutants have been intensively developed in past research; multi-air-pollutant exposure models, especially for particulate chemical species, have been however overlooked in Asia. Lack of an integrated model framework to calculate multi-air-pollutant exposure hinders the combined exposure assessment and the corresponding health assessment. This work applied the land-use regression (LUR) approach to develop an integrated model framework to estimate annual-average exposures of four major PM10 chemical species as well as four criteria air pollutants of PM10, PM2.5, NO2, and O3 in a typical high-rise and high-density Asian city (Hong Kong, China). Our integrated multi-air-pollutant exposure model framework is capable of explaining 91–97 % of the variability of measured air pollutant concentration, with the leave-one-out cross-validation R2 values ranging from 0.73 to 0.93. Using the model framework, the spatial distribution of the concentration of various air pollutants at a spatial resolution of 500 m was generated. The LUR model-derived spatial distribution maps revealed weak to moderate spatial correlations between the PM10 chemical species and the criteria air pollutants, which may help to distinguish their independent chronic health effects. In addition, further improvements in the development of air pollution exposure models are discussed. This study proposes an integrated model framework for estimating multi-air-pollutant exposure in high-density and high-rise urban areas, serving an important tool for combined exposure assessment and the corresponding epidemiological studies.

131 – Volatile Organic Compounds in Disposable Diapers and Baby Wipes in the US: A Survey of Products and Health Risks

Nan Lin[1,2], Stuart Batterman[3]
[1]Shanghai Jiao Tong University, Associate Professor, Shanghai, China. [2]University of Michigan, Research Fellow, Ann Arbor, USA. [3]University of Michigan, Professor, Ann Arbor, USA

Abstract
Many thousands of diapers are worn by young children and the elderly, who have thin and sensitive skin that is highly vulnerable to chemicals, including volatile organic compounds (VOCs) that may be ingredients of these products or present as inadvertent or residual components. The levels and potential health risks of VOCs in diapers have not been reported previously. In this study, 31 disposable hygiene products in the US market were selected based on market share and analyzed for 98 target VOCs using purge and trap sampling and thermal desorption/gas chromatography/mass spectrometer analysis. Exposures and risks were modeled using reasonable upper level exposure scenarios. Adult diapers contained the highest total target VOC concentration (median level of 23.5 μg/g), and the predominant VOCs were alkanes. In some diapers the estimated non-cancer risk from these VOCs was sometimes very large (hazard quotient of 1609). Baby diapers contained several known or suspected carcinogens, including benzene and 1,4-dioxane, and the lifetime cancer risk approached for some diapers approached 1 per million. Store-brand products had higher levels of VOCs than generic brands, and products labeled "organic" or "for sensitive skin" did not necessarily have lower levels. Our results show that toxic VOCs were found in all tested disposable diapers and wipes at trace levels, and risks from using some diapers in high use exposure scenarios are high enough to warrant additional attention and possibly corrective measures. We recommend eliminating and monitoring toxic ingredients, and disclosing all chemicals that may be in these products.

132 – POSTER: Assessment of $PM_{2.5}$ personal exposure of non- or passive smokers by using portable $PM_{2.5}$ monitor

Takashi Amagai[1], Ayana Komatsu[2], Kazushi Noro[3]

[1]University of Shizuoka, Professor, Shizuoka, Japan. [2]University of Shizuoka, Student, Shizuoka, Japan. [3]University of Shizuoka, Assistant Professor, Shizuoka, Japan

Abstract
Passive smoking is one of social problems. Smoking is known to emit a large amount of $PM_{2.5}$, and there are concerns about its health effects. $PM_{2.5}$ personal exposure concentrations of non- or passive smokers were measured in metropolitan areas in Japan in October to December 2022, after the outbreak of COVID-19. Portable PM monitor developed by Dr. Ishigaki was used for monitoring $PM_{2.5}$ concentration. We asked 5,390 participants to answer a questionnaire on passive smoking status on the web, and 194 of them (97 males and 97 females) were asked to spend 24 hours with a PM monitor. All 194 participants were non-smokers, and 80 of them answered that they were not exposed to passive smoking in their daily lives.

The arithmetic mean and standard deviation of $PM_{2.5}$ personal exposure concentrations in this study were 11.1 $\mu g{\cdot}m^{-3}$ and 14 $\mu g{\cdot}m^{-3}$, respectively. The median value was 7.7 $\mu g{\cdot}m^{-3}$. It was found that the distribution of $PM_{2.5}$ exposure concentration followed a logarithmic normal distribution. There was no significant difference in exposure levels between male and female. There were also no significant differences in exposure levels by age or occupation. Significant differences were found only in $PM_{2.5}$ exposure levels among groups exposed to passive smoking outdoors, indoors, both, and none during the study. This suggest that exposure to secondhand smoke indoors was associated with higher $PM_{2.5}$ exposure levels. $PM_{2.5}$ exposure due to second-hand smoke outdoors was not significant.

133 – Complex Interaction Effects of Air Pollution and Temperature on Blood Pressure: Evidence from the SALIA Cohort Study

Dayasri Ravi[1], Andreas Groll[2], Nidhi Singh[3], Tamara Schikowski[4]
[1]Department of Statistics, TU Dortmund University, Doctoral Student, Dortmund, Germany. [2]Department of Statistics, TU Dortmund University, Professor, Dortmund, Germany. [3]Department of Epidemiology, IUF-Leibniz Research Institute for Environmental Medicine, Postdoctoral Fellow, Düsseldorf, Germany. [4]Department of Epidemiology, IUF-Leibniz Research Institute for Environmental Medicine, Environmental Epidemiologist, Düsseldorf, Germany

Abstract
Background: Systolic (SBP) and diastolic blood pressures (DBP) have recently been closely associated with environmental factors such as temperature and air pollution. However, the interaction effect between these environmental factors in modeling blood pressure is often not considered. We examined the relationship between temperature and air pollution with blood pressure in women above 60 from heavily industrialized regions of Germany.

Methods: A generalized additive model with bivariate tensor splines was employed to capture the complex interaction effect of temperature and air pollutants such as particulate matter ($PM_{2.5}$), nitrogen dioxide (NO_2), and ozone (O_3) with SBP and DBP, mean arterial pressure (MAP), and pulse pressure (PP). The interaction was monitored over a lag period of up to 13 days.

Results: We observed a non-linear interaction effect of environmental factors ($PM_{2.5}$, NO_2, and O_3) on blood pressure. In warmer months with temperatures above 10°C, the impact of air pollutants was significantly associated with blood pressure. For example, an IQR-unit increase in $PM_{2.5}$ concentration (μg/m3) was associated with a 0.46% (95% CI: 0.06%, 0.98%) increase in SBP, a 0.67% (95% CI: 0.17%, 1.18%) increase in DBP, and a 0.6% (95% CI: 0.15%, 1.05%) increase in MAP.

Conclusion: We discovered an intricate synergistic relationship between temperature and air pollution on systolic and diastolic blood pressures. Considering the linear effect, we found a minimal statistically significant association between air pollutants and blood pressure. However, we discovered a highly significant association when considering the non-linear interactions between temperature and air pollution.

134 – The use of human biomonitoring data and guidance values to evaluate priority chemical substances under Canada's Chemicals Management Plan

Kristin Macey, Devika Poddalgoda, Natasha Chander, Innocent Jayawardene, Britannia Sitland, Sara Mohr, Scott Hancock

Health Canada, Evaluator, Ottawa, Canada

Abstract
Health Canada uses human biomonitoring (HBM) data and human biomonitoring health-based guidance values (HB2GVs) to assess the potential for harm to human health from priority chemical substances in commerce under the Canada's Chemicals Management Plan (CMP). With the availability of national-level HBM data in 2010 from the Canadian Health Measures Survey, the development of HB2GVs was given priority to support the health-based interpretation of HBM data. Health Canada has been involved in the development of HB2GVs for over 30 chemicals. These guidance values are included in the on-line repository of guidance values, referred to as the Human Biomonitoring Health-Based Guidance Value (HB2GV) Dashboard, recently launched by the International Human Biomonitoring (i-HBM) Working Group. In 2016, Health Canada published a science approach document that outlines considerations for using such information in human health risk assessments to characterize the potential for harm in the Canadian population as part of the chemicals management process. This talk will present an overview of human health risk assessments for several inorganic substances under Canada's CMP, such as lithium, silver, and tin, using HBM data and HB2GVs.

135 – Biomonitoring per- and polyfluoroalkyl substances (PFAS) and local fish consumption in susceptible Great Lakes populations, 2010 - 2020

Tara Serio[1], Zheng Li[2], Elizabeth Lewis-Michl[3], Ming Liu[4], Kurunthachalam Kannan[4,5], Wei Wang[4], Wan-Hsiang Hsu[4,5], Sanghamitra Savadatti[4], Carrie Tomasallo[6], Elizabeth Irvin[7], Angela Ragin-Wilson[8]
[1]Agency for Toxic Substances and Disease Registry, Associate Service Fellow, Atlanta, USA. [2]Agency for Toxic Substances and Disease Registry, Director, Office of Capacity Development and Applied Prevention Science, Atlanta, USA. [3]New York State Department of Health, Director, Division of Environmental Health Assessment, Albany, USA. [4]New York State Department of Health, Research Scientist, Albany, USA. [5]University of Albany School of Public Health, Research Scientist, Albany, USA. [6]Wisconsin Department of Health Services, Section Chief, Environmental Epidemiology and Surveillance, Madison, USA. [7]Agency for Toxic Substances and Disease Registry, Director, Office of Community Health Hazard Assessment, Atlanta, USA. [8]Agency for Toxic Substances and Disease Registry, Deputy Associate Director, ATSDR, Atlanta, USA

Abstract
Fish and other seafood are important dietary sources of nutrients. However, despite remediation efforts, the Great Lakes region continues to face harmful effects of historical pollution from municipal and industrial sources. From 2010 to 2020, the Agency for Toxic Substances and Disease Registry (ATSDR) conducted the Biomonitoring of the Great Lakes Populations program to assess exposure to legacy and emerging contaminants, including PFAS, in susceptible populations who consume fish caught in contaminated waterbodies.

The biomonitoring program included cross-sectional studies conducted collaboratively with multiple state health departments. The target populations were recruited independently using sampling strategies that were tailored for each specific population. Participants completed questionnaires focused on fish consumption and provided biological specimens. Serum PFAS levels were measured in seven angler and refugee populations in New York and Wisconsin. Multivariate regression was used to assess associations between serum PFAS concentrations and self-reported fish consumption from Great Lakes waters.

Serum perfluorooctane sulfonic acid (PFOS) was significantly elevated in five of the seven target populations compared to the U.S. population. Other PFAS were not elevated. Locally caught fish consumption was associated with serum PFOS in five of seven populations after adjusting for covariates. PFOS was also found to be associated with age and gender in four of the target populations. The results generated from the biomonitoring program helped to guide public health actions to reduce chemical exposures in vulnerable Great Lakes populations, including updated fish advisories and targeted education and outreach on the risks and benefits of fish consumption.

136 – POSTER: Establishing a Community Air Monitoring Network to Investigate Hyperlocal Exposure to Oil and Gas Pollution

Caroline Frischmon[1], Brittney Lu-Jones[2], Richard Parks[3], Bhavna Shamasunder[4], Jill Johnston[5], Michael Hannigan[6]
[1]University of Colorado Boulder, Graduate Student, Boulder, USA. [2]Redeemer Community Partnership, Community Organizer, Los Angeles, USA. [3]Redeemer Community Partnership, President, Los Angeles, USA. [4]Occidental College, Associate Professor, Los Angeles, USA.

[5]University of Southern California, Associate Professor, Los Angeles, USA. [6]University of Colorado Boulder, Professor, Boulder, USA

Abstract
The Las Cienegas oilfield in South Los Angeles sits among the most environmentally burdened communities in California. Community groups in South LA have long advocated for stronger environmental protections for those living within the oilfield. As part of an interdisciplinary team of community organizers, engineers, epidemiologists, and social scientists, we set up a dense network of multi-pollutant air sensors across South LA to better understand oil and gas exposure impacts.

Air monitoring in oil and gas communities typically lacks the spatial resolution needed to investigate air quality differences at a hyperlocal scale. These hyperlocal differences may arise from the effects of wind dispersion, location relative to extraction sites, and proximity to other pollution sources, such as highway traffic.

We established a monitoring network capable of discerning these differences using three key strategies: pre-deployment meteorological monitoring, strategic neighborhood canvassing, and the development of a multi-pollutant sensor, called a multi-pod. Meteorological monitors set up throughout the community revealed major differences in wind direction that informed the density of our multi-pod placement. Strategic community canvassing allowed us to identify multi-pod hosts living both upwind and downwind of an oil and gas development site in buffers of <250 meters, <600 meters, and <1000 meters. Finally, our unique low-cost multi-pods can quantify up to eight individual pollutants. Multiple pollutant sensor signals can help us apportion the cumulative pollution levels to specific sources, such as oil and gas or traffic, so we can better understand the air quality impacts specifically caused by oil and gas development.

137 – Examining Impacts of Cross-Validation Approaches on Machine Learning Model Performance for Ultrafine Particle Exposure Assessment

Sean Mueller[1], Prasad Patil[2], Jonathan Levy[3], Neelakshi Hudda[4], John Durant[5], Emma Gause[6], Breanna Van Loenen[7], Kevin Lane[2]
[1]Boston University, PhD Student, Boston, USA. [2]Boston University, Assistant Professor, Boston, USA. [3]Boston University, Professor, Boston, USA. [4]Tufts University, Assistant Professor, Medford, USA. [5]Tufts University, Professor, Medford, USA. [6]Boston University, Research Scientist, Boston, USA. [7]Boston University, Research Assistant, Boston, USA

Abstract
Introduction: Ambient ultrafine particles (UFP) vary substantially over space and time and are therefore difficult to model. Machine learning (ML) approaches have been used to predict the complex nonlinear associations between UFP, source characteristics, and meteorology. Cross-validation (CV) is an important process in ML model building and assessment, but there are multiple potential CV methods and the choice among them could impact model accuracy and generalizability. Our aim is to assess how different CV approaches can impact model performance and feature prediction.

Methods: UFP was measured at a stationary site in Chelsea, Massachusetts from 2014-2021. UFP measurements were merged with flight activity, road traffic, and meteorology. We used a random forest (RF) and multiple CV techniques including K-fold and rolling. Model hyperparameter optimization and performance were compared using Mean Absolute Error, Root Mean Square Error, and the coefficient of determination (R2). Feature importance was calculated using permutation methods.

Results: Models exhibited substantial variation by CV method. Using RF, the optimal number of predictors per split ranged from two to eighteen. K-fold CV methods had the best performance on training data, while rolling CV had the best performance on the test data when validating on an external dataset. Meteorological features were most important across all CV techniques, while transportation source feature importance varied across models.

Conclusion: We found large variation in model performance and feature prediction across CV methods, indicating the importance of proper selection of CV technique in evaluating ML models for air pollution prediction and source characterization.

138 – Variations in Occupational Farm Exposures are Associated with Shifts in the Indoor Home Microbiome

Kathryn Dalton[1], Mikyeong Lee[2], Ziyue Wang[1], Shanshan Zhao[3], Alison Motsinger-Reif[4], Stephanie London[4]
[1]NIEHS, Postdoc, RTP, NC, USA. [2]NIEHS, Staff Scientist, RTP, NC, USA. [3]NIEHS, Investigator, RTP, NC, USA. [4]NIEHS, Senior Investigator, RTP, NC, USA

Abstract
Farm work entails a heterogeneous mixture of exposures that vary considerably across farms and farmers. Farm work is associated with various health outcomes, both adverse and beneficial. One mechanism that farming exposures exert their effects is through the microbiome, including the indoor built environment microbiome. It is unknown how different occupational exposures shape the microbes in workers' homes.

We investigated whether farm work activities and self-reported pesticide use were associated with the indoor microbiome in the homes of 468 male participants in the Agricultural Lung Health Study, a nested case-control study of licensed pesticide applicators. Vacuumed dust from participants' bedrooms underwent whole-genome shotgun sequencing for microbiome assessment. We evaluated 6 different farm work tasks (hay, silage, feed, grains, soy, or fertilizer use) and 19 pesticide ingredients used in the past year.

All work tasks were associated with increased within-sample microbial diversity, with a positive dose-response for the sum of tasks (p=0.001). Tasks were associated with altered overall microbial compositions (weighted UniFrac p=0.001), and showed an increased abundance of specific microbes, including pathogens. Among the 19 pesticides, only glyphosate was associated with increased within-sample diversity (p=0.04). Nine pesticides were associated with altered microbial composition (p 0.001-0.04), while atrazine and pyrethroid were associated with decreased abundance of select microbes.

Our work demonstrates that specific farm exposures impact the microbiome inside homes, which can be meaningful for the health of workers and their families. It poses potential mechanistic insights by which occupational exposures influence health through the microbiome, offering possible future intervention targets.

139 – POSTER: Validation of the Polair3D Chemical Transport Model Over Quebec and Analysis of Emission Reduction Scenarios

Shoma Yamanouchi[1], Shayamilla Mahagammulla Gamage[1], Sara Torbatian[2], Jad Zalzal[2], Laura Minet[3], Audrey Smargiassi[4], Ying Liu[5], Youngseob Kim[6], Daniel Yazgi[7], Andrée-Anne Brown[8], Marianne Hatzopoulou[9]
[1]University of Toronto, Department of Civil and Mineral Engineering, Postdoctoral Fellow, Toronto, Canada. [2]University of Toronto, Department of Civil and Mineral Engineering, PhD Student, Toronto, Canada. [3]University of Victoria, Professor, Victoria, Canada. [4]Université de Montréal, Professor, Montreal, Canada. [5]Université de Montréal, Postdoctoral Fellow, Montreal, Canada. [6]CEREA, École des Ponts, Researcher, Marne-la-Vallée, France. [7]Swedish Meteorological and Hydrological Institute, Researcher, Norrköping, Sweden. [8]Ministère de l'Environnement et de la Lutte contre les changements climatiques, Researcher, Quebec City, Canada. [9]University of Toronto, Department of Civil and Mineral Engineering, Professor, Toronto, Canada

Abstract
Air pollution is a major health hazard, and while air quality overall has been improving in industrialized nations, pollution is still a major economic and public health issue. Chemical transport models (CTM) are valuable tools that aid in our understanding of the risks of air pollution both at local and regional scales.

In this study, the Polair3D CTM was set up over Quebec, Canada to assess the model's capability in predicting key air pollutant species over the region, at seasonal temporal scales and at regional spatial scales. Given the size of the country, and the province, the existing network of surface observation sites that monitor pollutants is far too sparse to be useful in conducting spatial variability analyses of air pollutants. Modeling the sources, chemistry, dynamic transport of atmospheric pollutants is crucial in understanding tropospheric pollution events and mitigating health impacts by identifying affected regions and sensitivity to various emissions. In this study, model was validated against surface observations, and was used to analyze changes in population exposure as a result of scenarios affecting traffic and industrial emissions.

The simulation by the model included 3 nested domains, at resolutions of 9x9km, 3x3km, and two 1x1km domains covering Montreal and Quebec City. We find that the model accurately captures the spatial variability and seasonal effects, and the 3km domain showed higher correlation than at the 1km domain. The performance of the Polair3D model was in line with other models over Canada, and comparable to Polair3D's performance over Europe.

141 – Silica Exposure Among Countertop Fabricators by Particulate Matter Size and Material

Allison Persing[1], Berkley Wood[2]
[1]RHP Risk Management, Senior Associate Health Scientist, Chicago, USA. [2]RHP Risk Management, Associate Environmental Health Scientist,

Chicago, USA

Abstract
The goal of this study was to investigate the particulates produced from cutting and polishing pieces of granite and quartz based on mass distribution and particle size distribution. Additionally, this study assessed exposures of respirable dust and respirable crystalline silica (RCS) during fabrication of granite and quartz with and without the use of a shroud.

A 400 ft.2 negative pressure enclosure was constructed, where both a slab of granite and quartz were fabricated. A shroud was in place on the angle grinder for one trial on each piece of stone and the shroud was removed during the second test on each piece of stone. During the trials, RCS concentration was measured with a 37 mm aluminum cyclone, number concentration was measured with an AeroTrak, and mass concentration was measured with a DustTrak.

It was evident that the shroud was an effective control in reducing the concentration of respirable dust below the PEL of 5 mg/m3 for both stone types during fabrication. However, the shroud was not an effective control, alone, to reduce the concentration of RCS below the OSHA PEL of 50 ug/m3 for each stone and activity.

For the overall particulate count basis, an average of 51 percent of total particles produced during all trials were less than or equal to 0.3 microns. Most particles produced during the activities were PM10, averaging 80 percent during cutting and 76 percent during polishing. The results of this study emphasize the importance of implementing controls during stone fabrication to mitigate risk.

144 – Introduction of Korea Air Pollutants EXposure (KAPEX) model

Kiyoung Lee[1], Wonho Yang[2], Haemin Park[1], Sohyun Kang[3], Ji Young Park[4]
[1]Seoul National University, Professor, Seoul, Korea, Republic of. [2]Daegu Catholic University, Professor, Daegu, Korea, Republic of. [3]Seoul National University, student, Seoul, Korea, Republic of. [4]Seoul National University, Research Professor, Seoul, Korea, Republic of

Abstract
Outdoor concentration is often used as a surrogate for population exposure in epidemiological studies. However, people stay most of their times indoors. The activity pattern should be considered in the estimation of personal exposure. The Korea Air Pollutants EXposure (KAPEX) model is being developed to predict population exposures to $PM_{2.5}$ and O_3. The input data for the model were 1) site and time specific outdoor concentrations from outdoor air pollution data of 511 ambient air monitoring stations in Korea, 2) detailed time activity patterns by regional survey of 20,000 subjects, and 3) the microenvironment-to-outdoor (M/O) concentrations of $PM_{2.5}$ and O_3 at seven microenvironments. The personal exposure by KAPEX model will be evauated by direct measurement of personal exposure. Field technicians simulated a total of 10 time-activity pattern scenarios for each season while carrying a real-time aerosol monitor and a miniaturized O_3 monitor. Personal exposure monitoring was conducted 40 person-days in winter, 50 person-days in spring, and 80 person-days in summer and autumn, respectively. For measurement in 2022, Seoul population-weighted exposure to $PM_{2.5}$ was 21.5 $\mu g/m^3$ in winter, followed by 15.0 $\mu g/m^3$ in summer, and 14.7 $\mu g/m^3$ in autumn, and 14.0 $\mu g/m^3$ in spring. The Seoul population-weighted exposure to O_3 was 10.5 ppb in spring, followed by 3.9 ppb in autumn, 3.8 ppb in summer, and 3.2 ppb in winter. The KAPEX model for $PM_{2.5}$ and O_3 could be used to estimate population exposure, identify determinants for high exposure groups and evaluate the effectiveness of control policies to reduce exposure.

145 – POSTER: Estimating Dust Loadings on Surfaces of Homes

Helena Solo-Gabriele[1], Cristina Fayad[2], Maribeth Gidley[3], Matthew Roca[4], Ryuichi Nitta[4], Panara Suraj[4], Sofia Suarez Barajas[4], Anabella Cordovez[4], Elizabeth Gabriele[4], Collette Thomas[4], Foluke Adelabu[5], Neema Olagbemiro[5], Paloma Beamer[6], Jenna Honan[7], Olusola Ogunseye[8], Alesia Ferguson[9]
[1]University of Miami, Professor, Coral Gables, USA. [2]University of Miami, Research Associate, Coral Gables, USA. [3]University of Miami, Research Scientist, Miami, USA. [4]University of Miami, Student, Coral Gables, USA. [5]North Carolina Agricultural and Technical State University, Phd Student, Greensboro, USA. [6]University of Arizona, Professor, Tucson, USA. [7]University of Arizona, Research Associate, Tucson, USA. [8]University of Arizona, Phd Student, Tucson, USA. [9]North Carolina Agricultural and Technical State University, Professor, Greensboro, USA

Abstract
Dust and soil within the home can serve as a source of contaminant exposure to children, where a greater dust loading is associated with higher transfer rates and, therefore, higher probabilities of exposures. The objective of this study was to measure dust loadings

(mass per surface area) on two types of surfaces in the home; surface wipes (from elevated surfaces such as tables, chairs, shelves) and vacuum samples (from floors). A total of 480 surface wipes and 76 vacuum samples were collected from 60 different homes with children (ages from 6 months to 6 years). Samples were collected from three regions (North Carolina, Arizona and Florida) and from different socio-demographic groups as documented through household surveys. The overall average dust loading from surface wipes was 35 μg/cm2 (coefficient of variation, COV = 2.3) and for vacuum samples was 130 μg/cm2 (COV = 1.7). The highest average values for surface wipes (51 μg/cm2) and vacuum samples (234 μg/cm2) were observed for Greensboro. Differences in dust loading were identified by surface type (e.g., wood, tile, laminate) and home characteristics (type of construction, home age, presence of pets). These results will be used as part of the Dust Ingestion childRen sTudy (DIRT) which documents child play behavior through video translation and will simulate soil/dust ingestion rates for children (from a range of socio-demographic groups) using a micro-level activity time series (MLATS) model.

146 – Use of Critically Important Antimicrobials (CIAs) In Poultry Health Clinics, Nangarhar province Eastern Afghanistan

Mohammad Bayer Darmel
Nangarhar University, Associate Professor, Nangarhar, Jalalabad, Afghanistan

Abstract
The international organizations (WHO, OIE and FAO) are developing new guidelines for safety of antimicrobials (AMs) and their use particularly Critical Important AMs (CIAs) of medical and veterinary importance. The data related to the use AMs is scanty related to Afghanistan in poultry sector where use of AMs is common drug used for health protection and growth support in poultry sector. This study was designed to determine the use of AMs particularly CIAs in Poultry Health Sector at Nangarhar province and colected data was compared with WHO lists under CIAs human concern. In the result total quantity of AMs used in treatment and prophylaxis was 25.482 tons, in one year which consist the AMs classes: Tetracyclines 22.27% (doxycycline 18.00 and oxytetracycline 4.27), Aminoglycosides 16.36% (neomycin), Macrolides 12.275% (Tylosine,7.55 and tilmycosin 4.72) fluoroquinolones 11.456% (Enrofloxacin 6.20 Ciprofloxacin,4.63 and norfloxacin 0.61) Aminopenicillins 10.367% (Amoxicillin), Polymyxins 8.075%(colistin), Amphencoles 7.72% (florophenicol) Sulphonamides 7.18%(Sulphadiazine 2.99, Sulphapyridazine, 3.35 Sulphacloropyridazine 0.68), Diaminopyridines 1.29 (trimethoprim), Quinolone 0.477% plorotiamules 0.42 %(tiamulin) antivirals 0.23% (Amantadine), lincosamides0.99% (lincomycin) and Aminocyclitoles 0.88 (Spectinomycin) respectively. The CIAs make 58. 5% where highest priority 31.81% (tylosin 7.55, tilmycosin 4.70, Colistin 8.075, ciprofloxacin 4.53, enrofloxacin 6.20 and norfloxacin 0.61) and high priority 26.697% (Neomycin16.36 and Amoxicillin 10.336) make total antimicrobials used by 78 poultry clinics. In conclusion, the high amount usage of AMs particularly CIAs of medical important can case a severe problem of public health concern and it is required to develop policies for the reduction of AMs particularly CIAs.

147 – Impact of Local Food Consumption on Exposure to Perfluorooctanoate and Perfluorooctane Sulfonate in a Contaminated Community in Northeastern Italy: A Scenario-based Approach

Hyeong-Moo Shin[1], Armando Olivieri[2], Mario Saugo[2]
[1]Baylor University, Associate Professor, Waco, USA. [2]College of Physicians of Vicenza, Physician, Vicenza, Italy

Abstract
Background: For residents in highly contaminated areas with per- and polyfluoroalkyl substances, the contribution from local food consumption is not well characterized. We evaluated the impact of local food consumption on perfluorooctanoate (PFOA) and perfluorooctane sulfonate (PFOS) exposures by using data from the most contaminated area of the Veneto Region, Italy.

Methods: We considered three exposure groups of residents that may have had different intake rates (IR, ng/kg-day) of PFOA/PFOS, due to their drinking water sources and diet choices: (1) those drinking water filtered with double granular activated carbon (GAC) and not consuming local foods (reference group); (2) those drinking GAC-filtered water and consuming local foods; (3) those drinking private well water and not consuming local foods. For each group, we applied a one-compartment pharmacokinetic model to compute time-dependent PFOA/PFOS serum concentrations and the time needed to reach a threshold level of PFOA/PFOS with recognized clinical significance.

Results: For those drinking filtered water and consuming local foods, their mean IR of PFOA/PFOS is 2.3 ng/kg-day, which is about one-quarter higher than the reference group, and it takes up to 3.4 more years for their PFOA/PFOS concentrations to reach the threshold of interest. Regardless of local food consumption, drinking contaminated private well water results in steady-state concentrations above the threshold.

Conclusion: Our study shows that unrestricted local food consumption can be an additional exposure pathway for those in a PFAS-contaminated community where a GAC filtration system is already in place in public water districts.

148 – Estimation of emission rates of black carbon while heating 17 commercial cooking oils

Mehdi Amouei Torkmahalleh[1], Mostafa Salmanimojaveri[2], Tomiris Madiyarova[3], Motahareh Naseri[3], Nadezhda Ushakova[3], Karina Yessengaziyeva[3], Gulnur Sultanova[4], Enoch Adoteyl[3], Gulnaz Zhemeney[5], Seyed Morteza Zamir[6], Dhawal shah[7]
[1]University of Illinois at Chicago, Assistant Professor, Chicago, USA. [2]Nazarbayev University, research asistant, Astana, Kazakhstan. [3]Nazarbayev, student, Astana, Kazakhstan. [4]Nazarbayev, Student, Astana, Kazakhstan. [5]Nazarbayev, Research asistant, Astana, Kazakhstan. [6]Tarbiat Modares University, Faculty, Tehran, Iran, Islamic Republic of. [7]Nazarbayev, Assistant Professor, Astana, Kazakhstan

Abstract
It has been scientifically demonstrated that cooking significantly contributes to indoor particulate matter (PM). A study found that cooking was the primary source of PM2.5 and submicron particles (measuring between 7 to 808 nm) among 20 indoor activities in 15 houses. The presence of the black carbon was reported in the cooking PM. The present study investigated emissions of black carbon from 17 heated pure cooking oils available in the US market. The process involved heating 100 ml of cooking oil in a beaker until the temperature reached 195oC, and continued for 20 minutes. The highest BC emission rate was observed in coconut oil, while sunflower oil emitted the lowest amount of BC. Other oils that emitted high levels of BC at 195°C include macadamia, flax, pumpkin, avocado, grapeseed, hazelnut, and sesame (high emitting oils).

On the other hand, almond, olive, walnut, corn, vegetable, canola, peanut, safflower, and sunflower oils emitted lower levels of BC at the same temperature (low emitting oils). The study found a significant difference in BC emissions between the high and low-emitting oil groups. In this study, we have developed a statistical model that estimates BC emission of the cooking oils using the oil temperature and smoke temperature of the oils. The highest correlation coefficient was found to be 0.61for the temperature range of 175- 185oC. The study suggests that exposure to cooking BC as well as the contribution of residential BC to ambient BC can be reduced by selecting a low-BC emitting oil.

149 – Associations between Personal Apparent Temperature Exposures and Asthma Symptoms in Children with Asthma

Linchen He[1], Shoshana Evans[2], Christina Norris[3], Karoline Barkjohn[4], Xiaoxing Cui[5], Zhen Li[6], Feng Li[6], Yinping Zhang[7], Marilyn Black[8], Michael Bergin[9], Jim Zhang[9]
[1]Lehigh University, Assistant Professor, Bethlehem, USA. [2]Lehigh University, Student, Bethlehem, USA. [3]University of North Carolina at Chapel Hill, Student, clnorris@email.unc.edu, USA. [4]US Environmental Protection Agency, Researcher, Research Triangle Park, USA. [5]Duke University, Student, Durham, USA. [6]Shanghai Jiao Tong University, Chef Physician, Shanghai, China. [7]Tsinghua University, Professor, Beijing, China. [8]Underwriters Laboratories, VP, Marietta, USA. [9]Duke University, Professor, Durham, USA

Abstract
Ambient temperature and relative humidity can affect asthma symptoms. Apparent temperature is a measure of temperature perceived by humans that takes into account the effect of humidity. However, the potential link between personal exposures to apparent temperature and asthma symptoms has not been investigated. We conducted a panel study of 37 asthmatic children, aged 5-11 years, during an early spring season (average daily ambient temperature: 14°C, range: 7-18°C). Asthma symptoms were measured 4 times for each participant with a 2-week interval between consecutive measurements using the Childhood Asthma-Control Test (C-ACT). Average, minimum, and maximum personal apparent temperature exposures, apparent temperature exposure variability (TV), and average ambient temperature were calculated for the 12 hours, 24 hours, week, and 2 weeks prior to each visit. We found that a 10 °C lower in 1-week and 2-week average & minimum personal apparent temperature exposures, TV, and average ambient temperature exposures were significantly associated with lower total C-ACT scores by up to 2.2, 1.4, 3.3, and 1.4 points, respectively, indicating worsened asthma symptoms. Our results support that personal apparent temperature exposure is potentially a stronger driver than ambient temperature exposures for the variability in asthma symptom scores. Maintaining a proper personal apparent temperature exposure, by approaches like using indoor air conditioning and a humidifier during the cold season, could be an effective strategy in personalized asthma management.

150 – Computational Modeling of the Biphasic Depletion of Ovarian Follicle Reserve and Environmental Effects on Ovarian Aging

Sarahna Azaria Moyd[1], Audrey J Gaskins[2], Shuo Xiao[3], Qiang Zhang[4]
[1]Emory University, PhD Candidate, Atlanta, USA. [2]Emory University, Assistant Professor, Atlanta, USA. [3]Rutgers University, Assistant Professor, Piscataway, USA. [4]Emory University, Associate Professor, Atlanta, USA

Abstract
Background: Exposure to chemotherapy drugs and environmental toxicants can accelerate follicular depletion potentially leading to adverse reproductive outcomes, including early menopause, primary ovarian insufficiency (POI), and infertility. The ovarian reserve exhibits bi-phasic depletion: constant decline from birth to mid-30s, followed by accelerated decline to senescence. While this pattern can be described mathematically, rarely is it modeled mechanistically.

Method: To address this knowledge gap, we generated a mechanistic model characterizing the dynamics of age-dependent follicle depletion to predict reproductive health effects of chemical exposures on ovaries. Specifically, we proposed a minimal mechanistic model, including (1) zero-order feedforward inhibition of primordial follicle activation by a local inhibitory factor secreted by NGF, and (2) high-gain feedback inhibition of primordial follicle activation by anti-Müllerian hormone (AMH) secreted by the growing follicles, or GF. These two regulatory processes optimize the supply of NGF and GF across the reproductive life span.
Results: Preliminary results show the size of the initial primordial follicle pool has a significant role in ensuing menopausal age, and unilateral ovariectomy minimally effects GF despite significant reduction in NGF. Simulations of transient chemotherapy suggest earlier age exposure may have greater effect on ovarian reserve and menopausal timing than later age.

Conclusion: Collective results have important implications for people seeking to reproduce, including sensitive groups such as those experiencing aging-related infertility. Future elaborations are warranted to integrate in vitro toxicity testing data to predict effects of select reproductive toxicants.

151 – Workplace exposure during producing and handling of nanoelectric devices, vacuum pump oils and graphene linked with occupational health risks

Ishtiaq Hussain[1], Faiza Naseem[2], Ikhtiar Uddin[3]
[1]National Institute of Vacuum Science And Technology, NCP Complex, QAU Campus, General Manager Occupational Health & Technical Services, Islamabad, Pakistan. [2]Rawalpindi Women University, Assistant Professor, Rawalpindi, Pakistan. [3]Eco-Health, Department of Environmental Sciences, PMAS Arid Agriculture University, Research Fellow, Rawalpindi, Pakistan

Abstract
Perilous workplace exposure typically occurs when recommended engineering safeguards and personal protection strategies are not properly implemented. At such sites, workers may be exposed to multiple contaminants by inhalation, cutaneous contact and gastroenteric pathways. A facility that processes vacuum pump oils and evaluates high-performance nanoelectronic devices and integrated circuits having graphene, carbon nanotubes, and semiconductor nanowires as building blocks underwent an extensive study to measure air contaminants. Samples from filters, sorbents, microscopy, and direct-reading instruments were used to obtain information. Clinical history of ailments and prevailing symptoms were recorded through detailed interviews of workers. We detected 5–54 times higher concentrations of respirable elemental carbon inside the facility than outdoors. Polycyclic aromatic hydrocarbons (PAHs) and carbon monoxide in breathing zone samples were up to 130 times higher. Estimates of fine/ultrafine iron-rich soot particles were significantly higher inside workplace than outdoor. Inhalation linked respiratory infections were highest among workers (X^2=6.7, $p<0.01$) performing fabrication processes related to device performance analysis followed by workers dealing with vacuum oils and nanomaterials where exacerbation of asthma and chronic obstructive pulmonary disease was significant (χ^2=5.8, $p<0.05$). Our study concludes that relative risks are likely to be significant for the exposures assessed at this particular workplace. Reducing inhalation and skin contact with harmful substances at workplace is crucial in enhancing respiratory wellbeing, and further investigations are necessary to broaden our understanding of treatments that could prevent untimely asthma and frequent bronchitis episodes.

152 – Predicting Monthly Community-level Radon Concentrations with Spatial Random Forest in the Northeast and Midwest United States

Longxiang Li[1], Brent Coull[2], Petros Koutrakis[3]
[1]Harvard Chan School of Public Health, Postdoctoral Fello, Boston, USA. [2]Harvard Chan School of Public Health, Professor, Boston, USA.

[3]Harvard Chan School of Public Health, Professor, Boston, USA

Abstract
United States Environmental Protection Agency recommended installing a mitigation system when indoor concentration of radon, a well-known radioactive gas, is above 148 Bq/m3. In response, tens of millions of radon measurements have been conducted in residential buildings in the past three decades either for disclosure or to initially evaluate the need for mitigation. These measurements, however, are currently underutilized to assess population radon exposure. Based on two relatively small radon surveys, Lawrence Berkeley National Laboratory developed a state-of-the-art national radon model. However, this model only provides coarse and invariant radon estimations, which limits the ability of epidemiological studies to investigate the health effects of radon. This study involved obtaining over 2.8 million short-term radon measurements from independent laboratories. Using these measurements, an innovative spatial random forest (SRF) model was developed based on geological, architectural, socio-economical, and meteorological predictors. The model was used to estimate monthly community-level radon concentrations for ZIP Code Tabulation Areas (ZCTAs) in the Northeastern and Midwestern regions of the United States from 2001 to 2020. Via cross-validation, we found that the community-level estimations were highly correlated with observations. The estimation errors declined quickly as the number of radon measurements in a community increased. When ≥15 measurements existed, the mean absolute error was 24.6 Bq/m3, or 26.5% of the observed concentrations (R2=0.70). Our study demonstrates the potential of the large amount of short-term radon measurements that have been obtained to accurately estimate longitudinal community-level radon exposures at unprecedented levels of resolutions and accuracy.

153 – Passive sampling for assessing time-weighted exposure estimation of reactive biocides

Jung-Hwan Kwon[1], Yerim Koo[2], Yeonjeong Ha[3], Wonsik Shin[4]
[1]Korea University, Professor, Seoul, Korea, Republic of. [2]Korea Unversity, Student, Seoul, Korea, Republic of. [3]Pyeongtaek University, Professor, Pyeongtaek, Korea, Republic of. [4]Korea University, Student, Seoul, Korea, Republic of

Abstract
The use of reactive biocides in diverse consumer products may cause health concerns. Unlike volatile and semi-volatile organic chemicals, reactive biocides decompose very quickly in the indoor environment, making it difficult to quantify their exposure concentration. Screening level mathematical models, which assume no degradation, often overestimate the exposure concentration. Although passive sampling is advantageous when assessing personalized average exposure for many indoor air pollutants, including persistent organic chemicals, this method has rarely been attempted for reactive biocides. In this study, passive samplers employing specific chemical reactions between reactive biocides and chemicals doped in polydimethylsiloxane (PDMS) have been developed. Redox dyes that change their colors after exposure to strong oxidants, such as chlorine and chlorine dioxide, were found to be useful in developing a ready-to-use passive sampler for indoor disinfectants. A passive sampler containing glutathione in the PDMS tube was also found to be eligible to detect atmospheric isothiazolinones that are often used for disinfection purposes. Levels of reactive biocides determined using the developed passive samplers under consumer exposure scenarios were much lower than those estimated using mathematical models assuming that those biocides are conservative. Therefore, the use of passive samplers should also be very helpful in determining a reasonable level of personalized time-weighted average exposure and avoiding any overestimations made at the screening exposure assessment stage using mathematical models.

154 – Exposure science as an important ally of the European Zero Pollution Ambition

André Conrad[1], Pia Splanemann[2], Judith Meierrose[2], Alena Bartonova[3], Till Weber[2], Ana Maria Scutaru[2], Ian Marnane[4]
[1]German Environment Agency (UBA), Head of Section, Berlin, Germany. [2]German Environment Agency (UBA), Scientist, Berlin, Germany. [3]NILU Norwegian Institute for Air Research, Research Director EU collaboration, Kjeller, Norway. [4]European Environment Agency (EEA), Expert, Copenhagen, Denmark

Abstract
The Zero Pollution Ambition (ZPA) is part of the European Green Deal and aims to reduce environmental pollution levels to avoid harms to human health and natural ecosystems by 2050. The Zero Pollution Monitoring and Outlook framework accompanies the ZPA, especially by providing up-to-date knowledge on pollutant exposures and measuring progress. In 2022, the European Environment Agency published the baseline Zero Pollution Monitoring Assessment (ZPMA) including a chapter on pollution and health with exemplary case studies (signals) communicating facts on pollution and health as well as emerging topics.

The current ZPMA contains health signals on air, noise, water, chemicals, and soil. For example, a signal on air pollutants shows the importance of room ventilation for reducing health-relevant exposures indoors. For chemicals, a health signal discusses regrettable substitution based on German human biomonitoring data for plasticizers. Another signal demonstrates how single-substance phthalate exposure assessments would have underestimated risks by up to 50 %. Moreover, the ZPMA contains cross-cutting stories providing further insight into the relevance of pollution across regulatory domains.

The ambitious ZPA vision can only be achieved by cooperation between policy areas. The multidisciplinary exposure science approach can substantially support the ZPA by integrated data and knowledge on health impacts and resources in view of an overarching monitoring and assessment. Exposure science can help closing knowledge and data gaps for the next ZPMA. By tapping into the dialog between academic communities, exposure science can make a difference on the way to a non-toxic environment for all.

Further Information: https://www.eea.europa.eu/publications/zero-pollution

156 – Receptor modeling for source apportionment of secondary aerosols with multiple time resolution data

Chun-Sheng Huang[1], Ho-Tang Liao[2], Chia-Yang Chen[3,4,5], Li-Hao Young[6], Ta-Chih Hsiao[7], Tsung-I Chou[8], Jyun-Min Chang[8], Kuan-Lin Lai[8], Chang-fu Wu[3,5]
[1]Institute of Environmental and Occupational Health Sciences, National Taiwan University, PhD student, Taipei, Taiwan. [2]Institute of Environmental and Occupational Health Sciences, National Taiwan University, Postdoctoral researcher, Taipei, Taiwan. [3]Institute of Environmental and Occupational Health Sciences, National Taiwan University, Professor, Taipei, Taiwan. [4]Institute of Food Safety and Health, National Taiwan University, Professor, Taipei, Taiwan. [5]Department of Public Health, National Taiwan University, Professor, Taipei, Taiwan. [6]Department of Occupational Safety and Health, China Medical University, Professor, Taichung, Taiwan. [7]Graduate Institute of Environmental Engineering, National Taiwan University, Professor, Taipei, Taiwan. [8]Institute of Environmental and Occupational Health Sciences, National Taiwan University, MS student, Taipei, Taiwan

Abstract
The receptor model of Positive Matrix Factorization (PMF) has been widely used to characterize the emission sources of fine particulate matter ($PM_{2.5}$). However, PMF results often retrieve an individual factor representing secondary aerosols, which makes it difficult to identify the associated primary emission sources. To address this issue, a two-stage PMF modeling approach was implemented to better characterize the sources of secondary aerosols.

For PMF inputs, the hourly $PM_{2.5}$ elemental and ionic compositions were collected from May to December 2021 in an urban area in Taiwan. Additional 24-h integrated filter samples were collected simultaneously, resulting in 56 daily samples that were analyzed for carbonaceous components. The hourly measurements were averaged into 3-h intervals to increase the weight of carbonaceous components in the computation of model solution. Multiple time-resolved $PM_{2.5}$ composition data were applied to a two-stage PMF model with distinct parameter settings for uncertainties on secondary aerosol-related species.

The results showed that the top three $PM_{2.5}$ contributors were aged emissions (contribution = 25%), oil combustion (20%) and biomass burning related (13%). Secondary aerosols were mainly distributed in factors of oil combustion (27%), aged emissions (24%) and nitrate-rich secondary aerosol (17%). Compared to the conventional one-stage PMF method, downweighting secondary aerosol-related species with the two-stage PMF approach was found to be more effective in linking them to primary emission sources. Two-stage PMF approach provides insights into the exploration of secondary aerosol origins, which can be beneficial for formulating $PM_{2.5}$ management strategies.

159 – POSTER: Innovative Method to Measure Mass of Dust on Children's Hands

Cristina Fayad Martinez[1], Maribeth Gidley[2], Matthew Roca[1], Ryuichi Nitta[1], Suraj Panara[1], Sofia Suarez Barajas[1], Anabella Cordovez[1], Elizabeth Gabriele[1], Collette Thomas[1], Helena Solo-Gabriele[3], Alesia Ferguson[4], Foluke Adelabu[5], Neema Olagbemiro[5], Paloma Beamer[6], Jenna Honan[7], Olusola Ogunseye[7]
[1]University of Miami, Student, Miami, USA. [2]University of Miami, Research Scientist, Miami, USA. [3]University of Miami, Professor, Miami, USA. [4]North Carolina A&T, Professor, Greensboro, USA. [5]North Carolina A&T, Student, Greensboro, USA. [6]University of Arizona, Professor, Tucson, USA. [7]University of Arizona, Student, Tucson, USA

Abstract
Dust is a heterogenous mixture of materials with potentially harmful chemicals. Children represent a vulnerable population to dust intake through ingestion due to their unique play behaviors. There are two general methods for estimating ingestion of a contaminant.

The first is based upon identifying a conservative tracer and measuring it in child biological samples (e.g., urine, and feces). The second method is through estimations of dust mass on children's hands, activity pattern modeling (hand or object to mouth transfers), and dust contaminant load. We provide an innovative method to directly measure the mass of dust accumulated on children's hands, overcoming some of the difficulties associated with traditional methods. Our method determines the volume of particles washed off children's hands using a Coulter Counter. To calculate the mass of dust on children's hands, the volume is then multiplied by the density of dust samples collected from the home through vacuuming. This study reports the mass of dust on hands obtained from different age groups as well as the particle size distribution from the dust collected. The majority of particles found on children's hand rinses had a diameter size <30 μm (D_{90}= 29.1 μm). The mean mass of dust per surface area calculated was 44.4 μg/cm^2, with a median of 11.2 μg/cm^2. These results are intended for integration to the Dust Ingestion childRen sTudy (DIRT) which includes a modeling effort to compute the soil/dust ingestion rates for young children aged 6 months to 6 years.

160 – POSTER: Exposure and risk assessment of active substances in biocidal products to disinfection worker

Gihong Min[1], Dongjun Kim[2], Jihun Shin[1], Jaemin Woo[2], Mansu Cho[3], Wonho Yang[3]
[1]Daegu catholic university, Ph.D. Student, Gyeongbuk, Korea, Republic of. [2]Daegu catholic university, MPH Student, Gyeongbuk, Korea, Republic of. [3]Daegu catholic university, Professor, Gyeongbuk, Korea, Republic of

Abstract
The importance of disinfection has increased due to the spread of infections such as COVID-19. Disinfection is highly recommended to prevent the spread of COVID-19. Disinfection workers' exposure and risk assessment for biocidal products has been insufficient. This study aimed to assess the exposure and risk of disinfection workers through questionnaires.

A total of 4,762 cases were obtained from a 10-day follow-up of 271 disinfection workers from June to September 2021. Inhalation and dermal exposure algorithms were used in ConsExpo. Applying ECHA's assessment factor based on the toxicity value, occupational exposure levels (OELs) were estimated. The hazard index (HI) for disinfection workers was calculated by dividing the inhalation concentration and dermal dose by the OELs. Logistic regression analysis was used to identify variables influencing health effects.

Public facilities had longer exposure to biocidal products and higher usage than apartments and detached houses. Disinfectant usage likely increased rapidly to prevent the spread of COVID-19 and maintain safe indoor environments. The highest HI was 6.50E+00 for ethanol, and for quaternary ammonium compounds (QACs) and benzalkonium chloride (BKC), the HI value was ≥1, indicating a potential health risk. The inverse associations found in regression analysis for the period of the disinfection worker occupation may be due to the healthy worker survival effect that causes downward bias.

This study suggests that disinfection workers using ethanol, BKC, and QACs could have potential health risks. Therefore, it would be necessary to control the exposure to active substances for disinfection workers using personal protective equipment.

161 – Sociodemographic characteristics and exposure factors affecting heavy metal concentrations in blood and urine of Korean adults

Gihong Min[1], Jihun Shin[1], Dongjun Kim[2], Jaemin Woo[2], Mansu Cho[3], Wonho Yang[3]
[1]Daegu catholic university, Ph.D student, Gyeongbuk, Korea, Republic of. [2]Daegu catholic university, Master course, Gyeongbuk, Korea, Republic of. [3]Daegu catholic university, Professor, Gyeongbuk, Korea, Republic of

Abstract
Human activities have led to elevated levels of exposure to lead (Pb), mercury (Hg), and cadmium (Cd). The consequences of the accumulation of heavy metals on human health are constantly being uncovered through research, and the demand for managing exposure levels to heavy metals is rising. This study aims to identify the factors affecting heavy metal concentration in Korean adults bodies and to produce the contribution rate of each exposure route.

This study used the raw data of the 4th Korean National Environmental Health Survey (KoNHES) from 2018 to 2020 in Korea. The questionnaire dietary intake frequency data was utilized to estimate the concentration of heavy metals in the body through exposure routes. The annual report of air quality in Korea was used to consider heavy metal concentrations in the air, and dermal exposure adapted for adsorption from dust and soil onto the skin.

Multiple regression analysis was conducted using sociodemographic factors and exposure characteristics as independent variables to determine their impact on exposure level to heavy metals. The concentration of heavy metals in the body was found to increase with age and was higher in smokers compared to non-smokers ($p<0.05$). Additionally, the physiologically based pharmacokinetic modeling (PBPK) model will be used to calculate the contribution rate of heavy metal concentrations in the body for each exposure route.

This study will to suggest management plan by producting the factors affecting the levels of heavy metals in Korean blood and urine and the contributoin rare of exposure pathways.

162 – Influence of climate-impacted storms on the spread of contaminants from a Superfund site: A case study at DZUS fasteners in New York

Lissa Soares[1], Dimitrios Bakatsias[2], Adrienne Popescu[2], Janroy Caraan[2], David Dukes[3], Carrie McDonough[4], Arjun Venkatesan[5], Christopher Gobler[6], Payson Long[7], Evelyn Hussey[8], James Kruegler[9], Michael Cruden[10], William Ottaway[11], Kevin Reed[12], Jaymie Meliker[6]
[1]Stony Brook University, Postdoctoral Scholar, Stony Brook, USA. [2]Stony Brook University, Undergraduate Student, Stony Brook, USA. [3]Stony Brook University, PhD Student, Stony Brook, USA. [4]Carnegie Mellon University, Assistant Professor, Pittsburgh, USA. [5]Stony Brook University, Research Assistant Professor, Stony Brook, USA. [6]Stony Brook University, Professor, Stony Brook, USA. [7]NY Department of Environmental Conservation, Environmental Engineer, Schnectady, USA. [8]NY Department of Environmental Conservation, Geologist, Schnectady, USA. [9]NY Department of Environmental Conservation, Environmental Engineer, Troy, USA. [10]NY Department of Environmental Conservation, Bureau Director, Remediation, Ballston Spa, USA. [11]NY Department of Environmental Conservation, Environmental Engineer, Albany, USA. [12]Stony Brook University, Associate Professor, Stony Brook, USA

Abstract
Introduction: In May, 2021, J. Alfredo Gómez, Director, Natural Resources and Environment, US Government Accounting Office, reported that 50% of Superfund sites are at risk of moderate flood hazard, 12% from storm surge, and 7% from sea level rise and concluded that EPA should take additional actions to manage risks from climate change at Superfund sites. We have created an academic-government partnership to better understand how storms may be impacting movement of contaminants near Superfund sites.

Methods: We report on New York Department of Environmental Conservation (NYDEC) data from a case study of Willetts Creek, West Islip, NY, just down creek from DZUS fasteners, a NY State Superfund site. Water and sediment sampling were conducted approximately yearly, including before and after Superstorm Sandy, October 2012. During Sandy, Willetts Creek flooded, which abuts a middle school, a high school, and the DZUS facility.

Results: In Apr 2013 after Sandy, 70 sediment samples were collected from Willetts Creek, 14 were above 90 mg/kg cadmium (Cd) with a maximum of 1600 mg/kg, and another 5 were above 9 mg/kg Cd, the site clean-up criterion level. The previous maximum sediment sampling concentration was 133 mg/kg. Similarly, in a monitoring well 3 meters below the surface, water concentrations increased from 64.4 μg/L Cd in Aug 2012 to 120 μg/L in Nov 2013.

Conclusion: Superstorm Sandy appears to have increased mobility of Cd from a Superfund site; increased monitoring at superfund sites is recommended to understand impacts of climate-related storms.

163 – Assessment of inhalation exposure to active substances in disinfectants for disinfection worker

Dongjun Kim[1], Jihun Shin[2], Gihong Min[2], Jaemin Woo[1], Mansu Cho[3], Wonho Yang[3]
[1]Daegu Catholic University, Master student, Gyeongsan, Korea, Republic of. [2]Daegu Catholic University, Ph.D student, Gyeongsan, Korea, Republic of. [3]Daegu Catholic University, professor, Gyeongsan, Korea, Republic of

Abstract
The utilization of disinfectants could be imperative in controlling viruses and bacteria. However, excessive use of disinfectants may pose a risk to human health. Disinfectant workers face a higher risk of exposure to the active substances of disinfectant during work, as compared to the public. Therefore, it is necessary to measure the inhalation exposure of active substances in disinfectants for disinfection workers.

The researcher simulated disinfection work to assess the inhalation exposure to active substances of disinfection workers. The target active substances were quaternary ammonium compounds (QACs), benzalkonium chloride (BKC), and isopropyl alcohol (IPA). The disinfection work simulation was performed considering the working time and the spraying method in the actual situation.

As a result of the exposure assessment performed for each substance, the personal exposure concentration of the disinfection worker on QACs was 0.10±0.05 mg/m3. The personal exposure concentrations of BKC and IPA were 0.02±0.01 mg/m3 and 224.12±74.79 mg/m3, respectively. The QACs had the potential to cause health risk, whereas BKC did not risk. The IPA had a high personal exposure concentration due to its high content and vapor pressure but had no health risk.

The inhalation exposure of disinfection workers and the subsequent risk assessment of the active substances in the disinfectant revealed that certain substances may have health risks. Therefore, it is necessary to have an appropriate protective method to reduce the risk due to active substance exposure, and it must be selected in consideration of the vapor pressure toxicity of the active substances.

164 – Determinations of Endocrine Disrupting Chemicals (EDCs) from Personal Care Products (PCPs) in Indoor Dusts in Taiwan

Shih-Wei Tsai[1], Jou-Chun Lung[2]
[1]National Taiwan University, Professor, Taipei, Taiwan. [2]National Taiwan University, Student, Taipei, Taiwan

Abstract
Personal care products (PCPs) are widely used daily. Within PCPs, various organic chemicals are added for different functions. However, these chemicals are primarily endocrine disruptors (EDCs). Although the determination of EDCs from PCPs in indoor dust has been gradually studied in the literature, more data in Taiwan still needs to be collected. Therefore, this study aimed to optimize a sensitive and effective pre-treatment and analytical procedure for simultaneous determinations of selected EDCs in indoor dust. The target EDCs comprised organic alkylphenols, phthalates, bisphenols, parabens, and triclosan, commonly contained or highly detected in PCPs. Sample pre-treatment procedure and instrumental analysis were conducted by MAE-HS-SPME (microwave-assisted extraction-headspace-solid phase microextraction) coupled with GC/MS. For indoor dust samples, the established experimental procedures were as follows: household dust was collected while particles with diameters smaller than 150 μm were filtered out by stainless mesh. Afterward, a 0.02 g dust sample spiked with known amounts of the surrogate was put in a 4 ml vial, and the vial was left overnight. Later, the sample vial was placed in microwave-assisted equipment with spiking 5mL of RO water. After extracting for 20 mins under 85°Cand 150W, filtration was followed. 2.5 ml of the sample was put into another 10 mL vial. For field dust samples in Taiwan, the concentration medians for the chemicals determined were 0.02~0.91 μg/g, including parabens, triclosan, and bisphenol A. Further exposure assessments will be conducted to address the health risk posed by exposure to EDCs from indoor dust in Taiwan.

166 – POSTER: Cumulative Risk Assessment of Environmental Phthalates Using Longitudinal Human Biomonitoring Data in Children: The Japan Environment and Children's Study (JECS)

Nayan Chandra Mohanto[1], Yuki Ito[2], Romanas Chaleckis[3], Sayaka Kato[4], Kyoko Minato[5], Kayo Kaneko[6], Michihiro Kamijima[7]
[1]Department of Occupational and Environmental Health, Nagoya City University Graduate School of Medical Sciences, PhD Student, Nagoya, Japan. [2]Department of Occupational and Environmental Health, Nagoya City University Graduate School of Medical Sciences, Associate Professor, Nagoya, Japan. [3]Department of Occupational and Environmental Health, Nagoya City University Graduate School of Medical Sciences, Assistant Professor, Nagoya, Japan. [4]Department of Occupational and Environmental Health, Nagoya City University Graduate School of Medical Sciences, Assistat Professor, Nagoya, Japan. [5]Department of Occupational and Environmental Health, Nagoya City University Graduate School of Medical Sciences, Research Technologist, Nagoya, Japan. [6]Department of Occupational and Environmental Health, Nagoya City University Graduate School of Medical Sciences, Fulltime Lecturer, Nagoya, Japan. [7]Department of Occupational and Environmental Health, Nagoya City University Graduate School of Medical Sciences, Professor, Nagoya, Japan

Abstract
The objective of this study was to assess cumulative health risks of phthalates in Japanese children using longitudinal human biomonitoring data at age of 1.5y, 3y, and 8y (n=447). Children were recruited from Aichi Regional Center of the Japan Environment and Children's Study. Human biomonitoring of 16 urinary metabolites of eight phthalates was performed through UPLC-MS/MS by assuring internal quality controls. Hazard quotient (HQ), and hazard index (HI) were calculated based on the tolerable daily intake (TDI), reference dose (RfD), and reference dose for anti-androgenicity (RfD_{AA}) approaches from estimated daily intake (EDI) of each phthalate. The highest median EDI was found for di-2-ethylhexyl phthalate (DEHP) at 3y-children (10.1 μg/kg b.w/day). DEHP, di-n-butyl phthalate (DnBP), and di-iso-butyl phthalate (DiBP) exhibited higher risks (HQ>1) than the cutoff level in a small proportion of children at each age based on the TDI and RfD_{AA} approaches; whereas HQ>1 was for DEHP based on RfD estimation. A significant number of children at each age were at cumulative risk of multiple phthalates beyond the cutoff level (HI > 1) based on TDI, RfD, and RfD_{AA} estimation, and the proportion was higher in 3y-children [TDI (19%), RfD (20%), and RfD_{AA} (61%)]. The frequency of HQ maximum was higher for DEHP followed by DnBP

and DiBP at all ages. DEHP, DnBP, and DiBP were priority phthalates in Japanese children irrespective of age. Compared to 1.5y- and 8y-children, 3y-children were more concerned in perspective to the cumulative health risks. Thus, evaluation of health outcomes is warranted.

167 – POSTER: Exposure to Incense Components and Compounds from Incense Burning

Saria Hoshino[1], Ryo Omagari[2], Wang Qi[3], Kazushi Noro[2], Takashi Amagai[4]
[1]Academic, Student, Shizuoka, Japan. [2]Academic, Assistant Professor, Shizuoka, Japan. [3]Academic, Assistant Professor, Kawasaki, Japan. [4]Academic, Professor, Shizuoka, Japan

Abstract
Incense has been used in sacred rituals and for enjoying fragrances. Exposure to incense smoke contributes and human health risks because it contains toxic compounds, such as volatile organic compounds, polycyclic aromatic hydrocarbons and aldehydes. Addition to these toxic compounds, combustion of incense sticks may emit fragrances, some of which may cause allergies. However, consumers cannot obtain detailed information of fragrances, since labeling fragrances in products is not mandatory in Japan. Here we show that use of incense stick is a source of atmospheric allergenic fragrances. Incense sticks contained seven types of fragrances. Incense smoke emitted twenty-four compounds, including six fragrances. Among the emitted fragrances, linalool, linalyl acetate, benzyl benzoate, and geraniol were allergenic. The relatively high concentrations of fragrances in the incense smoke were celestolide (<92.9 μg g^{-1}), linalool (<89.8 μg g^{-1}), tonalide (<83.1 μg g^{-1}) and linalyl acetate (<12.4 μg g^{-1}). Incense with high linalool content may cause allergic reactions because emission efficiency of linalool was high (33–68 %) and the autoxidation products of linalool are contact allergens. In addition, concentrations of benzene, toluene, formaldehyde, and furfural were <561 μg g^{-1}, <392 μg g^{-1}, <362 μg g^{-1}, and <193 μg g^{-1}, respectively. This study demonstrated that use of incense stick may cause allergic reaction owing to the exposure of fragrances emitted.

168 – POSTER: Prenatal Lead and Mercury Screening: Challenges and Efforts Towards a Sustainable Program

Eric Bind[1], Andrew Steffens[2], Veronica Chandra[3], Chang Ho Yu[1], Douglas Haltmeier[1], Zhihua (Tina) Fan[4]
[1]New Jersey Department of Health, Research Scientist 1, Ewing, USA. [2]New Jersey Department of Health, Research Scientist 2, Ewing, USA. [3]New Jersey Department of Health, Research Scientist 3, Ewing, USA. [4]New Jersey Department of Health, Service Director, Environmental and Chemical Laboratory Services, Ewing, USA

Abstract
Lead and mercury cross the placental barrier and harm babies' development, change behaviors, and can lead to other negative health outcomes. The NJDOH-NJ Biomonitoring Program (NJB) and University Hospital in Newark initiated standard-of-care prenatal lead and mercury screening for expectant women/newborns. Over 10,000 women/newborns have been screened since 2019. The results show that 20% (>1 ug/dL) and 50% (>1 ppb) are at risks of adverse health effects from lead and mercury exposures, respectively, with baby body burdens being proportional to the mother's. Further, exposed babies are more likely to have immigrant mothers (75% vs. 20%), abnormal newborn screening (31% vs. 5%), and ICU admittance (35% vs. 21%) than unexposed babies. The results suggest the importance of expanding the Program to more hospitals in NJ to confirm the findings.

NJB is working with three hospital networks to implement prenatal lead/mercury screening to improve patient outcomes. NJB's advocacy for prenatal screening includes joint efforts from NJB and NJ doulas and community partnerships to develop educational materials aimed at improving messaging and distribution and sharing data with other state health programs, academia, and hospital networks. State officials support this work, including the NJ Office of the First Lady which allocated funding for an ICP-MS in the state budget. Significant resources are required to expand the program statewide. NJB is also discussing third-party payer options with hospitals, phlebotomy services, and state officials.
This poster will cover the Program's success and challenges associated with developing a sustainable program.

169 – Associations between PFAS Contamination of US Community Water Systems and COVID-19 Mortality

Jahred Liddie[1], Marie-Abéle Bind[2,3], Mahesh Karra[4], Elsie Sunderland[5,6]
[1]Harvard T.H. Chan School of Public Health, Department of Environmental Health, PhD Student, Boston, USA. [2]Harvard University, Department of Statistics, Assistant Professor, Cambridge, USA. [3]Massachusetts General Hospital, Assistant Investigator, Boston, USA.

[4]Boston University, Pardee School of Global Studies, Assistant Professor, Boston, USA. [5]Harvard T.H. Chan School of Public Health, Department of Environmental Health, Professor, Boston, USA. [6]Harvard John A. Paulson School of Engineering and Applied Sciences, Department of Environmental Sciences and Engineering, Professor, Cambridge, USA

Abstract

Epidemiologic studies have documented associations between PFAS exposure and harmful effects on the immune system, with confirmation from animal studies. PFAS have also been identified as potential environmental risk factors for COVID-19 outcomes at individual and population levels, but empirical evidence is limited. Here we combine data on PFAS concentrations in U.S. community water systems (CWS) with county-level COVID-19 case records to examine associations between PFAS drinking water contamination and COVID-19 mortality. Our analyses leverage two datasets: one at the subnational level (including 4,868 CWS serving 543 U.S. counties) and one at the national level (including 4,798 CWS serving 1,677 counties). We use multilevel quasi-Poisson regressions to investigate associations between county-level measures of PFAS contamination in drinking water and COVID-19 mortality in 2020. Adjusting for several county-level factors, our estimates highlight a positive association between county-level drinking water PFAS contamination and COVID-19 mortality. In the subnational dataset, a one-standard-deviation (36%) increase in the proportion of CWS with concentrations of at least one PFAS over 5 ng/L was associated with a 7% increase [95% CI: 3, 11] in COVID-19 mortality. In the national dataset, a one-standard-deviation increase (12%) in the proportion of CWS with PFAS concentrations above the reporting limits was associated with a 4% increase [95% CI: 3, 6] in COVID-19 mortality. Our findings indicate that several PFAS may be important environmental risk factors for COVID-19 outcomes and should be systematically measured as part of water system monitoring efforts.

170 – Metabolomic Associations with an Exposure Mixture of Per- and Polyfluoroalkyl Substances in the Atlanta African American Maternal-Child Cohort

Donghai Liang[1], Kaitlin Taibl[2], Anne Dunlop[3], Dana Barr[4], P Barry Ryan[4], Parinya Panuwet[1], Kurunthachalam Kannan[5], Carmen Marsit[4], Dean Jones[3], Stephanie Eick[1]

[1]Emory Unviersity, Assistant Professor, Atlanta, USA. [2]Emory Unviersity, PhD Student, Atlanta, USA. [3]Emory University, Professor, Atlanta, USA. [4]Emory Unviersity, Professor, Atlanta, USA. [5]New York University, Professor, New York City, USA

Abstract

Background and Aim: Prenatal exposure to individual per- and polyfluoroalkyl substances (PFAS) are associated with maternal metabolic perturbations and adverse pregnancy outcomes. Despite out knowledge that individuals are exposed to multiple PFAS, the joint effects of PFAS on the human metabolome remains largely unknown. Here, we leveraged high-resolution metabolomics (HRM) data to identify the metabolic pathways and markers perturbed by exposure to a PFAS mixture during pregnancy.

Methods: Targeted assessment of perfluorooctanoic acid (PFOA), perfluorononanoic acid (PFNA), perfluorooctane sulfonic acid (PFOS), and perfluorohexane sulfonic acid (PFHxS), along with untargeted high-resolution metabolomics profiling were conducted on non-fasting serum samples collected from 286 pregnant African American people at 8–14 weeks gestation. We estimated the overall mixture effect using quantile g-computation, single pollutant effects were estimated using linear regression.

Results: Our analytic sample was comprised of a socioeconomically diverse group, with most receiving public health insurance. Relative to individual PFAS exposures (1,498 features $p<0.05$), we observed a larger number of significant metabolic features associated with the PFAS mixture (1,651 features $p<0.05$). Twelve metabolites associated with the PFAS mixture were confirmed with level-1 evidence, including linoleic acid, carnitine, valine, leucine, tyrosine, and uracil.

Conclusion: In one of the first studies of its kind, we demonstrate the feasibility and utility of using methods designed for exposure mixtures in conjunction with HRM to assess the synergistic effects of multiple PFAS on the maternal metabolome. We found that a simultaneous exposure to multiple PFAS during early pregnancy was associated with inflammatory and prooxidative pathways and metabolites.

171 – The Risk21 Framework: Applications in risk-based evaluation and risk communication

Edward Chikwana[1], Michelle Embry[2], Sandrine Deglin[3], Eliana Munarriz[4]

[1]Corteva Agriscience, Human Health Risk Assessor, INDIANAPOLIS, USA. [2]Health & Environmental Sciences Institute, Deputy Director, Washington DC, USA. [3]Health & Environmental Sciences Institute, HESI Senior Scientific Program Manager, Washington DC, USA. [4]University of Buenos Aires, Group Leader & Research Associate, Buenos Aires, Argentina

Abstract
The HESI Risk Assessment in the 21st Century (RISK21) project developed an integrated risk assessment framework that enhances efficiency and risk management. The project was initiated to develop a scientific, transparent, and efficient approach to risk assessment, relying on a problem formulation-based, exposure driven, tiered data acquisition. This RISK21 framework allows informed decisions on safety to be made when sufficient evidence is available, and it maximizes the ability to inform decisions and optimize resource usage. The program also developed a web-based tool that allows users to easily graph their risk assessment data and effectively communicate risk-based decisions, whether for a screening and prioritization purpose or a definitive risk assessment. This tool is freely available at www.risk21.org.

The utility and uptake of this approach and the web-tool has been demonstrated via several hands-on case study workshops that have engaged participants in real-world case examples across the world, including the US, Canada, China, Taiwan, Brazil, Argentina, Panama, Colombia, and Thailand. The approach and the associated web-tool are broadly applicable across disciplines for risk assessment, prioritization, communication, and outreach. Several case studies will be highlighted to demonstrate how the framework and RISK21 matrix tool can be applied to food safety, pharmaceutical, and agrochemical safety assessments to provide a clear picture of risk and drive specific additional data needs that inform business decisions. These case studies also demonstrate how the approach can be used as an effective communication tool by providing information in a format that is easily understood by multiple, technically-diverse stakeholders.

172 – POSTER: Phase-specific density and proximity metrics for oil and gas wells: Exposure estimates in the EXPERIVA study

Coreen Daley[1], Marc-André Verner[2], Élyse Caron-Beaudoin[3,4]
[1]Department of Physical and Environmental Sciences, University of Toronto Scarborough, PhD student, Toronto, Canada. [2]Department of Occupational and Environmental Health, School of Public Health, Université de Montréal, Associate Professor, Montréal, Canada. [3]Department of Physical and Environmental Sciences, University of Toronto Scarborough, Assistant Professor, Toronto, Canada. [4]Department of Health and Society, University of Toronto Scarborough, Assistant Professor, Toronto, Canada

Abstract
Background: Northeastern British Columbia represents one of Canada's key regions that produce natural gas. Hydraulic fracturing and horizontal drilling have triggered significant growth in the oil and gas sector which have provoked concerns regarding environmental and health consequences.

Objective: To investigate the association between various proximity-based exposure estimation metrics for conventional and unconventional wells, and measured VOCs in indoor air samples from the Exposures in the Peace River Valley (EXPERIVA) study.
Methods: 85 pregnant women participated in EXPERIVA. Passive indoor air samplers were analyzed for 47 VOCs. Wells legacy data were sourced from the British Columbia Energy Regulator. For each participant and accounting for 5km, 10km and no buffer distances around residences, Inverse Distance Weighted IDW) metrics were calculated to estimate gestational exposure to conventional and unconventional wells. Multiple linear regression models were used to test for associations between these different IDW metrics and indoor VOC.

Results: In EXPERIVA, densities ranged from 1-326 conventional and 1-578 unconventional wells for the 10 km buffer. Analyses indicate that: 1) all wells density metrics were positively associated with indoor air acetone, chloroform, decamethylcyclopentasiloxane and decanal; 2) unconventional wells density metrics were positively associated with indoor air chloroform and decamethylcyclopentasiloxane 3) Drilling and production-specific IDW metrics were positively associated with indoor air benzene, chloroform and decamethylcyclopentasiloxane.

Conclusion: Our results suggest that proximity to unconventional wells is associated with increased indoor air concentrations of certain VOCs. This research contributes to advancing our understanding of the impacts of oil and gas operations on air quality.

173 – POSTER: Integrating Multiple Data Sources to Generate an Emission Inventory for Residential Wood Burning in Quebec, Canada

Jad Zalzal[1], Ying Liu[2], Shayamila Mahagammulla Gamage[3], Shoma Yamanouchi[3], Laura Minet[4], Andree-Anne Brown[5], Audrey Smargiassi[6], Marianne Hatzopoulou[7]

[1]University of Toronto, PhD Candidate, Toronto, Canada. [2]Universite de Montreal, Postdoctoral Fellow, Montreal, Canada. [3]University of Toronto, Postdoctoral Fellow, Toronto, Canada. [4]University of Victoria, Professor, Victoria, Canada. [5]Ministry of the Environment and the Fight Against Climate Change, -, Montreal, Canada. [6]Universite de Montreal, Professor, Montreal, Canada. [7]University of Toronto, Professor, Toronto, Canada

Abstract
Background: Residential Wood Burning (RWB) is a major contributor to outdoor air pollution. However, emission inventories for this source have been generated using top-down approaches and are distributed based on population density, resulting in large uncertainties.

Methods: In this study, we developed a novel method that combines multiple data sources for generating a bottom-up emission inventory for RWB in Quebec, Canada. Gradient Boost and Nearest Neighbor models were used to estimate RWB activity potential for wood stoves and fireplaces separately at the dissemination area level (~400 to 700 people). The models used real estate data (~500,000 residences from Centris), and socioeconomic, demographic, and land use predictors from the census. A survey was also conducted with over 13,400 residences that use RWB as a heating source. Survey data was used to determine the amount of wood burned and the frequency of RWB appliance usage for different population clusters. The results of both stages were combined to develop a high-spatial/temporal resolution emission inventory for RWB.

Results: Residence type, income, education and building age were important predictors of RWB activity potential (R2 ~0.63-0.8). Residences owning wood stoves in rural areas used RWB more frequently (~70% daily in winter) and burned on average ~50% more wood compared to residences owning fireplaces or residences in urban regions.

Conclusion: This inventory will be used as input to a chemical transport model to assess the impact of RWB on air pollution. Our approach provides an accurate and detailed understanding of RWB emissions and their spatial distribution.

174 – Downscaling CTMs with Mobile Air Quality Measurements and Machine Learning: An Investigation of the Impact of Different Vehicle Fleets on Environmental Justice

Jad Zalzal[1], Laura Minet[2], Jeffrey Brook[3], Marianne Hatzopoulou[3]
[1]University of Toronto, PhD Candidate, Toronto, Canada. [2]University of Victoria, Professor, Victoria, Canada. [3]University of Toronto, Professor, Toronto, Canada

Abstract
Background: Determining the burden of different sources on air pollution-related disparities can be done through zero-out scenarios in Chemical Transport Models (CTMs), which are limited by their coarse spatial resolution and failure to capture intra-urban variability.

Methods: In this study, we downscaled CTM outputs (1Kmx1Km to 100mx100m) using mobile air quality measurements and land use machine learning (LU-ML) models. First, we generated downscaling factors for 100mx100m grids by combining LU-ML models and CTM outputs. Then, we trained a Gradient Boost model to update these factors for different scenarios based on local traffic predictors. We applied this method to investigate the impact of zero-out scenarios for light-duty vehicles, trucks, and buses on disparities in exposure to nitrogen dioxide (NO_2) and fine particulate matter ($PM_{2.5}$) in Toronto, Canada.

Results: The LU-ML models were able to predict NO_2 and $PM_{2.5}$ concentrations with Cross-Validated-R^2=0.77 and 0.9 respectively. The downscaling method improved the Pearson correlation between CTM concentrations and concentrations at reference stations from 0.33 to 0.77 for $PM_{2.5}$ and from 0.52 to 0.74 for NO_2. Our results indicate that coarse spatial resolutions tend to underestimate disparities in exposure to NO_2 by up to 30%. Additionally, the zero-out scenarios only resulted in a slight decrease in disparities for NO_2 (<10%). $PM_{2.5}$-related disparities on the other hand were not significant across scenarios and weren't affected by spatial resolution.

Conclusion: Our findings show the importance of high-spatial-resolution for environmental justice and suggest that downscaling techniques based on land-use regression are adequate tools for refining CTM outputs.

175 – Deciduous teeth from the New Hampshire Birth Cohort Study: Infant feeding and Pb in water predict dentine elements

Julia Anglen Bauer[1], Tracy Punshon[2], Brian P. Jackson[3], Felicitas B. Bidlack[4,5], Modupe O. Coker[6], Marc G. Weisskopf[7], Margaret R. Karagas[8]
[1]Dartmouth Geisel School of Medicine, Post doctoral Fellow, Lebanon, USA. [2]Dartmouth College, Research Scientist, Hanover, USA.

[3]Dartmouth College, Dartmouth Trace Element Lab Director, Hanover, USA. [4]The Forsyth Institute, Research Scientist, Cambridge, USA. [5]Harvard School of Dental Medicine, Scientist, Boston, USA. [6]Rutgers University, Assistant Professor, Newark, USA. [7]Harvard T.H. Chan School of Public Health, Professor, Boston, USA. [8]Dartmouth Geisel School of Medicine, Professor and Chair of Epidemiology, Lebanon, USA

Abstract
Objective: To estimate associations between maternal/child characteristics, environmental water concentrations during pregnancy, and element concentrations in shed deciduous teeth.

Methods: 152 mother-infant dyads were followed from the second trimester through the end of the first postnatal year from the New Hampshire Birth Cohort Study. During pregnancy and early infancy, we collected maternal and infant dietary and sociodemographic information via surveys, measured metals in home tap water, and later collected naturally shed teeth from child participants and measured weekly deposition of elements in dentine using LA-ICPMS. Multivariable linear mixed models were used to estimate associations between predictors and tooth metal concentrations.

Results: We measured 12 elements in dentine (Ba, Mn, Pb, Sr, Zn, Al, As, Cd, Cu, Hg, Li, and W). A doubling of Pb concentrations in water was associated with higher prenatal and postnatal dentine Pb [8.9% (95%CI: 3.2, 14.8%); 13.4% (95%CI: 5.0%, 22.4%)]. Exclusive formula feeding was associated with higher postnatal dentine element concentrations as compared to exclusive breast feeding at 4 months: Sr: 251% (95%CI: 54%, 699%); Ba: 120% (95%CI: 32%, 268%); Li: 73% (95%CI: 13%, 166%); Al: 136% (95%CI: -6.2%, 498%). A mix of formula and breast feeding was associated with higher elements but to a lesser degree than exclusive formula feeding, suggesting a dose response: dentine Sr: 72.9% (95%CI: 17.1, 155.2%), Ba: 26% (95%CI: -0.6%, 60%), Li: 47% (95%CI: 20%, 79%); Al: 4.8% (-32%, 63%) as compared to exclusive breastfeeding.

Conclusions: Pb in home tap water and any formula use were predictors of dentine elements.

176 – Toxic Metals Induced Thrombotic Risk and Altered Rheology in Traditional Bronze Casters and Environmental Cohorts in Bronze Locations in Benin City, South-South, Nigeria

James Osakue[1,2], Evarista Osime[3], Osaretin Igharo[4,5]
[1]Medical Laboratory Sciences, Basic Medical Sciences, University of Benin, Lecturer/Doctoral Student, Benin City, Nigeria. [2]Biomedical Toxicology and Chemical Safety Research Laboratory, University of Benin, Senior Biomedical Research Scientist, Benin City, Nigeria. [3]Medical Laboratory Sciences, Basic Medical Sciences, University of Benin, Professor, Benin City, Nigeria. [4]Medical Laboratory Sciences, Basic Medical Sciences, University of Benin, Associate Professor, Benin City, Nigeria. [5]Biomedical Toxicology and Chemical Safety Research Laboratory, University of Benin, Principal Investigator, Benin City, Nigeria

Abstract
Occupational and environmental exposures can result in different health complications. Haemostatic balance is necessary for normal vascular flow which is important for healthy living. Toxic metals accumulations have been associated with reduced fibrinolysis through various interactions like oxidative stress. Whereas, a balance between coagulation and fibrinolysis, is necessary for healthy blood flow, reduced fibrinolysis could lead to thrombotic conditions like ischemia, atherosclerosis, obstructive sleep apnoea (OSA), etc. The aim of this study was to evaluate tendencies to thrombosis among traditional Benin Bronze Casters (BBC), Environmental Cohorts (EC) in Benin bronze locations and Unexposed Participants (UP). 200 participants were recruited from Benin City, Nigeria comprising of 100 BBC, 40 age-and-sex matched EC and 60 UP. Occupational activities and anthropometric data of participants were collected using a structured questionnaire. Blood levels of toxic metals (Lead, chromium and cadmium and Mercury) were determined using Inductively Coupled Plasma Mass Spectrometry (ICP-MS). Whole blood (WBV) and plasma viscosities (PV) were evaluated using flow rate. Levels of Human Plasminogen activator inhibition-1(PAI-1) and α-2-antiplasmin (α2-AP) were also determined using enzyme linked immunosorbent assay (ELISA), Elabscience: E-EL-H2104 (PAI-1) and E-EL-HO968 (α2AP) technique. Data were analysed using ANOVA at α0.05. The levels of all toxic metals in the BBC and EC were significantly higher than UP ($P<0.05$). Also, the levels of WBV, PV, PAI-1and α2AP in BCC were significantly higher than Unexposed participants. This study revealed that BBC and their environmental cohorts are exposed to bronze borne toxicants which could put them at risk of thrombosis related complications.

177 – POSTER: From Linear Regression to Machine Learning: Comparing Performance of Multiple Intra-Urban Ultrafine Particles Models

Julien Vachon[1,2], Stéphane Buteau[3], Keith Van Ryswyk[4], Ying Liu[5], Marianne Hatzopoulou[6], Audrey Smargiassi[7,8]

[1]School of Public Health, University of Montreal, PhD Candidate, Montreal, Canada. [2]Center for Public Health Research (CReSP), University of Montreal and CIUSSS du Centre-Sud-de-l'Île-de-Montréal, Student Member, Montreal, Canada. [3]School of Public Health, University of Montreal, Assistant Professor, Montreal, Canada. [4]Air Health Science Division, Health Canada, Air Pollution Exposure Researcher, Ottawa, Canada. [5]School of Public Health, University of Montreal, Research Assistant, Montreal, Canada. [6]Department of Civil Engineering, University of Toronto, Associate Professor, Toronto, Canada. [7]School of Public Health, University of Montreal, Professor, Montreal, Canada. [8]Center for Public Health Research (CReSP), University of Montreal and CIUSSS du Centre-Sud-de-l'Île-de-Montréal, Researcher, Montreal, Canada

Abstract

Background: Limited data and high spatiotemporal variability of ambient ultrafine particles (UFPs, <100 nm) make them challenging to model. We compared the predictive performance of two statistical and eight machine learning (ML) methods in modelling ambient UFPs.

Methods: Daily and annual models were developed from UFP measurements from a 2019-2020 mobile monitoring campaign (1,770 100 m segments, 38 passages) in Quebec City (Canada), and using land use/cover types, roads and traffic, and meteorological predictors. Various road segment lengths were also considered (100/300/500 m). Statistical methods employed include stepwise multiple linear regression (MLR) and LASSO, while ML methods included Support Vector Machine, Kernel Ridge Regression, Bagging, Random Forest (RF), Gradient Boosting Machine, XGBoost, and Neural Networks. Nested cross-validation was used for hyperparameter tuning, model training, and assessing performance on unseen data.

Results: Daily UFP counts ranged from 99 to 217,675 particles/cm^{-3}, while annual averages ranged from 4,483 to 60,854 particles/cm^{-3}. For daily models, tree-based ML methods (e.g., RF, XGBoost) outperformed other methods (R^2 = 0.42-0.47 vs. R^2 = 0.07-0.38). For annual models, most ML methods outperformed statistical methods (R^2 = 0.74-0.87 vs. R^2 = 0.57-0.70). XGBoost showed the best overall performance at different road segment lengths (daily: R^2 =0.47-0.48, RMSE = 8,626-11,527; annual: R^2 = 0.79-0.87, RMSE = 2,087-2,124 particles/cm^{-3}).

Conclusion: ML methods can significantly improve the predictive performance of UFP models compared to statistical methods. MLR showed the worst overall performance, while XGBoost is recommended based on its performances and ease of use.

178 – German human biomonitoring health-based guidance values and approaches to interpret internal exposure levels

Marike Kolossa-Gehring[1], Holger M. Koch[2], Petra Apel[3], Claudia Roehl[4]

[1]German Environment Agency, Head of Section toxicology, health related environmental monitoring, Berlin, Germany. [2]Institute for Prevention and Occupational Medicine of the German Social Accident Insurance, Ruhr-University Bochum (IPA), Head of Section Human Biomonitoring, Bochum, Germany. [3]German Environment Agency, Senior scientist, Berlin, Germany. [4]State Agency for social Services Schleswig-Holstein (LAsD SH), Head of Section environmental midicine and toxicology, Neumünster, Germany

Abstract

The German Human Biomonitoring Commission (HBM-C) of the German Environment Agency (UBA) derives health-based guidance values (HBM-I and HBM-II values) for the risk assessment of chemicals (or metabolites) in human tissues (e.g. blood, urine) determined by human biomonitoring (HBM).

The HBM-I value represents the concentration of a substance in urine or blood at and below which no risk of adverse health effects is expected. The HBM-II value describes the respective concentration at and above which adverse health effects need to be expected. HBM values reflect current state of knowledge, are up-dated according to new data or assessments by European Agencies, and are derived in a single substance assessment. Selection of key studies and point of departure for derivation of the respective level are discussed extensively in the HBM-C, a panel of independent experts without conflict of interest. Discussion and validation integrate controversial views and increase acceptance of HBM values. A similar consultation mechanism has been implemented in the derivation of HBM guidance values (HBM-GVs) in HBM4EU and PARC, pan-European, EU-supported HBM activities.

Up to now HBM values had not been derived for non-threshold carcinogens. To fill this gap, a supplemental risk assessment concept for those carcinogens was most recently developed which is useful to derive quantitative dose descriptors of the internal body load. They allow to calculate for any concentration of a carcinogen, e.g. in urine or blood, an additional lifetime cancer risk, which can be used for the further evaluation of the internal exposure to carcinogens.

Further Information: https://www.umweltbundesamt.de/en/topics/health/commissions-working-groups/human-biomonitoring-commission/reference-hbm-values

179 – Assessing risk to phthalates in Human Health Exposure Analysis Resources (HHEAR) subjects according to Hazard Indices using the Human Biomonitoring Health-Based Guidance Value (HB2GV) Dashboard

Moira Bixby[1], Emily Spear[1], Chris Gennings[2], Susan L. Teitelbaum[2]
[1]Icahn School of Medicine at Mount Sinai Department of Environmental Medicine and Public Health, Biostatistician, NYC, USA. [2]Icahn School of Medicine at Mount Sinai Department of Environmental Medicine and Public Health, Professor, NYC, USA

Abstract
Low molecular weight (LMW) phthalates are believed to have additive adverse effects on health outcomes. The HB2GV Dashboard provides access to available biomonitoring health-based guidance values. The HHEAR Data Repository hosts human biomarker data from thousands of subjects. We assessed the risk posed to HHEAR subjects from LMW phthalates according to available HB2GVs. Urinary LMW phthalate concentrations from each child (0-18 years) and the most sensitive (lowest) HB2GVs were used to calculate per subject hazard quotients (HQs) per phthalate, and for cumulative risk HQs were summed to hazard indices (HIs). A HI >1 indicates concern for increased risk to adverse health endpoints. Using Kruskal-Wallis rank sum tests, HIs were compared across groups including sex, cohort, and age group (infant/child/adolescent). Nine HHEAR studies had phthalate biomarkers from 2,966 children (47% female; 9% adolescent, 85% children, 6% infant). Pertinent HB2GVs included monoethyl, mono-isobutyl, and mono-n-butyl phthalate. HHEAR subject's LMW phthalate HI median was 0.28, with 75th%tile=0.59, and 95th%tile=1.8, and 13% of the subjects had HI values > 1 indicating a level of concern. Significant differences in risk were observed by age group ($p<0.001$) (adolescents had highest average HIs; infants lowest), and across cohorts ($p<0.001$), but not by sex. Finally, 48% of subjects were under age 6, where guidance values have not been specified, suggesting that additional guidance is required for vulnerable populations. Additional phthalates, alongside other endocrine-disrupting chemicals, have not yet been assessed to capture the additive nature of this exposure. Future analyses will include additional endocrine disrupters.

180 – POSTER: A Comprehensive Approach to Health Assessment of Exposures from Petroleum Vapor Intrusion

tonia burk[1], Bradley Goodwin[1], Sandra Miller[2]
[1]ATSDR, Environmental Health Scientist, Atlanta, USA. [2]ATSDR, Environmental Engineer, Atlanta, USA

Abstract
ATSDR protects communities from harmful exposure to man-made and natural hazardous substances in the environment. Health assessment of petroleum vapors in indoor air from soil vapor intrusion involves screening subsurface contaminant concentrations for potential to migrate to indoor air. Uncertainties in evaluating complex migration pathways and petroleum mixtures make screening challenging. Environmental guidance presents vertical inclusion criteria (VIC) to screen in petroleum vapor intrusion (PVI) for further evaluation. VIC apply when high subsurface concentrations occur close to buildings and overwhelm the aerobic microorganisms that interrupt the PVI pathway. Otherwise, guidance recommends no further evaluation.

Other criteria may warrant further evaluation: chlorinated volatile plumes, preferential pathways, oxygen and soil moisture shadows, and special contaminants. Chlorinated plumes are easily detected if analysis includes chlorinated contaminants. The other four criteria are less well defined. This presentation shares a stepwise process for considering these criteria in PVI screening. ATSDR health-based comparison values are presented for special contaminants and contaminants in USEPA's petroleum vapor intrusion database.

The objective of this presentation is to share a comprehensive PVI approach that goes beyond the VIC and looks more critically at five other criteria. Considering all of these criteria increases the accuracy of PVI health conclusions which is key to conducting more protective public health assessments.

The findings and conclusions in this presentation have not been formally disseminated by the Centers for Disease Control and Prevention/the Agency for Toxic Substances and Disease Registry and should not be construed to represent any agency determination or policy.

181 – POSTER: Feasibility of a portable x-ray tube based KXRF system to measure lead in bone

Chandler Burgos[1], Thomas Grier[1], Maruf Khan[1], Marc Weisskopf[2], Kathryn Taylor[3], Aaron Specht[4]
[1]Purdue University, Graduate Student, West Lafayette, USA. [2]Harvard T.H. Chan School of Public Health, Professor, Boston, USA. [3]United States Army Research Institute of Environmental Medicine, Director SPHERE Repository, Boston, USA. [4]Purdue University, Assistant Professor, West Lafayette, USA

Abstract
X-ray Fluorescence (XRF) is commonly used to measure cumulative exposure to lead. For many years, a method called K-shell x-ray fluorescence (KXRF) had been utilized to detect the adverse health effects of metal exposure. Although the equipment had undergone numerous advancements, it is limited by its bulkiness, liquid nitrogen cooled germanium detectors, and radioisotope. A handheld L-shell XRF (pXRF) has been utilized recently as a replacement. This system has gained broad use for community level exposure assessment. However, this system cannot measure trabecular bone and has uncertainty linked to skin thickness. In this study, we developed a new portable K-shell XRF for bone lead measurement that would reap the benefits of both previous devices. The new KXRF utilizes a 140kV x-ray tube with two CdTe detectors. For this study, lead doped bone phantoms, cadaver bones, and goat bones were measured to test the limitations of the system. Using two CdZnTe detectors gave us a limit of detection (LOD) of 0.6 ug/g bone mineral for our new bone measurement system. This can be compared to the LOD for previous KXRF and pXRF at 2-3 ug/g and 2-10 ug/g respectively. Thus, our new measurement system can detect bone lead with a factor of 3 greater efficiency than previous approaches enabling it for measurements of nearly all environmentally exposed populations for cortical and trabecular measurements. The portable KXRF developed in this study allows for cost-effective, portable, in-field, biomonitoring of cumulative metal exposures – vital equipment to increase accessibility of exposure assessment.

182 – POSTER: Environmental injustice among measured sediment pollution within the Water Quality Portal

Patrick Hurbain[1], Dingsheng Li[2]
[1]University of Nevada, Reno, Graduate Student Assistant, Reno, USA. [2]University of Nevada, Reno, Assistant Professor, Reno, USA

Abstract
The Water Quality Portal (WQP) is a large publicly available water quality dataset maintained by the U.S. Environmental Protection Agency, the U.S. Geological Survey, and collaborating federal and local agencies. It contains measurements of various pollutants including heavy metals and organic compounds in environments pertain to water quality. Ten years (2010 - 2019) of available WQP sediment measurement data for lead, arsenic, cadmium, and mercury were extracted and curated based on relatively robust sample size compared to other pollutants recorded in WQP. This data was then paired with sociodemographic data from the U.S. Census Bureau at the county level to examine whether disparity in distributed pollution exists using state stratified multivariate generalized linear models. There are mixed results in potential disparity in exposure to these metals in U.S. counties with respect to racial makeup, income, and education attainment, although metal concentrations appear to be consistently elevated in counties with higher Black populations, and lower in counties with higher median incomes. Counties with increasing populations of college educated individuals appears positively associated with increasing metal concentrations. Many other pollutants are not collected across all counties, and sampling is not consistent across years in the WQP database. Data attrition inhibits the ability to analyze environmental justice in potential exposure to these environmental contaminants across years due to unbalanced sampling. Improved data coverage across more geographic regions with emphasis on consistent sampling over time would provide much needed understanding of environmental justice issues.

183 – POSTER: National-scale Estimation of Industrial Chemical Concentrations in Surface Waters from Wastewater Discharge using the National Hydrography Dataset

Craig Connolly[1], Bryan Groza[2], Nerija Orentas[1], Jason Todd[3]
[1]Environmental Protection Agency, Office of Pollution Prevention and Toxics, Biologist, Washington D.C., USA. [2]Environmental Protection Agency, Office of Pollution Prevention and Toxics, Exposure Assessor, Washington D.C., USA. [3]Environmental Protection Agency, Office of Chemical Safety and Pollution Prevention, Ecologist, Washington D.C., USA

Abstract
Humans and ecological species can be exposed to industrial chemicals in surface water from direct wastewater discharge. Under TSCA, EPA frequently estimates site-specific concentrations of these chemicals in receiving water bodies across the nation which requires detailed hydrologic information and representative national data sets. However, this information has been difficult to obtain and integrate into previous exposure modeling practices. Here we will present a proposed approach for estimating aqueous concentrations resulting from the wastewater discharge of industrial chemicals directly to surface waters using delineated stream networks and modeled stream flow values from the National Hydrography Dataset (NHD). Employing the NHD dataset allows for a comprehensive, efficient, and precise process to estimate aqueous chemical concentrations for a large number of site releases to surface waters at the national scale. The integration of the NHD presents an opportunity to estimate aqueous concentrations in downstream drinking water supplies, apply detailed hydrologic information to other exposure models, and readily update site-specific hydrologic information as new data sets become available.

DISCLAIMER: The views expressed in this abstract are those of the authors and do not necessarily represent the views or the policies of the U.S. Environmental Protection Agency.

184 – Lead Contamination in Drinking Water: A Review of National and Local Lead in Drinking Water Regulations in Illinois

Berkley Wood[1], Taylor Shughart[2]
[1]RHP Risk Management, Associate Health Scientist, Chicago, USA. [2]RHP Risk Management, Associate, Carlisle, USA

Abstract
The U.S. EPA Lead in Copper Rule (LCR) was published in 1991 with the goal of protecting public health by minimizing lead (Pb) and copper (Cu) in drinking water. The U.S. EPA notes that implementation of the LCR has resulted in a reduction of exposure to lead that can cause damage to the brain, red blood cells, and kidneys, particularly in children, even at low levels of exposure. Since its original release, there have been numerous revisions to the rule, with the most recent revision occurring in 2021. Locally, Illinois has implemented several lead-in-water reduction-focused regulations, such as Public Act 99-0922 requiring Illinois Schools (Pre-K through grade five) to test for lead in each "source of potable water" and the Lead Service Line Replacement and Notification Act, which requires a lead service line material inventory. The state estimated that, in 2019, there were 677,359 lead service lines connected to Illinois water systems.

With aging infrastructure and evolving regulations at the national and local level, we conducted a detailed review of current and anticipated regulations and programs such as the Lead Service Line Replacement Program and the Revised LCR and provide a detailed summary of regulations, both in Illinois and nationally. It is crucial to understand U.S. EPA and Illinois regulations, guidelines, and directives to address lead in drinking water at the national and local level with continued collaboration between scientific experts, community organizations, and government agencies to be better prepared to address requirements focused on reducing lead in drinking water.

185 – Community Liaison Perspectives on the NASEM PFAS Testing and Health Outcome Guidance

Sandy Wynn-Stelt
Great Lakes PFAS Action Network, co-chair, Belmont, USA
Abstract
The National Academies of Science, Engineering and Medicine Guidance on PFAS Exposure, Testing and Clinical Guidance published in 2022 was an important milestone for those impacted by PFAS contamination and was strengthened by the inclusion of community liaisons and several town halls to elevate the real-world challenges being experienced during interactions with clinicians. Strengths of the guidance include ethical framing of the problem, a clear summary of PFAS health effects with relative strengths of evidence, clear guidance for clinicians regarding the benefits and limitations of PFAS blood testing, and a list of specific clinical tests and screenings that can be useful for those with moderate and elevated PFAS blood levels. For individuals that have been impacted by PFAS contamination, this guidance has provided a framework for individuals and their physicians moving forward. Future updates to the guidance should include improved guidance on PFAS testing for breast milk and other food sources.

186 – Examining differences in extreme heat exposures for Black communities in New York City

Nadav Sprague[1], Stephen Uong[1], James Quinn[2], Katherine Keyes[3], Andrew Rundle[1]
[1]Department of Epidemiology, Columbia Mailman School of Public Health, PhD student, NEW YORK, USA. [2]Department of Epidemiology, Columbia Mailman School of Public Health, Geographer and Senior Geographic Information Systems (GIS) Analyst, NEW YORK, USA. [3]Department of Epidemiology, Columbia Mailman School of Public Health, Professor of Epidemiology, NEW YORK, USA

Abstract
Background: In the US, structural racism has resulted in neighborhoods with higher proportions of Black residents having more features that intensify and less that cool the local-heat environment. Blacktop surfaces and low altitudes are features that may heat local environments, whereas nature-cover and proximity to shorelines cool local environments. In this study of New York City (NYC)'s five boroughs, we investigate disparities in surface temperature exposure faced by Black communities.

Methods: Data from the 2017 American Community Survey, U.S Landsat-8 Analysis Ready Data on summer surface temperatures, and NYC Land Cover Dataset were aggregated to and analyzed at the census tract-level (n =2,098). Four cross-sectional linear regression

(LR) modeling strategies were used to estimate associations between tract-level percent of Black residents and surface temperature, adjusting for altitude, shoreline, and nature cover: overall NYC, borough-specific, community-district specific, and geographically weighted regression models.

Results: The overall NYC LR found a 1.87°F (95% CI: 1.52, 2.22) difference in surface temperature across the tract-level interquartile range of percent Black residents. Similarly, significant associations were found in every borough. However, both the NYC overall and borough-specific LRs had significant Moran's I scores, indicating the presence of spatial autocorrelation. Both the community-district LR and geographically weighted regression, which address the issue of spatial autocorrelation, identified specific locations (such as northwest Bronx, central Brooklyn, and uptown Manhattan) where these significant associations are concentrated.

Discussion: These results identify specific geographic areas for interventions that will mitigate disparities in heat exposure experienced by Black NYC residents.

187 – POSTER: Persistent Organic Pollutants Exposure (POPs) and Endogenous Hormones in Hispanic/ Latino Postmenopausal Women: The Hispanic Community Health Study/ Study of Latinos (HCHS/SOL)

Chibuzor Abasilim[1], Victoria Persky[2], Tessa Day[3], Konstantina Tsintsifas[4], Robert Sargis[5,6], Martha Daviglus[7], Sally Freels[8], Jianwen Cai[9], Brandilyn Peters[10], Gregory Talavera[11], Bharat Thyagarajan[12], Andreas Sjodin[13], Mark Davis[13], Richard Jones[13], Mary Turyk[2]
[1]Division of Epidemiology and Biostatistics, School of Public Health, University of Illinois Chicago, Postdoc, Chicago, USA. [2]Division of Epidemiology and Biostatistics, School of Public Health, University of Illinois Chicago, Professor, Chicago, USA. [3]Division of Epidemiology and Biostatistics, School of Public Health, University of Illinois Chicago, Epidemiologist, Chicago, USA. [4]Division of Epidemiology and Biostatistics, School of Public Health, University of Illinois Chicago, Doctoral Candidate, Chicago, USA. [5]Division of Endocrinology, Diabetes, and Metabolism, Department of Medicine, University of Illinois Chicago, Associate Professor, Chicago, USA. [6]Jesse Brown VA Medical Center, Associate Professor, Chicago, USA. [7]Institute for Minority Health Research, University of Illinois Chicago, Associate Vice Chancellor for Research, Chicago, USA. [8]Division of Epidemiology and Biostatistics, School of Public Health, University of Illinois Chicago, Associate Professor, Chicago, USA. [9]Department of Biostatistics, University of North Carolina at Chapel Hill, Professor, Chapel Hill, USA. [10]Department of Epidemiology and Population Health, Albert Einstein College of Medicine, Assistant Professor, New York, USA. [11]Department of Psychology, San Diego State University, Professor, San Diego, USA. [12]Department of Laboratory Medicine and Pathology, University of Minnesota, Professor, Minneapolis, USA. [13]Persistent Pollutants Biomonitoring Laboratory, Centers for Disease Control and Prevention, Atlanta, GA, Research Scientist, Atlanta, USA

Abstract
Background: POPs are endocrine disrupting chemicals implicated in the development of cardiometabolic diseases. While prior studies provide some evidence of hormone dysregulation with POPs exposure, findings are mixed and associations of POPs with hormones have not been explored in investigations using a heterogenous Hispanic/Latino population.

Methods: We studied 716 postmenopausal women free of diabetes at baseline examination from a subcohort of HCHS/SOL, the largest cohort study of US Hispanics. Women on medications related to hormones were excluded. Our study examined seven sex and thyroid-related hormones. Using multivariable linear regression adjusted for complex survey design and confounders, we estimated associations of baseline POPs exposures (continuous natural log transformed and quartiles) with baseline sex and thyroid-related hormones.

Results: Per 2.7-fold increase in POPs level; p,p'-dichlorodiphenyltrichloroethane (p,p'-DDT; β=0.07,95%CI:0.01,0.14) and polychlorinated biphenyl (PCB) group (3-methylcholanthrene inducers; β=0.12,95%CI:0.03,0.21) were associated with sex hormone binding globulin (SHBG) while oxychlordane (β=4.11,95%CI:0.13,8.09) was associated with triiodothyronine (T3). Per quartile increase in POPs level; various PCB groups (sum PCBs, dioxin-like, non-dioxin-like, and phenobarbital inducers) and organochlorine pesticides (OCPs; beta-hexachlorocyclohexane, hexachlorobenzene, oxychlordane, transnonachlor, and p,p'-dichlorodiphenyldichloroethylene) were also associated with SHBG while mirex was associated with T3 (β=2.78,95%CI:0.20,5.36) and thyroid stimulating hormone (TSH; β=-0.09,95%CI:-0.17,-0.01).

Conclusion: Oxychlordane and mirex were associated with thyroid-related hormones while various PCB groups and OCPs were associated with SHBG in postmenopausal women. No distinct associations were observed for other POPs with hormones. Hormonal dysregulation following POPs exposure involves multiple pathways and should be elucidated in future longitudinal studies.

188 – Prenatal Fluoride Exposure and Neurobehavioral Outcomes among Children Living in Los Angeles, California

Ashley J. Malin[1], Sandrah P. Eckel[2], Howard Hu[3], E. Angeles Martinez-Mier[4], Ixel Hernandez-Castro[5], Tingyu Yang[6], Shohreh F. Farzan[7], Rima Habre[2], Carrie V. Breton[8], Theresa M. Bastain[2]
[1]University of Florida, Assistant Professor, Gainesville, USA. [2]Keck School of Medicine of USC, Associate Professor, Los Angeles, USA. [3]Keck School of Medicine of USC, Professor & Flora L. Thornton Chair, Los Angeles, USA. [4]Indiana University School of Dentistry, Associate Dean for Global Engagement, Chairperson, Professor, Indianapolis, USA. [5]Keck School of Medicine of USC, Graduate Research Assistant, Los Angeles, USA. [6]Keck School of Medicine of USC, Statistician, Los Angeles, USA. [7]Keck School of Medicine of USC, Assistant Professor, Los Angeles, USA. [8]Keck School of Medicine of USC, Professor, Los Angeles, USA

Abstract
Background: Most of the United States (US) receives fluoridated water for dental caries prevention, with Los Angeles (LA) County being partially fluoridated. Higher prenatal and childhood fluoride exposure have been associated with poorer neurodevelopmental outcomes in Canadian and Mexican populations. However, associations of prenatal fluoride exposure with early childhood neurobehavioral health have not yet been examined in the US.

Participants and Methods: This study included 196 mother-child pairs from the Maternal and Developmental Risks from Environmental and Social Stressors (MADRES) cohort in LA. We examined associations of specific gravity-adjusted maternal spot urinary fluoride (MUFsg) during trimester three with index scores on the Child Behavior Checklist for Ages 1.5-5 at 36 months using linear and logistic regression. Models were adjusted for maternal age, education, ethnicity, pre-pregnancy BMI, marital status, household income, and child sex. Women who reported smoking (n=6) during pregnancy were excluded from analyses.

Results: Median (IQR) MUFsg was 0.76 (0.74) mg/L. Mean (SD) Total Problems, Internalizing and Externalizing T-scores were 47.55 (11.64), 47.05 (11.76), and 46.39 (10.61) respectively. There were 27 participants with Total Problems T-scores in the borderline clinical or clinical range. Each 1mg/L increase in MUFsg was associated with more than double the odds of having a Total Problems T-score in the borderline clinical or clinical range (OR=2.44, 95%CI=1.142, 5.193, p=0.021), and 3.6-point and 3.1-point higher Internalizing and Total Problems T-scores respectively (B=3.63, 95%CI=0.785, 6.472, p=0.013; B=3.13, 95%CI=0.33, 5.93, p=0.029).

Conclusion: Prenatal fluoride exposure may be a risk factor for childhood neurobehavioral problems in the US.

189 – POSTER: Persistent Organic Pollutants (POPs) Exposure and Incident Hypertension in Hispanic/ Latino Adults: The Hispanic Community Health Study/ Study of Latinos (HCHS/SOL)

Chibuzor Abasilim[1], Victoria Persky[2], Konstantina Tsintsifas[3], Robert Sargis[4,5], Maria Argos[6], Martha Daviglus[7], Sally Freels[6], Jianwen Cai[8], Brandilyn Peters[9], Gregory Talavera[10], Andreas Sjodin[11], Mark Davis[11], Richard Jones[11], Mary Turyk[2]
[1]Division of Epidemiology and Biostatistics, School of Public Health, University of Illinois Chicago, Postdoc, Chicago, USA. [2]Division of Epidemiology and Biostatistics, School of Public Health, University of Illinois Chicago, Professor, Chicago, USA. [3]Division of Epidemiology and Biostatistics, School of Public Health, University of Illinois Chicago, Doctoral Candidate, Chicago, USA. [4]Division of Endocrinology, Diabetes, and Metabolism, Department of Medicine, University of Illinois Chicago, Associate Professor, Chicago, USA. [5]Jesse Brown VA Medical Center, Associate Professor, Chicago, USA. [6]Division of Epidemiology and Biostatistics, School of Public Health, University of Illinois Chicago, Associate Professor, Chicago, USA. [7]Institute for Minority Health Research, University of Illinois Chicago, Associate Vice Chancellor for Research, Chicago, USA. [8]Department of Biostatistics, University of North Carolina at Chapel Hill, Professor, Chapel Hill, USA. [9]Department of Epidemiology and Population Health, Albert Einstein College of Medicine, Assistant Professor, Chicago, USA. [10]Department of Psychology, San Diego State University, Professor, San Diego, USA. [11]Persistent Pollutants Biomonitoring Laboratory, Centers for Disease Control and Prevention, Atlanta, GA, Research Scientist, Atlanta, USA

Abstract
Background: The impact of POPs during stages of hypertension development has implications for prevention and treatment but is not well understood. Nonlinear associations and confounding by other POPs during hypertension development has not been explored in many investigations.

Methods: We studied 1,065 adults free of hypertension and diabetes at baseline from a subcohort of HCHS/SOL. Hypertension was defined as anti-hypertensive medication use, SBP≥130mmHg or DBP>80mmHg and prehypertension as SBP≥120mmHg and ≤129mmHg and DBP<80mmHg. Using multivariable Poisson regression adjusted for complex survey design and confounders; we assessed associations of baseline POPs exposure (continuous natural log transformed and quartiles) with transition to prehypertension from normotension and to hypertension from prehypertension and normotension after 6 years of follow-up.

Results: Per 2.7-fold increase in POPs level; 2-(4-chlorophenyl)-2-(2-chlorophenyl)-1,1,1-trichloroethane (o,p'-DDT) was significantly associated with developing prehypertension (IRR=2.51,95%CI:1.08,5.80) and hypertension (IRR=1.92,95%CI:1.03,3.56) from normotension. Per quartile increase in POPs level; polybrominated diphenyl ethers (ΣPBDEs) and polychlorinated biphenyl (PCB) groups (dioxin-like and 3-methylcholanthrene inducers) were significantly associated with developing hypertension from normotension while per 2.7-fold increase in POPs level; p,p'-dichlorodiphenyldichloroethylene (p,p'-DDE; IRR=1.19,95% CI:1.03,1.37) and o,p'-DDT (IRR=1.56,95% CI:1.03,2.38) were significantly associated with developing hypertension from prehypertension.

Conclusion: o,p'-DDT but not PCBs were operative in the development of prehypertension while ΣPBDEs, o,p'-DDT, p,p'-DDE, and PCBs were associated with the development of hypertension. Our findings implicate diverse pathways for POPs in the development of hypertension and await confirmation in future studies.

190 – Bacterial Visualization Software Application (App) for Use During Extreme Flooding Events

Eric Hall
US Environmental Protection Agency, Physical Scientist, Durham, USA

Abstract
Recent reports indicate an increase in the projection of storms as a result of climate change (https://www.gfdl.noaa.gov/global-warming-and-hurricanes/). The 2022 hurricane season started May 15, in the Eastern Pacific Ocean, and June 1 in the Atlantic Ocean. In September 2022, the Atlantic hurricane season peaked with 5 storms. Cities in Puerto Rico and Florida were in the path of several storms (https://www.nesdis.noaa.gov/news/earth-orbit-2022-hurricane-season-begins). Climate change uncertainty has created the need to assess the distribution and concentration of microbial and chemical contaminants during flood events to inform public safety decisions. During these events, stormwater overwhelms local sewer systems, leading to potentially widespread fecal contamination. Overflow of combined sanitary and stormwater sewers presents acute infection risks to both emergency responders and resident populations. These events disproportionally impact underserved communities and communities with a higher composition of Black and Latino citizens. Because of elevated occurrences of gastrointestinal illness following urban floods, there is a need to accurately determine the distribution of these contaminated waters. Pathogens were discovered in the aftermath of Hurricane Katrina in the floodwaters that inundated 80% of the city (Pressley et al., 2006). The impact of harmful algal bacteria has also been documented during these storms (Phlips et al., 2020). Better tools for assessing the distribution of fecal pathogens and chemical contaminants, that can assess microbial risk exposure during flood events, are required. This project addresses this need by developing a novel application that can provide rapid microbial contaminant information, along with fecal source characterization, integrated into a geospatial platform.

191 – FRESSCA-Mujeres: Testing an Intervention to Protect Farmworker Women from Wildfire Smoke

Gina Solomon[1,2], Nayamin Martinez[3], Jeff Wagner[4], Brent Stephens[5], Isabella Kaser[6], Mohammad Heidarinejad[7], Aditya Singh[8], Ruben Rodriguez[9], Renata Valladares[10], Mingyu Wang[11], Stephanie Jarmul[12]
[1]Public Health Institute, Program Director, Oakland, USA. [2]University of California San Francisco, Clinical Professor, San Francisco, USA. [3]Central California Environmental Justice Network, Executive Director, Fresno, USA. [4]California Department of Public Health, Chief, Environmental Health Laboratory Branch, Richmond, USA. [5]Illinois Institute of Technology, Professor and Department Chair, Chicago, USA. [6]Public Health Institute, Project Manager, Oakland, USA. [7]Illinois Institute of Technology, Assistant Professor, Chicago, USA. [8]Illinois Institute of Technology, Doctoral Candidate, Chicago, USA. [9]Central California Environmental Justice Network, Community Organizer & Air Monitoring Specialist, Coalinga, USA. [10]Public Health Institute, Program Manager, Oakland, USA. [11]Illinois Institute of Technology, Graduate Student, Chicago, USA. [12]Office of Environmental Health Hazard Assessment, Environmental Scientist, Oakland, USA

Abstract
Agricultural workers in California's San Joaquin Valley are disproportionately exposed to wildfire smoke due to long work hours outdoors. Many farmworkers have evaporative ("swamp") coolers (ECs) in their homes. ECs bring in unfiltered outdoor air, creating an indoor hazard when wildfire smoke and heat coincide.

Filtration for Respiratory Exposure to wildfire Smoke in Swamp Cooler Air (FRESSCA) is a community-based participatory research project that responds to farmworker concerns by bringing together a multidisciplinary team to: 1) Design a filter for ECs that is effective and easy to install; and 2)Test the filter in the lab and in the homes of agricultural workers in the San Joaquin Valley.

In Year 2 of the study (2023), we are recruiting 50 farmworker women in non-smoking households with ECs, randomizing homes to filtration systems developed and tested during Year 1, and conducting continuous indoor/outdoor monitoring using Purple Air PA-II-SD monitors. This year of the study also includes air monitoring and urinary biomonitoring for polycyclic aromatic hydrocarbons (PAHs), volatile organic compounds (VOCs), metals found in wildfire smoke, and biomarkers of oxidative stress and inflammation. The biomonitoring will be done at baseline (spring), and during a wildfire event both after working outdoors and after sleeping indoors. Questionnaires will be used to assess potential exposures and health symptoms. If successful, the development of an effective filter for ECs could help protect thousands of low-income people in the Western U.S. exposed to the combined adverse effects of heat and wildfire smoke.

192 – Incorporating Health Impact Assessments (HIAs) into the Regional Environmental Health Data Visualization Interface (REHD-VI) for Use in Decision Making

Eric Hall
US Environmental Protection Agency, Physical Scientist, Durham, USA

Abstract
Environmental Justice (EJ) impact assessments and analyses are needed which combine software, databases, and environmental health indicators and population exposure considerations at the community level, to aid decision-makers in addressing cumulative impacts for permitting actions, non-regulatory actions, community-level, and regional planning activities. These include Health Impact Assessments (HIAs) to promote sustainable and healthy communities by assessing how a program, project, policy or plan impacts the health and well-being of a community (Bourcier et al., 2015). HIAs consider disproportionate impacts and inform decision-makers of potential outcomes. HIA Case Studies are available for review (https://www.epa.gov/healthresearch/epa-health-impact-assessment-case-studies). The Regional Environmental Health Data Visualization Interface (REHD-VI) is a readily available software application (app) that allows the user to select health data and relate them with known chemical and non-chemical stressors (https://edap.epa.gov/public/extensions/EHD_public/EHD_public.html). REHD-VI allows users to visualize tabular data and relationships, local 'hotspots', and provides insights into spatial interactions between health and exposure variables. REHD-VI allows the user to see relationships between data over many years, and can be displayed at the county and US census block group levels. Incorporating the HIA framework into REHD-VI will assist the US EPA in conducting consistent cumulative impact assessments that integrate EJ analysis and health impacts, while ensuring use of the most current information. The combined the HIA framework and REHD-VI can provide focus on specific pollution burdens, like PM2.5 and blood lead levels, and can have a real impact on reducing environmental exposure at the local community level.

194 – POSTER: A Roadmap for Efficient Systematic Reviews of Human Exposure Studies

Rebecca Nachman[1], Natalia Neal-Walthall[2], Peter Byrley[3], Rachel Shaffer[1]
[1]U.S. Environmental Protection Agency, Epidemiologist, Washington, DC, USA. [2]ORISE fellow for U.S. Environmental Protection Agency, Scientist, Washington, DC, USA. [3]U.S. Environmental Protection Agency, Scientist, Washington, DC, USA

Abstract
A key step of a systematic review for decision-making is the evaluation of individual studies. This evaluation informs how much weight each study will receive during the synthesis and summary of the evidence. Several tools and approaches exist for the evaluation of studies containing human exposure data. Despite the robustness of these frameworks, there is difficulty in applying them toward individual chemicals, because they are chemical-agnostic. A reviewer must have chemical-specific knowledge to effectively use these tools. In this presentation we introduce a roadmap for developing chemical-specific evaluation criteria that can supplement existing study evaluation tools.

To develop this roadmap, we identified multidisciplinary factors that are essential to evaluating the exposure domain. Principles of analytical chemistry, toxicokinetics, industrial hygiene, exposure sciences, and epidemiology informed the roadmap along with case examples from U.S. Environmental Protection Agency Integrated Risk Information System (IRIS) assessments and from the published literature.

We will first present the roadmap, which gives an overview of the multidisciplinary types of information that must be collected for a specific chemical(s) of interest. We then present case examples from systematic reviews conducted by EPA's IRIS program for three chemicals: hexavalent chromium, methylmercury, and ethylbenzene. For each example, we show how chemical-specific information outlined in the roadmap was integrated with the general IRIS study evaluation approach. These examples illustrate how the roadmap can be used to enhance the utility IRIS'S or other existing study evaluation tools and approaches to support a more efficient and robust systematic review process.

195 – POSTER: One-Step Cotinine Test to Assess Tobacco Use and Exposure

Jun Feng[1], Honest Achilihu[2], Tiffany Seyler[1], John Bernert[3], Lanqing Wang[4]
[1]Centers for Disease Control and Prevention, Research Chemist, Atlanta, USA. [2]Centers for Disease Control and Prevention, Chemist, Atlanta, USA. [3]Centers for Disease Control and Prevention, Guest Researcher, Atlanta, USA. [4]Centers for Disease Control and Prevention, Lead Research Chemist, Atlanta, USA

Abstract
Introduction: Disposable lateral flow test strips can provide a qualitative result in a few minutes. Cotinine is the most commonly measured tobacco exposure biomarker reflecting recent exposure to nicotine. A cotinine test strip with appropriate cutoff can differentiate tobacco users from nonusers, which provide qualitative result that is sufficient for some tobacco use and cessation studies.

Methods: Urine samples were tested with the cotinine test strips following the manufacture's protocol. We evaluated test strips with three different cutoffs, 10, 50, and 200 ng/mL, respectively. Cotinine concentrations for these samples were confirmed by two LC/MS/MS methods (High and Low) depending on the cotinine concentration. Data were analyzed in Excel and SAS.

Results: On average, classification by the strips with different cutoffs had a sensitivity of 98%, specificity of 94%, and accuracy of 96% in reference to results obtained by LC/MS/MS analysis. The positive predictive values are 92% and the negative predictive values are 98%. All the misclassified samples were relatively close to the cutoff for that strip. We applied the 50 ng/mL strip test to prescreen our samples for selection of the appropriate LC/MS/MS analysis method. Comparing to classification by self-report that we used previously, the misclassification rate reduced from 10% to <2%.

Conclusion: These test strips performed very well in our evaluations. They can be applied easily outside of the traditional testing laboratory to qualitatively assess nicotine exposure. Most importantly, cotinine test strips are faster and less expensive than the definitive LC/MS/MS analysis.

196 – Creating Built and Natural Environment Exposure Metrics by Applying Deep Learning Algorithms to 350 Million Street View Images across the United States from 2007-2020

Peter James[1], Marcia Jimenez[2], Pi-I Lin[3], Esra Suel[4], Jochem Klompmaker[5], Wenwen Zhang[6], Steve Hankey[7], Andrew Larkin[8], Perry Hystad[9]
[1]Harvard Medical School / Harvard TH Chan School of Public Health, Associate Professor, Boston, USA. [2]Boston University School of Public Health, Assistant Professor, Boston, USA. [3]Harvard Medical School / Harvard Pilgrim Health Care Institute, Research Scientist II, Boston, USA. [4]University College London, Lecturer, London, United Kingdom. [5]Harvard TH Chan School of Public Health, Postdoctoral Fellow, Boston, USA. [6]Rutgers University, Assistant Professor, New Brunswick, USA. [7]Virginia Polytechnic Institute and State University, Associate Professor, Blacksburg, USA. [8]Oregon State University, Assistant Research Professor, Corvallis, USA. [9]Oregon State University, Associate Professor, Corvallis, USA

Abstract
Background/Aims: Epidemiologic studies commonly use administrative and satellite-based spatial datasets to derive built and natural environment exposures. However, these top-down measures do not capture visual-level exposures. Deep learning applied to Google Street View (GSV) images offers opportunities to measure specific features of the built and natural environments at street-level that may be important to health. We aimed to develop nationwide, time-varying GSV measures that can be linked to Environmental influences on Child Health Outcomes (ECHO) consortium cohorts.

Methods: We created a 100m road network grid and used the closest GSV image to each grid point across all Metropolitan Statistical Areas in the United States from 2007-2020 (n=350 million GSV images). We classified each pixel in each GSV image into 150 different classifications by applying the open source PSPNet deep learning algorithm trained on the ADE20K dataset.
Results: We successfully derived measures of the percentage of pixels in each image that contained built features (e.g., buildings, sidewalks, benches) and natural features (e.g., trees, grass, plants). We then created nationwide annual raster surfaces (100m resolution) of these specific street-level built and natural features. GSV measures can be linked to any georeferenced health dataset (e.g., geocoded addresses, global positioning systems data) to create time-varying exposure metrics representing street-level built and natural features, providing novel opportunities for epidemiological studies.

Conclusion: We have created nationwide GSV exposure metrics that may provide novel insights on how built and natural environments influence human health.

197 – POSTER: An optimal method for determining the intraclass correlation coefficients of left-censored experimental data

Yukiko Nishihama[1,2], Yonghang Lai[3], Tomohiko Isobe[4], Shoji F. Nakayama[5]
[1]University of Tsukuba, Assistant professor, Tsukuba, Japan. [2]Japan Environment and Children's Study Programme Office, Health and Environmental Risk Division, National Institute for Environmental Studies, Collaborative Researcher, Tsukuba, Japan. [3]Japan Environment and Children's Study Programme Office, Health and Environmental Risk Division, National Institute for Environmental Studies, Research Associate, Tsukuba, Japan. [4]Japan Environment and Children's Study Programme Office, Health and Environmental Risk Division, National Institute for Environmental Studies, Chief Senior Researcher, Tsukuba, Japan. [5]Japan Environment and Children's Study Programme Office, Health and Environmental Risk Division, National Institute for Environmental Studies, Deputy Director, Tsukuba, Japan

Abstract
The concentrations of chemicals or their metabolites in biological samples are commonly used as surrogates of chemical exposure. When used in epidemiological studies, it is important to evaluate their reliability since it may introduce systematic bias resulting in misunderstanding and biased inference of the results. An intraclass correlation coefficient (ICC) has been used as a marker of reliability of a biomarker; however, the ICC has been initially proposed for evaluating inter-rater reliability and has not been assumed to be applied in an experimental study. In biomarker measurement, left censoring often occurs and its treatment has not been standardized yet.

In this study, we mathematically evaluated the effect of left-censoring on the estimated ICC and explored an optimal imputation method of left-censored data. Taking dialkylphosphate in children's urine as an example, 1) correlated data was simulated, 2) the data weas left-censored with different censoring rates and 3) the effect of different imputation method as examined. As a result, the Gibbs sampler-based left-censored missing value imputation approach (GSimp) outperformed the other methods including single value substitution. The ICC decreased when censoring rate increased even after GSimp imputation. We proposed a mathematical method to estimate a true ICC, i.e., the ICC would have been obtained from data with no left-censoring.

198 – IDSL.GOA: Gene Ontology Analysis for metabolomics - applications in exposomics

Dinesh Barupal[1], Priyanka Mahajan[2]
[1]Icahn School of Medicine at Mount Sinai, Assistant Professor, New York, USA. [2]Icahn School of Medicine at Mount Sinai, Programmer, New York, USA

Abstract
Blood metabolomics datasets contain signatures for metabolic processes which may be affected by exposure to a chemical hazard. Identification of these processes is useful for prioritizing and identifying chemical hazards for human health. For lists of genes, transcripts and proteins, Gene Ontology (GO) terms over-presentation analysis has become a standardized approach for biological interpretation. GO terms are not limited to predefined pathways but can also include relevant metabolic processes that are not included in pathway databases. Despite the several advantages of GO terms over traditional pathway maps, GO analysis has not been achieved for metabolomics datasets. To overcome this, we present a new knowledgebase and the online tool, Gene Ontology Analysis by the Integrated Data Science Laboratory for Metabolomics and Exposomics (IDSL.GOA) to conduct GO over-representation analysis for a metabolite list. The IDSL.GOA knowledge base covers 2,324 metabolic GO terms and associated 2,818 genes, 22,264 transcripts, 20,158 proteins, 1,482 EC annotations, 2,430 reactions and 2,212 metabolites. IDSL.GOA analysis for two large blood metabolomics datasets suggested that exposures to per- and polyfluoroalkyl substances (PFAS), including perfluorooctanoic acid (PFOA), perfluorohexane sulfonate (PFHxS) and perfluorooctane sulfonate (PFOS) can significantly affect the metabolic processes related to biogenic amines, sulfur amino acids and nucleotides metabolism. Interpreting metabolomics datasets in exposomics projects should not be limited to only pathway maps and can also leverage GO terms. IDSL.GOA provides a powerful tool for this purpose and can be accessed at https://goa.idsl.me/

199 – Effectiveness of Portable Air Cleaner in Reducing Exposure to Respiratory Infectious Aerosols in Learning Environments

Gen Pei[1], Parham Azimi[2], Donghyun Rim[3], Joseph G. Allen[4]
[1]Harvard T.H. Chan School of Public Health, Postdoctoral Fellow, Boston, USA. [2]Harvard T.H. Chan School of Public Health, Research Associate, Boston, USA. [3]Pennsylvania State University, Associate Professor, State College, USA. [4]Harvard T.H. Chan School of Public Health, Associate Professor, Boston, USA

Abstract
The utilization of portable air cleaner (PAC) is an important approach to reduce occupant exposure to respiratory viral aerosols in indoor environments. To maximize the PAC efficacy, science-based information is needed for determining the optimal PAC deployment location, depending on PAC characteristics, occupant placement, and room airflow pattern. We developed a Computational Fluid Dynamics (CFD) model to evaluate the PAC performance in reducing aerosol exposure in a classroom with a variety of PAC locations under different indoor airflow patterns. Our results illustrate PAC location as a critical factor affecting the PAC effectiveness. In the simulated classroom, positioning a PAC within 4 m distance to the infector can result in up to 50 % more reduction in occupant aerosol exposure than the case with a PAC at 8 m distance to the infector, suggesting the PAC placed closer to the infector has better aerosol removal performance. In absence of knowing who is infected, placing the PAC within the occupant zone (e.g., center of the classroom) should be considered, while the placement of PAC at back of the room or in the corner may not be ideal. Moreover, adjusting PAC flow discharge height to the occupant breathing level (e.g., 0.9-1.2 m for seated people) can improve its effectiveness, especially in displacement ventilated rooms. Our results also suggest that detailed airflow and aerosol transport analyses are needed before the application of multiple low-capacity PACs, which may lead to higher near-field aerosol exposure than using a single high-capacity PAC.

201 – Plastic waste generation and emissions from the domestic open burning of plastic waste in Guatemala

Michelle Bardales[1], Eri Saikawa[2], Mayari Hengstermann[3], Alexander Ramirez[4], John McCracken[5], Lisa Thompson[2]
[1]Emory University, Student, Atlanta, USA. [2]Emory University, Associate Professor, Atlanta, USA. [3]The Universidad del Valle de Guatemala, Anthropologist, Guatemala city, Guatemala. [4]The Universidad del Valle de Guatemala, Data Support, Guatemala city, Guatemala. [5]University of Georgia, Assistant Professor, Athens, USA

Abstract
Domestic, or household-level, open burning of plastic waste is a source of air pollutants and greenhouse gases that are often neglected in emission inventories. Domestic open burning is a considerable concern in Guatemala due to the lack of access to waste collection services, particularly in rural areas. This paper offers the first attempt to estimate emissions from the domestic open burning of waste at the city and departmental levels in Guatemala. Data were collected from the Xalap´an region of Jalapa, Guatemala and analyzed to determine the change in plastic waste generation over time as well as the socioeconomic factors that may affect the extent of plastic waste generation and burning. The annual per capita masses of plastic waste burned were used to estimate emissions from domestic open burning of plastic waste in the region of Xalap´an, the cities of Jutiapa and Guatemala city, and all 22 departments in Guatemala. Our results show that rural areas burn more waste domestically, likely because of a lack of access to waste collection, and 30.4% of OC, 24.0% of BC, 23.6% of PM2.5, and 2.4% of CO2 emissions in Guatemala may not be accounted for by excluding open plastic burning as a source.

202 – POSTER: Exposure Assessment of Isocyanates Emitted from Consumer Products

Yuna Kakimoto[1], Qi Wang[2], Ryo Omagari[3], Kazushi Noro[3], Yuichi Miyake[4], Takashi Amagai[5]
[1]Academic, Student, Shizuoka, Japan. [2]Academic, Researcher, Kawasaki, Japan. [3]Academic, Assistant professor, Shizuoka, Japan. [4]Academic, Associate professor, Yokohama, Japan. [5]Academic, Professor, Shizuoka, Japan

Abstract
Isocyanates are semi-volatile organic compounds used as raw materials for polyurethane foams, paints, and building materials. Exposure to isocyanates cause acute adverse health effects, such as irritation of the respiratory tract, skin, and eyes, inducing that asthma and sick house syndrome. Despite the potential presence of various sources indoors, studies for long-term exposures and risk assessment of isocyanates in the indoor environments are insufficient. Here we show that emissions of isocyanic acid (ICA) from consumer products may contribute on their indoor air concentrations. Toluene 2,4-di-isocyanate (TDI) and 4,4'-Diphenylmethane di-isocyanate (4,4'-MDI) were detected at relatively high concentrations in infant chairs, mattresses, and spray polyurethane forms, but no emissions were observed. ICA was found in and emitted from all consumer products. Emissions of ICA were in the range of 0.0536-1.37 ng g^{-1} day^{-1}). Isocyanates were detected in the range of 8.69-64.0 ng m^{-3} in the measurements of indoor air levels in six houses. In particular, ICA was detected at relatively high concentrations (7.69-62.8 ng m^{-3}). Relatively high I/O ratios were obtained in ICA (0.690-3.75), ethyl isocyanate (0.600-2.96), and propyl isocyanate (0.630-2.04). These results suggest that consumer products may be an indoor source of ICA. Margin of exposure of methyl isocyanate and TDI were between 100 and 1000 at most measurement sites, implying that they were possibilities of risk concern. Risk assessments in the indoor environment is required in future study.

203 – California's AB 617 - A Signature Environmental Justice Law for Community Air Protection: From the Context of Segmented Scientific Understanding of Cumulative Exposures and Cumulative Impacts with Environmental Justice Connections

Lily Wu
Cal EPA - Office of Environmental Health Hazard Assessment, Staff Toxicologist, Sacramento, USA

Abstract
California has long been on the national and international stage for environmental protection. As the disparate impacts of Cap and Trade with the state's AB 32 regulation (Global Warming Solutions Act of 2006) were becoming clearer - numerous community members, environmental justice advocates, multidisciplinary scientists, policy makers, and many others were adding to the growing data sets showing a need for policy change. Assembly Bill (AB) 617 was passed in 2017. AB 617 is a regulation which lead to the development a statewide strategy that emphasizes local plans to reduce emissions and the Community Air Protection Program within the state's air resources board. AB 617 sets up air monitoring plans and selection of high priority areas in need of air monitoring systems. It also mandates a statewide strategy to reduce emissions of greenhouse gas co-pollutants, which include toxic air contaminants (TACs) and criteria air pollutants (CAPs). The TACs and CAPS might affect the health of people living in communities highly burdened by environmental polluting sources. The objective of this presentation is to highlight lessons learned in program implementation, advance the understanding of cumulative exposures, cumulative risk assessments, and cumulative impacts. The impacts of climate change amid a recent global pandemic have highlighted the inequities of many systems. Now is the time to advance environmental equity, health equity, and community resiliency. In order to do so, we'll need to address the segmented way our environmental media are regulated (air, water, and soil) and realize the coordination necessary to address cumulative impacts.

204 – POSTER: PFAS Concentrations in Landfill Impacted Communities in North Carolina: A Pilot Study

Charlotte Robbins[1], Courtney Woods[2]
[1]University of North Carolina at Chapel Hill, Student, Chapel Hill, USA. [2]University of North Carolina at Chapel Hill, Faculty, Chapel Hill, USA

Abstract
Unlined landfills can serve as a major source of PFAS contamination into the surrounding environment and groundwater. Previous research has shown that landfills in North Carolina are more likely to be found in rural areas presenting a potential threat to public health, as many residents rely on private wells for drinking water. While several studies have assessed the presence of PFAS in landfill leachate, few studies have examined PFAS in groundwater and in private wells near landfills. This study aims to predict the potential risk landfill-related well water contamination in surrounding rural communities based on several community, hydrogeologic and landfill characteristics. We assess the concentrations of approximately 26 PFAS compounds found in four landfill-impacted communities across the state, including those hosting decades old municipal solid waste, pre-regulatory and construction and demolition landfills. Our team employs a community engaged approach to collect sample sand report findings. This poster presents our preliminary findings from this study.

205 – Understanding Dietary and Dust Contributions to Per- and Polyfluoroalkyl Substances (PFAS) Exposure among Pregnant Women and Newborns

Yerin Jung[1], Scott Bartell[2]
[1]University of California, Irvine, PhD Student, Irvine, USA. [2]University of California, Irvine, Professor, Irvine, USA

Abstract
Background: When drinking water is uncontaminated, exposure to PFAS is thought to occur primarily via ingestion of food and indoor dust.

Objectives: To understand the contribution of exposure routes in prenatal periods, this study examined whether published estimates of PFAS exposure from dust and diet provide reasonable predictions of PFAS serum concentrations in pregnant women and newborns.

Methods: This study estimated maternal and cord serum concentrations of four PFAS (PFOA, PFOS, PFHxS, and PFNA) based on published PFAS intake rates for food and indoor dust reported in the peer-reviewed literature, contact rates, and a one-compartment pharmacokinetic (PK) model using two different sets of parameters, and Monte Carlo simulation to account for variability/uncertainty.

The estimated serum concentrations for different exposure scenarios were then compared with measured serum levels reported in 23 published studies from 3 continents, measured from 1998 until 2016.

Results: Lower published estimates of dietary intake resulted in good prediction of maternal and cord serum concentrations. Higher published estimates of dietary intake overestimated maternal serum concentrations, especially for PFHxS and PFNA. Although some discrepancies exist among published estimates of indoor dust intake and PK parameters (half-life, volume of distribution and cord:maternal serum ratio) for PFAS, any combination of our selected estimates along with lower published dietary intake estimates are sufficient to provide reasonable estimation of published maternal and cord serum concentrations.

Significance: For communities without water contamination, PFAS serum concentrations are largely explained by dietary intake, though the exact contributions of individual foods remain poorly understood.

206 – Polychlorinated Naphthalenes and Polychlorinated Biphenyls in Street Dust and Surface Soils: Identifying Sources and Potential Exposures

Sung-Hee Seo, Stuart Batterman
University of Michigan, PhD, Ann Arbor, USA

Abstract
Polychlorinated naphthalenes (PCNs) and polychlorinated biphenyls (PCBs) are legacy contaminants that were produced and used in massive quantities for their electrical, cooling, lubricating and other properties. These persistence and toxic compounds were improperly disposed and are now broadly distributed in the environment, yet little data exists regarding urban levels. This study examines levels, distributions, and potential sources of these compounds in Detroit, Michigan. We collected 19 composite samples of surface soils and street dusts in residential and industrial areas, which were dried, sieved, extracted, and analyzed by gas chromatography/mass spectrometry for 32 PCN and 37 PCB congeners. Principal component and geospatial analyses helped identify mixtures and potential sources. The geometric mean concentrations of total PCNs and total PCBs were 56 (range: 7-1,933) and 422 (range: 30-18,636) ng/g dw, respectively. Among PCNs, the more chlorinated PCNs (tetra- to hexa-) were most common; PCB-59 was most prevalent congener. Among PCBs, hexa-congeners were most common; PCB-158 was dominant. PCN levels were highest near a waste pile at a gypsum facility, and PCB levels were highest in street dust at a scrap metal recycling facility. Soils and dusts in adjacent residential areas had similar compositions but lower concentrations. This study shows the presence of sources and "hotspots" of toxic compounds in urban and industrial settings, despite the phase-out of these chemicals many decades earlier. Sources should be identified and controlled to prevent further environmental dispersal and exposure. The facility type and land use history can help prioritize sampling strategies.

207 – POSTER: Assessment of pesticide exposure during the florescence and harvest periods using honeybees as bioindicator

Chien-Che Hung[1], Lih-Ming Yiin[2]
[1]Tzu Chi University, Post-doctoral researcher, Hualien, Taiwan. [2]Tzu Chi University, Pofessor, Hualien, Taiwan

Abstract
Honeybees are one of the important pollinators in the ecosystem; by pollinating, they may be exposed to a variety of pesticides, including insecticides, herbicides and fungicides. With the characteristics of certain foraging distances and the likelihood of exposures to pesticides, honeybees could serve as bioindicator of pesticide exposure in the environment. We conducted a study to evaluate the feasibility by collecting and analyzing worker bee samples from 15 apiaries located in east Taiwan during the florescence and harvest periods for fruits and betel nuts. Bee samples from the same hive boxes were composited, and the target pesticides were analyzed by gas or liquid chromatography, coupled with mass spectrometry. The preliminary result shows that more occurrences of sick or dead worker bees are present near the hive boxes during the florescence period than the harvest period (75% of the hive boxes vs. 26.6%). The pesticide analysis is consistent with the observation, showing that herbicides are more frequently detected than insecticides from healthy worker bees (33.7% vs. 15.9%), and from sick/dead worker bees (51.2% vs 22.7%). Glyphosate is the most frequently detected herbicide with the concentration up to approximately 750 ng/g of bees, whereas carbaryl and chlorpyrifos appear to be the most frequently detected insecticides with the concentrations lying among several hundred ng/g of bees. The finding suggests that use of herbicides could be more prevalent than that of insecticides; those pesticides need applying with caution during the florescence period to avoid unnecessary exposure to pesticides.

208 – Mediating role of serum vitamin D in the association between ambient air pollution and glucose metabolism in a nationally representative population of Korea

Habyeong Kang[1], Min Jae Ju[1], Yoon-Hyeong Choi[2]
[1]Korea University College of Health Science, Research Professor, Seoul, Korea, Republic of. [2]Korea University College of Health Science, Associate Professor, Seoul, Korea, Republic of

Abstract
Background: Previous studies have suggested a link between air pollution exposure and glucose metabolism. One possible mechanism for this association is reduced vitamin D synthesis, although this mediating role of vitamin D synthesis has not been fully understood. This study aimed to investigate the mediation by serum vitamin D on the association between ambient air pollution and glucose metabolism.

Methods: Among adults participating in the Korea National Health and Nutrition Examination Survey (KNHANES) 2009–2014, we included those having complete information on covariates, air pollutants concentrations estimated using the CMAQ model ($PM_{2.5}$, PM_{10}, CO, and NO_2; cumulative lag 0-29 days), serum vitamin D levels, and glucose metabolism outcomes (fasting blood glucose, FBG or hemoglobin A1c, HbA1c), yielding 26,124 and 15,191 participants for FBG and HbA1c, respectively. Survey-weighted linear regression and mediation analysis were used to examine the associations between air pollutants, serum vitamin D, and glucose metabolism outcomes, and the medicating effects of vitamin D.

Results: Air pollutants were positively associated with FBG or HbA1c. An interquartile range increase in PM_{10} (16.9 μg/m^3) was associated with a 1.69 ng/mL (95% CI: 0.62, 1.85) increase in FBG and a 0.12% (95% CI: 0.06, 0.12) increase in HbA1c. It was estimated that vitamin D mediated 38% (95% CI: 21, 47) and 22% (95% CI: 3, 23) of these associations of PM_{10} with FBG and HbA1c, respectively.

Conclusion: Our findings suggest that a significant portion of the association between air pollution and glucose metabolism may be mediated by reduced vitamin D synthesis.

209 – Fine particulate constituents and hospital admission for cardiovascular diseases among 183 major Chinese cities

Yi Zhang[1], Tiantian Li[2], Pei Gao[3]
[1]China CDC Key Laboratory of Environment and Population Health, National Institute of Environmental Health, Chinese Center for Disease Control and Prevention, Associate professor, Beijing, China. [2]China CDC Key Laboratory of Environment and Population Health, National Institute of Environmental Health, Chinese Center for Disease Control and Prevention, Professor, Beijing, China. [3]Department of Epidemiology and Biostatistics, School of Public Health, Peking University, Professor, Beijing, China

Abstract
Objective: This study aims to investigate the association between short-term exposure of $PM_{2.5}$ constituents and hospital admissions of cause specific CVD.

Methods: Time-series design was used. The daily concentration of $PM_{2.5}$ constituents during 2014-2017 were estimated by the Community Multi-scale Air Quality Model (CMAQ). Daily counts of city specific hospital admissions for CVD, ischemic heart disease (IHD), heart failure, ischemic stroke, and hemorrhagic stroke in 183 cities in China were extracted from the national Urban Employee Basic Medical Insurance database and summary of hospital discharge.

Results: Among the five $PM_{2.5}$ constituents, NO_3^- had the highest average daily concentration (12.98μg/m^3), followed by OC (9.43μg/m^3), SO_4^{2-} (8.62μg/m^3), NH_4^+ (6.93μg/m^3), and EC showed the lowest concentration (2.87μg/m^3). The cities with relatively higher concentrations of $PM_{2.5}$ constituents were mainly located in Beijing-Tianjin-Hebei, Yangtze River Delta region, Fenwei Plain and Chengdu-Chongqing area. The RR were 1.009 (95%CI: 1.005~1.012) for EC exposure per IQR increase, 1.005 (95%CI: 1.002~1.008) for SO_4^{2-} exposure, 1.006 (95%CI: 1.002~1.011) for NH_4^+. OC and NO_3^- were not found significant association with hospital admission for CVD. Among the specific diseases, the association between $PM_{2.5}$ constituents and hospital admission for ischemic stroke showed similar pattern to CVD. Only EC was found significantly associated with hospital admission for IHD (1.006, 95%CI: 1.003~1.010).

Conclusion: The risks of hospital admissions for CVD varied exposed to different $PM_{2.5}$ constituents. Exposed to EC had the highest risk to IHD and ischemic stroke, whereas all constituents were not significant associated with hemorrhagic stroke and heart failure.

210 – POSTER: Human Biomonitoring of Neonicotinoids and Neonicotinoid-like Compounds: Selection of Biomarkers of Exposure Based on Human Metabolism Studies

Sonja A. Wrobel[1], Daniel Bury[2], Stephan Koslitz[2], Heiko Hayen[3], Holger M. Koch[4], Thomas Brüning[5], Heiko U. Käfferlein[6]
[1]Institute for Prevention and Occupational Medicine of the German Social Accident Insurance, Ruhr University Bochum (IPA), PhD Student, Bochum, Germany. [2]Institute for Prevention and Occupational Medicine of the German Social Accident Insurance, Ruhr University Bochum (IPA), Researcher, Bochum, Germany. [3]Institute of Inorganic and Analytical Chemistry, University of Münster, Professor, Münster, Germany. [4]Institute for Prevention and Occupational Medicine of the German Social Accident Insurance, Ruhr University Bochum (IPA), Head of Lab, Bochum, Germany. [5]Institute for Prevention and Occupational Medicine of the German Social Accident Insurance, Ruhr University Bochum (IPA), Professor, Bochum, Germany. [6]Institute for Prevention and Occupational Medicine of the German Social Accident Insurance, Ruhr University Bochum (IPA), Section Head, Bochum, Germany

Abstract
Human biomonitoring (HBM) of neonicotinoids and neonicotinoid-like insecticides (NNIs) has mostly been performed using the respective parent compounds or their suspected breakdown metabolites 2-chloro-1,3-thiazole-5-carboxylic acid (2-CTA) or 6-chloronicotinic acid (6-CNA) as biomarker of exposure in urine. Nevertheless, human metabolism data is limited and knowledge on urinary excretion fractions (F_{ue}s) is almost inexistent. Therefore, the aim of this study was the quantitative investigation of the human metabolism and urinary excretion of the NNIs acetamiprid, clothianidin, flupyradifurone, imidacloprid, sulfoxaflor, thiacloprid, and thiamethoxam.

Each NNI was administered separately to male volunteers in controlled oral doses. The respective doses were based at current acceptable daily intake (ADI) levels. Complete and consecutive urine samples were collected over 48-hrs. All samples were analyzed by mass spectrometry following liquid or gas chromatographic separation.

We found extensive metabolization of NNIs containing a 6-chloropyridinyl moiety (acetamiprid, flupyradifruone, imidacloprid, thiacloprid). They were mainly excreted as oxidized metabolites with F_{ue}s ranging from 17% to 34% for some key metabolites. The 2-chlorothiazolyl-NNIs clothianidin and thiamethoxam were mainly excreted as the parent compounds. Despite their previous use as biomarkers, the group-specific breakdown metabolites 2-CTA and 6-CNA were of low metabolic relevance in humans with F_{ue}s <1%.

This study generates long needed quantitative data on urinary excretion of NNIs and provides valuable biomarkers of exposure for HBM. It shows that earlier NNI exposure assessments might be misleading due to insufficient knowledge on their human metabolism. The new set of exposure biomarkers will enable a robust exposure and risk assessment for NNIs via human biomonitoring.

212 – A Suspect Screening of Urinary Pesticides in the European General Population

Jelle Vlaanderen[1], Ilse Ottenbros[2], Priska Ammann[3], Erik Lebret[4], Carolin Huber[5], Arjen Lommen[6], Jean-Philippe Antignac[7], Pavel Cupr[8], Libor Sulc[8], Ondrej Mikes[8], Tamas Szigeti[9], Szilvia Középesy[9], Inese Martisone[10], Zanna Martinsone[10], Lasma Akulova[10], Olga Pardo[11], Sandra Fernandez[12], Clara Coscollá[12], Susana Pedraza-Diaz[13], Martin Krauss[14], Laurent Debrauwer[15], Kévin Wagner[16], Rosalie Nijssen[6], Hans Mol[6], Chiara Maria Vitale[8], Jana Klánová[8], Medea Imboden[3], Borja Garlito Molina[17], Jan-Paul Zock[18], Nuria León[12], Samuel Fuhrimann[3], Roel Vermeulen[4], Mirjam Luijten[18], Nicole Probst[3]
[1]Utrecht University, Associate Professor, Utrecht, Netherlands. [2]Utrecht University, PhD student, Utrecht, Netherlands. [3]Swiss TPH, Researcher, Basel, Switzerland. [4]Utrecht University, Professor, Utrecht, Netherlands. [5]UFZ, PhD Student, Leipzig, Germany. [6]Wageningen University and Research, Researcher, Wageningen, Netherlands. [7]INRAE, Professor, Nantes, France. [8]Recetox, Researcher, Brno, Czech Republic. [9]National Public Health Center, Researcher, Budapest, Hungary. [10]Rīga Stradiņš University, Researcher, Riga, Latvia. [11]Instituto de Salud Carlos III, Researcher, Majadahonda, Spain. [12]Foundation for the Promotion of Health and Biomedical Research of the Valencia Region, Researcher, Valencia, Spain. [13]nstituto de Salud Carlos III, Researcher, Majadahonda, Spain. [14]Helmholtz Centre for Environmental Research, Professor, Leipzig, Germany. [15]INRAE, Researcher, Nantes, France. [16]Toulouse University, Researcher, Toulouse, France. [17]niversitat Jaume I, Researcher, astelló de la Plana, Spain. [18]RIVM, Researcher, Bilthoven, Netherlands

Abstract
Humans are exposed to a mixture of pesticides through diet as well as through the environment. We conducted a suspect-screening based study to describe exposure to a set of pesticide biomarkers in six European countries (Latvia, Hungary, Czech Republic, Spain, Switzerland, and the Netherlands). We explored whether living in an agricultural area (compared to living in a peri-urban area), being a a child (compared to being an adult), and the season in which the urine sample was collected had an impact on the probability of detection of pesticides (-metabolites).
In total 2383 urine samples were collected from 1345 participants (including 525 parent-child pairs) and analyzed through harmonized suspect screening by five different laboratories. Fourty pesticide biomarkers relating to 29 pesticides were identified at high levels of

confidence in samples across all study sites. Most frequently detected were biomarkers related to the parent pesticides acetamiprid and chlorpropham. In 84% of the samples at least two different pesticides were detected. The median number of detected pesticides in the urine samples was 3, and the maximum was 13 pesticides detected in a single sample. We observed differences in the probability of detection of pesticide biomarkers among children compared to adults. No consistent effects were observed for living in an agricultural area or season of urine sampling.

This survey demonstrates the feasibility of conducting a harmonized European sample collection, combined with suspect screening to provide insight in the presence of exposure to pesticide mixtures in the European population.

213 – The Short-term Cardiopulmonary Impact of Cooking Particles: A Clinical Study

Motahareh Naseri[1], Seyedeh Mohadeseh Kazemitabar[2], Seyedeh Ayeh Esmaili Taleshi[3], Milad Malekipirbazari[4], Giorgio Buonanno[5], Luca Stabile[5], Flemming Cassee[6], Dhawal Shah[7], Mehdi Amouei Torkmahalleh[8]
[1]Nazarbayev University, PhD student, Astana, Kazakhstan. [2]Payam Noor University, student, Tehran, Iran, Islamic Republic of. [3]Mazandaran University of Medical Sciences, student, Sari, Iran, Islamic Republic of. [4]Chalmers University of Technology, Professor, Gothenburg, Sweden. [5]University of Cassino and Southern Lazio, Professor, Casino, Italy. [6]7National Institute for Public Health and the Environment, Professor, Bilthoven, Netherlands. [7]Nazarbayev University, Professor, Astana, Kazakhstan. [8]University of Illinois, Professor, Chicago, USA

Abstract
Individuals spend the vast majority of their time indoors, estimated to be 90%. Cooking is an essential activity that produces indoor ultrafine particles (UFPs) that can potentially impact human health. Inhalation of cooking UFPs can trigger inflammation and oxidative stress and induce changes in the autonomic nervous system. This study aims to investigate the impact of cooking UFPs on the human cardiovascular and respiratory systems of forty healthy non-smoker human subjects.

Study participants stayed in a fully furnished apartment for 48 hours consecutively. The first 24 hours was a control day, and the second 24 hours was the cooking (frying chicken and fries) day. The study participants' heart rate was monitored for 48 hours by attaching a portable ECG device to their chests. The Omron 10 blood pressure monitor measured systolic and diastolic blood pressure. A Bedfont NObreath FeNO Monitor and a peak flowmeter were utilized to measure fractional exhaled nitrogen oxide (FeNO) and the volume of air that can be blown out in a single quick burst, respectively. A NanoTracer and the RizgardPlus PM monitor (a low-cost) were used to measure the particle number and mass concentrations, respectively. The results showed that the diastolic blood pressure decreased significantly exactly after cooking and one hour after cooking; however, no statistical differences were observed for systolic blood pressure and heart rate. FeNO results showed statistically significant changes to lung function immediately after cooking until two hours post-exposure. The peak flow rate significantly increased exactly after cooking and 30 minutes after cooking.

214 – POSTER: Changes of Air Quality during the COVID-19 Pandemic and their Effects on Asthma Attacks in Japan

Masayuki Shima[1], Yoshiko Yoda[2]
[1]Hyogo Medical University, Professor, Nishinomiya, Japan. [2]Hyogo Medical University, Assistant professor, Nishinomiya, Japan

Abstract
Air pollution reductions during the COVID-19 pandemic were observed in many countries. We investigated the association between changes of air quality and asthma attacks in Himeji city in western Japan. We collected the daily average concentrations of fine particulate matter ($PM_{2.5}$) and nitrogen dioxide (NO_2) and daily maximum 8-hour average of ozone (O_3) from 2017 to 2022 from 9 air quality monitoring stations in Himeji city. The weekly numbers of asthma attacks were reported from 46 hospitals and clinics in the city. After the Japanese government declared the state of emergency on 7 April, 2020, NO_2 concentrations decreased by about 30% in April and May than the average of the past 3 years. Whereas, $PM_{2.5}$ concentrations had decreased from January by more than 25% compared with the past 3 years. The reduction may be due to the lockdown in China, because $PM_{2.5}$ concentrations in western Japan are influenced by the transport from the Asian continent. O_3 concentrations were unchanged during the period. The weekly numbers of asthma attacks dramatically decreased from April, and the decrease has continued thereafter. The associations between asthma attacks and the concentration of $PM_{2.5}$, NO_2, and O_3 were significant among elderly (≥65 years, β=0.032 [95% CI: 0.011, 0.052]), working adults (15–64 years, β=0.136 [95% CI: 0.016, 0.256]), and children (<15 years, β=0.045 [95% CI: 0.009, 0.082]), respectively. In conclusion, the decrease in asthma attacks may be partly caused by the lifestyle changes associated with the COVID-19 pandemic, but was also associated with air pollution levels.

216 – Per- and Polyfluoroalkyl Substances (PFAS) in Paired House Dust and Tap Water from United States Homes

Nikki DeLuca[1], Jason Boettger[2], Jeffrey Minucci[3], Kelsey Miller[4], Karen Bradham[5], James McCord[4], Peter Ashley[6], Warren Friedman[6], Eugene Pinzer[6], Elaine Cohen Hubal[7]
[1]USEPA, Postdoctoral Physical Scientist, Research Triangle Park, NC, USA. [2]USEPA, ORISE Participant, Research Triangle Park, NC, USA. [3]USEPA, Physical Scientist, Research Triangle Park, NC, USA. [4]USEPA, Chemist, Research Triangle Park, NC, USA. [5]USEPA, Research Physical Scientist, Research Triangle Park, NC, USA. [6]US HUD, Scientist, Washington, D.C., USA. [7]USEPA, Senior Science Advisor, Research Triangle Park, NC, USA

Abstract
House dust acts as an aggregate of chemicals in residential spaces and can be an important human exposure pathway for compounds such as per- and polyfluorinated alkyl substances (PFAS). The composition and levels of PFAS in house dust can be a function of the home's proximity to sources and/or the materials and products within it. PFAS in house dust may also be affected by tap water contamination, as this water may be used for household activities like cleaning and showering. U.S. Department of Housing and Urban Development (HUD) funded and conducted a survey of residential hazards called the American Healthy Homes Survey II in 2017. U.S. Environmental Protection Agency (EPA) collaborated on this effort and analyzed PFAS in house dust and tap water samples. Here we report results for 241 paired house dust and tap water samples from homes in 37 states. In house dust, 16 PFAS were measured with detection rates ranging from 97% (PFOA) to 9% (PFNS); in tap water, 13 PFAS were measured with detections ranging from 100% (PFBS) to 1% (PFNS). Four chemicals (PFOA, PFOS, PFHxS, PFHpA) were detected in at least 50% of the samples in both house dust and tap water. Correlations between paired house dust and tap water concentrations and geographic proximity to potential sources are analyzed to investigate possible links between PFAS in both media.

Disclaimer: The views expressed here are those of the author(s) and do not necessarily reflect the views or policies of the EPA nor HUD.

217 – Characterizing peak and mean VOC exposures associated with different nail salon services

Kimberly Anderson, Abbas Virji
NIOSH, Research Industrial Hygienist, Morgantown, USA

Abstract
Exposure assessments reported in the literature have focused on traditional acrylic manicures or salon spaces without characterizing the services or products offered; however, product formulations for different services and brands vary, resulting in different exposure profiles. The objective of this study was to estimate the short-term peak and average VOC exposures associated with different manicure types (lacquer, gel, dip, and acrylic) and product use from real-time data. Total volatile organic compounds (TVOC, photoionization detector, Tiger) and twenty-two individual compounds (Fourier Transform Infrared spectrometer, Gasmet Inc) were measured in the breathing zone while traditional lacquer nail polish, gel, dip, and acrylic manicures were simulated in an exposure chamber. Log-transformed concentrations were modeled with the fixed effects of tasks and covariates using a Bayesian spline model that accounted for non-stationary autocorrelation in real-time measurements and censored data. Task was the strongest predictor of average TVOC concentration followed by type of manicure. Gel manicures resulted in significantly higher average TVOC concentrations across all tasks compared to dip and lacquer manicures. Dip manicures resulted in the highest peak TVOC concentrations at 84 ppm compared to gel (45 ppm) and lacquer (10 ppm) manicures. Concentrations of butyl acetate, butyl acrylate, ethyl acetate, isopropyl acetate, and isopropanol varied significantly between the type of manicure being performed (dip, gel, or lacquer). These results demonstrate significant differences in short-term peak and average exposures associated with various nail-salon activities that are important for targeted intervention strategies to control exposures, but may be missed with the traditional sampling approaches.

218– POSTER: Determining Relationships Between Levels of Ethylene Oxide Hemoglobin Adducts and Smoking Biomarkers Among 6 Years and Older

James Pyon[1], Tebitha Mawokomatanda[2], James Durant[1], Benjamin Blount[3], Hubert Vesper[3]
[1]Centers for Disease Control and Prevention, Agency for Toxic Substances and Disease Registry, Office of Community Health Hazard Assessment, Environmental Health Scientist, Atlanta, USA. [2]Centers for Disease Control and Prevention, Agency for Toxic Substances and Disease Registry, Office of Community Health Hazard Assessment, Statistician, Atlanta, USA. [3]Centers for Disease Control and Prevention, National Center for Environmental Health, Division of Laboratory Sciences, Atlanta, USA

Abstract
Ethylene Oxide (EtO) is a gas used to sterilize certain food products and medical equipment and to produce other chemicals. EtO hemoglobin adducts (HbEO) can be used to evaluate occupational exposure to EtO. Community members near facilities that release EtO may have interest in using HbEO to evaluate recent exposures to EtO. Smoking can also produce HbEO, so confounding effects of smoking must be controlled when using HbEO to evaluate non-occupational EtO exposure. This analysis explores which smoking biomarker best predicts HbEO levels.

A one-third subsample of 5,447 participants (age six and older) in NHANES (2013-2016) had their whole blood analyzed for HbEO. Jackknife replicate weights, using NHANES design variables (masked strata and primary sampling units) and EtO subsample weights, were used with linear modeling to estimate model R-squared (R^2) with standard error (SE) for the following models: $Y_{HbEO} = X_{Cotinine}$, $Y_{HbEO} = X_{2CYEMA}$, and $Y_{HbEO} = X_{Cotinine} + X_{2CYEMA}$.

The R^2 value of the model with both biomarkers (0.529 SE 0.038) was greater than the R^2 of the model with 2-CYEMA (0.443 SE 0.057) and was greater than the R^2 of the model with cotinine (0.346 SE 0.039).

Using the combination of 2-CYEMA and cotinine may improve effectiveness of HbEO as a biomarker of EtO exposure in a non-occupational setting.

Disclaimer: The findings and conclusions in this abstract have not been formally disseminated by CDC/ATSDR and should not be construed to represent any agency determination or policy.

219 – POSTER: Evolution in disparity of PM2.5 pollution in China

Su Shi[1], Xia Meng[2]
[1]Fudan University, Student, Shanghai, China. [2]Fudan University, Associate professor, Shanghai, China

Abstract
Spatial disparity of air pollutants is one of key influential factors for environmental inequality. We quantitively evaluated the evolution of PM2.5 spatial disparity in China during 2013-2020, and investigated associations between PM2.5 spatial disparity and economic indicators. Differences in PM2.5 between more and less polluted cities declined over time, suggesting decreased absolute disparity. However, the more polluted cities in 2013 remained so in 2017 and 2020, and vice versa, indicating persistent relative disparity. PM2.5 pollution levels increased with higher GDP per capita in less-developed areas of China, but such negative effect weakened over time; while economic development tended to promote cleaner air in developed areas of China. Therefore, policies to improve air quality and promote economic development simultaneously are needed in China to reduce the disparity of air pollution and promote all people enjoy environmental equality.

220 – Hair as a Tool for Biomonitoring: Turning Weaknesses into Strengths

Ashna Lallmahomed[1], Fabien Mercier[2], Nathalie Bonvallot[3], Barbara Le Bot[3]
[1]Univ Rennes, EHESP, Inserm, Irset (Institut de recherche en santé, environnement et travail) – UMR_S 1085, F-35000 Rennes, France, PhD Student, Rennes, France. [2]Univ Rennes, EHESP, Inserm, Irset (Institut de recherche en santé, environnement et travail) – UMR_S 1085, F-35000 Rennes, France, Research engineer, Rennes, France. [3]Univ Rennes, EHESP, Inserm, Irset (Institut de recherche en santé, environnement et travail) – UMR_S 1085, F-35000 Rennes, France, Professor, Rennes, France

Abstract
Exposure assessment can be done directly via biological monitoring, and blood and urine are usually the preferred biological matrices to do so. However, hair is an innovative tool as it is easier to sample and store, and covers a wide and retrospective window of exposure, but its use is limited due to lack of data and the unknown limits of this matrix. This scoping review comprises literature data to identify relevant molecules to investigate organic contaminants in hair and a statistical analysis to correlate the physical-chemical properties of the selected relevant molecules, to their levels of concentration in hair. This review also explores hair properties and analytical procedures that can impact organic contaminant quantification. It groups articles published over the last ten years on European sample populations. Phthalates, pyrethroids and organophosphorus flame retardants were chosen to further study the interpretation of hair measurement for exposure assessment based on European priorities, the frequency of detection of these compounds in past studies, the diversity in the physical-chemical properties as well as the diversity in the exposure sources and pathways, and the

innovative aspect of studying these compounds in hair. Further data treatment then allowed to identify sample preparation methods (sampling, storage, washing) to question and adapt to improve the existing analytical methods.

221 – Property Estimations of Organophosphate Ester Transformation Products: Interpreting Environmental Fate with Atmospheric Chemical Space Plots

Cassandra Johannessen[1], Eric Wootton[2], J. Mark Parnis[3]
[1]Concordia University, Ms., Montreal, Canada. [2]McGill University, Mr., Montreal, Canada. [3]Trent University, Prof., Peterborough, Canada

Abstract
TCPP (Tris (2-chloroisopropyl) phosphate) and EHDP (2-ethylhexyl diphenyl phosphate) are two organophosphate ester flame retardants (OPFRs) found in relatively high abundance in air. Numerous transformation products (TPs), produced by heterogeneous atmospheric oxidation, have been identified for both TCPP and EHDP. However, there is limited understanding of the partitioning tendencies of these compounds. The primary objectives of this study were to estimate key physico-chemical partitioning properties of these two model OPFRs and gain a preliminary understanding of how the environmental fate of both chlorinated and non-chlorinated airborne OPFRs are altered upon heterogeneous transformation. Since experimentally determined partition ratios are not currently available for these chemicals, we have estimated their key physico-chemical properties using BIOVIA COSMOtherm. Using Chemical Space Plots, it is shown that heterogeneous oxidation has a significant impact on the partitioning behaviour of the various oxidation products. There is a general decrease in log KAW and increase in log KOA with increasing extent of oxidation, reflecting a lowering in vapour pressure and an increased tendency toward accumulation in condensed media. For model compartment volumes corresponding to an atmosphere containing secondary organic aerosols at the assumed liquid water content and organic aerosol levels of 10 µg m-3, this decrease results in shifts in EHDP transformation product equilibrium partitioning from the gas phase to the water-insoluble organic matter phase (WIOM), whereas those of TCPP shift from the gas phase to a roughly equal distribution between the water and WIOM phases.

222 – Effects of residential ventilation and air filtration interventions on asthma control associated with indoor air pollution among 51 adults with asthma in Chicago, IL

Insung Kang[1], Brent Stephens[2]
[1]Illinois Institute of Technology, Postdoctoral Research Associate, Chicago, USA. [2]Illinois Institute of Technology, Professor, Chicago, USA

Abstract
Here we present the Breathe Easy Study, a pseudo-randomized, parallel-group, intervention study design, conducted from July 2017 through March 2020 in Chicago, IL. We investigated the impacts of residential ventilation and filtration interventions on indoor air quality (IAQ) and asthma control, and the associations between IAQ and asthma control among 51 asthmatic adults. Three types of mechanical ventilation systems (e.g., continuous exhaust-only, intermittent supply-only, and continuous balanced systems with energy recovery ventilators) and, in eligible homes, improved central air filters, were installed midway through the study. A selection of indoor and outdoor pollutants and environmental conditions were measured quarterly in each home, and asthma control was assessed using Asthma Control Test (ACT) questionnaire monthly. Overall, home ventilation and filtration interventions reduced average normalized indoor concentrations for HCHO (-75%), CO2 (-38%), NO2 (-15%), PM1 (-47%), PM2.5 (-42%), and PM10 (-46%), and improved asthma control, with an approximately 2.7% (1.6-3.8%) increase in ACT scores. We found longitudinal associations between improved asthma control and a reduction in indoor concentrations of NO2 and particulate matter. Both the interventional effects and longitudinal associations varied by ventilation and filtration system type. Our results provide new empirical evidence that residential ventilation and filtration retrofits may modestly improve asthma control by improving IAQ, especially NO2 and particulate matter.

223 – treecompareR: an Open-source Software Package to Visualize Hierarchical Chemical Classifications

Paul Kruse[1], Caroline Ring[2]
[1]United States Environmental Protection Agency, ORISE Postdoctoral Fellow, Durham, USA. [2]United States Environmental Protection Agency, Computational Exposure Scientist, Durham, USA

Abstract
The open-source R package "treecompareR" offers a framework to visualize and explore chemical data with ClassyFire chemical

classifications. ClassyFire is a web tool that uses chemical structure to classify chemicals automatically in a hierarchical "tree of life" ontology (ChemOnt). The ClassyFire chemical classification framework is useful for evaluating the landscape of chemical data relevant to exposure, toxicity, and risk, as part of a cheminformatics approach in new approach methodologies for rapid chemical risk characterization.

The treecompareR package allows the user to:

- Visualize the ChemOnt ontology as a tree diagram
- "Prune" the tree: select and visualize branches of interest in more detail
- Assess and visualize similarity of ClassyFire classifications between two lists of chemicals
    1. Highlight tree branches according to list presence
    2. Identify portions of data sets that are highly similar or dissimilar to each other by employing hierarchical clustering in conjunction with heatmaps and a variety of similarity measures
- Visualize chemical data (e.g., median chemical concentration and presence of chemical in each survey across three National Sewage Sludge Surveys) on tree diagrams to explore relationships with chemical class

These capabilities will be demonstrated with visual examples from a case study analyzing biosolids and surface water monitoring data by chemical class.

We developed treecompareR, an open-source package for the R statistical-computing environment, built on the popular ggtree and ggplot2 R packages. While it is still under development, an initial release is planned for late 2023.

This abstract does not reflect U.S. EPA policy.

224 – Leveraging Crowd-Sourced Environmental Data to Inform Environmental Exposure Disparity

Tianjun Lu[1], Dulce Garcia[2], Armando Garcia[2]
[1]California State University, Dominguez Hills, Assistant Professor, Carson, USA. [2]California State University, Dominguez Hills, N/A, Carson, USA

Abstract
The COVID-19 pandemic has brought attention to the overall impact of air pollution exposure; however, its effects on the exposure disparity remain understudied. This study utilized crowd-sourced open data from the PurpleAir website, a widely used low-cost sensor network, and data from CalEnviroScreen 4.0, a tool for identifying disproportionate pollution burdens in California, US, to investigate air pollution exposure in Los Angeles County. The study compared Particulate Matter 2.5 (PM2.5) concentrations before and during the pandemic (March 2019-March 2020 vs. March 2020 vs. March 2021) across eight regions, with a focus on disadvantaged communities that often have high proportions of low-income and people of color residents.
The results revealed a notable decline in air pollution exposure during the study period; however, some disadvantaged communities experienced higher-than-average concentrations when lockdown measures were lifted, with a higher percentage of days exceeding the 35 µg/m3 threshold compared to non-disadvantaged communities. For example, in the San Fernando Valley region in 2019, the gap could be as high as 45%. This study confirmed that lower air pollution concentrations were observed in non-disadvantaged communities as compared to disadvantaged communities, indicating persistent disparities.

This study demonstrates the feasibility of leveraging crowd-sourced data to identify air pollution exposure disparities. The findings emphasize the need for targeted interventions and resilient urban planning to address the disproportionate burden of air pollution on vulnerable communities, especially during and after crises. Further research is warranted to refine and expand this approach to advance environmental justice efforts in diverse contexts.

225 – PFAS Exposure via Drinking Water, Biomonitoring and Report back for a Paper Mill Impacted Community

Rachel Bauer[1], Ankita Bhattacharya[1], Ying Guo[1], Heather Stapleton[2], Christopher Higgins[3], Sarah Choyke[4], John Adgate[5], Courtney Carignan[6]
[1]Michigan State University, Graduate Research Assistant, East Lansing, USA. [2]Duke University, Professor, Durham, USA. [3]Colorado School of Mines, Professor, Golden, USA. [4]Eurofins Environment Testing, Technical Director, Denver, USA. [5]University of Colorado School of Public Health, Professor, Aurora, USA. [6]Michigan State University, Assistant Professor, East Lansing, USA

Abstract
Per- and polyfluoroalkyl substances (PFAS) are a group of highly persistent man-made chemicals known for their water and stain repellant properties. In 2018, a state-wide public drinking water monitoring program identified elevated concentrations in a Michigan community (>1400 ppt PFOA+PFOS) that traced back to a landfill which had historically received paper-manufacturing waste. From June 2020 to November 2021, we conducted a biomonitoring study to characterize exposure to PFAS mixtures via drinking water, diet and the indoor environment. To participate, community members must have lived in our study area and consumed the water between 2005 - 2018. Samples and data collected during home visits included blood, drinking water, home grown produce and soil, indoor air and dust, and wristbands. Samples were shipped and analyzed for PFAS by analytical laboratories. We detected up to 14 and 23 different PFASs in water and serum, respectively, with median serum concentrations of PFOS, PFOA and PFHxS elevated compared to the general population as reported by NHANES.

Individual and overall study results were reported back to participants by mail. After reviewing their results, they were asked to complete and return a feedback survey on the clarity, structure and content of the results shared in the report. We reviewed and plan to incorporate the results from the feedback survey to improve future communication with communities. During this presentation we will share our report back of overall water and blood results, results of the feedback survey, and associations of PFAS serum concentrations with sociodemographic and water source variables.

226 – Effectiveness of portable air cleaners on indoor air quality and chronic obstructive pulmonary disease in high-risk U.S. military veterans in Chicago, IL: Study design, cohort characteristics, and preliminary analyses

Insung Kang[1], Brent Stephens[2], Mohammad Heidarinejad[3], Israel Rubinstein[4], Zane Elfessi[5], Kaveeta Jagota[6], Nancy Karpen[7]
[1]Illinois Institute of Technology, Postdoctoral Research Associate, Chicago, USA. [2]Illinois Institute of Technology, Professor, Chicago, USA. [3]Illinois Institute of Technology, Assistant Professor, Chicago, USA. [4]Jesse Brown Veterans Affairs Medical Center, Associate Chief of Staff, Chicago, USA. [5]Jesse Brown Veterans Affairs Medical Center, Emergency Medicine Pharmacist, Chicago, USA. [6]Jesse Brown Veterans Affairs Medical Center, Senior Research Coordinator, Chicago, USA. [7]Jesse Brown Veterans Affairs Medical Center, Registered Nurse, Chicago, USA

Abstract
This work describes a randomized, single-blind, placebo-controlled clinical trial designed to evaluate the effectiveness of portable air cleaners (PACs) in reducing indoor concentrations of particulate matter (PM) and improving chronic obstructive pulmonary disease (COPD) outcomes in a high-risk urban population of U.S. military veterans in Chicago, IL. The targeted recruitment is 80 participants with COPD, recruited through the Jesse Brown Veterans Affairs Medical Center. Half the participants will receive a blinded (placebo) PAC and half will receive a functional PAC, with a study duration of 1 year in each home. We assess baseline and end-line COPD-related health outcomes using respiratory questionnaires and clinical records, including COPD exacerbations, emergency room visits, 6-minute walk distance, oxygen saturation, etc. We deploy low-cost PM sensors to monitor the long-term effects of the interventions throughout the study. We also conduct housing condition assessments in each home to identify housing-related factors and environmental conditions that may contribute to COPD exacerbation. This work will describe the study design, cohort characteristics, and a preliminary result of cross-sectional associations between COPD outcomes in all participants with indoor PM concentrations and/or housing-related factors.

227 – POSTER: Assessing the quality of existing databases (secondary data) for surveillance of PRI in Illinois

Lupita Quintana
University of Illinois at Chicago, PhD Candidate, Chicago, USA

Abstract
Illinois lacks a public health surveillance program to track pesticide related illness (PRI) cases, limiting the ability to protect its residents and nearly 55,000 agricultural workers. A relatively inexpensive approach is to conduct data linkage and secondary analysis of existing health data that contain PRI cases. This allows enumeration of cases and targeting interventions, and serves as a model for inexpensive, state-based surveillance that can be replicated in other states. We used data from the Illinois Poison Center (IPC) and Illinois Hospital Discharge (HD) and Emergency Department (ED) databases between 2016-2021 using ICD-10 code T60.0 Toxic Effect of Pesticides to extract cases. A descriptive analysis was conducted. To determine the total number, we matched cases across these databases using a probabilistic record linkage method. We identified a total of 12,614 unique cases of PRI in Illinois, of which 970 linked across all databases. Most cases were male and affected children 17 and younger. Rodenticide poisonings were the most common in unintentional PRI/exposure of children under 5 and suicide attempts. Seasonality of pesticide exposure was observed, peaking in the Spring. There is a need to protect children from pesticides, especially at home. A detailed analysis of exposure-related factors that lead to PRI such as low supervision, education, and proper storage is needed to target interventions. The literature shows that while licensed

handlers are usually safe, farmworkers continue to be exposed to pesticides and hospitalized for work-related PRI. There is a need to improve the capture of occupational PRI data.

228 – An environmental study to investigate exposures to neonicotinoid insecticides using human biomonitoring strategies as part of the EIRE project

Alison Connolly[1], Darragh Doherty[2], Sonja A. Wrobel[3], Holger M. Koch[4], Daniel Bury[4], Marike Kolossa-Gehring[5], Heiko-Udo Käfferlein[4]
[1]University College Dublin, Assistant Professor, Dublin, Ireland. [2]University College Dublin, PhD, Dublin, Ireland. [3]Institute for Prevention and Occupational Medicine (IPA), PhD, Bochum, Germany. [4]Institute for Prevention and Occupational Medicine (IPA), Scientist, Bochum, Germany. [5]German Environment Agency (Umweltbundesamt), Scientist, Berlin, Germany

Abstract
Neonicotinoid insecticides and neonicotinoid-like compounds (NNIs) are the most extensively used class of insecticides worldwide, registered in 120 countries. In Europe, many NNIs have been banned or restricted, mostly due to their toxic effect on pollinators; however, there are increasing concerns about their potential adverse human health effects. Though this is a growing area of research, there is still a very limited number of human biomonitoring (HBM) studies investigating NNIs; thus, there is a dearth of information on the range and magnitude of human exposures to NNIs.

In Ireland, the EIRE 'nEonicotinoid Insecticide exposuREs' project aims to characterise acetamiprid exposures among workers and investigate NNI exposures among the general population. The study will use an HBM strategy to determine internal body burden of NNIs and their metabolites. The study will investigate population exposure by analysing samples collected in a previous project (i.e., IMAGE project). This previous study recruited 68 families from Ireland, including fathers, mothers and children and collected a morning void urine sample from each participant (n=224). The study aligned its protocols with the Human Biomonitoring for Europe (HBM4EU) initiative. These samples will be analysed using a developed and validated HBM method that quantifies seven NNIs (i.e. acetamiprid, clothianidin, flupyradifurone, imidacloprid, sulfoxaflor, thiamethoxam, thiacloprid), including nine of their specific metabolites derived from N-dealkylation and hydroxylation. Contextual information on diet, occupation, lifestyle and health will be analysed to investigate exposure sources.

Preliminary results will be presented at the conference, including the HBM results and future plans for the EIRE project.

229 – Improving Methodologies for Cumulative Impact Assessment: A Case Study of Health Risks in Communities on the Fenceline of Petrochemical Refineries in Southeast Pennsylvania

Andrea A. Chiger[1], Carolyn Gigot[1], Ellis S. Robinson[2], Mina Tehrani[2], Megan Claflin[3], Edward Fortner[4], Manjula R. Canagaratna[3], Scott Herndon[3], Tara I. Yacovitch[3], Mary Fox[5], Kirsten Koehler[6], Ana Rule[5], Thomas A. Burke[7], Peter F. DeCarlo[6], Keeve E. Nachman[6]
[1]Johns Hopkins University, PhD Candidate, Baltimore, MD, USA. [2]Johns Hopkins University, Postdoctoral Fellow, Baltimore, MD, USA. [3]Aerodyne Research, Principal Scientist, Billerica, MA, USA. [4]Aerodyne Research, Senior Scientist, Billerica, MA, USA. [5]Johns Hopkins University, Assistant Professor, Baltimore, MD, USA. [6]Johns Hopkins University, Associate Professor, Baltimore, MD, USA. [7]Johns Hopkins University, Professor Emeritus, Baltimore, MD, USA

Abstract
Cumulative impact assessment is key to characterizing health risks in fenceline communities, which face environmental pollution and challenging socioeconomic conditions. Current EPA methods consider mixtures, but only assess the most sensitive health endpoint for each chemical and do not quantify non-chemical stressors. We developed an approach that considers all known health endpoints associated with a chemical, which we demonstrate using data from our Hazardous Air Pollutant Monitoring and Assessment Project (HAP-MAP). Our transdisciplinary team used state-of-the-art mobile measurement techniques to estimate fenceline exposures to volatile organic compounds, providing higher spatial and temporal resolution than regulatory monitoring datasets such as EPA's Air Toxics Screening Assessment (AirToxScreen). We compiled hazard information on all non-cancer health endpoint categories associated with each measured chemical using ATSDR Toxicological Profiles and EPA's CompTox Chemicals Dashboard. Next, we employed a hazard quotient/index methodology to evaluate risks. We conducted literature reviews to investigate the effects of co-exposures to psychosocial stressors. Our approach found neurological, reproductive, and respiratory risks (hazard index >=1) in HAP-MAP fenceline communities, whereas no risks were identified using AirToxScreen or an assessment limited to sensitive endpoints only. Although we found evidence that co-exposures to psychosocial stress increase susceptibility to chemicals, we were unable to account for these impacts quantitatively. Instead, we propose lowering the allowable hazard index to 0.1 in vulnerable communities. Our results suggest that current approaches underestimate health risks in fenceline communities and highlight the need for transdisciplinary collaborations

and innovative approaches to improve methods for cumulative impact assessment.

230 – Comparing Human Biomonitoring Health-Based Guidance Values (HB2GVs): A Case Study of Di-(2-ethylhexyl) Phthalate (DEHP)

Tyler Pollock[1], Petra Apel[2], Antonia Calafat[3], Marike Kolossa-Gehring[4], Judy LaKind[5], Rosa Lange[6], Kristin Macey[7], Lisa Melnyk[8], Shoji Nakayama[9], Devika Poddalgoda[7], Annie St-Amand[10]
[1]Health Canada, Senior Scientist, Ottawa, Canada. [2]German Environment Agency, Scientific Officer, Berlin, Germany. [3]Centers for Disease Control and Prevention, Branch Chief, Atlanta, USA. [4]German Environment Agency, Section Head, Berlin, Germany. [5]Lakind Associates LLC, President, Maryland, USA. [6]German Environment Agency, Researcher, Berlin, Germany. [7]Health Canada, Senior Scientific Evaluator, Ottawa, Canada. [8]U.S. Environmental Protection Agency, Research Chemist, Cincinnati, USA. [9]National Institute for Environmental Studies, Deputy General Manager, Tsukuba, Japan. [10]Health Canada, Section Head, Ottawa, Canada

Abstract
The International Society of Exposure Science (ISES) International Human Biomonitoring (i-HBM) Working Group recently launched a free, online repository of biomonitoring guidance values referred to as the Human Biomonitoring Health-Based Guidance Value (HB2GV) Dashboard. The goal of the dashboard is to assist global human biomonitoring data users (e.g., risk assessors, risk managers) and human biomonitoring programs with a readily available compilation of guidance values for the general population. The dashboard contains around 600 HB2GVs for over 150 chemicals or their metabolites; although there are many different types of HB2GVs, most are Biomonitoring Equivalents (BEs), Human Biomonitoring (HBM-I and HBM-II) Values, or Human Biomonitoring Guidance Values (HBM-GVs). Therefore, there is a need to inform users of the relative differences among available guidance values and to help users identify the HB2GV that could be most suitable for their purposes. Here, we summarize the derivation of HB2GVs for a case study chemical, di-(2-ethylhexyl) phthalate (DEHP). We selected DEHP because there are over 30 unique HB2GVs available from the three most common guidance value types (BE, HBM-I, HBM-GV). We also compare the available HB2GVs with a focus on the differences among their derivation methods, relative quality and confidence, and interpretation. This approach can be applied to additional chemicals in the future to facilitate the use of HB2GVs and health-based interpretation of human biomonitoring data.

231 – The Use of State of the Science Meetings for Interdisciplinary Dissemination, Community Engagement and Improvement of Agriculture, Forestry and Fishing (AgFF) Related Research

Farah Arosemena-Murfee[1], Ashley McLeod-Morin[2], Tracy Irani[3], Ricky Telg[4], J. Glenn Morris, Jr.[5]
[1]University of Florida, Emerging Pathogens Institute, Assistant Scientist, Gainesville, USA. [2]University of Florida, Institute of Food and Agricultural Sciences, Associate Director of Strategic Communications, Gainesville, USA. [3]University of Florida, Institute of Food and Agricultural Sciences, Professor and Chair, Gainesville, USA. [4]University of Florida, Institute of Food and Agricultural Sciences, Professor, Gainesville, USA. [5]University of Florida, Emerging Pathogens Institute, Professor and Director, Gainesville, USA

Abstract
In 2018 the CDC/NIOSH Southeastern Coastal Center for Agricultural Health and Safety (SCCAHS) launched the annual State of the Science (SOS) meetings to better understand and enhance the uptake of evidence-based practices that improve the health and safety of agricultural, forestry, and fishing (AgFF) populations and spotlight unequal exposure and other emerging issues. The design of the SOS Meetings arose from the Center's Community Stakeholder Advisory Board report on the need for stronger dissemination of findings from across the Southeast U.S. and U.S. Caribbean regions. The SCCAHS has convened a total of five meetings 2018 – 2023, with the annual theme driven by community feedback on the importance to address topics on heat-related illness, mental health, global pandemics, health equity and climate change. The meetings have served to create a forum that addresses the effectiveness of practices and policies on AgFF communities, to encourage collaboration between communities and regional researchers, and to disseminate research findings more broadly to community stakeholders. The proposed presentation will provide insights on how to operationalize, conduct and implement a novel meeting approach to address occupational health issues facing the AgFF workforce and move the field forward to impact the associated unique burden and need identified by community stakeholders.

232 – Florida Red Tides and Preterm Birth: A Staggered Difference-in-Differences Analysis

Rain E. Freeman[1], Quynh Anh Nguyen[2,3], Jason L. Salemi[4], Catherine M. Bulka[5]
[1]College of Public Health, University of South Florida, Graduate Research Assistant, Tampa, FL, USA. [2]Department of Mathematics and

Statistics, College of Arts and Sciences, University of South Florida, Undergraduate, Tampa, FL, USA. [3]College of Public Health, University of South Florida, Undergraduate, Tampa, FL, USA. [4]College of Public Health, University of South Florida, Associate Professor, Tampa, FL, USA. [5]College of Public Health, University of South Florida, Assistant Professor, Tampa, FL, USA

Abstract

Background: Although Karenia brevis (K. brevis) 'red tide' events have been linked to respiratory and neurologic symptoms, there is limited research on pregnancy outcomes. This ecologic study leveraged the 2018 red tide event as a natural experiment to investigate cross-sectional effects on preterm birth along Florida's Gulf coast.

Methods: We obtained monthly K. brevis concentration estimates from a published report on historical blooms for sample locations within 5-km of the shoreline for two periods: June-September 2017 (when no red tide events occurred) and June-September 2018 (when red tides occurred). K. brevis cell concentrations were spatiotemporally assigned to coastal census tracts and linked with Florida Department of Health residential birth data. We conducted a staggered difference-in-difference analysis to compare trends in monthly preterm birth (<37 weeks) rates across 'exposed' (≥10,000 cells/Liter) and 'unexposed' (<10,000 cells/Liter) census tracts over time.

Results: The study population included 274 census tracts with 2,652 live births (10% preterm). A Wald pre-test of the parallel trends assumption found no significant difference in preterm birth rates between exposed and unexposed tracts before 2018 (p=0.83). During June-August 2018, red tides were associated with 3.4 fewer preterm births per 100 live births, but this difference was not statistically significant (95%CI: -8.0,1.3).

Conclusions: Our staggered difference-in-difference analysis found no association between red tides and preterm birth, however, we lacked spatiotemporally-refined measures of exposure at the individual-level, which may explain the null results. Future research on health effects of red tides should focus on improving methods for exposure assessment.

234 – Predictors of early life residential mobility in urban and rural Pennsylvania children with acute lymphoblastic leukemia and implications for environmental exposure assessment

Cassandra Clark[1], Joshua Warren[2], James Saiers[3], Xiaomei Ma[4], Michelle Bell[3], Nicole Deziel[2]
[1]Yale School of Public Health, Post-Doctoral Fellow, New Haven, USA. [2]Yale School of Public Health, Associate Professor, New Haven, USA. [3]Yale School of the Environment, Professor, New Haven, USA. [4]Yale School of Public Health, Professor, New Haven, USA

Abstract

Residential mobility can introduce exposure misclassification in pediatric epidemiology studies using birth address only. We examined whether residential mobility varies by sociodemographic factors and urbanicity/rurality among 400 children born in Pennsylvania between 2002-2015 and diagnosed with leukemia at 2-7 years. Mobility between birth and diagnosis addresses (obtained from state registries) were calculated as: whether a child moved, distance moved, and whether a mover changed census tract. We evaluated predictors of these measures in urban-born (n=357) and rural-born (n=43) children using chi-square and linear/logistic regression. Rural-born children were more likely to move than urban-born children (63% vs. 57%), and rural-born children moved farther on average (18.7 km) than urban-born children (14.8 km). In urban-born children, but not rural-born children, moving between birth and diagnosis was associated with Black maternal race, lower maternal education, participation in the Nutrition Program for Women, Infants and Children, and lower income (all c2 p<0.01). These associations may reflect real differences or sample size issues. For movers, no associations were observed between any studied factor and changing census tract or distance moved. Exposure misclassification from residential mobility may be more likely in rural-born children, who were more likely to move and move farther. However, exposures assigned at the census tract level may not differ because the frequency of movers changing census tract was similar across urban-born and rural-born. In summary, our results suggest that the impact of mobility on exposure misclassification may vary by urbanicity, sociodemographic status, and the spatial scale of the exposure assessment.

235 – Walking time to urban blue and green space, social inequalities, and health-related quality of life in children and adolescents

Gina Friedriszik[1,2], André Conrad[3], Christiane Bunge[4], Aline Murawski[5], Marike Kolossa-Gehring[3], Karl Eckert[6], Alice Schröder[7], Miriam Dross[7], Karla Romero Starke[8]
[1]German Environment Agency (UBA), Student, Berlin, Germany. [2]Chemnitz University of Technology (TU Chemnitz), Student, Chemnitz, Germany. [3]German Environment Agency (UBA), Head of Section, Berlin, Germany. [4]German Environment Agency (UBA), Scientist, Berlin, Germany. [5]German Environment Agency (UBA), Head of Subject Group, Berlin, Germany. [6]German Environment Agency (UBA), Scientist,

Dessau-Roßlau, Germany. [7]German Environment Agency (UBA), Head of Section, Dessau-Roßlau, Germany. [8]Technical University Dresden (TU Dresden), Section Leader, Dresden, Germany

Abstract
Blue and green space exposure is central for urban environmental health. This study of the "Advancing the New European Bauhaus: Sustainable Mobility and Resilient Urban Spaces for a Better Quality of Life" (AdNEB) project investigates associations between socio-economic position (SEP) and proximity to blue and green space. Associations between walking times and health-related quality of life (HRQL) in children are evaluated.

Population-representative data of the German Environmental Survey for Children and Adolescents 2014-2017 (GerES V) and the German Health Interview and Examination Survey for Children and Adolescents (KiGGS Wave 2) was analyzed: self-reported walking times to the nearest blue and green space, socio-demographic characteristics, family situation, SEP, and migration background. HRQL is assessed using the KIDSCREEN-27 instrument.

Preliminary results show that about half of the children and adolescents aged 3-17 years living in German urban areas (20,000+ inhabitants) can reach the nearest green space within 5 minutes of walking, while approximately a quarter live the same walking time away from the next blue space. Both fractions are larger for high SEP participants. Multivariate analyses investigate associations between walking times to urban green and blue space, socio-demographic characteristics, and HRQL.

Our study adds to the knowledge of nature-based solutions for healthy urban environments. It further demonstrates the role of interdisciplinary cooperation between exposure and social science for driving sustainable development of communities by addressing environmental health.

We thank all study participants and the RKI for cooperation and data sharing. Funding of AdNEB by the German Environment Ministry (BMUV) is gratefully acknowledged.

237 – Demographic Characteristics and Behavior Patterns Predict Young Infants' Exposure to Parabens

Elizabeth Boxer[1], Yilin Zhong[2], Jessica Levasseur[2], Sharon Zhan[3], Brian Eichner[4], Heather Stapleton[5], Kate Hoffman[6]
[1]Duke University, PhD Student, Durham, USA. [2]Duke University, Durham, USA. [3]Duke University, Research Project Manager, Durham, USA. [4]Duke School of Medicine, M.D. / Associate Professor, Durham, USA. [5]Duke University, Professor, Durham, USA. [6]Duke University, Assistant Professor, Durham, USA

Abstract
Parabens are widely used as antimicrobials in various products, including personal care products and pharmaceuticals. While previous studies demonstrate paraben exposure is ubiquitous, data investigating infants' exposure is limited. We sought to characterize infants' exposure to parabens and identify factors associated with higher exposure levels. We also evaluated whether parabens are associated with markers of immune function. Families enrolled in the Children's Immune Response Study between 2016-2018. Parents completed questionnaires, providing information on demographic characteristics and behaviors, and urine samples were collected when infants were 1 to 3 months old (n=72) and 12 months old (n=29), with 18 infants evaluated at both time points. Blood samples were collected at 12 months. Methylparaben, ethylparaben, and propylparaben were detected in >70% of urine samples, and concentrations ranged several orders of magnitude (specific-gravity-corrected medians: MPB=25.44; PPB=3.55; EPB=0.90 ng/mL). Butylparaben was detected less frequently (<50%). Urinary paraben concentrations were lower than those reported for older children and adults; however, we did not find statistically significant differences in paraben concentrations by age. Correlations between measurements taken at both time points were poor, suggesting multiple measurements are needed to capture cumulative exposure information. We observed differences in exposure by race, ethnicity and socioeconomic status; non-Hispanic Black infants and infants whose parents completed less education had higher paraben exposure. Recent lotion usage strongly predicted paraben exposure in 1-3 month-olds, suggesting behavioral changes may be a feasible exposure reduction strategy. Preliminary analyses do not support associations between parabens and antibody response to tetanus and diphtheria vaccines.

238 – Creating a consistent dataset of daily fine spatial scale PM2.5 concentration estimates for the western US 2008-2018 for understanding population exposures to wildfire smoke

Colleen Reid[1], Ellen Considine[2]
[1]University of Colorado Bouler, Assistant Professor, Boulder, USA. [2]Havard T. H. Chan School of Public Health, Doctoral Student, Boston, USA

Abstract
In most of the U.S., fine particulate matter (PM2.5) air pollution is improving, however, in much of the western US, increases in PM2.5 due to emissions from more frequent and intense wildfires have offset these gains in air quality. Although there is significant evidence of health impacts of PM2.5 exposures in general, many questions remain about health impacts of wildfire PM2.5. Addressing these questions is hampered by lack of consistent estimates of PM2.5 exposures across space and time that incorporate wildfire smoke. To improve population exposure assessment of PM2.5 across space and time, researchers are increasingly using statistical methods to "blend" information from multiple data sources to better estimate PM2.5. Some studies have created daily fine-resolution estimates of PM2.5 for the whole US, but they tend to perform poorly in the western US. In this talk, we will present a dataset we have created using machine learning that combines various data sources (satellite, meteorological, monitoring, land use and more) to estimate daily PM2.5 estimates at the census tract, ZIP code, and county levels during 2008-2018. Our methods improve upon previous models by use of a more extensive monitoring station network, which captures more spatial locations and proximity to wildfires. We have also validated our estimates using independent measurements from mobile smoke monitors deployed by the US Forest Service. Relative to another widely-used and high-coverage machine learning-derived PM2.5 dataset, our estimates better capture the high PM2.5 concentrations in places affected by wildfire smoke.

239 – POSTER: Prenatal Exposure to Perfluoroalkyl and Polyfluoroalkyl Substances and Neurobehavioral Development in Early Childhood

Jeong Weon Choi[1], Jiwon Oh[2], Deborah Bennett[3], Irva Hertz-Picciotto[3,4], Rebecca Schmidt[5], Hyeong-Moo Shin[6]
[1]Department of Environmental Science, Baylor University, Postdoc, Waco, USA. [2]Department of Public Health Sciences, University of California, Postdoc, Davis, USA. [3]Department of Public Health Sciences, University of California, Professor, Davis, USA. [4]UC Davis MIND (Medical Investigations of Neurodevelopmental Disorders) Institute, Professor, Davis, USA. [5]Department of Public Health Sciences, University of California, Associate Professor, Davis, USA. [6]Department of Environmental Science, Baylor University, Associate Professor, Waco, USA

Abstract
Background: Prenatal exposure to per- and polyfluoroalkyl substances (PFAS) may adversely affect child neurobehaviors; however, findings from epidemiologic studies remains inconsistent.

Methods: Participants were 176 mother–child pairs from MARBLES (Markers of Autism Risk in Babies – Learning Early Signs), a high-risk autism spectrum disorder (ASD) cohort. We quantified nine PFAS in prenatal maternal serum. Child's neurobehavioral development at 2-4 years was assessed by the Child Behavioral Check List (CBCL) for ages 1.5-5. We used multiple linear regression models to estimate changes in CBCL T-scores per log2-unit increase in prenatal PFAS concentrations. We adjusted for covariates in our model, including child diagnosis for ASD.

Results: Overall, associations of prenatal PFAS serum concentrations with CBCL T-scores were null. When stratified by child's sex, perfluorononanoic acid (PFNA) was associated with higher internalizing (ß=3.90, 95% CI: 0.63-7.16) and externalizing scores (ß=4.45, 95% CI: 1.10-7.79) in boys and perfluorohexane sulfonate was associated with higher externalizing scores (ß=3.20, 95% CI: 0.02-6.39) in girls.

Conclusions: Our study showed that prenatal exposure to some PFAS had negative effects on child neurobehavioral problems when analysis was stratified by child's sex. However, our results should be interpreted with caution due to small sample size and because children with ASD and other neurodevelopmental concerns have shown more behavioral problems in our cohort.

240 - POSTER: A fossil fuel exposure assessment in a North American preconception cohort study

Mary Willis[1], Erin Campbell[2], Jonathan Buonocore[1], Erin Polka[3], Amelia Wesselink[1], Lauren Wise[4]
[1]Boston University, Assistant Professor, Boston, USA. [2]Boston University, Research Assistant, Boston, USA. [3]Boston University, Research Data Analyst, Boston, USA. [4]Boston University, Professor, Boston, USA

Abstract
Background: In the United States (U.S.), fossil fuel emissions from petroleum, natural gas, and coal made up ~80% of primary energy consumption in 2021. Pollutants from fossil fuel sector emissions have been associated with adverse pregnancy outcomes in previous studies. Little is known about the distributions and health effects of exposures to co-occurring fossil fuel sector activity.

Methods: We used U.S. data from Pregnancy Study Online, an internet-based preconception cohort study of couples attempting pregnancy. Using participants' residential address at baseline, we integrated public data sources on fossil fuel sector activity to assess exposure to key potential hazards: oil and gas development, coal mines, fossil fuel power plants, underground storage tanks, highways/ expressways, and railroads. We calculated participants' exposures by measuring proximity to each source and categorizing into health-relevant distances from the source based on existing literature. We described co-occurring exposures and disparities by sociodemographic characteristics.

Results: Among 14,295 participants, 91.3% were exposed to ≥1 fossil fuel sector activity at a health-relevant distance, and 59.2% were exposed to ≥2. The most common co-occurring exposures were from underground storage tanks and fossil fuel power plants (n=3,662 participants). For instance, when we examined sociodemographic characteristics, lower income participants (<$50,000/year) were more likely to be exposed to fossil fuel sources, except highways and power plants, compared to participants with higher incomes.

Conclusion: In this geographically diverse cohort, we observed that a large proportion of participants were exposed to multiple fossil fuel sector activities. Future work will examine how co-occurring exposures influence reproductive outcomes.

241 – Evaluation of Urinary Limonene Metabolites as Greenness Biomarkers

Zhengzhi Xie[1], Saurin Sutaria[2], Jin Chen[2], Hong Gao[3], Daniel Conklin[4], Rachel Keith[5], Sanjay Srivastava[4], Aruni Bhatnagar[4], Pawel Lorkiewicz[6]
[1]University of Louisville, Research Scientist, Louisville, USA. [2]University of Louisville, Postdoctoral Associate, Louisville, USA. [3]University of Louisville, Research Associate, Louisville, USA. [4]University of Louisville, Professor, Louisville, USA. [5]University of Louisville, Associate Professor, Louisville, USA. [6]University of Louisville, Assistant Professor, Louisville, USA

Abstract
Exposure to plants (greenness) has been linked with improved physical and mental health. Existing studies, however, are observational and lack individual-level data on exposure to greenness and outcome. Quantitative measures of greenness exposure (i.e., biomarkers of greenness) are needed in order to establish cause-and-effect relationships between greenspaces and human health. Limonene is one of the most abundant biogenic volatile organic compounds emitted by plants, so we hypothesized that urinary limonene and/or its metabolites may serve as urinary greenness biomarkers.

We used liquid chromatography-high resolution mass spectrometry to analyze urine samples from eight human volunteers after inhaling limonene or visiting a local park/green environment (e.g., forest bathing). The data were acquired on Waters Synapt XS UPLC-QTOF-MS and processed using Waters UNIFI software. An integrated library-guided analysis (ILGA) workflow was developed to perform data analysis and biomarker discovery and identification. Kinetic parameters were estimated using nonlinear mixed effect models for 18 detected metabolites that significantly increased in urine post-limonene inhalation. The three most abundant metabolites, namely two isomers of uroterpenol glucuronide (LMN-8,9-O-GlcA) and dihydroperillic acid glucuronide (DHPA-GlcA) were also elevated after a forest bathing episode (4hr). The identities and structures of these three metabolites were validated using stable isotope tracing and tandem mass spectral comparison. Our data suggest that urinary metabolites of limonene, especially LMN-8,9-O-GlcA and DHPA-GlcA, may be useful as biomarkers of greenness exposure at individual level. However, more research is needed to evaluate the specificity of these biomarkers and their suitability in cross-sectional studies.

242 – Characterization of organophosphate ester flame retardants and plasticizers in the LIFECODES Fetal Growth Study

Alexa Friedman[1], Paige A. Bomarrito[2], Barrett M. Welch[3,4], David E. Catonwine[5], Maria Ospina[6], Antonia M. Calafat[7], John Meeker[8], Thomas F. McElrath[9], Kelly K. Ferguson[10]
[1]National Institute of Environmental Health Sciences, IRTA Postdoctoral Fellow, Durham, USA. [2]National Institute of Environmental Health Sciences, Staff Scientist, Durham, USA. [3]University of Nevada, Reno, Assistant Professor, Reno, USA. [4]National Institute of Environmental Health Sciences, Special Volunteer, Durham, USA. [5]Brigham and Women's Hospital, Assistant Professor, Boston, USA. [6]Center for Disease Control, Research Chemist, Alanta, USA. [7]Center for Disease Control, Chief of the Organic Analytical Toxicology Branch, Alanta, USA. [8]University of Michigan, Professor, Ann Arbor, USA. [9]Brigham and Women's Hospital, Associate Professor, Boston, USA. [10]National Institute of Environmental Health Sciences, Principal Investigator, Durham, USA

Abstract
Organophosphate ester flame retardants and plasticizers (OPEs) have increased in use since the early 2000s. Studies examining temporal trends and predictors of OPE exposure are limited, especially among pregnant people. We examined temporal trends (from 2007 to 2018) and predictors of OPE biomarker concentrations.

We used repeated measured of seven OPE biomarkers from urine samples in pregnant people from the LIFECODES Fetal Growth Study (n=900). We examined OPE biomarker concentrations and their variability across pregnancy. Within-subject geometric means with 95% confidence intervals (CIs) were calculated and we assessed temporal trends in OPE biomarker concentrations and associations with sociodemographic characteristics.

Five OPEs were highly detected and there was moderate to high variability of OPE biomarkers across pregnancy. We observed heterogenous temporal trends in biomarkers across the study period. Bis(1-chloro-2-propyl) phosphate (BCIPP), increased from 0.10 ng/mL (95% CI: 0.08, 0.11) in 2007-2008 to 0.21 ng/mL (95% CI: 0.13, 0.32) in 2017 – 2018 whereas bis(1,3-dichloro-2-propyl) phosphate concentrations (BDCIPP) increased from 0.52 ng/mL (95% CI: 0.45, 0.59) in 2007 – 2008 to 0.96 ng/mL (95% CI: 0.78, 1.17) in 2013 – 2014, then declined. BDCIPP was 59% higher among non-Hispanic Black participants and 50% higher among Hispanic participants compared to non-Hispanic White participants. OPE biomarker concentrations were higher among younger individuals, participants with higher pre-pregnancy body mass index and lower educational attainment.
OPE biomarkers displayed heterogenous trends from 2007 – 2018 and concentrations varied according to sociodemographic factors. Future studies examining sources and health effects of OPE exposure are warranted.

243 – POSTER: Reconstructing Historical Exposures to Oil and Gas Development at Critical Windows of Exposure for Reproductive Outcomes in a Preconception Cohort Study

Erin Campbell[1], Amelia Wesselink[2], Nicole Deziel[3], Ann Aschengrau[4], Lauren Wise[4], Mary Willis[5]
[1]Boston University School of Public Health, Research Assistant, Boston, USA. [2]Boston University School of Public Health, Research Assistant Professor of Epidemiology, Boston, USA. [3]Yale University School of Public Health, Associate Professor of Epidemiology, New Haven, USA. [4]Boston University School of Public Health, Professor of Epidemiology, Boston, USA. [5]Boston University School of Public Health, Assistant Professor of Epidemiology, Boston, USA

Abstract
Background: A key challenge in exposure science is building exposure measurements across the lifespan, as specific windows of development may be critical for long-term health. Records of oil and gas development location are available across time, presenting a unique opportunity to reconstruct historical exposure measures.

Methods: We leveraged United States (U.S.) data from Pregnancy Study Online (PRESTO) in 2013-2022, an online preconception cohort study of pregnancy planners. Using historical city and town boundaries from the U.S. Census, we estimated the density and presence of active oil and gas development sites across three time points in participants' lives: birth, age 15, and study enrollment (i.e., preconception). We calculated the percentage of participants who were exposed at each life stage and calculated Spearman correlation coefficients among the exposure measures.

Results: Among 8,047 participants (median age=30) with residential history data, 50.8% resided in the same state at all three time points. The prevalence of oil and gas development exposure was 13.1% at birth, 11.1% at age 15, and 9.9% at study enrollment. Spearman correlations were highest between birth and age 15 (0.55) and lower between age 15 and enrollment (0.38) as well as birth and enrollment (0.29). In total, 3.2% of participants were exposed to oil and gas development at all three time points.

Conclusion: This study demonstrates a framework for reconstructing a lifetime of exposures to the oil and gas industry, facilitating the identification of critical exposure windows in epidemiologic analyses.

244 – Aluminum Cookware as a Source of Lead Exposure in the United States

Katie Fellows[1], Stephen G Whittaker[2], Shar Samy[3]
[1]Hazardous Waste Management Program, Environmental Scientist, Seattle, USA. [2]Hazardous Waste Management Program, Research Services Program Manager, Seattle, USA. [3]Environmental Health Laboratory and Trace Organics Analysis Center, University of Washington, Director, Seattle, USA

Abstract
Lead poisoning investigations in King County, Washington revealed that refugee children have a high prevalence of elevated blood lead levels. We previously described how in-home investigations of these cases found that cookware used by refugees often contained thousands of parts per million (ppm) of lead. The goal of this study was to expand our investigation to include cookware used by other communities, as well as cookware manufactured in the US.

We acquired 25 aluminum cookpots from online marketplaces, seven of which were made from the aluminum alloy Hindalium. We conducted extensive XRF analyses on the cookware and used our previously published leachate method to determine the extent to which lead migrates from the cookware under simulated cooking and storage conditions. We compared the lead levels in leachate to the FDA's recommended maximum daily intake for lead (Interim Reference Level, IRL). We conducted the same analyses on 17 cookpots manufactured from stainless steel, and one made of brass.

Many aluminum cookpots contained very high lead levels (500 to 1,000 ppm), particularly those made from Hindalium. In the leachate experiments, all Hindalium cookware exceeded IRLs, and one Hindalium cookpot exceeded the infant IRL by nearly 1,400-fold. Many aluminum cookpots that were anodized also exceeded IRLs, as did the brass cookpot.

Aluminum cookware commonly available in the US is a potential source of lead exposure. While the previous investigation focused on refugees, this study revealed that long-term residents are also at risk of lead poisoning from aluminum cookware purchased in the US.

245 – Application of Air Dispersion Modeling in Exposure Estimation of Soil Fumigants for California Communities

Jing Tao
California Office of Environmental Health Hazard Assessment, Research Scientist IV, Oakland, USA

Abstract
Soil fumigants are used to control pests living in the soil and that are harmful to plant growth and crop production. They are highly volatile and are applied as a gas or form a gas shortly after their application. They can disperse beyond the application sites and into the ambient air in nearby communities. Soil fumigants can be associated with acute and chronic adverse health effects in humans and inhalation is the most common route of exposure.

We are interested in investigating communities' exposure to soil fumigants in ambient air and their association with adverse health outcomes in populations living nearby. For this type of study, reliable exposure data is necessary. In this exploratory study, we are looking at the relationship between communities' emergency department visits for respiratory-related illnesses and acute exposure to one of the major soil fumigants, 1,3-dichloropropene (1,3-D). For the study design, we use American Meteorological Society (AMS)/EPA Regulatory Model (AERMOD) to perform air dispersion modeling and generate continuous exposure data. We perform modeling for 52 selected Zip Code Tabulation Areas (ZCTAs) in California and set the population weighted centroid of ZCTAs as modeling receptors. AERMOD modeling generate 24-hr average concentrations of 1,3-D at receptors for each day of a 5-year study period, which provides the exposure estimates needed for the study. This case study demonstrates that air dispersion modeling with appropriate data inputs can be a useful tool to estimate communities' exposure for epidemiological studies on pesticides.

248 – Get the Lead Out: A community-driven approach to characterizing urban soil lead levels near a battery smelter

Jill Johnston[1], mark! Lopez[2], Venezia Ramirez[3], Oscar Reyes Zapien[4], Francisca Castro[5], Bhavna Shamasunder[6]
[1]University of Southern California, Associate Professor, Los Angeles, USA. [2]East Yard Communities for Environmental Justice, Organizer, Commerce, USA. [3]University of Southern California, Research Coordinator, Los Angeles, USA. [4]University of Southern California, Student Fellow, Los Angeles, USA. [5]Occidental College, Coordinator, Los Angeles, USA. [6]Occidental College, Associate Professor, Los Angeles, USA

Abstract
Lead (Pb), a potent neurotoxicant, may persist indefinitely in soil, and is frequently detected in dangerous concentrations near lead-acid battery smelters. In urban Los Angeles, CA, one such smelter operated adjacent to densely populated residential neighborhoods without proper permits. As a result of community organizing, this site is now the largest soil cleanup project in the state's history. Over 4000 residential properties have received soil remediation.

The Get The Lead Out Study is an on-going community-academic research collaboration with East Yard Communities for EJ that seeks to build community resilience to systematically address Pb contamination. We aimed to characterize soil Pb in yards that received a state-funded cleanup and in neighborhoods adjacent to the cleanup. We recruited residents to complete a survey and collected surface soil samples, using a novel geospatial application. Deidentified samples were dried and analyzed with an XRF.

We found that the majority of homes that received a soil cleanup still had soil Pb concentrations exceeding the goal of <80 ppm and over 25% of homes had at least one sample > 400 ppm. Further, highly elevated soil Pb concentrations are present in neighborhoods adjacent beyond the current eligible area with 22% of samples >400 ppm. Results were sent to participants and a community forum was held.

Our findings generated data responsive to community questions and identified continued health hazards. Our community-driven research spawned renewed discussion with state agencies to revise cleanup protocols, expand blood lead testing, resample cleaned homes and improve workers protections.

249 – Maternal Metabolome as a Biological Intermediary of Periconceptional Ambient Temperature Exposure and Preterm Birth

Kaitlin R. Taibl[1], Priyanka A. Bhanushali[2], Anne L. Dunlop[3], Noah Scovronick[4], Stephanie M. Eick[4], Howard H. Chang[5], Dana Boyd Barr[3], P. Barry Ryan[3], Youran Tan[1], Jeremy A. Sarnat[5], Carmen Marsit[6], Dean P. Jones[3], Donghai Liang[4]
[1]Emory University, PhD Student, Atlanta, Georgia. [2]Emory University, MPH Student, Atlanta, Georgia. [3]Emory University, Professor, Atlanta, Georgia. [4]Emory University, Assistant Professor, Atlanta, Georgia. [5]Emory University, Associate Professor, Atlanta, Georgia. [6]Emory University, Rollins Distinguished Professor of Research, Atlanta, Georgia

Abstract
Background/Aim: Thermal stress is associated with adverse birth outcomes, but the biological effects in pregnant people are poorly understood. Here, we profiled the maternal metabolome to characterize the molecular signatures of periconceptional ambient temperature exposure and preterm birth (PTB; <37 completed gestation weeks).

Methods: Participants were enrolled in the Atlanta African American Maternal-Child Cohort between 2014-2018 (N=330), including 66 with PTB. Maternal serum samples collected during early/middle pregnancy (6-18 weeks gestation) were analyzed by untargeted high-resolution metabolomics. Mean daily maximum ambient temperature was estimated at the Atlanta Hartsfield Jackson International Airport for three exposure windows: 12 weeks preconception to conception, conception to early/middle pregnancy, and 12 weeks preconception to early/middle pregnancy. Metabolome-wide association studies (MWAS) were conducted with regression models adjusted for maternal age, education, BMI, parity, substance use, and season of conception. We identified and measured molecular signatures of the exposure-outcome relationship with the meet-in-the-middle framework.

Results: Four metabolic pathways that collectively involve glucogenic (arginine, aspartate, glycine, serine), proteinogenic (alanine, proline, threonine), and redox (cysteine, methionine) amino acids were enriched across MWAS for the exposure windows and PTB ($p<0.05$). Of the 6 metabolites confirmed with Level-1 confidence, methionine was an intermediate marker for the effect of periconceptional ambient temperature on PTB ($p<0.05$).

Conclusion: Our results provide novel mechanistic information by which higher temperatures might increase the odds of PTB, a leading cause of death for African American infants. Additional research is needed to determine the impact of climate change on maternal-child health.

250 – POSTER: Association between Organophosphate Ester Exposure and Insulin Resistance with Glycometabolic Disorders among Older Chinese Adults 60–69 Years of Age: Evidence from the China BAPE Study

enmin ding
China CDC Key Laboratory of Environment and Population Health, National Institute of Environmental Health, Chinese Center for Disease Control and Prevention, visiting staff, Beijing, China

Abstract
Background: OPEs are common endocrine-disrupting chemicals, and OPE exposure may be associated with type 2 diabetes (T2D). However, greater knowledge regarding the biomolecular intermediators underlying the impact of OPEs on T2D in humans are needed to understand biological etiology.

Objectives: We explored the associations between OPE exposure and glycometabolic markers among older Chinese adults 60–69 years of age to elucidate the underlying mechanisms using a multi-omics approach.

Methods: This was a longitudinal panel study comprising 76 healthy participants who lived in Jinan of northern China. We measured a total of 17 OPEs in the blood, 11 OPE metabolites in urine, and 4 glycometabolic markers . The blood transcriptome and serum/urine metabolome were also evaluated. An adverse outcome pathway (AOP) was established to determine the biomolecules mediating the associations.

Results: Exposure to five OPEs and OPE metabolites (trimethylolpropane phosphate, triphenyl phosphate, tri-iso-butyl phosphate, dibutyl phosphate, and diphenyl phosphate) was associated with increased levels of glycometabolic markers. The mixture effect analysis further indicated the adverse effect of OPE mixtures. Multi-omics analyses revealed that the endogenous changes in the transcriptional and metabolic levels were associated with OPE exposure. The putative AOPs model suggested that triggers of molecular initiation events with subsequent key events, including disruptions in signal transduction pathways and biological functions, may constitute the diabetogenic effects of OPEs.

Discussion: OPEs are associated with the elevated risk of T2D among older Chinese adults 60–69 years of age. Implementing OPE exposure reduction strategies may help reduce the T2D burden among these individuals.

251 – POSTER: Impact of Mobile Monitoring Design on Epidemiologic Inference – A Case Study of Ultrafine Particles and Cognitive Function

Magali Blanco[1], Adam A Szpiro[2], Lianne Sheppard[3]
[1]University of Washington, Postdoctoral Scholar, Seattle, USA. [2]University of Washington, Associate Professor, Seattle, USA. [3]University of Washington, Rohm & Haas Endowed Professor in Public Health Sciences, Seattle, USA

Abstract
Mobile monitoring is increasingly used to measure traffic-related air pollution (TRAP), including ultrafine particles (UFP). We investigated the impact of campaign design on epidemiologic inference, using UFPs and cognitive function as a case study. We leveraged an extensive campaign in greater Seattle to assess TRAP exposures for the Adult Changes in Thought (ACT) cohort, a community-based study investigating the aging brain. Monitoring was conducted during a single year on all days of the week and most hours of the day, including a total of 7,806 visits (278 sites × ~28 visits). We developed UFP exposure prediction models from all the data as well as random subsampling designs (30 campaigns each) including: fewer total visits (3,336, 1,800); visits concentrated during business or rush hours; and visits concentrated in 1-3 seasons. We assessed prediction accuracy and adjusted associations between UFP exposure and cognitive function at baseline (Cognitive Abilities Screening Instrument-Item Response Theory [CASI-IRT]) in ACT (N= 4,385). Prediction accuracy for the exposure model using all data was good (R^2=0.8), with some degradation for fewer visits, fewer seasons, and rush hours designs (median R^2~0.65-0.75) and significant degradation for business hours (R^2~0.35). Using the all-data exposure model, the adjusted mean baseline CASI-IRT score decreased by -0.012 (95% CI: -0.022, -0.003) per additional 1,000 pt/cm^3 UFP. The fewer stop and season designs produced similar health effect estimates (median difference within ~25%), whereas business and rush hour designs produced more biased, attenuated estimates (median difference~40-75%). Monitoring design is critical for accurate epidemiologic inference.

252 – Exposure Profile and Cumulative Risk Assessment of Bisphenol A and Its Substitutes in Taiwanese – Taiwan Environmental Surveys for Toxicants 2013

Po-Chin Huang[1], Yu-Jung Lin[2], Hsin-Chang Chen[3], Wan-Ting Chang[4], HSIN-I HUANG[1], Jung-Wei Chang[5]
[1]National Institute of Environmental Health Sciences, National Health Research Institutes, Associate Investigator, Miaoli County, Taiwan. [2]National Institute of Environmental Health Sciences, National Health Research Institutes, Postdoctoral, Miaoli, Taiwan. [3]Department of Chemistry, Tunghai University, Assistant Professor, Taichung City, Taiwan. [4]National Institute of Environmental Health Sciences, National Health Research Institutes, Research Assistant, Miaoli, Taiwan. [5]Institute of Environmental and Occupational Health Sciences, School of Medicine, National Yang Ming Chiao Tung University, Associate Professor, Taipei, Taiwan

Abstract
Background and Aims: With the restriction of bisphenol A (BPA) use, concerns have arisen regarding its substitutes, such as bisphenol S (BPS) and bisphenol F (BPF). This study aimed to identify the exposure characteristics and estimate the cumulative risk of bisphenol and their substitutes in the general Taiwanese population.

Methods: We collected urine samples including 366 participants (271 adults and 95 minors) from Taiwan Environmental Survey for Toxicants 2013. The bisphenol and their substitutes in urinary samples were converted into daily intake (DI) values. The health-based guidance values (TDI, RfD) were used to calculate to hazard quotients (HQs). The hazard index (HI) is the cumulative summation of HQs for each compound.

Results: BPF was determined to be the most important exposure rather than BPA in our participants. The median levels of BPA in adults (7.96 μg/L) were significantly higher than those found in minors (4.83 μg/L). Female adults had higher BPS DI values (0.69 ng /kg/day) than male adults (0.49 ng/kg/day). In contrast, males had higher DI values for BPF (1.15) and BPS (0.26) than females (0.57, 0.20, respectively) in the minor group. While the HI values in this study were lower than 1, the HI values in adults were also higher than in minors.

Conclusion: We found that exposure profiles and risk of bisphenol analogues in Taiwanese varied by age and sex. Further research is warranted on exposure sources and chronic health effects.

253 – Co-use characteristics of consumer chemical products for aggregate exposure assessment in Korean population

Miyoung Lim[1], Kiyoung Lee[2]
[1]Seoul National University, Institute of Health and Environment, Research Professor, Seoul, Korea, Republic of. [2]Seoul National University, Graduate School of Public Health, Professor, Seoul, Korea, Republic of

Abstract
The use of consumer chemical products can cause adverse health effects to human body. Aggregate exposure through various products should be considered to safety management of consumer chemical products. The co-use characteristics of consumer chemical products is useful information for refined aggregate exposure assessment. This study investigated co-use patterns and characteristics of consumer chemical products by Korean population groups considering demographic and occupational characteristics. The target consumer chemical products included a total of 18 types, consisting of eight types of cleaning products, five types of automotive care products, and five types of surface protection products. The survey results from consumer chemical product exposure factor database, which targeted a total of 10,000 respondents aged 15 years and older throughout South Korea, was used. The 64.7% of respondents reported using one or more products, with an average of 2.35 ± 1.62 products. Among the high use group that reported co-using five or more cleaning products (n = 231), 25.1% were housewives or domestic workers, while among the high use group that reported co-using three or more automotive care products (n = 158), 31.0% were office workers. The most common combination of two co-used products was household bleach and glass cleaner (14.6%), followed by household bleach and mold stain remover (11.7%), and household bleach and drain cleaner (9.5%). The most common combination of three co-used products was household bleach, glass cleaner, and mold stain remover (5.0%). These co-use characteristics will be useful as exposure scenario for aggregate exposure assessment of consumer chemical products.

254 – Understanding the interrelationship among human behaviors, building and flood characteristics, mold growth, and respiratory health after Hurricanes Ian and Ida

Parham Azimi[1], Zahra Keshavarz[2], Christian Caballero[2,3], Maryam Pakdehi[4], Zhengxiao Yan[3], Joseph Allen[5], Ebrahim Ahmadisharaf[6]
[1]Harvard University, Research Associate, Boston, USA. [2]Harvard University, Research Assitant, Boston, USA. [3]Florida State University, Student, Tallahassee, USA. [4]Florida State University, Ph.D. Student, Tallahassee, USA. [5]Harvard University, Associate Professor, Boston, USA. [6]Florida State University, Assistant Research Professor, Tallahassee, USA

Abstract
Mold growth in buildings affects human respiratory health. Flooded buildings are susceptible to mold growth. There is currently a lack of understanding of interrelationships among flood characteristics, building and HVAC system properties (e.g., ventilation rates), human behaviors (e.g., time spent in homes), and vulnerability to mold growth (e.g., asthma symptoms) in the built environment, particularly in residential buildings. We studied the aftermath of Hurricanes Ida (2021) and Ian (2022) and collected critical data required for examining human respiratory health risks associated with mold growth in 60 water-damaged residential buildings across the US. We explored a comprehensive range of factors influencing mold growth in water-damaged buildings using a combination of questionnaires, home inspections, and existing datasets were used. Our preliminary results demonstrated that maximum flood height, number of days spent at home per week, and number of steps toward remediation of dampness in buildings are the top three factors influencing respiratory health indicators. Using air cleaners, the number of mitigation measures taken after hurricanes, and the age of the roof were three

important factors in presence of mold in studied buildings. The collected data and identified relationships will be used to guide building designers and occupational health scientists to establish resilient indoor environments, provide a foundation to develop flood-induced mold growth and asthma risk models, assist public health officials and emergency managers to have a better understanding of indirect health-related impacts of floods and support the development of timely strategies for disaster management in population centers.

255 – Compound Risk of Air Pollution and Heat Days and the Influence of Wildfire by SES across California, 2018–2020: Implications for Environmental Justice in the Context of Climate Change

Shahir Masri[1], Yufang Jin[2], Jun Wu[3]
[1]University of California Irvine, Associate Specialist, Irvine, USA. [2]University of California Davis, Professor, Davis, USA. [3]University of California Irvine, Professor, Irvine, USA

Abstract
Major wildfires and heatwaves have begun to increase in frequency throughout much of the United States, exacerbating air pollution and other public health risks. With climate change and increasing global average temperatures, the frequency of major wildfires, heat days, and air pollution episodes is projected to increase, resulting in the potential for compounding risks. In this study, we processed daily particulate matter (PM) data from over 18,000 low-cost PurpleAir sensors, along with gridMET daily maximum temperature data and government-compiled wildfire perimeter data from 2018–2020 in order to examine the occurrence of compound risk (CR) days (high temperature and high PM2.5) at the census tract level in California, and to understand how such days have been impacted by the occurrence of wildfires. We also examined the extent to which CR days were correlated with socioeconomic factors at the census tract level. Results showed census tracts with a higher frequency of CR days to have statistically higher rates of poverty and unemployment, along with high proportions of child residents and households without computers. The frequency of CR days and elevated daily PM2.5 concentrations was strongly related to the occurrence of nearby wildfires, with over 20% of days with sensor-measured average PM2.5 > 35 µg/m3 showing a wildfire within a 100 km radius and over two-thirds of estimated CR days falling on such days with a nearby wildfire. Findings from this study are important to policymakers and public health officials as it relates to establishing climate resilient, healthy communities.

256 – Race and Street-Level Firework Legalization as Primary Determinants of July 4th Air Pollution across Southern California

Shahir Masri[1], Leonel Flores[2], Jose Rea[3], Jun Wu[4]
[1]University of California Irvine, Associate Specialist, Irvine, USA. [2]MPNA, Community Organizer, Santa Ana, USA. [3]MPNA, Director, Santa Ana, USA. [4]University of California Irvine, Professor, Irvine, USA

Abstract
Air pollution is a major public health threat that is associated with asthma, cardiovascular disease, respiratory disease and all-cause mortality. Among the most important acute air pollution events occurring each year are celebrations involving fireworks, such as the 4th of July holiday in the United States. In this community-engaged study, academic partners and residents collaborated to collect indoor and outdoor PM2.5 concentration measurements in the disadvantaged city of Santa Ana, California, using low-cost AtmoTube sensor devices before, during and after the July 4th firework celebration, while also examining July 4th data extracted from the PurpleAir sensor network across over a hundred other cities in southern California. Average outdoor PM2.5 concentrations on July 4th were found to be three-to-five times higher than baseline, with hourly concentrations exceeding 160 µg/m3 . Outdoor averages were roughly 30% to 100% higher than indoor levels. The most polluted cities exhibited 15-times higher PM2.5 levels compared with the least contaminated cities and were often those where household-level fireworks were legal for sale and use. Race/ethnicity was found to be the leading predictor of July 4th-related air pollution across three counties in southern California, with greater PM2.5 being associated with higher proportions of Hispanic residents and lower proportions of White residents. The findings from this study underscore the importance of environmental justice as it relates to firework-related air pollution exposure, and the critical role city[1]and county-level firework policies play in determining exposure.

257 – Associations of PM2.5 constituents with blood lipid homeostasis in adults aged 40-89 years: the mediation role of thyroid hormones

Jianlong Fang, Xiaoming Shi

National Institute of Environmental Health, Chinese Center for Disease Control and Prevention, Professor, Beijing, China

Abstract
The relationship between PM2.5 constituents and blood lipids has not been fully illustrated and the potential mechanisms remain unclear. We explored the associations of high levels of PM2.5 constituents with hyperlipidemia and 10 biomarkers of lipid profiles and further investigate the underlying mechanism of thyroid hormones on the associations. A multi-center cross-sectional study of 1827 aged 40-89 years was carried out in 10 cities. Linear mixed-effect regression was conducted to quantify associations of PM2.5 and its constituents with lipid profiles and interactions of PM2.5 constituents and thyroid hormones on lipid profiles. Mediation analysis was further used to quantify the potential mediation effects of thyroid hormones. The levels of OC, EC, Ca2+, K+, Mg2+, Ga, Co, Mn, Pb, Se, Ti, Tl, V, and Zn were positively and SO42-, NH4+, Cr, Cu, Fe, and As were negatively associated with at least one lipid biomarker (except for ApoA1/ApoB). The odds ratios for hyperlipidemia associated with Cl-, Cu, Pb, and Ti were 1.146 (95% CI: 1.006, 1.035), 0.866 (95% CI: 0.764, 0.982), 1.074 (95% CI: 1.003, 1.149), and 1.090 (95% CI: 1.002, 1.185), respectively. A majority of PM2.5 constituents had joint effects of thyroid hormones on blood lipid profiles. T4 mediated 42.53% and 13.58% of NH4+-associated decreases of HDL-C and ApoB, respectively. T4 mediated 4.78% of Cu-associated decreases of TG .T3 mediated 25.83% and 11.78% of Cr-associated decreases of ALB and ApoA1, respectively. Our findings suggest that thyroid hormones play an important role in the regulation of blood lipids induced by PM2.5 constituents.

258 – Occupational Exposure Limits Recommended by The Japan Society for Occupational Health

Yuki Ito[1], Michihiro Kamijima[2]
[1]Nagoya City University Graduate School of Medical Sciences, Associate professor, Nagoya, Japan. [2]Nagoya City University Graduate School of Medical Sciences, Professor, Nagoya, Japan

Abstract
The Japan Society for Occupational Health (JSOH) recommends Occupational Exposure Limits (OELs) as reference values for preventing adverse health effects on workers caused by occupational exposure to chemical substances, continuous or intermittent noise, impulsive or impact noise, heat stress, cold stress, whole-body vibration, hand-arm vibration and time-varying electric, magnetic and electromagnetic fields, as well as ultraviolet and ionizing radiation. To date, OELs have been set for approximately 200 chemical substances. The Recommendations for OELs of JSOH are published annually in Environmental and Occupational Health Practice, one of the official journals of the JSOH.

The Recommendation for OELs committee currently consists of a chair, five special advisors, 32 regular members, and eight regular and four temporary drafters. They must declare the conflicts of interest prior to deliberation. Candidates for the committee members are selected in consultation with the chair and regular members based on their research achievements or their expertise in a specific area. The meeting is held 4–5 times a year on Saturdays. The activity is voluntary and the travel expenses to attend the face-to-face meeting are generally not supported by the JSOH. One-way fares are covered only for those whose draft proposals are discussed in that meeting.

There are several working groups within the committee such as OELs based on biological monitoring and carcinogenicity classification. Most members are assigned a task to draft documentation about one or more environmental factors including chemicals, metals, vibration, noise, etc. We will introduce how the JSOH committee derives the OELs.

259 – Pesticide Exposure Induced Epigenetic Changes in Sperm: Implications for Offspring Health

Xuan Ma[1], Bingqian Wang[1], Weiyue Hu[1], Yankai Xia[2]
[1]School of Public Health, Nanjing Medical University, /, Nanjing, China. [2]School of Public Health, Nanjing Medical University, Professor, Nanjing, China

Abstract
Recently, paternal pesticide exposure has received increasing concern due to its potential adverse effects on offspring health. However, the underlying mechanism remains elusive. Sperm acts as a critical memory carrier that records environmental stress through epigenetic changes, and thus might further affect offspring health. Among various pesticides, we focused on glufosinate-ammonium (GLA), a widely used herbicide with emerging concern over its reproductive toxicity. Adult male C57BL/6J mice were administered 0.2 mg/kg·day GLA for 5 weeks, and then copulated to obtain offspring. Neurobehavioral tests showed decreased social novelty, learning and memory in 5-week-old GLA offspring. Then, we focused on sperm to uncover memory transmission from father to offspring. After

examination on fertility, testis histology and semen quality in the GLA group, we found no significant abnormality in GLA mice. Next, we performed deep sequencing to identify DNA methylation, transcriptionally active (H3K4me3 and H3K27ac) and repressive (H3K27me3 and H3K9me3) histone modifications, and mRNA transcripts in sperm. Next generation sequencing showed alterations of these epigenetic marks and extensive transcription inhibition in sperm, which were mainly enriched in synapse organization related pathways. Consistently, histology analysis showed decreased dendrite complexity and spine density in hippocampus and prefrontal cortex of offspring, indicating abnormal synapse organization. Together, with little effect on fertility, testis histology or semen quality, GLA predominantly affected sperm epigenome and transcriptome, which might lead to the synapse organization abnormalities in hippocampus, thereby explaining decreased memory in offspring. Epidemiological studies and further risk managements on GLA are required to protect farmers and residents.

260 – Estimation of premature death by exposure to PM2.5 using advanced exposure assessment

Jihun Shin[1], Gihong Min[1], Dongjun Kim[2], Jaemin Woo[2], Mansu Cho[3], Wonho Yang[3]
[1]Daegu Catholic University, Ph.D student, Gyeongsan, Korea, Republic of. [2]Daegu Catholic University, Master student, Gyeongsan, Korea, Republic of. [3]Daegu Catholic University, professor, Gyeongsan, Korea, Republic of

Abstract
Until now, the concentrations measured at the air quality monitoring stations (AQMS) has been used for the population exposure of PM2.5 at an area. However, it does not accurately assess the exposure of the whole population and sub-populations. The population exposure at target area could be assessed using a sensor-based monitoring instrument (SAMI) with AQMS and be estimated the premature death.

The outdoor PM2.5 concentrations measured AQMS and SAMI were collected in Seoul, Korea. In addition, Geographic Information System (GIS) and machine learning models were developed, and the population exposure was estimated by dividing them into regional units and grids (1km×1km). The number of premature deaths will be calculated by the time-weighted average concentration (PWAC) and disease prevalence provided by the government.

As a result of comparing the three kinds of GIS interpolation methods and the machine learning models using the random forest and support vector machine, high accuracy at a coefficient of determination (R2) of 0.83 was shown. The root mean squared error (RMSE) was 10.1 μg/m3 and mean absolute error (MAE) was 6.1 μg/m3. Analyzing and comparing PWAC using this model, the number of premature deaths would be higher than the values using air quality.

PM2.5 was closely related to climate change, which significantly affects human exposure. Therefore, an accurate forecasting methodology will be needed. SAMI could assist AQMS in advancing the prediction of PM2.5 exposure concentrations. If SAMI operates a real-time monitoring and prevention system, it will be expected to reduce the number of premature deaths.

261 – Development of Spatiotemporal Resolution of PM2.5 concentration Using Sensor-based Air Monitoring System and Geographic Information System

Woo Jaemin[1], Shin Jihun[2], Min Gihong[2], Kim Dongjun[1], Cho Mansu[3], Yang Wonho[3]
[1]Daegu Catholic University, Master student, Gyeongsansi, Korea, Republic of. [2]Daegu Catholic University, phD Students, Gyeongsansi, Korea, Republic of. [3]Daegu Catholic University, Professor, Gyeongsansi, Korea, Republic of

Abstract
Sensor-based Air Monitoring Stations (SAMS) could provide high-resolution air quality data by offering more detailed mapping of areas that Air Quality Monitoring Stations (AQMS) could not reach. It is enhancing the precision of estimating PM2.5 concentration levels for areas that have not been directly measured, thus enabling more accurate assessment of exposure.

The study period was from September 30th to October 2nd, 2019 in Guro-gu, Seoul, Korea. Four models were applied to assess the applicability of SAMS and to visualize the temporal and spatial distribution of PM2.5 levels. Model 1 used only concentrations measured in an AQMS in Guro-gu, and Model 2 used the AQMS concentrations installed near Guro-gu along with Model 1. Model 3 used only SAMS, while Model 4 used all data, including the measured values of Models 1-4. Assuming the PM.2.5 concentrations measured at a SAMS which was located in the center of Guro-gu is true value, the PM2.5 concentrations estimated using QGIS spatial interpolation techniques were compared. Those SAMSs were used at seven points (S1 to S7) according to distance.

Model 3 and Model 4 could accurately estimate the unmeasured points with a higher coefficient of determination (R2) compared to the other Models. As the distance from AQMS increased, the R2 between the observed and estimated values decreased from 0.89 to 0.29 on S1 to S7.

The auxiliary installation of SAMS could resolve regional concentration imbalances, allowing for accurate estimation of pollutant concentrations and improved risk assessment for the population.

262 – Exposure Assessment of PAH by Combined Measurements of Indoor Air, Dust, Diet, and Urine: A Family-based Panel (CAFE-P) Study

Suhyun Park[1], Na-Youn Park[2], Jae-Eun Lim[3], Hyo-Bang Moon[4], Jeongim Park[5], Younglim Kho[6], Kyungho Choi[7,8]
[1]Department of Environmental Health Sciences, Graduate School of Public Health, Seoul National University, PhD student, Seoul, Korea, Republic of. [2]Department of Health, Environment & Safety, Eulji University, PhD student, Seongnam, Korea, Republic of. [3]Department of Marine Science and Convergent Technology, College of Science and Convergence Technology, Hanyang University, PhD student, Ansan, Korea, Republic of. [4]Department of Marine Science and Convergent Technology, College of Science and Convergence Technology, Hanyang University, Professor, Ansan, Korea, Republic of. [5]Department of Environmental Health Sciences, College of Natural Sciences, Soonchunhyang University, Professor, Asan, Korea, Republic of. [6]Department of Health, Environment & Safety, Eulji University, Professor, Seongnam, Korea, Republic of. [7]Department of Environmental Health Sciences, Graduate School of Public Health, Seoul National University, Professor, Seoul, Korea, Republic of. [8]Institute of Health and Environment, Seoul National University, Professor, Seoul, Korea, Republic of

Abstract
Polyaromatic hydrocarbons (PAHs) can enter the human body through various pathways. Traditional exposure assessments typically focus only on carcinogenic PAHs, while biomonitoring covers mostly non-carcinogenic PAHs.

We employed a family-based panel of Korean participants (CAFÉ-P), with 48 families of 124 participants. Panel was recruited from two mid-sized adjacent cities of Cheonan and Asan in Korea and sampled in August (summer) and October (autumn) 2020, as well as in January (winter) 2021. We measured 9 carcinogenic and 6 non-carcinogenic PAHs in indoor dust, air, and diet, and 20 metabolites, primarily non-carcinogenic, in urine to evaluate exposure of PAHs.

The exposure estimates showed that naphthalene had the highest exposure, followed by anthracene, fluorene, phenanthrene, and benzo(a)pyrene for both children and parents. Between parents and children, naphthalene metabolite showed a strong correlation, while fluorene metabolite showed a strong correlation between mother and child. Diet was the most important exposure source for non-carcinogenic PAHs, including naphthalene, acenaphthylene, fluorene, and phenanthrene, while indoor dust was an important exposure source for carcinogenic PAHs, such as BaA, BbF, and BghiP.

The present comprehensive assessment of PAHs indicates that approximately 10% of the participants were at potential risk due to PAHs exposure. Diet was identified as the most significant contributor, explaining about 95% of the total exposure.
This research was supported by Korea Ministry of Food and Drug Safety(20162MFDS112) and Korea Environment Industry & Technology Institute (KEITI) through Technology Development Project for Safety Management of Household Chemical Products, funded by Korea Ministry of Environment(MOE)(2022002970004).

263 – Exposure and Risk Assessment of Urinary Trans, Trans-Muconic Acid in School-aged Children in the Vicinity of a Petrochemical Complex in Central Taiwan

Po-Chin Huang[1], Vinoth Kumar Ponnusamy[2], Karthikeyan Prakasham[3], Wan-Ting Chang[4], Po-Keng Cheng[5]
[1]National Institute of Environmental Health Sciences, National Health Research Institutes, Associate Investigator, Miaoli, Taiwan. [2]Department of Medicinal and Applied Chemistry, Kaohsiung Medical University, Associate Professor, Kaohsiung City, Taiwan. [3]PhD Program in Environmental and Occupational Medicine, College of Medicine, Kaohsiung Medical University, Ph.D candicate, Kaohsiung City, Taiwan. [4]National Institute of Environmental Health Sciences, National Health Research Institutes,, Research Assistant, Miaoli, Taiwan. [5]Department of Finance and Cooperative Management, National Taipei University, Assistant Professor, New Taipei City, Taiwan

Abstract
School-aged children living near large petrochemical factories may have an increase risk of exposure to benzene. We aimed to investigate the urinary concentrations of trans, trans-muconic acid (t,t-MA) in school-aged children residing near a petrochemical complex and estimate their benzene exposure dose.

We used Taiwan Petrochemical Complex Cohort for Children (TPE3C) consisting of school-aged children (6–13 years old) living near large petrochemical factories in central Taiwan from 2013 to 2014. We recruited a total of 297 children from five elementary schools and measured urinary t,t-MA levels in all participants. We used multiple regression analysis to investigate potential risk factors of urinary t,t-MA level of children in five schools.

Median urinary t,t-MA levels were 64.1(A), 61.0(B), 59.4(C), 42.4(D), and ND (not detected)(E) ng/mL, respectively. Median estimated benzene daily intake were 11.1(A), 15.3(B), 14.8(C), 11.7(D), and 0.1(E) μg/kg/day, respectively. Our findings indicated that the distance between the school location and the petrochemical complex (farther the school: $\beta=-0.29$, 95% CI =-0.57 to -0.01, $p<0.05$), and the age of the children (older: $\beta=-3.64$, 95% CI =-5.89 to -1.39, $p<0.05$) were important risk factors. Young children attending schools in proximity to the petrochemical complex may be more sensitive to benzene exposure, and their health shall be followed. After adjustment for confounders, we observed a decrease in urinary t,t-MA levels in school-aged children as the distance between the elementary schools and the petrochemical complex increased.

264 – Identification of Pathogenic Bacteria in Public-Use Facilities using NGS Analysis

Jihyeon Kim[1], Sungchul Seo[1], Juchan Hwang[2], Bo Yeon Kwon[1], kyungmin kim[1], hyosun lee[1]
[1]Seokyeong University, Researcher, Seoul, Korea, Republic of. [2]Institute Pasteur Korea, Researcher, Seoungnam, Korea, Republic of

Abstract
The purpose of this study is to select and prioritize microorganisms for focused management in public-use facilities by analyzing the microbiome of samples collected from surfaces and air, taking into account their prevalence and harmfulness. This will provide foundational data for the development of management strategies, including exposure level reduction, prediction, and alerts.

To collect samples from the air in public-use facilities, total floating bacteria were collected using a MAS-100eco from December 2021 to February 2022. After incubation according to the indoor air quality process test standards, the colony was scooped and stored in a PBS solution. Additionally, to collect surface samples, Swab kits were used to vigorously wipe the surfaces for 1-2 minutes from September to December 2021, and the samples were stored in DNA/RNA shield.

The facilities were classified into 'Sensitive-Use Facilities', 'Transportation-Related Facilities', 'Temperature-Controlled Facilities', and 'Other Facilities' based on their characteristics. Analysis of the ratio of pathogenic and non-pathogenic bacteria revealed that 30% of the bacteria were potentially pathogenic.

'Rothia mucilaginosa', which was identified as a pathogenic bacterium in 'Sensitive-Use Facilities', has recently been recognized as an opportunistic pathogen. 'Pseudomonas putida', one of the representative bacteria that cause catheter-related opportunistic infections, was found specifically in 'Other Facilities'. While a variety of pathogenic bacteria were found to be distributed, most of them were opportunistic pathogens that do not cause infections in healthy individuals.

265 – A Random Forest Model for $PM_{2.5}$ Personal Exposure Assessment for a Chinese Cohort

Yanwen Wang[1], Renjie Chen[2], Haidong Kan[2], Tiantian Li[3]
[1]National Institute of Environmental Health, Chinese Center for Disease Control and Prevention, Associate Professor, Beijing, China. [2]Fudan University, Professor, Shanghai, China. [3]National Institute of Environmental Health, Chinese Center for Disease Control and Prevention, Professor, Beijing, China

Abstract
Background: Errors in air pollution exposure assessment are often considered a major limitation in epidemiologic studies. However, it is difficult to obtain accurate exposure on cohort populations due to the often prohibitively expensive costs. Personal exposure estimation models are used in lieu of personal exposures, but still suffer from the issues of availability and accuracy. We aim to establish a personal $PM_{2.5}$ exposure assessment model for a cohort population, and assess its performance by applying our model on cohort subjects.

Methods: We analyzed data from representative sites selected from the Sub-Clinical Outcomes of Polluted Air in China (SCOPA-China) cohort study, and established a random forest model to estimate $PM_{2.5}$ personal exposure. Parameters obtained from questionnaires and outdoor and meteorological monitoring sites. We also applied the model among subjects recruited in the above project within the same area and study period to estimate the reliability of the model.

Results: The established model showed good fit with an R^2 of 0.81. Ambient $PM_{2.5}$ contributed the most to personal exposure concentrations, and indoor passive smoking, meteorological parameters, and durations in different microenvironments also ranked high in feature importance. The model application results showed similar patterns with empirically measured data, supporting the performance of our established model.

Conclusions: Our pilot study provided a validated and feasible modeling approach to assess personal $PM_{2.5}$ exposure for large cohort populations, and the preliminary application among cohort subjects proved satisfactory. The promising model framework can improve $PM_{2.5}$ exposure assessment accuracy for future environmental health studies of large populations.

266 – Relationships between indoor air quality and type of store in underground shopping districts in Korea

Jeonghoon Kim[1], In-Keun Shim[2], Jongchun Lee[2]
[1]National Institute of Environmental Research, Researcher, Incheon, Korea, Republic of. [2]National Institute of Environmental Research, Senior researcher, Incheon, Korea, Republic of

Abstract
Most stores in underground shopping districts (USDs) are open or semi-open types and share air with passages in Korea. However, indoor air quality (IAQ) may be vary depending on selling merchandises in stores. This study aimed to examine relationships between IAQ and store types in USDs.

A total of 128 stores in 10 USDs were examined. IAQ, including fine particle ($PM_{2.5}$), aldehyde, volatile organic compound (VOC) concentrations were measured in store, passage, and outdoor. Store were categorized as clothing, fashion accessory, and food services. Environmental factors, including CO_2 levels, in each store were recorded.

While $PM_{2.5}$ concentrations were higher at outdoor, aldehyde, VOC, and total VOC (TVOC) concentrations were higher in passages or stores than those in outdoor. In 10 USDs, average store to passage ratios of $PM_{2.5}$ and individual aldehyde and VOC concentrations were ranged from 1.06 to 4.93. Among the stores, the formaldehyde and TVOC concentrations were higher in clothing and/or fashion accessories while $PM_{2.5}$ concentrations were higher in food services. Other individual compounds, including propionaldehyde, hexaldehyde, benzene, n-heptane, toluene, n-octane, ethylbenzene, xylene, d-limonene, n-undecane, and n-dodecane concentrations were associated with store types. Most aldehyde and VOCs were positively associated with CO_2 concentrations in the multivariable regression models.

Although stores shared air with passages, IAQ were different by area or store types. Aldehyde and VOC concentrations were associated with CO_2 concentrations in stores. Findings suggested USD-level operation of air conditioning system considering CO_2 levels in stores might be helpful to improve IAQ in stores at USDs.

268 – POSTER: Contribution of personal care products and diet to paraben exposure among young adults of Korea: A Crossover Intervention Study

Aram Lee[1], Sohyeon Choi[2], Jae-Hyun Kim[2], Na-Youn Park[3], Younglim Kho[4], Kyoungho Choi[5], Hyo-Bang Moon[6], Jeongim Park[7]
[1]Soonchunhyang University, Postdoctoral researcher, Asan, Korea, Republic of. [2]Soonchunhyang University, Researcher, Asan, Korea, Republic of. [3]Eulji University, Researcher, Seongnam, Korea, Republic of. [4]Eulji University, Professor, Seongnam, Korea, Republic of. [5]Seoul National University, Professor, Seoul, Korea, Republic of. [6]Hanyang University, Professor, Ansan, Korea, Republic of. [7]Soonchunhyang University, Professor, Asan, Korea, Republic of

Abstract
Parabens are widely used as preservatives in cosmetics, personal care products (PCPs), and food, but are suspected as endocrine disruptors. However, the major sources of paraben exposure are not well characterized. In this study, we designed a crossover intervention study for cosmetics, PCPs, and diet, with 27 young adults. During the 6-day intervention period, participants were instructed to use paraben-free cosmetics and PCPs for the entire period and were provided with a "paraben-free meal" on days 3-4, followed by a

"regular meal" on days 5-6. Urine samples were collected from the participants on days 0, 3, 5, and 7 to quantify the effects of the intervention. MeP, EtP, OH-MeP, and OH-EtP concentrations were measured in urine using LC-MS/MS. Before intervention, the GM of urinary MeP, EtP, OH-MeP, and OH-EtP concentrations were 6.05, 76.9, 4.79, and 1.87 ng/mL, respectively. No significant differences were observed in urinary paraben concentrations in the urine collected after cosmetics and PCPs intervention. Following the intervention of a paraben-free meal, urinary EtP and OH-EtP levels significantly reduced by 79.7% and 66.5%, respectively. On day 7, after returning to a routine diet, urinary EtP and OH-EtP levels significantly increased, while MeP and OH-MeP levels did not change after the paraben-free meal intervention but significantly increased on day 7. Our study supports that diet is a major source of EtP exposure among young adults in Korea. This work was supported by KEITI through Technology Development Project for Safety Management of Household Chemical Products, funded by MOE(2022002970004).

269 – POSTER: A passive sampling method using silicone wristband for determination of polycyclic aromatic hydrocarbons and their halogenated derivatives

Eri Tagawa[1], Qi Wang[2], Takashi Amagai[3], Yuichi Miyake[4]
[1]Yokohama National University, Student, Yokohama, Japan. [2]University of Shizuoka, Assistant Professor, Shizuoka, Japan. [3]University of Shizuoka, Professor, Shizuoka, Japan. [4]Yokohama National University, Associate Professor, Yokohama, Japan

Abstract
Polycyclic aromatic hydrocarbons (PAHs) and their halogenated derivatives (XPAHs) are emitted by anthropogenic activities with incomplete combustion. The PAHs and XPAHs are partitioned between gaseous and particulate phases. Some PAHs and XPAHs have adverse effects for human health such as carcinogenicity and mutagenicity. The measurements of individual exposures to PAHs and XPAHs are needed to assess their human health risks. A recent study showed that silicone wristband could be a useful tool as passive sampler to measure individual exposures to PAHs because of no requirement of electric power. However, there is a lack of knowledge on the passive sampling method using silicone wristbands, especially on target chemicals and environmental effects. The aims of this study were: (1) to evaluate the relationships between the concentrations of PAHs and XPAHs measured by active sampling and their collection amounts by passive sampling using silicone wristbands; (2) to assess the effects of distribution phase for PAHs and XPAHs to sampling rate; (3) to assess the wind effects during the sampling. In this study, we revealed that silicone wristbands enable to collect the PAHs and XPAHs with a wide range of concentrations compared with previous studies. In addition, we showed that the concentrations of 24 PAHs and 27 XPAHs in air could be estimated from their collection amounts in silicone wristbands.

270 – Sampling for improving air quality monitoring for Buffalo African-American community

eunhye yoo
University at Buffalo, associate professor, buffalo, USA

Abstract
To better understand the contexts of air pollution exposures at the community level, including who are the most disproportionately affected by which air pollutant and what are the major contributors to air pollution, we increase localized and concurrent community-specific air quality data by deploying low-cost ambient air pollution sensors. Meanwhile, we maximize the utility of limited resources, i.e., i.e., the fixed number of low-cost sensors and the cost for their maintenance, being available and maintainable at the community level.

To achieve this goal, we develop an optimal spatio-temporal sampling strategy that collects PM2.5 measurements at multiple times a day at selected sampling sites over a period of time, capturing both spatial and temporal variability of PM2.5 concentrations. Specifically, we develop a spatial sampling design based on a utility function that combines prediction uncertainty and risk factor criteria including the inputs from the community and spatially varying risk of adverse health outcomes, such as the relative risk of children's hospitalizations for asthma. The prediction uncertainty is quantified through a spatial data fusion approach based on a variant of kriging methods that efficiently tackle data with different spatial resolutions and measurement errors. We conducted an experimental study based on realistically simulated data.

The preliminary results indicate that the proposed sampling strategy improves the performance, quantified by the population root mean square error, in comparison to simple random sampling and the population Nash-Sutcliffe model efficiency coefficient. However, the prediction uncertainty suggests that further calibration is needed to improve the redundancy.

271 – Source Tracking of Heavy Metals from Blood and Urine of Residence near Industrial Area

Wonho Yang[1], Jihun Shin[2], Gihong Min[2], Dongjun Kim[3], Jaemin Woo[4], Chaekwan Lee[5]
[1]Daegu Catholic University, Professor, Kyongsansi, Korea, Republic of. [2]Daegu Catholic University, Ph.D. student, Kyongsansi, Korea, Republic of. [3]Daegu Catholic University, Master student, Kyongsansi, Korea, Republic of. [4]Daegu Catholic University, Master studentr, Kyongsansi, Korea, Republic of. [5]Inje University, Professor, Pusan, Korea, Republic of

Abstract
Residence living around industrial complexes could be expossed to hazadous agents such as heavy metals and may be suffer health risks. The heavy metals such as cadmium (Cd) are categorized as class I carcinogenic pollutants are of concern for their ability to cause injury, neurotic disorder, and lung cell damage. The purpose of this study identifies the pollution level and probable sources of heavy metals in blood and urine for residence neary industrial area. The study area is one of the most important industry-based area with a large steel mill in Korea. There are about 100 residents living near a large steel mill. Target heavy metals were Cd, Ni, and Pb, which will be measured in air, soil, foods, and drinking water. Simultaneously those metals will be analyzed in blood and urine for residents. The questionnaire dietary intake frequency data was utilized to estimate the concentration of heavy metals. To track the exposure route and sources, principal cluster analysis (PCA) was used with the correlation or regression analysis methods. The potential exposure pathways for target heavy metals can be in three ways ; (a) ingestion, (b) inhalation, and (c) dermal absorption. Exposure will be computed to calculate the average daily dose, which will be compared with their levels of blood and urine. It will suggest the methodology that source contributions and exposure routes that affect the levels of heavy metals in blood and urine, although this study is currently underway.

272 – Identifying Mixtures of Exposure in a Young Population – Results from the German Environmental Survey for Children and Adolescents (GerES V)

Liselot Koelman[1], Aline Murawski[2], Rosa Lange[1], Nina Vogel[1], Marike Kolossa-Gehring[3]
[1]German Environment Agency, Scientist, Berlin, Germany. [2]German Environment Agency, Head of Subject Group, Berlin, Germany. [3]German Environment Agency, Head of Section, Berlin, Germany

Abstract
The German Environmental Survey for Children and Adolescents (GerES V) was conducted between 2014–2017, and analysed 107 substances in urine and blood in partially overlapping subgroups (N = 485–2294), yielding population-representative data on exposure to environmental contaminants in Germany. So far, exposure to these contaminants has mostly been assessed for individual substances or groups of substances such as phthalates. Further analyses are now aimed to identify the mixtures of substances to which the young population has been exposed to. The 'Exposure Load' is a metric introduced by Willey et al. (2021) that counts the number of substances measured above a certain threshold. By using this measure, we aimed to identify typical mixtures of exposure for certain socio-demographic groups. To maximise the number of substances while retaining as large a sample size as possible, a subset of 29 substances was selected including metals, phthalates, PFAS, pesticides, aprotic solvents, and fragrances. The Exposure Load with median as threshold, ranged between 7 to 29 (median: 20) with no distinct differences between sex, age groups, or socio-economic status. Our results suggest two conclusions: (1) all individuals were exposed, i.e. were among the respective higher exposed group for at least some environmental contaminants, and (2) a large amount of exposure is due to ubiquitous contaminants, and individual behaviour seems to affect this type of exposure only to a rather small degree.

273 – POSTER: Fire Fighter Union Perspective: Urgent Need for Exposure Research & Interventions regarding the chemicals that harm us

Neil McMillan[1], Dan Whu, MD, MPH[2]
[1]International Association of Fire Fighters, Director of Science & Research, Ottawa, Canada. [2]International Association of Fire Fighters, Chief Medical Officer, Washington D.C., USA

Abstract
As the Director of Science & Research and the Chief Medical Officer for the International Association of Fire fighters we urge the environmental health research and regulatory community to partner with us in understanding the mixture of work-related exposures among our membership and taking effective action to protect them and their families. While most fire fighters maintain a high standard of physical health, they are disproportionately afflicted with higher rates of many cancers that are often diagnosed at an earlier age and/or more aggressive forms. Cancer incidences for fire fighters are up to 5 times higher than the general population with 74-95% of

firefighter line of duty deaths (LODD) attributable to cancer. Children of firefighters are a devastating 27 times more likely to be diagnosed with cancer compared to their peers. These rates are so alarming that we have begun to take a central role in understanding and reducing our exposures to the hazards that contribute to these diseases. We played a central role in opposing the use of PBDE flame retardants in favor of alternative forms of fire protection that don't risk the health of firefighters and the public, and more recently under the leadership of President Ed Kelly, we have begun to take action against PFAS in fire fighting foam, protective gear and other exposure pathways fire fighters face. We will share perspectives and concerns of firefighters on these topics, highlighting important gaps in understanding and potential opportunities for partnership.

274 – Codeveloping tailored interventions to reduce exposure to fine particulate matter ($PM_{2.5}$) and improve asthma-related health

Amy McCarron[1], Sean Semple[2], Christine Braban[3], Vivien Swanson[2], Colin Gillespie[4], Heather Price[5]
[1]University of Stirling, PhD Researcher, Stirling, United Kingdom. [2]University of Stirling, Professor, Stirling, United Kingdom. [3]UK Centre for Ecology and Hydrology, Atmospheric Composition Change Group Leader, Edinburgh, United Kingdom. [4]Scottish Environment Protection Agency, Air Modelling Unit Manager, Stirling, United Kingdom. [5]University of Stirling, Senior Lecturer, Stirling, United Kingdom

Abstract
$PM_{2.5}$ (fine particulate matter ≤2.5μm in diameter) is a key air pollutant responsible for acute and long-term health impacts. People with asthma, being more vulnerable to the effects of air pollution, are commonly advised to avoid exposure to air pollution, however, most people are unaware of the quality of the air they breathe and therefore cannot make informed exposure-minimising behaviour changes. Wearable air quality sensors can provide high-resolution air quality data across microenvironments and thus have the potential to inform effective, tailored exposure reduction behaviours.

$PM_{2.5}$ personal exposure data were collected by 32 participants over 1-week at baseline and for 1-week post-intervention using wearable sensors whilst participants simultaneously kept a time-activity diary documenting their microenvironments, activities, respiratory health and medication usage. Participants were randomised into control arm (n=16) or intervention arm (n=16), whereby intervention participants received written and verbal personalised air quality data feedback to inform tailored codeveloped interventions.

Most codeveloped interventions took place within the home microenvironment (n=15), and participants opted primarily for personalised prompts and cues to trigger the intervention behaviour. Preliminary results show that, on average, intervention arm participants reduced their $PM_{2.5}$ personal exposure by 2.8 ± 5.5μg/m^3 compared with control arm participants who reduced their exposure by 2.5 ± 3.5 μg/m^3. We found no clinically significant difference in self-reported asthma quality of life in either arm.

This feasibility study has demonstrated that personalised air quality data can raise awareness of exposure to air pollution across microenvironments and can inform effective, codeveloped, tailored interventions to minimise personal exposures.

275 – Development of a Measuring Vest for Epidemiological and Occupational Exposure Monitoring Using Low-Cost Miniature Sensors

Fidelis Theinert[1], Sam Aerts[2], John Bolte[3]
[1]The Hague University of Applied Sciences, Senior researcher, Delft, Netherlands. [2]The Hague University of Applied Sciences, Associate professor, Delft, Netherlands. [3]The Hague University of Applied Sciences, Professor, Delft, Netherlands

Abstract
A measuring vest prototype was developed for epidemiological and occupational exposure monitoring. It has a modular design and consists of a microprocessor and logging system, a battery pack, and a wired network of easy-to-replace sensor nodes. Currently, the vest contains inexpensive, lightweight micro-electromechanical systems (MEMS) sensors that can detect aerosols, temperature, and other environmental parameters. Sensors for particulate matter or certain chemicals can also be integrated into the vest with minimal effort. In addition, small, low-cost radiofrequency (RF) electromagnetic field (EMF) sensors are being developed that can be easily added to the vest, in particular when the required antennas are small, which is the case when measuring the frequency bands being introduced by the fifth generation of telecommunications (5G). The sensor nodes can be controlled via different interfaces, and the logging system software can collect data from all sensor nodes at a rate of up to one sample per second (along with RTC timestamp and GNSS coordinates). The locally stored data can also be retrieved by a smartphone or a central server. The measuring vest has the potential to improve epidemiological or occupational exposure monitoring. Survey results can be verified and hotspots or individual

patterns of adverse habits can be identified. It can also function as an intervention tool, alerting the user to potentially harmful environments and situations. In the future, the vest will be combined with an inner garment monitoring vital signs and different sensor technologies will be explored to ensure data accuracy and privacy while keeping costs low.

276 – Variability of the human serum metabolome over three months in the EXPOsOMICS Personal Exposure Monitoring study

Jelle Vlaanderen[1], Max Oosterwegel[2], Dorina Ibi[3], Lützen Portengen[3], Gerard Hoek[1], Nicole Probst-Hensch[4], Sonia Tarallo[5], Alessio Naccarati[5], Medea Imboden[6], Ayoung Jeong[6], Nivonirina Robinot[7], Andre Amaral[8], Erik van Nunen[2], John Gulliver[9], Debbie Jarvis[10], Marc Chadeau-Hyam[10], Paolo Vineis[10], Roel Vermeulen[11], Pekka Keski-Rahkonen[7]
[1]Utrecht University, Associate Professor, Utrecht, Netherlands. [2]Utrecht University, PhD student, Utrecht, Netherlands. [3]Utrecht University, Researcher, Utrecht, Netherlands. [4]Swiss TPH, Professor, Basel, Switzerland. [5]IIGM, Researcher, Turin, Italy. [6]Swiss TPH, Researcher, Basel, Switzerland. [7]IARC, Researcher, Lyon, France. [8]Imperial College London, Researcher, London, United Kingdom. [9]University of Leicester, Professor, Leicester, United Kingdom. [10]Imperial College London, Professor, London, United Kingdom. [11]Utrecht University, Professor, Utrecht, Netherlands

Abstract
Metabolomic techniques such as high-resolution liquid chromatography mass spectrometry (LC-HRMS) are increasingly used in exposome studies to study the interaction between non-genetic factors and the metabolome. To reliably and efficiently link detected compounds to exposures and health phenotypes in such studies it is important to understand the variability in metabolome measures. We assessed the within- and between-subject variability of untargeted LC-HRMS measurements in 298 non-fasting human serum samples collected at two occasions from 157 subjects. Samples were collected circa 107 days apart as part of the multicenter EXPOsOMICS Personal Exposure Monitoring study. In total 4294 metabolic features were detected. For 206 features a chemical identity was established with high confidence. These features referred to 184 unique compounds.

The median intraclass correlation coefficient (ICC) across all metabolic features was 0.51 (IQR: 0.29) and increased to 0.64 (IQR: 0.25) for the 184 uniquely identified compounds. For this group, the median ICC marginally changed (0.63) when we included common confounders (age, sex and body-mass-index) in the regression model. When grouping compounds by compound class, the ICC was largest among glycerophospholipid (median ICC 0.70), and steroids (0.67). ICCs varied substantially within chemical class.

Our results suggest that the metabolome as measured with untargeted LC-HRMS is moderately stable over 100 days for more than half of the features to reflect average levels across this time period. Variance across the metabolome will result in differential measurement error across the metabolome which needs to be considered in interpretation of metabolome results.

277 – POSTER: Identification of volatile organic compounds (VOCs) in the spray-type household products using non-target analysis

Hyejin Yun[1], Jeonghoon Park[2], Chungsik Yoon[3], Kiyoung Lee[3], KYUNG-DUK ZOH[3]
[1]Seoul National University, MS Student, Seoul, Korea, Republic of. [2]Seoul National University, PhD Candidate, Seoul, Korea, Republic of. [3]Seoul National University, Professor, Seoul, Korea, Republic of

Abstract
The use of volatile organic compounds (VOCs) in consumer products is a source of potentially harmful chemical exposures to humans. In this study, we analyzed VOCs in of four different types of consumer products, including cleaner, coating agent, deodorizer, and repellent, by non-target screening. Non-target screening was performed on the extracts using a headspace GC-MS instrument with a NIST library and Agilent Unknown Analysis software. A total of 298 VOCs were identified in the samples, of which a wide range of VOCs, including alkanes, monoterpenoids, and cycloalkanes. Among VOCs identified by non-target screening, 15 compounds were confirmed with the standards, implying the reliability of analysis. The identified VOCs were characterized as potentially hazardous based on their toxicity and/or detected frequency. Some of these compounds are known or suspected to be endocrine disruptors or fragrance allergens, implying the importance of non-target screening to identify potential health risks associated with VOCs contained in the consumer products.

278 – The impact of wildfire smoke on exposure measurement error of fine particulate matter

Yisi Liu[1], Roxana Khalili[1], Yan Xu[1], Nathan Pavlovic[2], Crystal McClure[2], Fred Lurmann[3], Karl O'Sharkey[1], Jane Cabison[4], Marisela Rosales[5], Nathana Lurvey[6], Deborah Lerner[7], Jill Johnston[8], Shohreh Farzan[8], Theresa Bastain[8], Carrie Breton[9], Genevieve Dunton[9], Rima Habre[8]
[1]University of Southern California, Postdoc Researcher, Los Angeles, USA. [2]Sonoma Technology, Inc., Scientist, Petaluma, USA. [3]Sonoma Technology, Inc., Chairman, Petaluma, USA. [4]University of Southern California, Research Coordinator Supervisor, Los Angeles, USA. [5]University of Southern California, Research Coordinator, Los Angeles, USA. [6]Eisner Health, Medical Director, Culver City, USA. [7]Eisner Health, Chief Medical Officer, Culver City, USA. [8]University of Southern California, Associate Professor, Los Angeles, USA. [9]University of Southern California, Professor, Los Angeles, USA

Abstract

Air pollution epidemiology studies use ambient concentrations as proxy measures for personal air pollution exposure of outdoor origin, potentially introducing exposure measurement error. This is especially a concern in wildfire studies since human behaviors may change on smoky days. We aim to understand the impact of wildfire smoke on measurement error for $PM_{2.5}$.

We collected 4-day continuous personal $PM_{2.5}$ among 63 pregnant women in the MADRES cohort during the 1st and 3rd trimester and 4-6 months postpartum using the RTI MicroPEM. Ambient $PM_{2.5}$ exposure was estimated at residences using inverse distance squared weighted interpolation from regulatory air monitors. Daily wildfire smoke exposure was assessed using CalFire data for Southern California, Hazard Mapping System (HMS) smoke density contours, and wildfire-$PM_{2.5}$ concentrations estimated with the Hybrid Single-Particle Lagrangian Integrated Trajectory (HYSPLIT) model. We analyzed relationships between daily ambient and personal $PM_{2.5}$ exposures on wildfire and non-wildfire days using mixed effect models.

A total of 605 person-days were measured with a mean (SD) daily personal, ambient, and wildfire-$PM_{2.5}$ exposure of 19.1 (38.9), 11.4 (5.2), and 1.0 (3.9) $\mu g/m^3$, respectively. Ambient $PM_{2.5}$ was less correlated with personal $PM_{2.5}$ on low (Spearman r=0.26) or medium (r=-0.26) smoke days compared to no smoke days (r=0.40) from HMS data. On days with wildfire-$PM_{2.5}$ > 0$\mu g/m^3$, ambient $PM_{2.5}$ had a weaker relationship with personal $PM_{2.5}$ as compared to days with no wildfire smoke (-1.80 [-3.04, -0.53] $\mu g/m^3$).

Our findings highlight potentially higher exposure measurement error on days with wildfire smoke, suggesting the importance of considering exposure averting behaviors in wildfire studies.

279 – POSTER: 12 Years of Intraindividual Trends of Perfluoroalkyl Acids (PFAAs) Among Women of Child-Bearing Age

Anders Glynn[1], Jonathan Benskin[2], Sanna Lignell[3], Irina Gyllenhammar[4]
[1]Swedish University of Agricultural Sciences, Professor, Uppsala, Sweden. [2]Stockholm University, Professor, Stockholm, Sweden. [3]Swedish Food Agency, PhD, toxikologist, Uppsala, Sweden. [4]Swedish Food Agency, PhD, toxicologist, Uppsala, Sweden

Abstract

Few studies have looked at changes in PFAA concentrations in repeated serum samples from women of reproductive age during an extended time period. Serum from women (median age 29 years, N=44-57) were analyzed for PFOA, PFNA, PFDA, PFUnDA, PFHxS and PFOS, starting 3 weeks after their first pregnancy and ending 12 years later. Correlations between concentrations measured at the first time point and 4 years later were significant (Spearman´s r=0.38-0.74), with the weakest correlations for PFOA and PFNA (r<0.5). Correlations between concentrations at the first time point and 8 and 12 years later were weaker, r being <0.39 at the 12-year sampling. Among mothers sampled for the first time 1996-1999, and again after 12 years, average PFOA and PFOS concentrations decreased while concentrations of PFNA, PFDA, PFUnDA and PFHxS increased. For mothers initially sampled from 2004, PFOA and PFOS concentrations decreased during the follow-up, whereas the patterns for the other PFAAs varied. PFDA initially decreased to a steady-state later in the follow-up period. PFNA did not change markedly during follow-up, PFUnDA increased towards a plateau, and PFHxS decreased during the follow-up. Thus, the intraindividual trends varied with the time period studied, probably partially influenced by timing of international measures to limit production/use of the different PFAAs. The change in trend of PFHxS was due to remediation of drinking water contamination during the latter part of the study period. Giving birth during follow-up had a strong influence on serum concentrations, on average decreasing 30-50% per child-birth.

280 – POSTER: Seasonal and Location-Dependent Alterations in the Placental Transcriptome Associated with Prenatal Pyrethroid Exposure

Yewei Wang[1], Karen Hermetz[2], Amber Burt[2], Elizabeth Kennedy[1], Corina Lesseur[3], Dana Barr[4], Ke Hao[5], Jia Chen[5], Carmen Marsit[4]
[1]Emory University, Postdoctoral Fellow, Atlanta, USA. [2]Emory University, Research Specialist, Atlanta, USA. [3]Icahn School of Medicine at Mount Sinai, Assistant Professor, New York, USA. [4]Emory University, Professor, Atlanta, USA. [5]Icahn School of Medicine at Mount Sinai, Professor, New York, USA

Abstract
Background: Prenatal exposure to pyrethroids is linked to adverse health effects in early life. Proper placental function is critical to development. This study aimed to explore the impact of prenatal pyrethroid exposure on the placental transcriptome, stratified by season and location to inform its effect on function.

Methods: The study of Asian Women and their Offspring's Development and Environmental Exposures (SAWASDEE) recruited pregnant farm-working women from two agricultural districts in the Chiang Mai province of Thailand from 2017 to 2019. Maternal urinary pyrethroid metabolites were measured during three trimesters respectively in 248 participants, specifically 3-phenoxybenzoic acid (PBA), cis-3-(2,2-Dichlorovinyl)-2,2-dimethylcyclopropane carboxylic acid (CDCCA), and trans-3-(2,2-Dichlorovinyl)-2,2-dimethylcyclopropane carboxylic acid (TDCCA). Individual metabolites were adjusted for urinary creatinine. The placental transcriptome was analyzed using RNA-Seq. Generalized linear regression was employed to identify the location and season-specific differentially expressed genes (DEGs) and associations between placental DEGs and the sum of each metabolite. Pathway analysis was performed to examine the potential mechanisms associated with these DEGs.

Results: We identified 253 DEGs by location and 25 were also associated with prenatal pyrethroid metabolites. 3130 DEGs were associated with the season and 27 were also associated with pyrethroids. These DEGs involved erythropoietin signaling, pulmonary fibrosis idiopathic signaling, oxidative phosphorylation, and neurodevelopment-associated pathways.

Conclusions: Our findings demonstrate that the placental transcriptome undergoes seasonal and location-dependent changes following prenatal exposure to pyrethroids. These alterations in gene expression may have significant implications for fetal neurodevelopment and other pathologic processes, highlighting the importance of monitoring environmental exposures during pregnancy.

281 – Trace metal concentrations around mining deposits : pathway to residential environmental media

Sabine Guerin, Laurence Lethielleux, Morgane Salomon
Ineris, Research engineer, Verneuil-en-Halatte, France

Abstract
Trace metal exposure around closed mining sites remains a public health concern. The transfer of particulates from soil sources, such as mining deposits, to outdoor and indoor air and dust is a complex mechanism that involves various influencing parameters. This study seeks to identify parameters that can explain the exposure pathway between mining sources and residential environmental media. Data were collected from various environmental compartments such as soil, indoor and outdoor air, indoor dust, and atmospheric deposit.

Eighteen trace metals and metalloids were analyzed in at least one sampled media including lead, arsenic, and cadmium. Environmental concentrations were compared to background values when available. Regression analyses were used to derive slopes between yard soil and indoor dust concentrations and between yard soil concentrations and distance to the source.

Lifestyle questionnaires regarding house age, presence of pets, and cleaning techniques were available for 30 dwellings and helped understand outliers and atypical data identified when performing statistical analysis. Indoor dust was collected in those dwellings with a wipe and a vacuum, and those two types of sampling were compared to assess their correlation. We faced challenges during the data analysis process such as a lack of quantified data and different sampling duration depending on the method used.

Despite these challenges, this study provides valuable insights into the exposure to trace metals around closed mining sites in France regarding sampling strategy and collecting additional data on factors influencing particulate pathway.

282 – Exposure Investigation of Community Exposure to Particulate Matter and Silica Dust near a Silica Sand Processing Facility, Valley, Stevens County, Washington State

James Durant[1], Bradley Goodwin[2], Aaron Young[1], Debra Gable[3]
[1]Agency for Toxic Substances and Disease Registry, Office of Community Health Hazard Assessment, Environmental Health Scientist, Atlanta, USA. [2]Agency for Toxic Substances and Disease Registry, Deputy Director, Atlanta, USA. [3]Agency for Toxic Substances and Disease Registry, Office of Community Health Hazard Assessment, Environmental Health Scientist, Seattle, USA

Abstract
In 2016, ATSDR conducted an exposure investigation to measure community exposure to respirable crystalline silica (RCS) and particulate matter (PM) near a silica sand processing facility in Valley, Washington.

ATSDR sampled air at 10 locations (8 outdoors, 2 indoors) around the facility, including a nearby elementary school. RCS was sampled outdoors using a 24-hour filter with a cutpoint of 4 microns, followed by gravimetric and X-ray diffraction analysis. PM was collected indoors and outdoors using environmental beta attenuation monitors (E-BAMs) with size selective inlets of 10, 4, and 2.5 microns. PM4 was used as a surrogate for RCS because the RCS sampler was too loud to run indoors. The percent of RCS contained in PM4 ranged from 0–31%.

Mean RCS concentrations at three locations (max 4.56 micrograms per cubic meter (μg/m3)) were higher than the California reference exposure level (3 μg/m3). The highest indoor mean PM4 E-BAM concentration was 4.9 μg/m3. The highest outdoor mean PM4 was 16.6 μg/m3. Mean PM10 concentration (30.7 μg/m3) was higher than the World Health Organization (WHO) air quality guideline. The highest mean PM2.5 (10.3 μg/m3) was also above WHO guidelines.

Silica and PM increased with proximity to the facility and when winds were from the direction of the facility. ATSDR identified potential steps for reducing exposures to silica and PM.

The findings and conclusions in this presentation have not been formally disseminated by the Agency for Toxic Substances and Disease Registry and should not be construed to represent any agency determination or policy.

283 – POSTER: Association between exposure to environmental chemical mixtures and hypertension in a Wuhan Elderly population

Ruifang Li, Yang Shen
Huazhong University of Science and Technology, PhD, Wuhan, China

Abstract
Since most studies focused on the single-chemical effects, our understanding of the combined effects of multiple pollutants on hypertension in an elderly population is limited. The purpose of this study is to ascertain the potential adverse health effects exerted by a wide range of pollutants on hypertension in an elderly population.

We investigated the associations between multiple-pollutant exposure and hypertension in a cross-sectional study of 961 elderly participants from Wuhan, China. Seven classes of environmental chemicals including phthalates, polychlorinated biphenyls (PCBs), polycyclic aromatic hydrocarbons (PAHs), organochlorine pesticides (OCPs), organophosphate pesticides (OPPs), synthetic pyrethroids (SPs), and herbicides were analyzed in serum samples. Both single-pollutant (logistic regression adjusting for covariates) and multi-pollutant approaches (hierarchical Bayesian kernel machine regression and Quantile g-computation) were used to estimate the associations between multiple exposures and hypertension.

In addition to discovering the relationships between individual chemicals and thyroid hormones, hierarchical BKMR was performed for variable selection at the chemical-group level. Important variables of chemical groups were selected if the group PIPs were > 0.5. Among the 7 chemical groups, the herbicide group was the most important group for hypertension (PIP = 0.89), which showed a monotonic positive association. OCPs and OPPs showed monotonic increasing trends (PIP = 0.72 and 0.75 respectively). The PAHs and SPs were selected as important groups for hypertension (PIP = 0.70 and 0.64 respectively), with a U-shape dose-response curve. These results were also replicated in the Quantile g-computation.

The results suggest that exposure to environmental chemical mixtures is related to hypertension.

284 – Fine Particulate Matter Exposure During Childhood is Associated with Altered Developmental Trajectories of Core Neurocognitive Brain Networks

Clara Zundel[1], Cole Brokamp[2,3], Julia Evanski[1], Samantha Ely[4], Leah Gowatch[5], Amanpreet Bhogal[5], Patrick Ryan[6,7], Hilary Marusak[8]
[1]Wayne State University School of Medicine, Postdoctoral Research Fellow, Detroit, USA. [2]University of Cincinnati College of Medicine, Associate Professor, Cincinnati, USA. [3]Cincinnati Children's Hospital Medical Center, Associate Professor, Cincinnati, USA. [4]Wayne State University School of Medicine, Translational Neuroscience PhD Student, Detroit, USA. [5]Wayne State University School of Medicine, Research Assistant, Detroit, USA. [6]University of Cincinnati College of Medicine, Professor, Cincinnati, USA. [7]Cincinnati Children's Hospital Medical Center, Professor, Cincinnati, USA. [8]Wayne State University School of Medicine, Assistant Professor, Detroit, USA

Abstract
Air pollution exposure has been associated with adverse mental health outcomes. Youth may be particularly susceptible given ongoing brain development. However, the neurobiological mechanisms underlying the association between air pollution and mental health remain unclear. We examine the impact of fine particulate matter ($PM_{2.5}$) on mental health symptoms and resting-state functional connectivity (rsFC) of the salience network, a network involved in attentional orienting to emotional and sensory stimuli, during the transition into adolescence.

The sample consisted of baseline and two-year follow-up data from 5,389 youth in the Adolescent Brain Cognitive Development Study (M+SD=9.92+0.62 years; 46% female). One-year annual ambient $PM_{2.5}$ concentrations during late-childhood were estimated using hybrid ensemble spatiotemporal modeling. The Child Behavioral Checklist was used to evaluate mental health symptoms. rsFC between the salience and other large-scale networks was estimated using functional neuroimaging. Linear mixed models, adjusting for potential confounders (e.g., income, greenspace), were used to examine $PM_{2.5}$-by-time interactions on rsFC and symptoms.

A $PM_{2.5}$-by-time interaction was significant for salience-default mode network rsFC such that youth with higher $PM_{2.5}$ exhibited increased rsFC over time while youth with lower $PM_{2.5}$ exhibited decreased rsFC over time (β=0.003, CI = (0.002, 0.004)). A $PM_{2.5}$-by-time interaction was observed for internalizing symptoms, such that all youth exhibited a decrease in symptoms over time, however in youth with higher $PM_{2.5}$, this decrease occurred at a faster rate (β= -0.196, CI = (-0.360, -0.033)).

PM2.5 exposure during late-childhood alters the development of core neurocognitive networks which may impact trajectories of mental health symptoms across adolescence.

285 – POSTER:

Sohyeon Choi[1], YoungJun Jeon[2], Aram Lee[3], Inae Lee[4], Jae-Eun Lim[5], Nayoun Park[6], Kyungho Choi[7], Hyo-Bang Moon[8], Younglim Kho[9]
[1]Soonchunhyang University, Asan, Republic of Korea, Ph.D Student, Asan, Korea, Republic of. [2]Soonchunhyang University, Asan, Republic of Korea, master student, Asan, Korea, Republic of. [3]Soonchunhyang University, Asan, Republic of Korea, Ph.D, Asan, Korea, Republic of. [4]Seoul National University, Seoul, Republic of Korea, Ph.D, Seoul, Korea, Republic of. [5]Hanyang University, Ansan, Republic of Korea, Ph.D Student, Ansan, Korea, Republic of. [6]4Eulji University, Republic of Korea, Ph.D Student, Seongnam, Korea, Republic of. [7]Seoul National University, Seoul, Republic of Korea, Professor, Seoul, Korea, Republic of. [8]Hanyang University, Ansan, Republic of Korea, Professor, Ansan, Korea, Republic of. [9]4Eulji University, Republic of Korea, Professor, Seongnam, Korea, Republic of

Abstract
Phthalates and alternative phthalates are well known endocrine-disrupting chemicals and can be exposed through plastic products and food packages. This study aimed to analyze plasticizer and their metabolites from dietary and urine samples and investigate their correlations. Forty-eight households (124 individuals) were recruited. Duplicate diet samples for 3 days were collected and urine samples were collected before bedtime on the day of dietary sample collection. Twenty-one plasticizers in dietary samples were analyzed using GC-MS/MS. Thirty-seven metabolites in urine samples were analyzed using UHPLC-MS/MS. Twelve plasticizers were detected at a detection rate (range 0.017~14.1ng/g dw) of ≥70% in dietary samples, and the concentration (median) ranged from 0.73 to 105.56ng/g dw. Twenty-two metabolites were detected at a detection rate (range 0.018~0.463ppb) of ≥70%, and the concentration (median) ranged from 0.23 to 42.25ppb. Typical phthalate plasticizers in dietary samples and their metabolites were shown positive correlation (DMP&MMP p=0.03; DEP&MEP p=0.02; DBP&MnBP p=0.01; BBzP&MBzP p<0.001). Association between alternative plasticizer and their metabolites were positively associated (DiNP&OH-MiNP p<0.001; cx-MMeHP p<0.001; DEHTP&5cx-MEPTP p<0.001; 5OH-MEHTP p<0.001; 5oxo-MEHTP p<0.001; DEHA&MEHA p<0.001; DINCH&cx-MINCH p<0.001; OH-MINCH p<0.001). This study observed significant positive correlation between plasticizers in dietary samples and their metabolites, indicating that exposure may occur through food consumption. This research was supported by a grant (21162MFDS074) from Ministry of Food and Drug Safety in 2023. This work was supported by KEITI through Technology Development Project for Safety Management of Household Chemical Products, funded by MOE(2022002970004).

286 – POSTER: ALAD polymorphisms and exposure to lead in Slovenian men

Anja Stajnko[1], Janja Snoj Tratnik[1], Neža Palir[2], Agneta Annika Runkel[1], Darja Mazej[1], Milena Horvat[3], Ingrid Falnoga[4]
[1]Jozef Stefan Institute, PhD, Ljubljana, Slovenia. [2]Jozef Stefan Institute, MsC, Ljubljana, Slovenia. [3]Jozef Stefan Institute, Prof., Ljubljana, Slovenia. [4]Jozef Stefan Institute, Assoc. prof, Ljubljana, Slovenia

Abstract
The genetic susceptibility to low-level lead exposure in general populations is poorly investigated and mostly limited to the individual single nucleotide polymorphism (SNP) rs1800435 in the delta-aminolevulinic acid dehydratase gene (ALAD). In the present study, we aimed to investigate possible associations between 10 selected ALAD SNPs with blood lead concentrations (BPb) in 281 adult men from Slovenia, with geometric mean Pb levels of 19.6 ng/g (range: 3.86 – 84.7 ng/g). The possible genetic influence was tested based on the level of SNP haplotypes, individual SNPs and two-SNP combinations using multiple linear regression analyses. While there was no significant association at the level of SNPs haplotypes, individually, the presence of variant alleles of rs1800435 (C>G) and rs1805312 (C>G) SNPs resulted in a significant 13 % decrease in BPb. Furthermore, homozygous carriers of rs1139488 (C>G) variant allele showed a significant 20 % BPb increase compared to carriers of the common allele. However, the addition of individual SNPs in the model explained only around 1 additional % point of BPb variability. Contrarily, SNPs combination analyses identified 6 two-SNP combinations which significantly explained 3 - 22 additional % points in BPb variability; the highest was for the rs1800435-rs1139488 and rs1139488-rs1805313 combinations. As such we demonstrated that the identification of individuals with specific combinations of ALAD SNPs explains a larger part of Pb variability and as such, after the confirmation in other populations, might represent a better susceptibility biomarker in Pb exposure than individual SNPs.

287 – Smart Home Air Quality Systems for Community Led Research

Daniel Morrish[1], Gustavo Olivares[2], Ian Longley[2], Elizabeth Somervell[2]
[1]NIWA, Air Quality Technician, Auckland, New Zealand. [2]NIWA, Air Quality Scientist, Auckland, New Zealand

Abstract
Measuring indoor air quality has become of interest to many over the past few years. The team at NIWA have been developing the Smart Home Air Quality (SHAQ) system. It's a portable, configurable, real-time, remotely accessible, user-friendly system using low-cost indoor air quality sensors and Home Assistant home automation. Initially designed for behaviour change studies in community spaces, SHAQs include window and door sensors, low-cost CO2 and particulate matter (PM) sensors, and dashboards designed to inform and prompt action when stale air is detected in a space. SHAQs are also being used in studies for biomass burning for home heating, including portable air cleaners and outdoor PM sensors to allow the characterisation and modification of the relationship between indoor and outdoor air. Preliminary findings from these use cases, and potential future use cases and technological developments will be explored.

288 – POSTER: Modelling Spatial & Temporal Variability of Air Pollution in an Area of Unconventional Natural Gas Operations

Miranda Doris[1], Coreen Daley[2], Jad Zalzal[1], Romain Chesnaux[3], Laura Minet[4], Mary Kang[5], Élyse Caron-Beaudoin[6], Heather MacLean[7], Marianne Hatzopoulou[7]
[1]University of Toronto, PhD Student, Toronto, Canada. [2]University of Toronto Scarborough, PhD Student, Scarborough, Canada. [3]University of Quebec at Chicoutimi, Professor, Chicoutimi, Canada. [4]University of Victoria, Professor, Victoria, Canada. [5]McGill University, Professor, Montreal, Canada. [6]University of Toronto Scarborough, Professor, Scarborough, Canada. [7]University of Toronto, Professor, Toronto, Canada

Abstract
Background: Despite the growing unconventional natural gas production industry in northeastern British Columbia, Canada, few studies have explored the air quality implications on human health in nearby communities. Researchers who have worked with pregnant women cohorts in this area have found higher levels of volatile organic compounds (VOCs) in the indoor air and tap water of their homes in association with higher density and closer proximity to oil and gas wells.

Methods: To inform ongoing exposure assessments, we develop land use regression models to predict air pollutant concentrations at the homes of pregnant women living near hydraulic fracturing activities. The regression models were developed at three temporal scales for 18 air pollutants using the existing monitoring network. Predictor variables for the regression models were developed using oil and gas production activities, including well operations, facilities, pipelines, road segments and annual emissions data. The developed models were used to predict ambient concentrations at cohort participants' homes and across the study domain.

Results: The models predicting monthly, bi-annual, and annual mean concentrations explained 23% to 96%, 54% to 98%, and 71% to 97% of the variability in air pollutant concentrations, respectively. The models predicted high concentrations of VOCs for cohort participants compared to values collected by the existing monitoring network.

Conclusion: The elevated concentrations of VOCs highlight the need for additional monitoring of hydraulic fracturing-related air pollutants near areas of residence. These models will help investigate associations between prenatal exposure to air pollutants related to natural gas extraction and adverse health outcomes.

289 – Developing, and Piloting in the Columbia Basin, a Prioritization Model and Screening Approach for PFAS Contamination in Fish Tissue

Elaine Cohen Hubal[1], Nicole Deluca[1], Dylan Wallis[2], Ashley Mullikin[1], Peter Brumm[3], Ashley Zanolli[4]
[1]USEPA, Scientist, RTP, USA. [2]ORISE, Scientist, RTP, USA. [3]USEPA, Scientist, Helena, USA. [4]USEPA, Scientist, Portland, USA

Abstract
Jurisdictions are currently challenged with efficiently identifying and characterizing the extent of per- and polyfluoroalkyl substances (PFAS) contamination and human exposure to PFAS. While PFAS contamination can be presumed near well-studied sources, the number of potential sources in larger regions, unknown facilities' PFAS use, and uncertain fate and transport, make sampling prioritization overwhelming and complex. In the Columbia Basin, particularly in and around the region's tribal lands, the availability of PFAS occurrence data is limited and exposure is less characterized than in other parts of the United States. This project seeks to develop and pilot a cost-effective approach to investigate and monitor for the presence and extent of PFAS contamination, with a focus on identifying PFAS contamination in fish tissue. In phase one, a geospatial prioritization workflow was developed that leverages existing PFAS measurements in fish tissue, along with information on potential sources of PFAS including industry facilities, fire training sites, and landfills. Machine learning and statistical models were employed to predict areas in natural waters of the Columbia Basin in which high PFAS concentrations are likely to occur in fish tissue, but sampling has not yet confirmed. Based on these model predictions, information on population vulnerability, and regional partner fish collection activities, additional fish tissue data will be generated in phase two of this pilot study. A novel tiered analytical chemistry strategy will be applied to screen for PFAS in these collected specimens and results will be used to evaluate the geospatial model predictions.

290 – A Low-cost Sensor Node for RF-EMF Exposure in 5G Networks Based on Requirements Identified in Horizon Europe NextGEM Project

Erdal Korkmaz[1], Stephan Littel[2], Sam Aerts[3], Marco Spirito[4], John Bolte[5]
[1]The Hague University of Applied Sciences, Senior researcher, Delft, Netherlands. [2]The Hague University of Applied Sciences, Junior researcher, Delft, Netherlands. [3]The Hague University of Applied Sciences, Associate professor, Delft, Netherlands. [4]Delft University of Technology, Professor, Delft, Netherlands. [5]The Hague University of Applied Sciences, Professor, Delft, Netherlands

Abstract
With the roll-out of the fifth generation of telecommunications (5G) networks, new transmission protocols and new frequency bands are introduced, which alter and may increase our everyday exposure to electromagnetic fields (EMFs). As it remains unclear whether there are any biological effects related to this exposure, large-scale and long-term monitoring of the EMF through networks of low-cost sensor nodes is a key objective of the four Horizon Europe projects on EMF exposure and health, such as NextGEM (Next Generation Integrated Sensing and Analytical System for Monitoring and Assessing Radiofrequency Electromagnetic Field Exposure and Health).

In this study, (design) requirements and (calibration and validation) recommendations are identified for the development of low-cost next-generation EMF measuring equipment, specified for different use cases (e.g., use on the body, on vehicles, or on infrastructure). Furthermore, based on the identified issues, a sensor node was developed to monitor specifically the exposure in four frequency bands used for 5G in the Netherlands, aiming to improve the sensitivity, accuracy, and stability of previously developed sensors. The sensor node consists of three stacked Printed Circuit Boards (PCBs) with carefully chosen off-the-shelf components. It has been validated with in-lab measurements for both continuous wave signals and modulated signals, and for both cases, and for all frequency bands, an error less than 1 dB for a dynamic range of 40 dB was obtained. Finally, in-situ validation measurements in the vicinity of 5G base stations are planned, as well as a thorough comparison to current state-of-the-art measuring equipment.

291 – Prenatal Exposure to Organophosphate Esters (OPEs) is Associated with Childhood Obesity in the Environmental influences on Child Health Outcomes (ECHO) Cohorts

Alicia Peterson[1], Cathleen Yoshida[2], Stacey Alexeeff[3], Jennifer Ames[4], Lyndsay Avalos[5], Emily Barrett[6], Theresa Bastain[7], Deborah Bennett[8], Lisa Croen[5], Anne Dunlop[9], Juanran Feng[10], Monique Hedderson[11], Julie Herbstman[12], Margaret Karagas[13], Cindy McEvoy[14], Thomas O'Connor[15], Megan Romano[16], Sheela Sathyanarayana[17], Susan Schantz[18], Rebecca Schmidt[19], Anne Starling[20], Leonardo Trasande[21], Tracey Woodruff[22], Qi Zhao[23], Yeyi Zhu[24], Assiamira Ferrara[5]
[1]Kaiser Permanente Northern California, Postdoctoral Research Fellow, Oakland, USA. [2]Kaiser Permanente Northern California, Data Reporting and Analytics Consultant III, Oakland, USA. [3]Kaiser Permanente Northern California, Research Scientist II, Oakland, USA. [4]Kaiser Permanente Northern California, Research Scientist I, Oakland, USA. [5]Kaiser Permanente Northern California, Research Scientist III, Oakland, USA. [6]Rutgers School of Public Health, Associate Professor of Biostatistics and Epidemiology, Piscataway, USA. [7]University of Southern California, Associate Professor, Los Angeles, USA. [8]University of California, Davis, Professor, Davis, USA. [9]Emory University School of Medicine, Associate Research Professor, Atlanta, USA. [10]Kaiser Permanente Northern California, Data Reporting and Analytics Consultant V, Oakland, USA. [11]Kaiser Permanente Northern California, Associate Director, Women's and Children's Health Section, Oakland, USA. [12]Colombia University, Professor, New York, USA. [13]Geisel School of Medicine at Dartmouth, Professor, Lebanon, USA. [14]Oregon Health & Science University, Professor, Portland, USA. [15]University of Rochester Medical Center, Professor, Rochester, USA. [16]Geisel School of Medicine at Dartmouth, Assistant Professor, Lebanon, USA. [17]University of Washington, Professor, Seattle, USA. [18]University of Illinois at Urbana-Champaign, Professor, Urbana, USA. [19]University of California, Davis, Associate Professor, Davis, USA. [20]University of North Carolina at Chapel Hill, Assistant Professor, Chapel Hill, USA. [21]New York University School of Medicine, Professor, New York, USA. [22]University of California, San Francisco, Professor, San Francisco, USA. [23]University of Tennessee Health Science Center, Associate Professor, Memphis, USA. [24]Kaiser Permanente Northern California, Research Scientist III Research Fellow, Oakland, USA

Abstract
Background: OPEs are ubiquitous chemicals with suspected endocrine-disrupting properties in utero and potential impacts on future health. We assessed the association of prenatal OPEs with childhood obesity risk within diverse ECHO cohorts.

Methods: Fifteen cohorts in ECHO (4,720 mother-child pairs) were included. OPE metabolite concentrations were measured in prenatal urine samples collected in 2006-2020. Metabolites with >50% detection were included. DPHP, DBUP/DIBP and BDCPP (>85% detection) were log2-transformed and modeled continuously, whereas BCPP, BCETP and BBOEP (53%-69% detection) were modeled as high (≥median), low (<median), and non-detected. Childhood obesity was determined using age-and-sex-specific body mass index (BMI) ≥95th percentile during infancy (0.5-1.9 years, n=3676), early-childhood (2.0-4.9 years, n=3560) and mid-childhood (5.0-10.0, years, n=2084). Modified Poisson regression models within life stages were adjusted for urinary dilution, cohort, maternal age, race/ethnicity, education, pre-pregnancy BMI, and parity.

Results: Childhood obesity prevalence was approximately 18% in each life stage. Increases in risk of obesity were observed for per log increase of DPHP in infancy (RR=1.03, 95% CI [1.02-1.05]) and mid-childhood (RR=1.04 [1.03-1.06]), BDCPP in infancy (RR=1.03 [1.01-1.04]) and DBUP/DIBP in mid-childhood (RR=1.06 [1.01-1.11]). As compared to children with non-detected prenatal levels, children with BCETP or BBOEP levels ≥ median had increased obesity risk in infancy (RR=1.07 [1.00-1.15]) or in mid-childhood (RR=1.24 [1.10-1.39]), respectively, whereas children with BCPP ≥ median had reduced risk in early-childhood (RR=0.81 [0.73-0.91].

Conclusions: Modest positive associations were observed between prenatal OPE exposures and early life obesity, warranting further investigation and examination of possible effect modification by population subgroups.

292 – Perfluoroalkyl Ether Acids in Dust From Households Near a Fluorochemical Manufacturing Facility, Fayetteville, North Carolina

Kauleen S. Proctor[1], Jane A. Hoppin[2], Sharon Zhang[3], Heather M. Stapleton[4], Detlef R.U. Knappe[2], Nadine Kotlarz[5]
[1]North Carolina State University, Research Assistant, Raleigh, USA. [2]North Carolina State University, Professor, Raleigh, USA. [3]Duke University, Research Project Manager, Durham, USA. [4]Duke University, Professor, Durham, USA. [5]North Carolina State University, Research Scholar, Raleigh, USA

Abstract
Exposure to per- and polyfluoroalkyl substances (PFAS) via household dust may be an important exposure pathway but PFAS levels in dust are not well characterized. In 2019, we collected settled dust samples from 82 GenX Exposure Study homes near a fluorochemical manufacturing facility in Fayetteville, North Carolina. Perfluoroalkyl ether acids (PFEAs), a subgroup of PFAS, were emitted from the facility to air and water. Dust samples were sieved to <150 microns, extracted and analyzed for 48 PFAS, including 12 PFEAs, by liquid chromatography-mass spectrometry.

Preliminary results for 10 of the dust samples revealed that several PFAS were commonly detected in dust, including PFEAs. Nine PFEAs (PEPA, PFO2HxA, GenX, PFO3OA, PMPA, Nafion byproduct 2, PFO4DA, PFMOAA, PFO5DoA) were detected in ≥6 of the 10 samples; PEPA and PFO2HxA were detected in all samples. Of PFEAs detected, PMPA, PFO2HxA, and GenX had the highest median concentrations (8 ng/g, 5 ng/g, and 4 ng/g, respectively). PFEA concentrations were lower than diPAP homologues and trifluoroacetic acid (TFA), which were detected in all samples (6:2/8:2 diPAP median=150 ng/g, 8:2 diPAP=90 ng/g, 6:2 diPAP=80 ng/g, TFA=30 ng/g). diPAPs are frequently detected in dust whereas TFA is not as widely reported in dust. Two PFAS of health concern are PFOA and PFOS. PFOA was detected in all samples (median=4 ng/g); linear PFOS was detected in 9 samples (median=2 ng/g). These results suggest occupants were exposed both to PFEAs and other PFAS via dust in their indoor environment.

293 – POSTER: A Novel Method for Capturing Windows of Association Between Spatiotemporal Exposures and DNAm on an Epigenome-Wide Scale

Milan Parikh[1], Erika Rasnick Manning[2], Liang Niu[3], Stephen Colegate[2], Andrew Vancil[2], Kelly Brunst[3], Cole Brokamp[4,5]
[1]Division of Biostatistics and Epidemiology, Cincinnati Children's Hospital Medical Center, Research Scientist, Cincinnati, USA. [2]Division of Biostatistics and Epidemiology, Cincinnati Children's Hospital Medical Center, Biostatistician, Cincinnati, USA. [3]Department of Environmental & Public Health Sciences, University of Cincinnati College of Medicine, Associate Professor, Cincinnati, USA. [4]Division of Biostatistics and Epidemiology, Cincinnati Children's Hospital Medical Center, Associate Professor, Cincinnati, USA. [5]Department of Pediatrics, University of Cincinnati College of Medicine, Associate Professor, Cincinnati, USA

Abstract
Associations between exposures and DNA methylation (DNAm) at individual methylation sites are traditionally characterized using epigenome-wide association studies (EWAS). However, EWAS can't account for temporally varying exposures like daily particulate matter with an aerodynamic diameter less than 2.5 microns (PM2.5). The specific timing of the effects of temporal exposures can be captured using distributed lag models (DLMs), but they haven't been used in combination with high-dimensional -omic data. Here, we propose a novel screening method for epigenome-wide DLMs that selects DNAm sites based on the presence of sensitive temporal windows, rather than on averaged exposures. We simulated neonatal DNAm with windows of association defined in 0.1% of 829,012 methylation sites based on prenatal PM2.5 concentrations from 455 subjects enrolled in the Cincinnati Combined Childhood Cohorts. We used traditional EWAS methodology to screen for DNAm sites associated with average exposures and used our proposed method to screen for DNAm sites that had a significant temporal window of association with the exposure. After using false discovery rate correction, the DLM was more sensitive than EWAS (0.782 vs 0.693) and was less specific (0.994 vs 1.000). Our proposed method is more sensitive when capturing DNAm sites with early (1.0 vs 0.46) and late (1.0 vs 0.88) susceptible windows of association with PM2.5 at the expense of a higher false positive rate. Regardless, our novel method provides a previously unavailable way to sensitively screen epigenomic, genomic, or other -omic datasets for associations with exposures localized to specific time periods.

294 – How the European Joint Programme HBM4EU supports the European Green Deal

Marike Kolossa-Gehring[1], Antje Gerofke[2], Philipp Weise[2], Nina Vogel[2], Rosa Lange[2]
[1]German Environment Agency, Head of section toxicology, health related environmental monitoring, Berlin, Germany. [2]German Environment Agency, -, Berlin, Germany

Abstract
The European Joint Programme HBM4EU, a consortium of 116 partners from 30 countries, created innovation at the science-policy interface. The work programme was designed to answer open research questions relevant for chemical policy defined by national risk assessing and managing agencies as well as EU-institutions for priority substances, i.a. phthalates. The impact of phthalates including exposure levels of the strictly regulated phthalates was assessed based on the analyses of harmonized data from the EU-wide HBM4EU Aligned Studies, existing Human Biomonitoring (HBM) data of 29 studies covering all European regions and Israel, exposure time trend analyses, identification of exposure factors, modelling approaches, analyses of occupational exposure in an E-waste study, newly developed and broadly consented HBM guidance values, systematic literature reviews on health effects, and a mixture risk assessment. Traditional biomarkers of effect and Adverse Outcome Pathways (AOPs) were inventarized and new ones identified.

Finally, policy relevant research questions were answered and diverse communication materials such as Policy Briefs, Fact Sheets, and videos were developed.

For the group of phthalates, the main message is that mixture exposure of children and adolescents to five reprotoxic phthalates is still so high, that for about 17% health impacts cannot be ruled out. The HBM4EU data supply the baseline against which the success of the

Zero Pollution ambition in the European Green Deal will be measured.

The project has received funding from the European Unions' Horizon 2020 research and innovation Programme under grant agreement No 733032 HBM4EU.

Further Information: https://www.hbm4eu.eu

295 – Large Scale Application of Low-Cost PM Sensors At The Workplace

Eelco Kuijpers[1], Sander Ruiter[2], Delphine Bard[3], John Saunders[3], Jean-Philippe Gorce[3], Nick Warren[3], Fanti Giacomo[4], Cody Wolfe[5], Emanuele Cauda[5], Anjoeka Pronk[1]
[1]TNO, Researcher, Utrecht, Netherlands. [2]TNO, Researcherq, Utrecht, Netherlands. [3]HSE, Researcher, Buxton, United Kingdom. [4]University of Insubria, Researcher, Varese, Italy. [5]NIOSH, Researcher, Pittsburg, USA

Abstract
The burden of occupational diseases is estimated between 5-7% of the total global mortality. Reducing work-related exposures provides a starting point for prevention but understanding where, why, and when these exposures occur remains a challenge. Sensors offer new possibilities for characterizing exposure, but adoption is in its infancy.

We explored the performance of low-cost particulate matter (PM) sensors for early warning of peak exposures and accurate time-weighted average exposures in six different sectors with in total 42 personal measurements. Measurements were collected with low-cost sensors (Plantower 5003/7003), high resolution monitors (APS/AeroTrak/GRIMM) and traditional gravimetrical sampling (respirable/inhalable). In addition, contextual information was collected using high resolution sensors (e.g. location) and time-integrated observations.

Highest concentrations were observed in sectors with welding fume, flour dust and quartz exposure. When comparing low-cost sensors with reference instruments both particle peaks and particle number concentrations were accurately detected (R^2 >0.61, R^2 = 0.81, respectively), while mass concentration measurements were moderately accurate (R^2 = 0.52). With high resolution worker location data and Kriging techniques, exposures were visualized in time and space. In addition, contextual information was successfully determined using sensors (e.g. for start and stop times of welding using UV sensors), guiding the interpretation of peak exposures.

In conclusion, low-cost sensors can contribute in understanding when, where and why exposure occur. Sensors can contribute in the early detection of peaks of exposure and help in the management of exposures to reduce the burden of occupational diseases.

296 – Dietary PFAS Exposure from Home Grown, Raised and Locally Captured Foods in a Midwestern PFAS-Impacted Community

Ankita Bhattacharya[1], Courtney Carignan[2], Rachel Bauer[1], Christopher Higgins[3], Heather Stapleton[4], Sarah Choyke[5], Jason Conder[6], Shenghui Zhang[7]
[1]Michigan State University, Doctoral Student, East Lansing, USA. [2]Michigan State University, Assistant Professor, East Lansing, USA. [3]Colorado School of Mines, Professor, Golden, USA. [4]Duke University, Professor, Durham, USA. [5]Colorado School of Mines, Post Doctoral Fellow, Golden, USA. [6]Geosyntec consultants, Principal Scientist, Costa Mesa, USA. [7]Duke University, Research Project Manager, Durham, USA

Abstract
Dietary exposure to PFASs has been raised as a concern among communities with PFAS-impacted water. Elevated concentrations have been identified in livestock and wildlife as well as soils and produce from home gardens in impacted areas. While elevated PFASs in drinking water clearly contribute to elevated exposures, less is known about the contribution of home-produced and locally captured foods. Therefore, we conducted a detailed exposure assessment for a Midwestern community where drinking water contamination was discovered in 2018 (>1400 ng/L PFOA+PFOS). During the growing seasons of 2020/21, we collected samples of produce and soil from the gardens of 27 homes as well as chicken eggs (n=2) and venison (n=3). Data were collected for each sample and participants completed a detailed exposure questionnaire. Samples were tested for 48 PFASs using LC-MS/MS (animal products) and LC-QTOF-MS (soil and produce) with improved methods designed to minimize bias during sample extraction and non-targeted suspect screening was performed on a subset of five soil samples. Modestly elevated concentrations of PFOS were found in eggs (3.5 ng/g) and venison (14 ng/g). PFOS was detected infrequently at low levels in produce (Max = 0.20 ng/g) with PFBA at the highest concentrations, perhaps reflecting greater plant uptake of shorter chain PFAS and/or ongoing low levels of PFBA in tap (irrigation) water. In addition to sharing

details of these findings, we will report exposure estimates for home-produced and locally captured foods using home-specific PFAS concentration and consumption frequency data.

297 – Please, sir, I want some more! How epidemiologists can give risk assessors what they want.

Carol Burns
Burns Epidemiology Consulting, President, Thompsonville, USA

Abstract
Like the quotable line in the Charles Dickens story of Oliver Twist, human health risk assessors are hungry for more. First, they wanted human data. Then they wanted more specific exposure information. Then, they promulgated quality considerations (as in the US EPA Office of Pesticide Programs 2016 Framework). It is clear that regulators are seeking opportunities to integrate epidemiology data into risk assessments. However, epidemiologists are faced with barriers of time, cost and ethics to provide more robust data that can be translational for decision making. Added to this challenge, few investigations are conducted with risk assessment in mind. While the solutions to these difficulties are complex, there are opportunities in the short term for epidemiologists to provide information which can be more informative in risk assessment. Three examples will be discussed that only require additional analyses, using data already collected. These include evaluating the shape of the exposure-response curve, harmonizing the exposure categories, and describing the direction and magnitude of the error(s). These approaches are intended to begin to bridge the gap from observational studies to risk assessment. While it will not resolve underlying data inadequacies, better awareness of what regulators want will help to fill their "bowl".

298 – Tiered Approach to Estimate Chemical Concentrations in Ambient Air and General Population Exposures

Albana (Alba) Bega, Kevin Vuilleumier, Kiet Ly, Yousuf Ahmad, Benjamin Kunstman, Edward (Ed) Lo
U.S. EPA, Env. Scientist/Exposure Assessor, D.C., USA

Abstract
The Office of Pollution Prevention and Toxics (OPPT) evaluates the risk of industrial chemical exposures to human health and the environment under the Toxic Substance Control Act (TSCA) across a variety of exposure pathways and routes. The ambient air is one such pathway and may present a primary exposure pathway of concern for the general population.

Therefore, it is important for exposure assessors to have an approach to quantitatively evaluate chemical releases, exposures, and associated risk estimates via the ambient air pathway as part of the TSCA risk evaluation process. This presentation will describe a tiered approach to quantitatively estimate concentrations in ambient air following industrial chemical releases using reasonably available data, information, and existing models. The presentation will include a description of the methodology, strengths and limitations, and the general decision-making framework for moving from tier to tier. The tiered approach begins by estimating a high-end exposure scenario to estimate ambient air concentrations and associated exposures/risks that are not specific to a facility, source category, or condition of use. Each tier of analysis considers and incorporates more specific and detailed information which may include a central tendency exposure scenario, evaluating multiple years of release data, site specific inputs, and other inputs or data to further characterize the potential exposures and associated risks to a level where conclusions related to risks can be drawn.

DISCLAIMER: The views expressed in this abstract are those of the authors and do not necessarily represent the views or the policies of the U.S.EPA.

299 – Targeted and Non-Targeted Screening for Seasonal Environmental Contamination Levels in Durham, NC Surface Waters using Liquid Chromatography-High Resolution Mass Spectrometry

Imari Walker-Franklin[1], Nateja Hale[2], Porché L. Spence[3], Julia Apoian[1], Omar E. Christian[3], Carresse Gerald[3]
[1]RTI International, Research Scientist, Research Triangle Park, USA. [2]North Carolina Central University, Student, Durham, USA. [3]North Carolina Central University, Faculty, Durham, USA

Abstract
Monitoring environmental pollutants in waterways is commonly done using targeted analysis for a small subset of chemical space. However, with limited data on the chemical contaminant profile of surface waters in Durham, NC, a combination of targeted, suspect

screening, and non-targeted analysis was opted for as the more efficient approach for sites of interest. In this study, we analyzed three sites in Durham, NC to understand variability of pollutants during the summer to winter months. Water samples were filtered, solid-phase extracted, and analyzed by both targeted and non-targeted analysis on a UHPLC-Orbitrap-MS/MS. In total, over 1306 features (1043 Positive Mode, 303 Negative mode) were detected in the samples.

Compounds confirmed with an analytical standard (Level 1 confidence) include caffeine, sucralose, and PFOS. Caffeine, a wastewater tracer, had concentrations ranging 2.28 to 13.2 ng/L, with highest concentrations detected in July and September. Principal component analysis indicated the largest difference between all features detected was by sample time point and not sample location. There were 40 compounds with a spectral library match to commercial and in house databases (Level 2 confidence; 29 in positive and 11 in negative mode). Mass lists were further used for tentative identification of other environmental contaminants. Occurrence of PFAS chemicals, pesticides, plastic additives, and personal care products were noted in the surface waters with differences over time. Some of these contaminants were known hazards to the aquatic environment. The implications of this work highlight the necessity for year-round regional monitoring of contamination in surface waters.

300 – POSTER: Human Biomonitoring of Homosalate – Applicability of Toxicokinetic Data for the Oral Pathway for the Assessment of Dermal Exposures

Katharina Ebert[1], Peter Griem[2], Tobias Weiss[3], Thomas Brüning[4], Heiko Hayen[5], Holger M. Koch[6], Daniel Bury[7]
[1]Institute for Prevention and Occupational Medicine of the German Social Accident Insurance, Institute of the Ruhr University Bochum (IPA), PhD Student, Bochum, Germany. [2]Symrise AG, Principal Expert Toxicology, Holzminden, Germany. [3]Institute for Prevention and Occupational Medicine of the German Social Accident Insurance, Institute of the Ruhr University Bochum (IPA), Head of Department, Bochum, Germany. [4]Institute for Prevention and Occupational Medicine of the German Social Accident Insurance, Institute of the Ruhr University Bochum (IPA), Director, Bochum, Germany. [5]Institute of Inorganic and Analytical Chemistry, University of Münster, University Professor, Münster, Germany. [6]Institute for Prevention and Occupational Medicine of the German Social Accident Insurance, Institute of the Ruhr University Bochum (IPA), Scientific Head of Laboratory, Bochum, Germany. [7]Institute for Prevention and Occupational Medicine of the German Social Accident Insurance, Institute of the Ruhr University Bochum (IPA), Senior Researcher, Bochum, Germany

Abstract
Homosalate (HMS) is a UV filter used worldwide in sunscreens at maximum concentrations of 10–15%, but a reduction of the safe maximum dose was recently suggested on EU level. Human biomonitoring for exposure assessment is therefore considered desirable, however, the substance has presented several challenges requiring closer investigation.

HMS is used as a mixture of cis- and trans-isomers. Previous studies did not provide any data on isomer ratios, although it is well-known that toxicokinetics of isomers may differ strongly. Indeed, in an oral metabolism study, we showed that bioavailability of cis-HMS was one order of magnitude lower than that of trans-HMS, and urinary excretion of cis-HMS-derived metabolites even two orders of magnitude lower than those derived from trans-HMS. These results emphasize the necessity to consider both isomers separately in exposure and risk assessment.

Dermal bioavailability is influenced by many different factors, such as product formulation, skin thickness etc., thereby impeding estimation of HMS uptake. In contrast, oral uptake allows a precise determination of the internal dose. However, the applicability of oral data may be questioned when dermal exposure is likely the major uptake route. Consequently, we performed a dermal sunscreen application study in four volunteers to investigate this.

We found that the ratios between excreted metabolite and parent HMS amounts were similar between the oral and dermal uptake routes. Therefore, we deem urinary excretion fractions determined after oral exposure suitable to estimate dermal uptake of HMS and thus recommend their use in human biomonitoring for assessment of total exposure.

301 – Management of Smoke Haze Exposure Among Preschool Children in Intensive Burning Area in Upper Northern Thailand

Warangkana Naksen[1], Tippawan Prapamontol[2], Nattapon Pansukul[3], Nathaporn Thongjan[4]
[1]Faculty of Public Health, Chiang Mai University, Lecturer, Chiang Mai, Thailand. [2]Research Institute for Health Sciences, Senior reseracher, Chiang Mai, Thailand. [3]Faculty of Public Health, Chiang Mai University, PhD student, Chiang Mai, Thailand. [4]Faculty of Science, Chiang Mai University, PhD student, Chiang Mai, Thailand

Abstract
In Upper Northern Thailand, smoke haze caused by open burning, especially the burning of crop residues, is a significant source of air pollution. Residents' health can be negatively impacted by air pollution, particularly vulnerable populations such as young children. According to scientific data, exposure to air pollutants is linked to some adverse health outcomes in young children, including abnormal birth outcomes, infant mortality, damaged lung function, asthma, and cancer, a higher risk of developing neurological abnormalities and childhood obesity.

In this study, we aim to monitor urine 1-hydroxypyrene (1-OHP), a biomarker of PAH exposure, in preschool children in Chiang Dao District, Ching Mai Province, an active open-burning area in Upper Northern Thailand, in order to examine the air pollutant exposure. Additionally, we concentrate on lowering exposure in these young children by using the findings from this study to create the proper community guidelines or plans or policies through community participation approach.

In total, 256 preschool school students (2–5 years old) were selected from 4 preschools in Chiang Dao District. The parents were interviewed to complete a questionnaire including information about demographic data, birth data, diet, and environmental exposure history. Spot urine samples of preschool school students were collected and analyzed for 1-OHP using high-performance liquid chromatography with a fluorescence detector. The result of urinary 1-OHP will be communicated to individuals, communities, and stakeholders. Some interventions, such as community education, a low-cost air pollutant monitoring system, and a clean room for preschools, will be further implemented and evaluated.

302 – Estimating Bias Introduced by Within-Individual Autocorrelation in a Case-Crossover Design: Simulating Daily Fine Particulate Matter and Pulmonary Exacerbation Cases in Cystic Fibrosis Patients

Stephen Colegate[1], Marepalli Rao[2], Rhonda Szczesniak[3,2], Cole Brokamp[4,5]
[1]Cincinnati Children's Hospital Medical Center, Postdoctoral Research Fellow, Cincinnati, USA. [2]University of Cincinnati College of Medicine, Professor, Cincinnati, USA. [3]Cincinnati Children's Hospital Medical Center, Professor, Cincinnati, USA. [4]Cincinnati Children's Hospital Medical Center, Associate Professor, Cincinnati, USA. [5]University of Cincinnati College of Medicine, Associate Professor, Cincinnati, USA

Abstract
Cystic fibrosis (CF) is an autosomal recessive disorder with recurring episodes of declines in lung function called pulmonary exacerbation (PEx) events. The case-crossover study design is one of the major epidemiological approaches for estimating relationships between transient exposures and acute outcomes. We evaluate associations between exacerbations and air pollution exposure with a case-crossover design and examine how frequent PEx events from the same individual can bias its conclusions. We simulated PEx cases with a known effect size and implemented the case-crossover design to study associations between daily ambient air pollution exposure and PEx prevalence. We compared the odds ratio of multiple cases contributed per person and excluded PEx cases within 7, 14, 21, and 28 days against an ideal approach where one person contributes exactly one PEx case in an immensely generated cohort. We found PEx cases resulting from the same individual with a constant baseline risk are autocorrelated, of which a washout period alleviates. The median absolute percentage error from the true effect odds ratio is smallest with a 14-day washout (5.8%) compared to a 7-day (6.1%), 21-day (6.0%) and 28-day (6.4%) period. Increasing the washout from 14 days to 21 and 28 days removes 27.7% and 55.0% useable PEx cases, respectively. We recommend a 14-day washout for a case-crossover design to balance the bias from the true odds ratio and the precision of the estimated odds ratio in PEx case-crossover studies. Broader implications for development and application of the case-crossover design in other biomedical research contexts are discussed.

303 – Estimating Chemical Concentrations Downstream from Surface Water Releases for Use in Drinking Water Exposure Assessments

Bryan Groza[1], Jason Todd[2], Craig Connolly[3], Susanna Wegner[4]
[1]US EPA, Physical Scientist, Washington, DC, USA. [2]US EPA, Ecologist, Washington, DC, USA. [3]US EPA, Biologist, Washington, DC, USA. [4]US EPA, Toxicologist, Washington, DC, USA

Abstract
Under the Toxic Substances Control Act (TSCA), the Office of Pollution Prevention and Toxics (OPPT) assesses exposure to chemicals across exposure pathways, including via drinking water, to support risk evaluations. For chemicals released to surface water that can contaminate drinking water, monitoring data may only be available for wastewater at releasing facilities, particularly for chemicals not regulated under the Safe Drinking Water Act. When monitoring data are limited, OPPT relies on models to estimate concentrations in drinking water. To assess exposure at the tap in these circumstances, the analysis requires: releasing facility location data; a process for

following the chemical from the point of release as it travels downstream; and consideration of chemical properties, environmental conditions, and treatment processes. OPPT is developing strategies for conducting national-scale chemical assessments and modeling concentrations in drinking water derived from upstream surface water release information to inform TSCA risk evaluations. These approaches include leveraging database connectivity between U.S. EPA reporting products, high resolution hydrologic data, and developing new tools to enable efficient and replicable processing of chemical releases. This presentation will describe one available approach, as well as considerations of computational strengths and limitations.

Disclaimer: The authors views expressed here do not necessarily reflect views or policies of U.S. EPA

304 – POSTER: An integrated nanosensor-smartphone-cloud platform for point-of-care biomonitoring of human exposure to pesticides

Hussian Maanaki[1,2], Letice Bussiere[1,2], Aleksandr Smirnov[3], Xiuxia Du[4], Thomas Arcury[5], Phillip Summers[6], Landon Bulter[7], Carey Pope[8], Anna Jensen[9], Komal Madamwar[10], Jun Wang[11,4]
[1]NanoDiagnostic Technology LLC, Research Associate, Kannapolis, USA. [2]University of North Carolina at Charlotte, PhD student, Charlotte, USA. [3]University of North Carolina at Charlotte, Research Assistant Professor, Charlotte, USA. [4]University of North Carolina at Charlotte, Professor, Charlotte, USA. [5]Wake Forest University School of Medicine, Professor, Winston-Salem, USA. [6]Wake Forest University School of Medicine, Scientist, Winston-Salem, USA. [7]Oklahoma State University, PhD student, Stillwater, USA. [8]Oklahoma State University, Professor t, Stillwater, USA. [9]North Carolina Farmworkers Project, Director, Benson, USA. [10]University of North Carolina at Charlotte, MS student, Charlotte, USA. [11]NanoDiagnostic Technology LLC, President, Kannapolis, USA

Abstract
We present an integrated point-of-care resistive nanosensor-smartphone-cloud platform (NSCP) for simple, reagentless, rapid, and sensitive biomonitoring of exposure to organophosphate (OP) pesticides from a minimal finger-stick blood. Due to acute or chronic OP poisoning resulted from the cholinesterase inhibition by OP pesticides, biomonitoring is critical for exposure assessment and diagnosis through measuring exposure biomarkers acetylcholinesterase (AChE) and butyrylcholinesterase (BChE) in blood during pre/post-exposure. The NSCP takes advantage of the transport properties of polyaniline nanofibers (PAnNFs) and carbon nanotubes (CNTs) to provide a sensitive Chitosan/PAnNF/CNT nanocomposite film on a gold interdigitated electrode (AuIDE). The principle of the NSCP is based on AChE and BChE facilitated hydrolysis of acetylcholine (ACh) and butyrylcholine (BCh) pre-dried on a glass fiber pad, respectively, generating protons that dopes the polyaniline resulting in increasing conductance across an AuIDE. As such, changes in conductance can be used to quantitate cholinesterase activity and assess acute/chronic pesticide exposure. Furthermore, a mobile app and cloud infrastructure were developed for this nanosensor for processing resistive data, displaying, managing, tracking, and sharing testing results. Under optimal conditions, the nanosensor demonstrated a wide dynamic range of 0.25-18.0 U/mL AChE and 0.1-5.0 U/mL BCHE in blood, respectively, and excellent reproducibility with CV of < 4%. The NSCP was validated against a radiometric method utilizing in-vitro pesticide-spiked blood samples and blood samples from farmworkers. Results demonstrated excellent correlation between two methods (r-squared of ~0.96). This platform provides a new avenue for simple, rapid, sensitive, and sensitive biomonitoring of exposure to pesticides.

305 – POSTER: Aligned and Defined: Capturing and Analyzing Real-Time Residential Environmental Exposure Data and Asthma Symptoms

Emily Cramer[1], Luz Huntington-Moskos[2], Kamal Eldeirawi[3], Sharmilee Nyenhuis[4], Kathryn Krueger[5], Olivia Bimbi[6], Barbara Polivka[7]
[1]Children's Mercy Kansas City, Associate Professor, Kansas City, USA. [2]University of Louisville School of Nursing, Associate Professor, Louisville, USA. [3]University of Illinois-Chicago, Associate Professor, Chicago, USA. [4]University of Chicago, Associate Professor, Chicago, USA. [5]University of Kansas School of Nursing, Program Manager, Kansas City, USA. [6]University of Illinois-Chicago, Graduate Student, Chicago, USA. [7]University of Kansas School of Nursing, Associate Dean for Research and Professor, Kansas City, USA

Abstract
Purpose: We will discuss the methods used in a real-world study to integrate sensor technology with an ecological momentary assessment (EMA) and assess real-time exposures on asthma symptoms.

Background: The proliferation of sensor data enables new opportunities to examine the real-time impact of environmental exposures. By linking air quality monitors to EMA apps, researchers can collect self-report data from participants at the moment of exposure and examine the real-time effect on asthma symptoms.

Methods: An EMA app was linked to an indoor air quality monitor and programmed to send push notifications when elevated levels of VOCs or PM2.5 were detected. Surveys attached to the push notifications asked participants to report their asthma symptoms and environmental exposures. During the study period, 40 participants completed 2 weeks of monitoring and reporting.

Results: Linking technologies required detailed specifications and meticulous testing. During the testing period, fluctuations in VOCs/PM2.5 caused unnecessarily frequent EMA push notifications within a short period of time. Specifications in the EMA app required balance between participant burden and the reporting interval of the device. Real-time, 5-minute, and 15-minutes averages were tested for feasibility. To analyze the data, definitions were created to identify the appropriate time window and calculate the frequency and duration of exposure, and frequency and severity of reported symptoms.

Conclusions: Strategies to align data across diverse platforms must balance accuracy and feasibility. Analytic strategies rely on detailed specifications and definitions to translate large-scale sensor data into correlational models with self-report data.

306 – Assessment of Commuting Exposure to $PM_{2.5}$ and the Commuting to Ambient $PM_{2.5}$ Ratio for Different Commuting Modes in Taiwan and Indonesia

Ming-Chien Mark Tsou[1], Shih-Chun Candice Lung[2]
[1]Research Center for Environmental Changes, Academia Sinica, Postdoctoral Research Fellow, Taipei, Taiwan. [2]Research Center for Environmental Changes, Academia Sinica, Research Fellow, Taipei, Taiwan

Abstract
Commuting exposure to $PM_{2.5}$ is one of the concerned issue in Asian countries due to the severe outdoor air pollution. The aim of this study is to monitor the commuting $PM_{2.5}$ exposure during the rush hours for various commuting modes (walking, bicycle, scooter, sedan, and bus) in Taipei, Taiwan and Bandung, Indonesia using the low-cost sensing devices. Studies were carried out on 2018 and 2019. Total of 35 and 32 subjects were recruited in Taiwan and Indonesia, respectively. Each of them monitored the commuting exposures in rush hours for 1 to 7 days for 2 seasons. Results indicated that the commuting $PM_{2.5}$ exposures depended on different commuting modes and cities. In Taipei, compared to other commuting modes, the incremental contribution to commuting $PM_{2.5}$ exposure was highest in the scooter mode. However, the contribution to commuting $PM_{2.5}$ exposure was highest in the bicycle mode in Indonesia. In addition, both in Taiwan and Indonesia, the sedan mode had the highest reduction the commuting $PM_{2.5}$ exposure among all commuting modes. Finally, the $PM_{2.5}$ commuting to ambient ratios (C/A ratios) were also calculated for different commuting models. These C/A ratios can be used to estimate the commuting $PM_{2.5}$ exposures in rush hours by the ambient $PM_{2.5}$ data according to the commuting modes.

307 – Evaluating Precision Medicine Tools in Cystic Fibrosis for Racial and Ethnic Fairness

Stephen Colegate[1], Anushka Palipana[2], Rhonda Szczesniak[3,4], Cole Brokamp[5,6]
[1]Cincinnati Children's Hospital Medical Center, Postdoctoral Research Fellow, Cincinnati, USA. [2]Cincinnati Children's Hospital Medical Center, Research Fellow, Cincinnati, USA. [3]Cincinnati Children's Hospital Medical Center, Professor, Cincinnati, USA. [4]University of Cincinnati College of Medicine, Professor, Cincinnati, USA. [5]Cincinnati Children's Hospital Medical Center, Associate Professor, Cincinnati, USA. [6]University of Cincinnati College of Medicine, Associate Professor, Cincinnati, USA

Abstract
Cystic fibrosis (CF) is often characterized by frequent episodes of acute decline in lung function called pulmonary exacerbation (PEx) events. Clinical and place-based markers were collected from the CF Foundation Patient Registry (CFFPR) from individuals aged >=6 years (2003 – 2018) to develop a classification model to accurately estimate the probability of future PEx events. We characterize group-level fairness of the model, with respect to self-reported race and ethnicity, by implementing receiver operating characteristic (ROC) curves on k-fold cross-validation model PEx predictions and their three-, six-, and twelve-month outcome. We screened predictors (and outcomes) by minority population status based on their association with race and ethnicity to identify reasons for discriminatory performance. Individuals identifying as African American have lower average sensitivity and specificity (60.4% and 61.5%, respectively) than those who identify as white (62.4% and 65.0%, respectively) or from another race (63.2% and 62.0%, respectively) with lower performance across all three timeframes. We found no significant difference in model performance between Hispanic (AUC: 67.5%) and non-Hispanic (AUC: 67.8%) groups for every timeframe. We found key factors such as CF mutation, proximity to the healthcare center, and differences in clinical utilization and treatment to explain these discrepancies. Racial and ethnic differences in precision care algorithm accuracies in CF exist and, although unintentional, can lead to prediction algorithms that could replicate systemic racism in the data and setup used to create it. Researchers should proactively address group-fairness in their designs and models to account for racial and ethnic discrimination.

308 – Human Health Risk and Risk-Benefit of Fish Consumption During Pregnancy in Suriname: Considering Mercury, Selenium, and Docosahexaenoic Acid Intake

Lissa Soares[1], Firoz Abdoel Wahid[2], Maureen Lichtveld[3], Wilco Zijlmans[4], Arti Shankar[5], Ashna Hindori-Mohangoo[6], Paul Ouboter[7], Kenneth Wan Tong You[8], Clementino Djakiman[9], Martin Shafer[10], Jeffrey Wickliffe[11]
[1]Stony Brook Medicine, Postdoctoral Associate, Stony Brook, USA. [2]University of Pittsburgh School of Public Health, Visiting Assistant Professor, Pittsburg, USA. [3]University of Pittsburgh School of Public Health, Dean, Pittsburg, USA. [4]Anton de Kom University of Suriname, Pediatrician, Paramaribo, Suriname. [5]Tulane University School of Public Health and Tropical Medicine, Professor, New Orleans, USA. [6]Foundation for Perinatal Interventions and Research in Suriname (PeriSur), Principal Investigator, Paramaribo, Suriname. [7]NeoWild, Chairman/Director, Paramaribo, Suriname. [8]Anton de Kom University of Suriname, Assistant Researcher, Paramaribo, Suriname. [9]Anton de Kom University of Suriname, Chemist, Paramaribo, Suriname. [10]University of Wisconsin-Madison & Wisconsin State Laboratory of Hygiene, Associate Scientist, Madison, USA. [11]University of Alabama at Birmingham School of Public Health, Professor and Chair, Birmingham, USA

Abstract
Fish consumption is an important source of neurotoxic methylmercury exposure and a dietary source of neuroprotective nutrients which may counteract methylmercury's neurotoxicological mechanism(s). In Suriname, the Caribbean Consortium for Research in Environmental and Occupational Health (CCREOH) conducted a dietary survey among methylmercury-exposed pregnant women to assess the risk of impaired neurodevelopment in the child.

A food frequency questionnaire measuring species-specific fish intake was administered to 1135 pregnant women upon enrollment into the CCREOH study (2016-2020). We measured mercury, selenium, and docosahexaenoic acid concentrations in commonly consumed marine and freshwater species (N=100; n≥5/species). Maternal fish intake patterns and related intake estimates of mercury, selenium, and docosahexaenoic acid were assessed using three approaches; a) US Environmental Protection Agency Reference dose (USEPA-RfD) for chronic oral exposure to methylmercury (0.1 µg mercury/kg bodyweight/day), b) FAO/WHO classification system, and c) selenium health benefit value.

Fish was consumed among 96% of CCREOH-participants residing in major coastal cities and 100% of those residing in riverside villages. Considering the diet reported by those in the coastal cities, 37% of probabilistic simulations exceeded the USEPA-RfD, and 1.6% resulted in subsequent loss of Intelligence Quotient points in the child. Among those residing in riverside villages, these values were 95% and 15%, correspondingly. Selenium Health Benefit values of consumption patterns were consistently positive.

Public health action is recommended to reduce methylmercury exposures. Public health research is recommended to assess the implications of nutrient levels in fish and how they may mitigate adverse effects on neurodevelopment.

309 – A Novel Model System to Estimate Near Real-time Personal Exposure to Hyperlocal Air Pollution – Highlights of Ongoing Testing and Development Phase

Jibran Khan[1,2,3], Ole Hertel[4], Thomas Ellermann[5], Matthias Ketzel[6,7], Steen Solvang Jensen[5]
[1]Department of Environmental Science, Aarhus University, Roskilde, Denmark, Assistant Professor, Roskilde, Denmark. [2]Danish Big Data Centre for Environment and Health (BERTHA), Department of Environmental Science and Department of Public Health, Aarhus University, Denmark, Assistant Professor, Roskilde, Denmark. [3]King's College London, United Kingdom, Visiting Researcher, London, United Kingdom. [4]Department of Ecoscience, Aarhus University, Roskilde, Denmark, Pro Dean, Roskilde, Denmark. [5]Department of Environmental Science, Aarhus University, Roskilde, Denmark, Senior Researcher, Roskilde, Denmark. [6]Department of Environmental Science, Aarhus University, Roskilde, Denmark, Professor, Roskilde, Denmark. [7]Global Centre for Clean Air Research (GCARE), University of Surrey, Guildford, United Kingdom, Visiting Professor, Guildford, United Kingdom

Abstract
Understanding personal exposure to air pollution is crucial for protecting public health and developing effective mitigation strategies. However, most published studies have used wearable sensors, requiring significant time and resources, or stochastic and probabilistic modelling approaches at coarser spatial resolutions to address this issue. In addition, model systems to estimate near real-time personal exposure to air pollution at high spatial resolution using physics and chemistry-based dispersion modelling are hardly seen. None to our knowledge in the literature. Therefore, this research aims to develop a novel model system to estimate near real-time exposure to hyperlocal air pollution. As a demonstration city, the new system is under development for the Danish Capital, Copenhagen. In short, the system tracks individuals' location in real-time using the Global Positioning System (GPS) and a PostgreSQL database with its spatial extension, PostGIS. Then, pre-processed and clean GPS coordinates are used to estimate air pollution levels (NO2, PM10, PM2.5, BC and PNC) using the Danish DEHM/UBM/AirGIS, a state-of-the-art and extensively validated human exposure modelling system. Finally,

the estimated pollution levels are displayed to the user in near real-time via a web portal. The oral presentation shall reflect the system description and operation, validation tests, strengths, limitations and possible future directions.

310 – Integrating Indoor Air Quality Sensors and Ecological Momentary Assessment to Examine the Real-time Impact of Residential Environmental Exposures on Asthma Symptoms

Kathryn Krueger[1,2], Barbara Polivka[3], Emily Cramer[4,5], Kamal Eldeirawi[6], Sharmilee Nyenhuis[7], Luz Huntington-Moskos[8], Olivia Bimbi[9]
[1]University of Kansas Medical Center, Program Manager, Kansas City, USA. [2]Rutgers University, Doctor of Public Health Student, Newark, USA. [3]University of Kansas Medical Center, Associate Dean for Research, Kansas City, USA. [4]Childrens Mercy Research Institute, Research Faculty, Kansas City, USA. [5]University of Missouri-Kansas City, Associate Professor, Kansas City, USA. [6]University of Illinois Chicago, Associate Professor, Chicago, USA. [7]University of Chicago, Associate Professor, Chicago, USA. [8]University of Louisville, Associate Professor, Louisville, USA. [9]University of Illinois Chicago, PhD Student, Chicago, USA

Abstract
Background: Effects of residential VOCs and PM2.5 exposures on adults with asthma remain unknown. Furthermore, current methods used to assess the residential environment don't capture real-time data on modifiable environmental exposures nor their effect on asthma symptoms.

Purpose: This feasibility study integrated indoor air quality monitoring and ecological momentary assessment (EMA) to evaluate type, frequency, and level of environmental exposures and asthma symptoms.

Methods: We recruited adults with asthma who previously reported high use of disinfectant/cleaning products (5 times/week or greater) and have poorly controlled asthma with oversampling of minorities. Participants received an Awair® Omni™ indoor air quality monitor to measure VOCs and PM2.5. EMA data were collected using a smartphone application. Participants received scheduled, random, and air quality event-triggered EMA notifications to assess asthma symptoms, environmental exposures, and mitigation strategies. Participants also completed acceptability, appropriateness, and feasibility survey items.

Results: 40 participants completed the study with 33% minority and ages 27-74 years. On average, participants received 19 event-triggered EMA surveys during the 14-day study period, with an average survey response rate of 77%. Air quality data revealed that 98% of participants had elevated VOCs and PM2.5 in their residences. The most common exposures reported with elevated VOCs and/or PM2.5 were cooking (16.4%), pets (11.1%), cleaning products (6.2%), and dusting (3.3%).

Conclusion: Findings will support a powered study to address the impact of home environmental exposures on asthma symptoms and develop tailored, home-based asthma interventions responsive to the changing home environment and impact home routines.

311 – POSTER: Chemical Uptake into Silicone Wristbands in a Natural Experimental Setting

Samantha Samon[1], Kate Hoffman[2], Nicholas Herkert[3], Heather Stapleton[4]
[1]Duke University, Postdoctoral Associate, Durham, USA. [2]Duke University, Assistant Research Professor, Durham, USA. [3]Duke University, Researcher, Durham, USA. [4]Duke University, Professor, Durham, USA

Abstract
Silicone wristbands are a noninvasive exposure assessment tool and an alternative to traditional biomonitoring; however, questions about their utility remain as validation studies are ongoing. This study sought to evaluate the experimental rate of uptake of 114 organic pollutants from several chemical classes (i.e., polychlorinated biphenyls (PCB), organophosphate esters (OPEs), polybrominated diphenyl ethers (PBDEs), brominated flame retardants (BFR), phthalates, pesticides, and polycyclic aromatic hydrocarbons (PAHs) over a five-day period. Adult participants (n=10) wore five silicone wristbands at the beginning of the sampling period and removed one wristband each day. By day five, 21 chemicals were detected in at least seven participants' wristbands. Uptake rates for these compounds were calculated using the geometric mean wristband concentration for each day worn. For many compounds, uptake into silicone wristbands was linear over time, but uptake rates differed by chemical class: PAH (7.71 pmol/g WB/day), PBDE (9.60 pmol/g WB/day), BFR (12.5 pmol/g WB/day), OPE (46.8 pmol/g WB/day), pesticides (50.9 pmol/g WB/day), and phthalates (5280 pmol/g WB/day). Uptake rates also varied among individuals, likely due to the impact that changing microenvironments and differing environmental chemical levels have on wristband concentrations. Additionally, while the number of target analytes detected did increase the longer wristbands were worn, the detection frequencies did not increase drastically over the sampling period, and several compounds were

detected in all day one wristbands; indicating a shorter sampling period may be sufficient to capture information about chronic exposures for some chemicals.

312 – Forecasting Gene Transfer Potential to Minimize Spreading of Bacterial Antibiotic Resistance in the Soil Environment

Itza Mendoza-Sanchez[1], Abigail Perez-Valdespino[2], Brandon Hardiman[3]
[1]Texas A&M University, Assistant Professor, College Station, TX, USA. [2]Instituto Politecnico Nacional, Associate Professor, Mexico City, Mexico. [3]Texas A&M University, Graduate Student, College Station, TX, USA

Abstract
Horizontal gene transfer (HGT) is a significant mechanism for the spread of genetic elements between bacteria in the environment. This process can lead to the acquisition of novel antibiotic resistance from environmental origins to human pathogens, posing a significant threat to public health. Our study considers that water-borne pathogens and residual levels of antibiotics are introduced into the soil environment via irrigation with treated wastewater. Despite the importance of HGT in the spread of antibiotic resistance, current methods for quantification of HGT during irrigation with treated wastewater do not take into account key factors such as antibiotic chemical structure, soil characteristics, and antibiotic selection pressure. The main goal of this study is to quantify the frequency of horizontal transfer of genes that encode for antibiotic resistance from a model water-borne pathogen (Escherichia coli) to soil bacterial communities. The frequency of HGT was quantified by conducting conjugal transfer experiments under different soil-antibiotic conditions, i.e. antibiotic concentrations and soil physicochemical characteristics. We predicted high-risk conditions using a process-based model that predicts acquisition of antibiotic resistance genes among soil bacteria as a function of horizontal gene transfer frequency, concentration of pathogenic bacteria, expected concentration of antibiotic absorbed in the soils, and antibiotic selection pressure. The findings of this investigation will contribute to a better understanding of the mechanisms of HGT in the soil environment and their impact on human health.

313 – Heat in Combination with Occupational and Environmental Exposures in an Agricultural Population

Rietta Wagoner[1,2], Nicolas Lopez-Galvez[3], Paloma Beamer[4], Robert Canales[5], Isaiah casarez[2], Leslie Farland[6], Frank von Hippel[4], Melissa Furlong[6]
[1]University of Arizona, PhD Candidate, Tucson, USA. [2]San Diego State University, Research Scientist, San Diego, USA. [3]San Diego State University, Assistant Professor, San Diego, USA. [4]University of Arizona, Professor, Tucson, USA. [5]George Washington University, Associate Professor, Washington, USA. [6]University of Arizona, Assistant Professor, Tucson, USA

Abstract
Agricultural workers in the US-Mexico border region are exposed to a combination of environmental and occupational stressors in addition to heat on a daily basis, increasing the risk for morbidity/mortality. For example, as core body temperature rises, acute kidney injury (AKI) can occur. The aim of this study is to assess the relationship between heat, environmental and occupational exposures, and AKI in agricultural workers.

A cohort of adult males working in agriculture and non-agricultural sectors in Sonora, Mexico was recruited in spring 2019 and followed through the summer. A questionnaire was administered at enrollment. Weather and anthropometric measurements were recorded. Heat exposure was estimated for each participant. AKI biomarker neutrophil gelatinase-associated lipocalin (NGAL) was measured in urine. All samples were collected once in spring and once in summer. We compared agricultural workers with non-agricultural workers and evaluated changes over time.

Agricultural workers more often spoke an Indigenous language as their primary language, had lower education rates and pay, and higher rates of migration compared to non-agricultural workers. NGAL levels in agricultural workers were significantly higher in the summer than in the spring, suggesting that NGAL varies by season and that heat exposure may affect kidney functioning.

Elevated heat can affect levels of exposure and toxicity of contaminants through physical changes and modifications in absorption, distribution, and metabolism, further complicating kidney functioning. As the number of hot days and heat waves is expected to increase globally, it is expected that vulnerable populations will experience increased heat-related morbidity and mortality.

314 – Investigating the spatiotemporal decay of PM 2.5 from 444 Indian cities using satellite remote sensing

Jonathan Gingrich[1], Varun Katoch[2], Alok Kumar[2], Sagnik Dey[3], Josh Apte[4]
[1]Dordt University, Assistant Professor of Engineering, Sioux Center, USA. [2]Indian Institute of Technology Delhi, Project Scientist, Delhi, India. [3]Indian Institute of Technology Delhi, Institute Chair Professor, Delhi, India. [4]University of California Berkeley, Associate Professor of Civil and Environmental Engineering, Berkeley, USA

Abstract
Ambient air pollution is a chief environmental risk factor for mortality worldwide and is responsible for nearly 1.6 million deaths each year in India attributed to exposure to ambient air pollution. It has been observed that particulate matter is a regionally dispersed pollutant rather than a locally dispersed pollutant. We attempted to determine the extent of the regional variation of PM2.5 in India by combining a daily, 1km x 1km satellite PM2.5 estimation for all of India as cited in Dey et al. 2020 with the "Universe of Cities" dataset developed by Angel et al. (2016). We developed buffers concentric annular buffers every 10 km from the city center to 100 km for 444 cities in India, dividing the cities into "small", "medium", and "large" cities and separating the cities located in the Indo-Gangetic Plain. Using these buffers, we aggregated the average concentration of each buffer and plotted the average decay curve. From this analysis, we determined that larger cities have a larger decay (~8% from city center to 100 km away on an annual basis), but the median overall decay is only 6% from the city center on an annual basis. When dividing the cities between the IGP and non-IGP, we observe that concentrations in areas further away from the city center are higher than they are within the city, suggesting that in the IGP especially, a regional focus on PM2.5 reduction is needed as opposed to the current individual city plans.

315 – Prenatal exposure to organophosphate ester flame retardants and behavioral outcomes in early childhood: The Environmental influences on Child Health Outcomes (ECHO) Program

Jiwon Oh[1], Lesliam Quiros-Alcala[2], Xuan Li[3], Kurunthachalam Kannan[4], Wenjie Lyu[5], Lyndsay Avalos[6], Emily Barrett[7], Theresa Bastain[8], Jessie Buckley[9], Giehae Choi[10], Patricia Cintora[11], Lisa Croen[12], Anne Dunlop[13], Akhgar Ghassabian[14], Rima Habre[8], Julie Herbstman[15], Ixel Hernandez-Castro[16], Irva Hertz-Picciotto[17,18], Margaret Karagas[19], Donghai Liang[20], Thomas O'Connor[21], Sheela Sathyanarayana[22], Susan Schantz[23], Rebecca Schmidt[17,18], Julie Schweitzer[17,18], Anne Starling[24,25], Heather Volk[26], Megan Woodbury[27], Tracey Woodruff[28], Qi Zhao[29], Yeyi Zhu[6], Deborah Bennett[17]
[1]University of California Davis, Postdoctoral Fellow, Davis, USA. [2]Johns Hopkins Bloomberg School of Public Health, Assistant Professor, Baltimore, USA. [3]Johns Hopkins Bloomberg School of Public Health, Biostatistician, Baltimore, USA. [4]New York University Grossman School of Medicine, Professor, New York, USA. [5]New York University Grossman School of Medicine, Lead Lab Analyst, New York, USA. [6]Kaiser Permanente Northern California, Research Scientist, Oakland, USA. [7]Rutgers School of Public Health, Associate Professor, Piscataway, USA. [8]University of Southern California, Associate Professor, Los Angeles, USA. [9]Johns Hopkins Bloomberg School of Public Health, Associate Professor, Baltimore, USA. [10]Johns Hopkins Bloomberg School of Public Health, Postdoctoral Fellow, Baltimore, USA. [11]University of Illinois Urbana-Champaign, Postdoctoral Fellow, Urbana, USA. [12]Kaiser Permanente Northern California, Senior Research Scientist, Oakland, USA. [13]Emory University School of Medicine, Assistant Professor, Atlanta, USA. [14]New York University Grossman School of Medicine, Associate Professor, New York, USA. [15]Columbia University Mailman School of Public Health, Professor, New York, USA. [16]University of Southern California, PhD Candidate, Los Angeles, USA. [17]University of California Davis, Professor, Davis, USA. [18]UC Davis MIND (Medical Investigations of Neurodevelopmental Disorders) Institute, Professor, Sacramento, USA. [19]Dartmouth Geisel School of Medicine, Professor, Lebanon, USA. [20]Emory University Rollins School of Public Health, Assistant professor, Atlanta, USA. [21]University of Rochester, Professor, Rochester, USA. [22]University of Washington and Seattle Children's Research Institute, Professor, Seattle, USA. [23]University of Illinois Urbana-Champaign, Professor, Urbana, USA. [24]University of North Carolina at Chapel Hill, Assistant Professor, Chapel Hill, USA. [25]University of Colorado Anschutz Medical Campus, Adjunct Assistant Professor, Aurora, USA. [26]Johns Hopkins University, Associate Professor, Baltimore, USA. [27]Northeastern University, Postdoctoral Fellow, Boston, USA. [28]University of California, San Francisco, Professor, San Francisco, USA. [29]University of Tennessee Health Science Center, Associate Professor, Memphis, USA

Abstract
Background/Aim: Organophosphate ester (OPE) flame retardants have been increasingly used in consumer products after phase-out of polybrominated diphenyl ethers due to toxicity concerns. OPE exhibit neurotoxic effects in animal studies, but epidemiological evidence on the effects of prenatal OPE exposures on child neurodevelopment remains inconclusive. We examined whether prenatal OPE exposures were associated with behavioral outcomes in early childhood.

Methods: We included 2664 mother-child dyads from 11 cohorts in the Environmental influences on Child Health Outcomes Program. Nine OPE biomarkers were quantified in a spot or first morning urine sample collected during mid- to late-pregnancy. Child behaviors were assessed using the Child Behavior Checklist preschool (ages 1.5-5) version. To examine associations between prenatal maternal urinary OPE biomarker concentrations and internalizing, externalizing, or total problems T-scores, we performed covariate-adjusted linear regression with generalized estimating equations to account for clustering at cohort level.

Results: Bis(1-chloro-2-propyl) phosphate (BCPP) was associated with increases in total problems (adj-β for low vs. non-detect=1.20, 95% CI: 0.21, 2.18; adj-β for high vs. non-detect=1.02, 95% CI: 0.06, 1.97), externalizing behaviors (adj-β for low=1.17, 95% CI: 0.22, 2.12), and internalizing behaviors (adj-β for high=0.87, 95% CI: 0.14, 1.60). The low exposure category of bis(butoxyethyl) phosphate (BBOEP), vs. non-detect, was associated with increases in total problems (adj-β=1.17, 95% CI: 0.23, 2.11), externalizing behaviors (adj-β=1.11, 95% CI: 0.22, 2.00), and internalizing behaviors (adj-β=1.03, 95% CI: 0.10, 1.97).

Conclusions: In this large, diverse U.S. population, prenatal exposure to several OPEs was associated with greater behavioral problems in early childhood.

316 – A participatory approach to designing and implementing a pilot occupational health intervention for the nail salon community in the Greater Philadelphia area

Tran Huynh[1], Duong Nguyen[2], Nga Vu[3], Lucy Robinson[4], Emily Nguyen[5], Nancy Nguyen[6], Amy Carroll-Scott[4], Igor Burstyn[4]
[1]Drexel University, Assistant Professor, Philadelphia, USA. [2]University of Pennsylvania, Research Coordinator, Philadelphia, USA. [3]Vietlead, Health Projects Senior Coordinator, Philadelphia, USA. [4]Drexel University, Associate Professor, Philadelphia, USA. [5]Vietlead, Health Projects Manager, Philadelphia, USA. [6]Vietlead, Co-Executive Director, Philadelphia, USA

Abstract
Background: The nail salon industry in the US comprises mostly immigrant-owned, small mom-and-pop salons that employ primarily first-generation immigrant workers from Asia. Because of the cultural and language barriers, both owners and workers may not avail themselves of the occupational safety resources. We formed an academic-community partnership to co-design a feasibility study and multi-level occupational health intervention targeting Vietnamese-speaking salon owners, workers, and community-based organization.

Methods: The intervention for each salon included 1) two-hour in-person training covering chemical safety, infection control, musculoskeletal prevention, and workers' rights for both the owners and their employees, 2) a tailored recommendation report for the owner, and 3) check-ins with the owners. Community partner was trained to deliver the in-language intervention. Baseline and post-intervention individual data about health symptoms and behaviors, as well as personal chemical exposures were collected and analyzed.

Results: A total of 44 participants from 12 consented salons enrolled in the study. Analysis of the differences between post-and pre-intervention showed a tendency toward reduction in self-reported symptoms in the respiratory system, skin, and eyes, neurotoxicity score, as well as some chemical exposures. We could not rule out seasonality as an explanation for these trends. Increase in self-efficacy was observed post-intervention.

Conclusions: Our study demonstrated a successful academic-community partnership to engage community members in the intervention study. While the intervention effects from feasibility study should be interpreted with caution, our preliminary results indicated that our community-based intervention is a promising approach to reduce work-related exposures and illnesses in vulnerable nail salon worker population.

317 – Evaluating Non-Targeted Analysis methods for comprehensive screening and identification of specific organic chemical tracers in soil and dust to estimate children's exposure

Natalia Soares Quinete[1], Luciana Dias Cappelini[2], Olutobi Ogunbiyi[3], Mymuna Monem[4], Emily Mejias[5], Daniel Bagner[6], Piero Gardinali[7], Florence George[6]
[1]Florida International University, Assistant Professor, North Miami, USA. [2]Florida International University, Postdoctoral Researcher, North Miami, USA. [3]Florida International University, PhD Candidate, North Miami, USA. [4]Florida International University, PhD Student, Miami, USA. [5]Florida International University, Recruitment Coordinatior, Miami, USA. [6]Florida International University, Associate Professor, Miami, USA. [7]Florida International University, Professor, North Miami, USA

Abstract
Children are vulnerable to environmental exposure of contaminants due to their small size, lack of judgement skills, as well as their proximity to dust, soil, and other environmental sources. A better understanding about the types of contaminants that children are exposed to or how their bodies remove these compounds is needed. In this study, we have implemented and optimized a methodology based on non-targeted analysis (NTA) for comprehensive assessment of chemicals in dust, soil, and in the diet. To identify specific

tracers of soil and dust ingestion by young children (from 6 months to 6 years) to improve calculations on the ingestion rates needed to estimate health risks, families with children between 6 months and 6 years of age of underrepresented groups were recruited in the greater Miami area. Samples of soil, indoor dust, food, water, and urine were provided by the caregivers, prepared by different techniques (involving online SPE, ASE, USE, QuEChERs), and analyzed by liquid chromatography-high resolution mass spectrometry (LC-HRMS). Post-processing was performed using the small molecule structure identification software, Compound Discoverer (CD) v. 3.3 and identified features were plotted using Kendrick mass defect plot and Van Krevelen diagrams to show unique patterns in different samples and regions of anthropogenic compound classifications. The performance of the NTA workflow was evaluated in terms of accuracy, precision, selectivity, and sensitivity. Common features detected in each matrix were prioritized, providing detailed information on organic contaminants ingested by young children and potential tracers in soil and dust are proposed.

318 – Air Quality Life Index (AQLI): A Global Interactive Tool Communicating the Health Impact of PM2.5

Nishka Sharma[1], Aarsh Batra[2], Christa Hasenkopf[3]
[1]Energy Policy Institute at the University of Chicago (EPIC), Research & Policy Manager, AQLI and Air Quality Programs at EPIC, Chicago, USA. [2]Energy Policy Institute at the University of Chicago (EPIC India), Research Associate, Air Quality Life Index (AQLI), Delhi, India. [3]Energy Policy Institute at the University of Chicago (EPIC), Director, AQLI and Air Quality Programs at EPIC, Munich, Germany

Abstract
Air Pollution is one of the greatest health threats facing humanity today. But, how can we effectively communicate its impact on human health to the public, specifically the implications it has on our life spans? How do we effectively nudge people to take action on air pollution?

Air Quality Life Index (AQLI), an interactive map tool, fills this gap by estimating the relationship between PM2.5 pollution and Life Expectancy, allowing users to visualize the gain in Life Expectancy they could experience if their community met the latest World Health Organization (WHO) Annual PM2.5 guideline, national standards, or another standard.

It does so by leveraging results from a pair of large cohort quasi-experimental studies set in China. The results of the studies are combined with detailed global population and satellite derived PM2.5 data to estimate the impact of particulate matter on life expectancy across the globe.

This presentation will share the latest findings of global air quality and its impact on human health, as calculated by AQLI. It will also discuss methods AQLI uses to communicate about air pollution to the public, including AQLI's interactive map tool, providing a hyperlocal view of air pollution in a given region and how it compares to other regions and across time. The presentation will also discuss how various public stakeholders have used this information to nudge their communities, cities, and countries to action.

319 – Identification of Novel Exposure Biomarkers for Short-term and Long-term Exposure to Woodsmoke Among Firefighters

Xiangtian Wang[1], Yan Lin[2], Emily Craig[3], Tenley Weil[3], Heather Stapleton[4], Junfeng(Jim) Zhang[4]
[1]Duke Unversity, Ph.D candidate, Durham, USA. [2]Duke Unversity, Postdoctoral Associate, Durham, USA. [3]Duke University, Ph.D student, Durham, USA. [4]Duke Unversity, Professor, Durham, USA

Abstract
Drastically increased frequency and intensity of wildfires due to climate change have placed firefighters at increased risk of woodsmoke exposure. Polycyclic aromatic hydrocarbons (PAHs) are hazardous compounds released during incomplete combustion. Urinary measures of hydroxylated metabolites of PAHs have been widely used to quantify short-term exposure to generic combustion sources; however, there is a lack of short-term and long-term exposure biomarkers specific to woodsmoke. This study aims to identify woodsmoke exposure biomarkers.

We hypothesize that urinary metabolites of methyl-PAHs are good short-term exposure biomarkers of woodsmoke. To test this hypothesis, we conducted a controlled study of eight healthy adults exposed to wood fires for four hours and collected urine samples before, 2, 4, 8, 12, and 24 hours after exposure. While none of the hydroxy-PAHs were increased after exposure, urinary 1-carboxylic pyrene, a metabolite of 1-methyl-pyrene, was significantly increased 2.24-fold (95% confidence interval (CI)= 1.18 – 5.77) after 4 hours of exposure, and returning to baseline levels 24 hours after exposure, suggesting urinary 1-methyl-pyrene is a short-term exposure biomarker of woodsmoke.

We further hypothesize that benzo[a]pyrene (BaP) hemoglobin adducts may serve as longer-term exposure biomarkers of fires. We measured two BaP-tetrol enantiomers (hydrolysis products of BaP-hemoglobin adducts) in red blood cells (RBC) from 29 Durham, NC. firefighters and 30 pregnant women residing in Rochester, NY. The detection rates of two types of BaP-tetrols among firefighters were 36.9% and 47.5%, while none were detected among pregnant women. These results suggest measures of BaP-tetrol may serve as fire exposure biomarkers.

320 – Prenatal Exposure to Fine Particulate Matter is Associated with Severe Respiratory Distress in Term Newborns

Markey Johnson[1], Lauren Mazur[2], Mandy Fisher[3], William D. Fraser[4,5], Liu Sun[6], Perry Hystad[7], Chintan K. Gandhi[8]
[1]Water and Air Quality Bureau, Health Canada, Research Scientist, Ottawa, Canada. [2]Department of Pediatrics, Penn State University College of Medicine, Researcher, State College, USA. [3]Environmental Health Sciences and Research Bureau, Researcher, Ottawa, Canada. [4]Université de Sherbrooke, Physician, Sherbrooke, Canada. [5]Centre de recherche du Centre Hosptialier Universitaire de Sherbrooke (CHUS), Physician, Sherbrooke, Canada. [6]Water and Air Quality Bureau, Health Canada, Researcher, Ottawa, Canada. [7]School of Biological and Population Health Sciences, Oregon State University, Professor, Corvallis, USA. [8]Department of Pediatrics, Penn State University College of Medicine, Professor, State College, USA

Abstract
Respiratory distress is the leading cause of neonatal morbidity and mortality worldwide, and prenatal exposure to air pollution exposure is associated with adverse long-term respiratory outcomes. However, the impact of air pollution on neonatal respiratory distress is not well characterized. We examined associations between prenatal exposures to fine particular matter (PM2.5) and nitrogen dioxide (NO2) with physician-diagnosed respiratory distress and related outcomes in term neonates.

We utilized data from the Maternal-Infant Research on Environmental Chemicals Study, a prospective pregnancy cohort spanning 10 Canadian cities. PM2.5 and NO2 were estimated using land use regression and satellite-derived models coupled with ground-level monitoring. Associations between air pollution and respiratory outcomes were explored using hierarchical logistic regression models adjusting for maternal and infant covariates.

Among 1321 neonates in our analyses, 7% experienced respiratory distress. Interventions included: oxygen therapy (6%), assisted ventilation (2%), and systemic antibiotics (3%); 2% of neonates received multiple interventions and 4% were admitted to neonatal intensive care units (NICUs). Median PM2.5 and NO2 concentrations during pregnancy were 8.8 μg/m3 and 18.0 ppb, respectively. Prenatal air pollution exposures were not associated with respiratory distress, oxygen therapy, or NICU admissions. However, PM2.5 exposures were strongly associated with severe respiratory distress indicators (assisted ventilation and multiple interventions) and systemic antibiotics. NO2 was also weakly associated with those outcomes.

Prenatal PM2.5 exposures increased the risk of severe respiratory distress among term newborns. These findings support the development and prioritization of public health and prenatal care strategies to increase awareness and minimize prenatal air pollution exposures.

321 – POSTER: Using Colorimetric Wipes to Characterize Lead Surface Levels in Construction Workers' Homes and Cars

Diana Ceballos[1], Cynthia Wu[2], Marie Bermudez[3], Jocelyn Fimbres[4], Julia Noguchi[3], Shane Whitacre[5], Nicholas Basta[6], Noah Buncher[7], Jonathan Levy[8], Junenette Peters[9]
[1]University of Washington, Assistant Professor, Seattle, USA. [2]University of Washington, Grad student, Seattle, USA. [3]Boston University, Research Assistant, Boston, USA. [4]Boston University, Grad student, boston, USA. [5]Ohio State University, Research Assistant, Columbus, USA. [6]Ohio State University, Professor, Columbus, USA. [7]Boston University, Assistant Professor, Boston, USA. [8]Boston University, Professor, Boston, USA. [9]Boston University, Associate Professor, Boston, USA

Abstract
Lead home inspections require a lead-certified technician using an expensive X-ray-fluorescent (XRF) analyzer and performed after a confirmed lead-poisoned child. Colorimetric wipes that change color from yellow to red in the presence of lead provide a visual and economic assessment of lead on surfaces; however, colorimetric wipes are not used in home investigations. This presentation's objectives are to 1) characterize lead levels in surfaces from 9 homes and 7 cars of Boston area families with someone working in a lead-exposed construction job and living with a child, 2) compare color readings from the colorimetric wipe results with ICP-MS

laboratory analysis. Of 80 home surfaces, 58% showed orange/red color, suggesting the presence of lead, including surfaces in contact with work items or window seals. In homes, the redder wipes included surfaces in the kitchen followed by the entrance to the home, the living room, the bedroom, and the laundry room. Of 24 car surfaces, 71% were orange/red, indicating a high likelihood of finding lead in the cars. The redder wipes in cars sampled included surfaces in the trunk, followed by the back and front seats. Surface wipe color readings were statistically significantly correlated with laboratory analysis results (Kendall's τ =0.42); 69% of wipes changed color in the presence of lead above method detection (18ug). Of the 8.6% of samples above 40ug/ft2 guideline, 71% changed color. Given their relatively low cost and immediate results, we conclude that colorimetric wipes may be a suitable alternative for home lead inspections.

322 – Advancing Exposure Science with Online Chemical Ionization Mass Spectrometry

Nusrat Jung[1], Brandon Boor[2]
[1]Purdue University, Assistant Professor, West Lafayette, Indiana, USA. [2]Purdue University, Associate Professor, West Lafayette, Indiana, USA

Abstract
Recent innovations in chemical ionization mass spectrometry (CIMS) offer exciting new opportunities for advancing exposure science research. Accurate characterization of human exposure to gas-phase volatile and semi-volatile chemicals is challenging. People are often exposed to complex mixtures of chemical contaminants that vary in space and time. Such mixtures can include hundreds of compounds that exhibit wide variations in chemical volatility and reactivity. Concentrations can span from less than a ppt to hundreds of ppm. Many emission sources in residential and occupational environments are inherently episodic and change rapidly over short time-scales. This study will introduce how CIMS can be used to improve non-targeted human exposure assessment to volatile and semi-volatile chemicals. CIMS involves soft ionization of chemical contaminants in a drift tube, followed by their separation and detection with a high-resolution time-of-flight mass spectrometer. CIMS instrumentation can detect hundreds of chemical contaminants in the air in real-time with ppt limit of detection and excellent sensitivity. Variable reagent ions (H_3O^+, O_2^+, NO^+, NH_4^+, Br^-, I^-) can be used to detect chemicals in the gas-phase across a wide range in volatility (VOCs, IVOCs, SVOCs, LVOCs). Results from recent laboratory- and field-based exposure measurements with CIMS at Purdue University will be shared. Such measurements demonstrate that CIMS is a robust analytical tool that facilitates non-targeted human exposure assessment to complex chemical mixtures produced by highly transient emission sources. Furthermore, CIMS data can be integrated with mass balance-based exposure models to quantify chemical-specific emission factors, inhalation intakes and cumulative doses, and intake fractions.

323 – Airborne Fragrance Chemicals Exposure in Nail Salons

Diana Ceballos[1], Xianqiang Fu[2], Thomas Webster[3], Chunrong Jia[4]
[1]university of WA, Assistant Professor, Seattle, USA. [2]University of Memphis, Research Assistant, Memphis, USA. [3]Boston University, Professor, Boston, USA. [4]University of Memphis, Professor, Memphis, USA

Abstract
The indoor air of nail salons is full of volatile organic compounds (VOCs), many of which have fragrances. Little is known about the chemicals used as fragrance ingredients in nail products as they are often considered trade secrets. This study aimed to identify fragrance chemicals and their sources and exposures in nail salons. Indoor air samples were collected in seven nail salons in the Greater Boston Area between November 09, 2016—June 28, 2017. Personal samples were simultaneously collected from ten nail workers during work time. Follow-up area sampling was performed in two salons one year after the initial visits. All the area and personal samples were collected using thermal desorption (TD) tubes and analyzed on a TD-gas chromatography/mass spectrometry (GC/MS) system for 55 target compounds. Sixteen compounds were detected in area samples, and they belonged to terpenes, alcohols, carbonyls, ethers, and esters. 2-Butoxyethanol, d-limonene, benzyl alcohol, and α-pinene had the highest mean concentrations of 76.6, 53.1, 16.4, and 8.7μg/m3, respectively. The concentrations displayed limited spatial variation within a salon but moderate variation over time. Personal concentrations of most compounds were highly correlated with area concentrations (Spearman 0.69-0.92). Area and personal concentrations did not show significant correlations with the ventilation rate. The factor analysis revealed common sources, including nail products, cleaning agents, and personal care products. This study reveals a subset of fragrance chemicals in nail salons and calls for future research on a full spectrum of these chemicals and their health effects among nail workers.

324 – Dioxin Biomarker Serum Levels and Determinants in Old Crow, a Western Arctic Community within Canada

Ashlyn Simpson[1], Mallory Drysdale[2], Mylène Ratelle[3], Jeremy Brammer[4], Mary Gamberg[5], Brian Laird[6]
[1]University of Waterloo, Graduate Student, Waterloo, Canada. [2]University of Waterloo, Alumni, Waterloo, Canada. [3]University of Waterloo, Adjunct Professor, Waterloo, Canada. [4]Vuntut Gwitchin First Nation, Fish and Wildlife Manager, Old Crow, Canada. [5]Gamberg Consulting, Environmental Consultant, Whitehorse, Canada. [6]University of Waterloo, Associate Professor, Waterloo, Canada

Abstract
Dioxins, furans, and dioxin-like PCBs are a group of persistent and toxic chemicals that are known to have human health effects at low levels. Dioxins and furans occur through industrial emissions or natural sources. Dioxin-like PCBs were formerly used in electrical applications before they were banned internationally (2004). These chemicals are widespread environmentally as they can contaminate air and travel distances > 100 km from any known emission source. Community concern surrounding the safety of traditional foods prompted a human biomonitoring project in Old Crow, Yukon Territory (2019). Through collaborative community engagement, dioxins and like-compounds were identified as contaminants of concern in traditional foods.

Blood samples (n=54) collected in Old Crow were used to measure exposures to seven dioxins, ten furans, and four dioxin-like PCBs. 1,2,3,6,7,8-HxCDD, 1,2,3,7,8,9-HxCDD, 1,2,3,4,6,7,8-HpCDD, OCDD, 2,3,4,7,8-PeCDF, 1,2,3,6,7,8-HxCDF, PCB 126, and PCB 169 were detected in >50% of samples. Among these congeners, the only elevated congener was PCB 169, ~2x elevated compared to the Canadian national average. No significant sex-based differences in exposure were observed. Generally, the concentrations of the detected congeners increased with age, except for 1,2,3,4,6,7,8-HpCDD. Multiple linear regression models were used to assess the potential sources of these exposures from commonly consumed traditional foods.

This research establishes the first dioxin and like-compound levels in Old Crow, furthering the information available on chemical exposures in the Arctic. Further research could be directed towards the investigation of PCB 169 exposure sources and continued monitoring of exposure levels and determinants over time.

325 – A Novel Method for Quantitation of Serum Terpenes as a Measure of Terpene Exposure

Lalith Silva[1], Cody Newman[2], Deepak Bhandari[3], Ben Blount[4]
[1]CDC, Research Chemist, Atlanta, USA. [2]CDC, Chemist, Atlanta, USA. [3]CDC, Lab Chief, Atlanta, USA. [4]CDC, Branch Chief, Atlanta, USA

Abstract
Introduction: Volatile terpenes are a class of volatile organic hydrocarbons commonly produced by vegetation and released into the atmosphere. These compounds are responsible for the scents of pine forests, citrus fruits, and flowers. Humans can be exposed to terpenes through different routes, including ingestion (e.g., eating terpene-containing foods) and inhalation (e.g., inhaling smoke, scents, or e-cigarette aerosols). We have developed a novel, high-throughput method for quantifying six terpenes (α-pinene, β-pinene, β-myrcene, 3-carene, limonene, and β-caryophyllene) in human serum to aid investigations of human exposure to terpenes.

Methods: This method employs headspace sampling with solid-phase microextraction (SPME) coupled with gas chromatography-tandem mass spectrometry to detect and quantify five monoterpenes and one sesquiterpene in the low parts-per-trillion to low parts-per-billion range. Our method can provide essential biomonitoring data to establish baseline exposure levels for terpenes in humans. We applied this method to quantify serum levels in over 2500 National Health and Nutrition Survey (NHANES) 2013-2014 surplus samples of participants ages 6 and older.

Results: The method is rugged, accurate, precise, and sensitive. The intraday and interday variability (percent error) of the method are ≤2% and ≤11%, respectively. In addition, this method showed excellent recovery in human serum (between 85 and 115% for all analytes). The assay precision ranges between 4.0% and 11%. Finally, the limits of detection ranged between 0.032 and 0.162 µg/L. Of the NHANES 2013-2014 surplus samples, α-pinene, β-pinene, and limonene had unweighted mean serum levels of 0.090, 0.094, and 1.460 µg/L, respectively. These analytes were not significantly associated with smoking status, as determined by serum cotinine or marijuana use.

Conclusion: The serum terpenes method is a sensitive, rugged, and reliable way to quantify this volatile class of organic compounds. It has been used to establish evaluate exposure levels of terpenes in the U.S. population based on NHANES 2013-2014 surplus sera. This data will be used as a baseline reference in future studies to help understand the impact of terpene exposure on human health and overall well-being.

Disclaimer: The findings and conclusions in this presentation have not been formally disseminated by the Centers for Disease Control and Prevention and should not be construed to represent any agency determination or policy.

326 – Healthcare Workers' Experiences Protecting their Families during the COVID-19 Pandemic

Diana Ceballos[1], Diana Vasquez[2], Julia Noguchi[3], Jonnathan Levy[4], Jennifer Grief Green[5], Lina Maria Ceballos Ochoa[6], William Baker[7], Elissa Perkins[7], Jessica Leibler[8]
[1]University of Washington, Assistant Professor, Seattle, USA. [2]Boston University, Graduate student, Boston, USA. [3]Boston University, Grad student, Boston, USA. [4]Boston University, Professor, Boston, USA. [5]Boston University, Associate Professor, Boston, USA. [6]EAFIT, Professor, Medellin, Colombia. [7]Boston Medical Center, Professor, Boston, USA. [8]Boston University, Assistant Professor, Boston, USA

Abstract
Purpose: This qualitative study aimed to characterize the facilitators and challenges healthcare workers faced in preventing SARS-CoV-2 transmission from work to household members and how these might inform preparations for future pandemics.

Methods: 23 remote semi-structured interviews were conducted with frontline hospital workers (i.e., 15 clinicians and 8 non-clinicians) in the summer of 2021. Participants were asked about personal precautions to prevent transmission during work, when leaving work, and during their commute and personal mitigation activities within the home. By applying journey maps to represent experiences before, during, and after work shifts, challenges and facilitators were identified from March 2020 through October 2021.

Results: Major facilitators included COVID-19 vaccines and testing, better information regarding virus transmission, and adequate PPE. Significant challenges were a lack of lockers, changing rooms, and showers at work. A significant area for improvement is the leaving work stage due to the lack of designated areas for end-of-day disinfection.

Conclusion: By exploring participants' experiences regarding preventing take-home exposures, a deeper understanding of psychosocial and environmental factors associated with occupational exposure to SARS-CoV-2 among healthcare workers was gained. To better prepare for future emergencies, our findings suggest preventing the spread of COVID-19 from healthcare settings.

327 – Developmental, Behavioral, and Environmental Determinants of Infant Dust Ingestion

Brandon Boor[1], Alexander Laskin[2], Karen Adolph[3], Laura Claxton[1], Satya Patra[4], Emily Halpern[4], Paige Thompson[4], Orit Herzberg[5], Meghan Kalvey[6], Brian Magnuson[4]
[1]Purdue University, Associate Professor, West Lafayette, Indiana, USA. [2]Purdue University, Professor, West Lafayette, Indiana, USA. [3]New York University, Professor, New York, New York, USA. [4]Purdue University, Ph.D. Student, West Lafayette, Indiana, USA. [5]New York University, Research Scientist, New York, New York, USA. [6]New York University, Ph.D. Student, New York, New York, USA

Abstract
This project elucidates determinants of indoor dust ingestion in 6- to 24-month-old infants (age range for major postural and locomotor milestones). Our 5 specific objectives test: (1) whether the frequency and characteristics of indoor mouthing events change with age and motor development stage for different indoor micro-environments; (2) how home characteristics and demographic factors affect indoor dust mass loading and dust toxicant concentration; (3) how dust transfer between surfaces is influenced by dust properties, surface features, and contact dynamics; and (4) contributions of developmental, behavioral, and socio-environmental factors to dust and toxicant-resolved dust ingestion rates. And (5), the creation of a shared corpus of video, dust, toxicant, and ingestion rate data to increase scientific transparency and speed progress through data reuse by the broader exposure science community. Our preliminary findings provide improved estimates of indoor dust ingestion rates in pre-sitting to independently walking infants and characterize inter-individual variability based on infant age, developmental stage, home environment, and parent behaviors. Dust transport experiments with a robotic contact simulator and mass balance modeling provide new mechanistic insights into the factors that affect the migration of dust from the floor to mouthed objects to an infant's mouth. The documentation of indoor dust mass loadings, size distributions, and toxicant concentrations allows for predictive understanding of the effects of the indoor environment on potential ingestion of hazardous toxicants. The shared corpus will enable data reuse to inform future research on how dust ingestion contributes to infants' total exposure to environmental toxicants.

328 – POSTER: Accuracy of self-reported distance to nearest unconventional oil and gas well in Pennsylvania, Ohio, and West Virginia residents and implications for exposure assessment

Cassandra Clark[1], Joan Casey[2], Michelle Bell[3], Desiree Plata[4], James Saiers[3], Nicole Deziel[5]
[1]Yale School of Public Health, Post-Doctoral Fellow, New Haven, USA. [2]University of Washington, Assistant Professor, Seattle, USA. [3]Yale

School of the Environment, Professor, New Haven, USA. [4]Massachusetts Institute of Technology, Associate Professor, Cambridge, USA. [5]Yale School of Public Health, Associate Professor, New Haven, USA

Abstract
Self-reported distances to industrial sources have been used in environmental epidemiology as proxies for potential exposure to environmental hazards and indicators of awareness and perception of pollutant sources. Unconventional oil and gas development (UOGD) can emit environmental pollutants and has been associated with adverse physical and mental health outcomes. In a study of 303 residents of Pennsylvania (n=92), Ohio (n=157), and West Virginia (n=54, all participants surveyed between August 2018-November 2020), we compared self-reported distance to nearest UOG well to the geographic information system (GIS)-calculated distance using Cohen's Weighted Kappa coefficient. Agreement was low (Kappa=0.18), and self-reports by Ohioans (39% accuracy) were more accurate than West Virginia residents (22%) or Pennsylvania (13%) residents ($p<0.05$). Individuals living closer to UOG wells tended to overestimate the distance to nearest well, whereas people living farther tended to underestimate. Of the demographic characteristics studied (age, race, ethnicity, household income, educational attainment), the only one related to reporting accuracy was educational attainment; residents with 12-16 years of education were more likely to estimate accurately (31.3% of the group) as compared to those with <12 (16.7%) or >16 years (16.7%). Understanding differences between objective and subjective measures of UOG well proximity could inform studies of perceived exposures or risks and could provide insights into awareness of the impacts of distance-related policies, such as setbacks.

329 – Formaldehyde Exposure and Epidemiological Evidence of Adverse Health Effects in Children

Laura Magaña[1], Luoping Zhang[2]
[1]UC Berkeley, PhD Student, Berkeley, USA. [2]UC Berkeley, Adjunct Professor Emeritus, Berkeley, USA

Abstract
Formaldehyde is a ubiquitous environmental contaminant in the building environment, such as wood-processed products, household goods and disinfectants, and clothing. Children are more susceptible to formaldehyde exposures during their development periods due to increased respiratory capacity and constant hand-to-mouth activity. They also spend much time indoors, in which formaldehyde concentrations are higher than outdoors. There is growing evidence of the association between formaldehyde exposure and children's upper respiratory outcomes, particularly asthma. However, there is limited information regarding formaldehyde exposure and non-respiratory products, including reproductive, neurodevelopmental, and cancer risks. Therefore, we conducted a literature review to identify and consolidate additional non-respiratory evidence regarding formaldehyde exposure during early life stages. A comprehensive literature search followed PRISMA 2020 guidelines to report our findings. A total of 12 studies were reviewed and evaluated with four components, formaldehyde exposure, cytogenic markers, carcinogenic risk, and neurodevelopmental outcomes. We concluded that there is mounting evidence of toxicity for associations between formaldehyde exposure and increased cancer risk, neurodevelopmental toxicity, and adverse birth outcomes. These epidemiological data also confirm that formaldehyde exposure causes other adverse health effects in children and has a significant association with asthma. Environmental regulation of indoor formaldehyde exposure and children's health outcomes may be of increasing concern and research interest.

330 – Participant perspectives related to individual chemical exposure report-back approaches in three environmental health studies

Julia Varshavsky[1], Marley Zalay[2], Jessica Trowbridge[3], Gabriel Jones[4], Tracey Woodruff[5], Asa Bradman[6], Brenda Eskenazi[7], Kim Harley[8], Katherine Kogut[9], Sara Hoover[10], Duyen Kauffman[11], Phil Brown[12], Julia Brody[13], Rachel Morello-Frosch[8]
[1]Northeastern University, Assistant Professor, Boston, USA. [2]University of California, Berkeley, PhD Student, Berkeley, USA. [3]University of California, San Francisco, Research Scientist, San Francisco, USA. [4]University of California, Berkeley, Student, Berkeley, USA. [5]University of California, San Francisco, Professor, San Francisco, USA. [6]University of California, Merced, Professor, Merced, USA. [7]University of California, Berkeley, Distinguished Professor, Berkeley, USA. [8]University of California, Berkeley, Professor, Berkeley, USA. [9]University of California, Berkeley, Study Coordinator, Berkeley, USA. [10]Office of Environmental Health Hazard Assessment, Research Supervisor, Oakland, USA. [11]Office of Environmental Health Hazard Assessment, Community Organizer, Oakland, USA. [12]Northeastern University, Professor, Boston, USA. [13]Silent Spring Institute, Director, Newton, USA

Abstract
Communicating results in biomonitoring and environmental exposure studies has become more common in recent years, due to 1) heightened researcher awareness about study participant expectations for report-back, and 2) assuaged concerns among researchers about potential risks of reporting environmental health data with inherent uncertainties around exposure sources and

health implications. Despite increased report-back and evidence of clear benefits for study participants, challenges remain regarding communicating results among populations with limited resources and capacity for mitigating environmental exposures. We compared different report-back approaches across three studies, with distinct designs that included low-income pregnant people of color and childcare workers in urban and rural communities of the San Francisco Bay Area and Salinas Valley, California. We examined commonalities and differences and observed an overall positive impact of report-back across studies, although differences in the overall study design, population characteristics, and approach to report-back (i.e., active vs. semi-active) played a role in facilitating participant understanding of their exposure results and mitigation strategies. All participants wanted their results, and many wanted to learn about exposure reduction strategies. However, geographic location, demographic characteristics, and levels of engagement with researchers, influenced results comprehension, environmental health literacy, and participant views of individual vs. community-level action for reducing exposure. This analysis contributes to the development of best practices for communicating chemical exposure results in future studies among potentially susceptible and/or highly exposed subpopulations.

331 – Protecting Patient Privacy by Matching a Synthetic, Place-Based Dataset to Real World Geolocations

Andrew Vancil[1], Anushka Palipana[2], Rhonda Szczesniak[3], Cole Brokamp[4]
[1]Cincinnati Children's Hospital Medical Center, Biostatistician/Epidemiologist, Cincinnati, USA. [2]Cincinnati Children's Hospital Medical Center, Research Fellow, Cincinnati, USA. [3]Cincinnati Children's Hospital Medical Center, Professor-Faculty, Cincinnati, USA. [4]Cincinnati Children's Hospital Medical Center, Associate Professor, Cincinnati, USA

Abstract
Data privacy continues to be an important consideration in many areas of research, including exposure assessment. While many researchers may have access to valuable, protected health information, it can be challenging to share important findings while maintaining the privacy of those involved. In this study, we present a novel approach to adapt a private place-based model (cohort n=1019) to be used publicly without sharing any sensitive information, namely the home zip code of the participant. Given a model with specific, Zip Code Tabulation Area (ZCTA)-level geomarker inputs and a clinically relevant output, we performed distance matching with those geomarkers to find geographically different, but statistically identical ZCTAs. The matched ZCTAs have geomarker values within 1% of the original home ZCTAs, giving us a simulated, geographical realization of real place-based data. From there, we can attach the model prediction to an area that has no relation to the original protected location of the participant. We found that all participants had at least one matched ZCTA within 1% of their home ZCTA and 97% of all ZCTAs had at least one participant match. The median number of participants that matched to each ZCTA was 40 while the median number of ZCTAs matched to each patient was 3,218 (33,120 matchable ZCTAs). In practice, we can take this another step and operationalize this protocol to create an interactive web tool available to clinicians and patients. In taking this adaptable approach, researchers can safely share their place-based research without exposing protected health information.

332 – Using Silicone Passive Samplers to Better Understand Personal Chemical Exposures of Structural Firefighters

Emily Bonner[1], Carolyn Poutasse[2,3], Walker Poston[4], Sara Jahnke[5], Christopher Haddock[6], Lane Tidwell[7], Kaley Adams[8], Peter Hoffman[9], Kim Anderson[10]
[1]Oregon State University, Graduate Fellow, Corvallis, USA. [2]Oregon State University, Graduate, Corvallis, USA. [3]CalEPA, Air Resources Board, Sacramento, USA. [4]NDRI, Chief Executive Officer, Kansas City, USA. [5]NDRI, Chief Operating Officer, Leawood, USA. [6]NDRI, Senior Scientist, Leawood, USA. [7]Oregon State University, Research Assistant, Leawood, USA. [8]Oregon State University, Research Assistant, Corvallis, USA. [9]Oregon State University, FSES Assistant Director, Corvallis, USA. [10]Oregon State University, Professor, Corvallis, USA

Abstract
Complex occupational hazards exist for firefighters, including mixtures of organic chemicals in smoke. These mixtures can increase cancer risk, which is a major health concern in the firefighting community. We recruited 57 structural firefighter participants from two fire departments in Kansas City to wear silicone dogtags around their necks as passive samplers. Separate tags were worn by each firefighter while on- and off-duty, each for 30 days total. Forty three polychlorinated biphenyls (PCBs), 63 polycyclic aromatic hydrocarbons (PAHs), 21 volatile organic compounds (VOCs), and 43 flame retardants were quantified using GC/MS. We hypothesized chemical concentrations were greater in paired on-duty versus off-duty tags, and in on-duty tags from the department with more calls per month. We are also exploring questionnaire data (e.g. rank, years firefighting) in relation to exposure. Certain flame retardants and PAHs were elevated in on-duty versus off-duty dogtags, and at the high call volume station. The difference in on- and off-duty concentrations and detections was highly individual for both PCBs and VOCs. Furthermore, 40% (n=30) of high call volume station

firefighters and 15% (n=27) of low call volume station firefighters had detectable PCB exposures, providing some support for association between higher call volume and greater exposures. There was no evidence of differing levels of VOC exposures between departments. This study expands the current data on firefighter chemical exposures with diverse analytes. Furthermore, the ease of sample collection and data produced in this study demonstrate the value of silicone passive sampling in assessing personal occupational chemical exposures.

333 – Comparison and Fairness of Commonly Used Traffic Pollution Measures

Andrew Vancil[1], Cole Brokamp[2]
[1]Cincinnati Children's Hospital Medical Center, Biostatistician/Epidemiologist, Cincinnati, USA. [2]Cincinnati Children's Hospital Medical Center, Associate Professor, Cincinnati, USA

Abstract
Traffic-related air pollution is a key component in a broad spectrum of contemporary, multidisciplinary research. While many researchers value including such a measure, not all have the expertise to examine the quality and fairness of the measure chosen. Currently, there are two common, "off the shelf" measures of traffic pollution: the EPA's Environmental Justice (EJ) Screen traffic proximity and the US Department of Transportation's Average Annual Daily Traffic (AADT). In this examination, we set out to assess both the accuracy and fairness of these two measures by introducing a local, parcel-based measure of highway proximity as our gold standard. With this measure, we can accurately estimate the proportion of people living in close proximity to a highway. We compared measures across three different geographies of increasing size: census block group, census tract, and zip code tabulation area (ZCTA). By comparing the mismatch and resulting regression coefficients between the two standard measures and our own measure, we observed that the AADT moving truck density tends to be more strongly correlated (R^2=0.6) with our measure than the EJ Screen traffic proximity (R^2=0.48) while also maintaining fairness across varying levels of racial makeup and community deprivation. The EJ Screen tended to overestimate pollution exposure in more deprived areas (R^2=-0.28) and areas with a higher proportion of black residents (R^2=0.22). We encourage other researchers to explore their own local measure for traffic related pollution exposure, while also suggesting that using AADT when that is not available will be the most equitable alternative.

334 – POSTER: Analyzing high-throughput assay data to advance the rapid screening of environmental chemicals for reproductive toxicity

Julia Varshavsky[1], Juleen Lam[2], Courtney Cooper[3], Patrick Allard[4], Jennifer Fung[5], Ashwini Oke[3], Ravinder Kumar[3], Josh Robinson[3], Tracey Woodruff[5]
[1]Northeastern University, Assistant Professor, Boston, USA. [2]Cal State East Bay, Assistant Professor, Hayward, USA. [3]University of California, San Francisco, Researcher, San Francisco, USA. [4]University of California, Los Angeles, Professor, Los Angeles, USA. [5]University of California, San Francisco, Professor, San Francisco, USA

Abstract
Rapid screening of environmental chemicals for reproductive toxicity is required to advance predictive toxicology and chemical assessment. High-throughput (HTP) assays have been used as platforms for rapid assessment of reproductive toxicants; however, additional germline assays and method validation would further advance knowledge regarding fertility and reproductive health. We assessed yeast (S. cerevisiae) and roundworm (C. elegans) HTP assays related to germline function and other reproductive health endpoints by screening toxicity of 125 environmental chemicals and modeling each dataset using a streamlined, semi-automated benchmark dose software (BMDS) approach. We then extracted and modeled mammalian in vivo data from the Toxicological Reference Database (ToxRefDB, Version 2.0). We potency-ranked the data set from each evidence stream by BMD and evaluated rank order correlation between datasets using Kendall's Tau Correlation. Finally, we constructed a prediction model using machine learning to incorporate model descriptors (including data on QSAR, chemical properties, etc.) and tested model performance in predicting time to half-max (yeast assay). We compared agreement of the coefficient of determination (R2) and mean absolute error (MAE) between training and testing sets (80/20 split) to evaluate the model's predictive power. We found good correlation between data sets and great promise in predicting toxicological outcomes with reasonable accuracy (Training R2: 0.97, testing R2: 0.61). The agreement between the training set and the testing set showed that the algorithm has a great potential of making accurate predictions outside its calibration domain and could play a significant role in predicting reproductive toxicity of environmental chemicals.

335 – A Novel Method for Quantitation of Serum Terpenes as a Measure of Terpene Exposure: Disclaimer: The findings and conclusions in this presentation have not been formally disseminated by the Centers for Disease Control and Prevention and should not be construed to represent any agency determination or policy

Lalith Silva[1], Cody Newman[2], Deepak Bhandari[3], Ben Blount[4]
[1]CDC, Research Chemist, Atlanta, USA. [2]CDC, Chemist, Atlanta, USA. [3]CDC, Lab Chief, Atlanta, USA. [4]CDC, Branch Chief, Atlanta, USA

Abstract
Introduction: Volatile terpenes are a class of organic hydrocarbons commonly produced by vegetation and released into the atmosphere. These compounds are responsible for the scents of pine forests, citrus fruits, and flowers. Humans can be exposed to terpenes through different routes, including ingestion and inhalation. We developed a novel, high-throughput method for quantifying six terpenes (α-pinene, β-pinene, β-myrcene, 3-carene, limonene, and β-caryophyllene) in human serum to aid investigations of human exposure to terpenes.

Methods: This method employs headspace sampling with solid-phase microextraction (SPME) coupled with gas chromatography-tandem mass spectrometry to detect and quantify six terpenes in the low parts-per-trillion to low parts-per-billion range. Our method can provide biomonitoring data to establish baseline exposure levels for terpenes in humans. We applied this method to quantify levels in over 2500 National Health and Nutrition Survey (NHANES) 2013-2014 surplus samples of participants ages 6 and older.

Results: The method is rugged, accurate, precise, and sensitive. Of the NHANES 2013-2014 samples, α-pinene, β-pinene, and limonene had unweighted mean serum levels of 0.090, 0.094, and 1.460 µg/L, respectively. These analytes weren't significantly associated with smoking status, as determined by serum cotinine or marijuana use.

Conclusion: The serum terpenes method is sensitive, rugged, and reliable. It has been used to establish exposure levels of terpenes in the U.S. population based on NHANES 2013-2014 sera. This data will be used as a baseline reference in future studies to help understand the impact of terpene exposure on human health.

336 – Non-targeted Chemical Analysis and In Vitro Epithelial Cell Line Analysis of Microplastics Dosed in Simulated Lung Leachates and Extracts

Imari Walker-Franklin[1], Zahkura Eastman[1,2], Kirstyn Tober[1], Leslimar Rios-Colon[1,3]
[1]RTI International, Research Scientist, Research Triangle Park, USA. [2]North Carolina Central University, Student, Durham, USA. [3]North Carolina Central University, Research Scientist, Durham, USA

Abstract
Human exposure to microplastics as an air pollutant can occur due to proximity to recycling facilities (shredding of polyethylene) or recycled by-products (tire rubble in soccer fields). The risk of microplastics as an air pollutant can be attributed to the physical characteristics of the inhaled particles, the ability of these small particles to release potentially hazardous polymer-associated chemicals (PACs), and the toxicological responses within the lungs. In this study, we utilized non-targeted analysis to examine the release of PACs from polyethylene spheres (PE), shredded high-density polyethylene fragments (HDPE), and tire rubble. All microplastics were placed in deionized water, simulated lung fluid, and methanol, to determine readily leachable PACs from microplastic concentrations ranging 0.001 g/mL to 0.1 g/mL. PACs structurally annotated included flame retardants, adhesives, plasticizers, UV absorbers, and antioxidants. Additionally, epithelial cell models were used to determine the toxicological response of microplastic leachates. Alveolar carcinoma cells (A-549 cells) were treated with cell culture medium previously incubated for 7 days with 0.001 g microplastics/mL. Cell proliferation and metabolic activity were assessed using an Incucyte live-cell imager and an MTT assay over 5 days. The cells incubated with leachate of tire rubble particles showed decreased proliferation and a statistically significant decrease in metabolic activity compared to the media-only control. Leachate of PE microspheres caused a decreased metabolic activity in cells; fragmented HDPE demonstrated no changes in cell proliferation or metabolic activity. These findings have implications for the human health risk posed by microplastics as global and regional air pollutants.

337 – Portable Air Cleaner Usage and Particulate Matter Exposure Reduction in an Environmental Justice Community: A Pilot Study

Alina McIntyre[1], Madeleine Scammell[2], Patrick Kinney[3], Kiran Khosla[4], Layne Benton[5], Roseann Bongiovanni[6], Chad Milando[7]

[1]Boston University School of Public Health, PhD Candidate, Boston, USA. [2]Boston University School of Public Health, Associate Professor, Boston, USA. [3]Boston University School of Public Health, Professor, Boston, USA. [4]Boston University School of Public Health, Research Assistant, Boston, USA. [5]GreenRoots, Inc., Lead Environmental Justice Organizer, Chelsea, USA. [6]GreenRoots, Inc., Executive Director, Chelsea, USA. [7]Boston University School of Public Health, Research Scientist, Boston, USA

Abstract
Particulate matter (PM) exposure is associated with premature death and adverse health impacts. Research assessing high-quality and long-term strategies to reduce indoor PM exposure in Environmental Justice (EJ) communities is limited. Through a five-month pilot study, we leveraged an academic-community partnership to assess Portable Air Cleaner (PAC) usage and household PM exposure in Chelsea, Massachusetts (MA).

We recruited eight Chelsea households where at least one family member had an asthma diagnosis. QuantAQ MODULAIR sensors measured PM1, PM2.5, PM10, temperature and relative humidity. HOBO PlugLoad Data Loggers measured Austin Air HealthMate Filter (PAC) usage. We calculated daily and hourly average PM, temperature, relative humidity, and PAC use for each participant, and used regression models to describe the impact of PAC use on PM concentrations.

After PACs were installed, all PM levels decreased, though PAC usage varied. Daily average $PM_{2.5}$ concentrations decreased by 11.1-24.3%, with post-installation hourly averages ranging from 2.4-9.8 ug/m^3. After adjusting for time trends, temperature, relative humidity and outdoor $PM_{2.5}$, hourly average indoor $PM_{2.5}$ decreased by -0.64 ug/m^3 (95% CI: -0.99, -0.29) when PACs were in use (versus not in use). Using a stronger PAC flow setting decreased hourly $PM_{2.5}$ concentrations more than a weaker setting, by 2.25 ug/m^3 (95% CI: 1.31, 3.39).

Findings support future investigation of household-level interventions to reduce PM and other indoor sources of air pollution. We demonstrate PAC efficacy, though high variability in PAC usage emphasizes the need for supplementary engagement to ensure that PAC interventions lead to reduced exposures and improved respiratory health.

338 – Enabling State Adoption of Non-Targeted Analysis (NTA) to Address Pressing Public Health Needs: Maryland, Minnesota, and California Leading the Way

Timothy Buckley[1], Jon Sobus[1], James McCord[2], Jacqueline Bangma[2], Denise MacMillan[3], Antony Williams[4], Sinisa Urban[5], Sophie Greene[6], Kaila Hanson[7], Rosie Rushing[7], Erin Newman[8], Wendy Linck[9], Erica Kalve[9], Matthew Small[10], Heather Whitehead[11]
[1]US EPA/ORD/CCTE, Senior Scientist, RTP, USA. [2]US EPA/ORD/CEMM, Chemist, RTP, USA. [3]US EPA/ORD/CCTE, Chemist, RTP, USA. [4]US EPA/ORD/CCTE, Chemistt, RTP, USA. [5]Maryland Dept. Hlth., Division Chief, Baltimore, USA. [6]MN Pollution Control Agency, Env Scientist, St. Paul, USA. [7]MN DOH, Project Coordinator, St. Paul, USA. [8]US EPA Region 5, Env Scientist, Chicago, USA. [9]CA Water Boards, Senior Eng Geologist, Sacramento, USA. [10]US EPA Region 9, Regional Science Liaison, Richmond, USA. [11]ORISE, Postdoctoral Fellow, RTP, USA

Abstract
Environmental health concerns over PFAS and other emerging contaminants are spurring state investment in, and adoption of, non-targeted analysis (NTA) methods. States are using NTA in conjunction with targeted methods to evaluate PFAS occurrence beyond the few dozen chemicals for which targeted methods exist. This is important since the chemical class is known to number in the thousands and broader contamination is suspected. States are motivated to provide a more comprehensive assessment of PFAS to inform public health protection strategies related to source attribution, exposure estimation, risk assessment, and risk management. However, there are significant barriers to state adoption of NTA including instrument cost, analytical expertise, and availability of data processing workflows and tools. Working toward a vision of "making non-targeted the new targeted", EPA/ORD has been anticipating these barriers and working to develop methods, tools, databases, and data processing workflows to enable NTA adoption. To better assess barriers and evaluate developed tools, EPA/ORD and regional offices (3,5,9) partnered with three states (MD, MN, and CA) leading efforts to deploy NTA to address differing scenarios of environmental health concern. This presentation will provide an overview of lessons learned through this collaboration, state motivation toward NTA, study plans, barriers to implementation, and strategies for overcoming those barriers. Together, this will provide a roadmap and best practices for EPA engagement as other states and regions develop an interest in pursuing NTA. Disclaimer: The views expressed in this article are those of the authors and do not represent the US EPA.

339 – Prenatal Exposures to Organophosphate Ester Metabolites and Infant Motor Development in the MADRES Cohort

Ixel Hernandez-Castro[1], Sandrah P. Eckel[2], Kurunthachalam Kannan[3], Morgan Robinson[4], Helen B. Foley[5], Brendan Grubbs[6], Deborah Lerner[7], Nathana Lurvey[7], Laila Al-Marayati[6], Rima Habre[2], Genevieve F. Dunton[8,9], Shohreh F. Farzan[10], Max T. Aung[10], Carrie V. Breton[8],

Theresa M. Bastain[2]
[1]Department of Population and Public Health Sciences, Keck School of Medicine, University of Southern California, Graduate Researcher, Los Angeles, USA. [2]Department of Population and Public Health Sciences, Keck School of Medicine, University of Southern California, Associate Professor, Los Angeles, USA. [3]Wadsworth Center, Professor, New York, USA. [4]Wadsworth Center, Researcher, New York, USA. [5]Department of Population and Public Health Sciences, Keck School of Medicine, University of Southern California, Postdoctoral Researcher, Los Angeles, USA. [6]Department of Obstetrics and Gynecology, Keck School of Medicine, University of Southern California, MD, Los Angeles, USA. [7]Eisner Health, MD, Los Angeles, USA. [8]Department of Population and Public Health Sciences, Keck School of Medicine, University of Southern California, Professor, Los Angeles, USA. [9]Department of Psychology, University of Southern California, Professor, Los Angeles, USA. [10]Department of Population and Public Health Sciences, Keck School of Medicine, University of Southern California, Assistant Professor, Los Angeles, USA

Abstract
Background: Organophosphate esters (OPEs) used as flame retardants and plasticizers are suspected neurotoxicants that may impact the development of motor skills. We evaluated overall and sex-specific associations between prenatal OPE metabolites concentrations and infant motor development.

Methods: We measured third trimester urinary concentrations of nine OPE metabolites in 329 MADRES participants. Child gross and fine motor development at 6, 9, 12, and 18-months were assessed with the Ages and Stages Questionnaire-3, using instrument-specific cutoffs for possible motor developmental delay. Specific gravity adjusted OPE metabolites were modeled continuously (log-transformed) if detected in >60% of participants; otherwise, they were modeled dichotomously (detect/non-detect). We performed longitudinal random effects mixed models and cross-sectional logistic regressions for each OPE metabolite and dichotomized fine and gross motor development category (above/below cutoff), adjusting for key covariates. Sex-specific effects for each metabolite were assessed using a statistical interaction term and fully stratified models.

Results: Thirty-one and 23% of children had gross and fine motor scores within the ASQ-3 at-risk categories at least once across infancy. A doubling in prenatal DPHP exposure was associated with 26% increased risk of fine motor delays (OR=1.26,p=0.04) and 30% increased risk at 6-months (OR=1.30,$p<0.05$). Significant interactions by infant sex were observed for associations of DPRP with gross motor development ($p<0.05$) and BCIPP with fine motor development (p=0.02), with greater risk of motor delays among female infants.

Conclusion: Prenatal OPE metabolites were associated with adverse impacts on infants' motor development, with evidence of sex-specific effects, warranting further examination of OPEs neurotoxicity.

340 – Barriers and opportunities for air quality and extreme heat information sharing: perspectives from decision-makers in two environmental justice communities

Alina McIntyre[1], Leila Heidari[1], Michael Hagen[2], Roseann Bongiovanni[3], Bianca Bowman[4], Patricia Fabian[5], Patrick Kinney[6], Madeleine Scammell[5]
[1]Boston University School of Public Health, PhD Candidate, Boston, USA. [2]Boston University School of Public Health, Research Assistant, Boston, USA. [3]GreenRoots, Inc., Executive Director, Chelsea, USA. [4]GreenRoots, Inc., Climate Justice Coordinator, Chelsea, USA. [5]Boston University School of Public Health, Associate Professor, Boston, USA. [6]Boston University School of Public Health, Professor, Boston, USA

Abstract
Air pollution and extreme heat exposure are leading causes of cardiorespiratory outcomes that disproportionately impact lower income people of color. Decision-maker awareness of these exposure and health risks is varied, community-specific, and dependent on data access and quality. Research on decision-maker perspectives regarding environmental exposures, specifically air pollution and heat, is limited.

We used semi-structured interview methodology and directed content analysis to examine air quality and heat perceptions, related data resources, and barriers to information sharing among 19 decision-makers serving residents in two environmental justice communities: Chelsea and East Boston, Massachusetts (MA).
Eighty percent of decision-makers were concerned about poor air quality, however, fewer than half were aware of Criteria Air Pollutants, National Ambient Air Quality Standards, or how their community compared to these established metrics. All decision-makers expressed concern about extreme heat, citing more "reliable" data sources such as the National Weather Service. Many commented that risk communication for different exposures is siloed and lacks a cumulative impact perspective. Leaders in the non-profit and local government sectors noted that existing environmental data sources are exclusionary; hard to access, interpret, and share with community members. Opportunities to overcome data sharing barriers are outlined, including the use of community-based methods, open-data portals, and creative science communication.

This study highlights critical challenges decision-makers face in understanding and sharing environmental health information in environmental justice communities, including improved data access and science communication tailored to local populations. Key insights highlighted the importance of sustained collaborations between academic, government, healthcare, and non-profit partners.

341 – Exposure to metal-containing aerosols emitted from material extrusion 3D printing with thermoplastic and composite materials

Qian Zhang[1], Rodney Weber[2], Todd Luxton[3], Derek Peloquin[4], Eric Baumann[5], Marilyn Black[6]
[1]UL Research Institutes, Research Scientist, Marietta, USA. [2]Georgia Institute of Technology, Professor, Atlanta, USA. [3]U.S. Environmental Protection Agency, Research Chemist, Cincinnati, USA. [4]U.S. Food and Drug Administration, Chemist, Cincinnati, USA. [5]U.S. Environmental Protection Agency, Chemist, Research Triangle Park, USA. [6]UL Research Institutes, VP and Executive Director, Marietta, USA

Abstract
Material extrusion 3D printing has been widely used in indoor environments that include industrial, educational, and residential spaces. 3D printing is an emerging source of micro- and nano-plastics. Studies have shown that 3D printing with thermoplastics emits high levels of ultrafine particles and could pose a hazard when inhaled. However, chemical composition of the emitted particles that influences health impacts, particularly metal contents, has not been well characterized. This study used ICP-MS to analyze metal (and metalloid) compositions of thermoplastic and composite filaments in addition to metals in the particles emitted during printing with these filaments. We found that Si had the highest concentrations and largest emission yields, followed by crustal metals like Na, Mg, and Al. Heavy metals like Cr, Mn, Cu, Zn, As, Pb, Cd, and Co, were also detected in particles, though with lower concentrations. We found that crustal metals tended to have higher partitioning factors from thermoplastic filaments to emitted particles, while bronze and stainless-steel powder in composite filaments were less likely to transfer from filaments to particles. We also found that the printer itself could be the source of metals like B, As, Mn, and Pb that were only detected in particles. In addition, we used an indoor exposure model to estimate heavy metal concentrations in non-industrial environments and found that metal exposure is not likely to exceed current guidelines and standards. However, estimated total inhalable fine particles could be higher than typical indoor levels and exceed ambient air quality standards for some print conditions.

343 – Prenatal and Pre-Conception Exposure to Pesticides and Associations with Childhood ADHD

Melissa Furlong[1], Paloma Beamer[2], Beate Ritz[3]
[1]University of Arizona, Assistant Professor, Tucson, USA. [2]University of Arizona, Professor, Tucson, USA. [3]University of California in Los Angeles, Professor, Los Angeles, USA

Abstract
Prenatal biomarkers of organophosphate (OP) and pyrethroid pesticide metabolites have been associated with ADHD behaviors in United States cohorts, although European cohorts report null associations for OPs. These studies used general metabolites of multiple pesticides; thus, there are no reports of specific pesticides with ADHD. We performed a large-scale study of pre-conception and prenatal exposure to specific OP, pyrethroid, and carbamate pesticides with childhood ADHD. We linked 1,029,398 Arizona Medicaid enrollees from 2008-2018 to birth certificates, and applied a validated algorithm to identify 36,229 children with ADHD. We linked pesticides reported in Arizona's Pesticide Use Registry to birth residence, and constructed exposure metrics for living within 500 meters of an application during a 90-day pre-conception window and each trimester.

We also considered any exposure during the entire prenatal period. We used logistic regression and controlled for sex, birth year, race/ethnicity, marital status, birth county, maternal education, conception season, and any pesticide exposure within 500 meters, with false discovery rate (FDR) adjustment. In trimester-specific models, oxydemeton-methyl during pre-conception (OR 2.86, 95% CI 1.74, 4.72), and acephate during the second trimester (OR 1.39, 95% CI 1.17, 1.64), were associated with ADHD at FDR<0.05. For any prenatal exposure, two OPs (chlorpyrifos; bensulide), two pyrethroids (zeta-cypermethrin; cyfluthrin), and carbamates (oxamyl; carbamates as a class), were positively associated with ADHD. The OPs acephate and oxydemeton-methyl have been banned in the European Union since the early 2000s. This may explain discrepancies in findings for OPs between populations, and implicates specific pesticides in ADHD etiology.

344 – Personal Monitoring of Ambient Environmental Exposures while Cycling on a Vegetated Trail versus the Road in an Urban Setting

Kevin Lanza[1], Brendan Allison[2], Baojiang Chen[3], Ethan Hunt[1], Katie Burford[4], Yuzi Zhang[4], Leigh Ann Ganzar[5], Preston Wilson[6], Timothy Keitt[6]
[1]UTHealth School of Public Health in Austin, Assistant Professor, Austin, USA. [2]University of Texas at Austin, Doctoral Student, Austin, USA. [3]UTHealth School of Public Health in Austin, Associate Professor, Austin, USA. [4]UTHealth School of Public Health in Austin, Doctoral Student, Austin, USA. [5]Professional Data Analysts, Senior Evaluator, Austin, USA. [6]University of Texas at Austin, Professor, Austin, USA

Abstract
Introduction: Health/environmental benefits of cycling are well-established, yet less is known about the impact of where one cycles on their ambient environmental exposures. We determined how cycling on a vegetated trail versus road influenced environmental exposures in an urban setting.

Methods: This cross-sectional study occurred in Austin, Texas, at a ~16km vegetated trail and a comparable road route. Twice daily (7:30-8:30, 17:30-18:30) on one weekday and weekend day per route in September 2021, we rode bicycles with devices measuring air temperature (HOBO MX2302A), light intensity (HOBO MX2202), PM2.5 and VOCs (Atmotube Pro), and noise (Zoom F2). After time matching data at one-minute averages, we conducted linear mixed effects modeling to test the association between route type and environmental exposures, adjusting for weekdays and afternoons.

Results: On average, the trail had lower levels of temperature (-0.7°C), light intensity (-252lx), and anthropogenic noises (-16dB) and higher levels of PM2.5 (+2.1μg/m³), VOCs (+0.1ppm), and natural noises (+3.0dB) than the road. In adjusted models, the trail was associated with different levels of temperature (β = -0.68; 95% CI [-1.15, -0.22]), light intensity (β = -0.36; 95% CI [-0.47, -0.24]), PM2.5 (β = 0.31; 95% CI [0.18, 0.44]), VOCs (β = 0.11; 95% CI [0.06, 0.17]), natural noises (β = 0.25; 95% CI [0.05, 0.45]), and anthropogenic noises (β = -0.89; 95% CI [-0.99, -0.79]) than the road ($p<0.05$).

Conclusions: Disparities in environmental exposures between a vegetated trail and road may help inform the design of urban cycling infrastructure for health and well-being.

345 – Citizen Science Approaches for Air Quality Characterization of Localized Sources in Urban Neighborhoods

Weaam Jaafar[1], Junshi Xu[2], Emily Farrar[1], Cheol-Heon Jeong[2], Arman Ganji[2], Greg Evans[3], Marianne Hatzopoulou[3]
[1]University of Toronto, Ph.D. Candidate, Toronto, Canada. [2]University of Toronto, Research Associate, Toronto, Canada. [3]University of Toronto, Professor, Toronto, Canada

Abstract
Background: In Toronto, city planners are relying urban densification to accommodate high population growth. This process has led to many "infill developments" entailing numerous construction projects impacting local communities. Understanding how neighborhoods are impacted by these localized sources can be achieved by implementing citizen science approaches.

Methods: To assess fine particulate matter ($PM_{2.5}$) levels, citizen scientists placed low-cost sensors on balconies in a neighborhood with dense construction activity. Community members were also given a survey to understand their perception of air quality in their neighborhood.

Results: Balcony sensors followed the same trend temporally, and higher $PM_{2.5}$ concentrations were measured at the west, where most construction sites are located. $PM_{2.5}$ levels at balconies on the west end ranged between 9 and 12 μg/m³ while the range was between 5 and 7 μg/m³ towards the east end. The 96 survey responses echoed the impacts of construction sites as revealed by $PM_{2.5}$ levels. Most respondents felt that air quality was poor and attributed this effect mostly to construction activities and heavy traffic. Individuals who felt 'very much affected' by air pollution also felt 'extremely disturbed' by construction projects. When asked to envision their ideal neighborhood, the top choices were good air quality, community safety, and greenspace.

Conclusion: Involving participants in the study provides them with greater knowledge of air quality. Additional surveys and focus groups will be conducted to track changes in perception after sharing results with the participants. This, coupled with air pollution data will provide insights on impacts of localized sources.

346 – POSTER: Glyphosate exposure in the United States general population

Maria Ospina[1], Lee-Yang Wong[2], Andre Schuetze[3], Pilar Morales-Agudelo[3], Meghan Vidal[2], Antonia Calafat[4]
[1]CDC, Lead Research Chemist, Atlanta, USA. [2]CDC, Statistician, Atlanta, USA. [3]CDC, Chemist, Atlanta, USA. [4]CDC, Branch Chief, Atlanta, USA
Abstract
Background: Glyphosate is the most widely used herbicide worldwide. The applied volume of glyphosate herbicides has grown about 100-fold in the United States since its introduction in the 1970s. Exposure to glyphosate in the American public has not been well characterized.

Methods: We used ion chromatography isotope dilution tandem mass spectrometry to measure glyphosate in urine from participants aged 3 years and older from the 2013-2016 National Health and Nutrition Examination Survey. We calculated distribution percentiles and estimated glyphosate exposures by age group, sex, and race/ethnicity. Survey weights and design variables were used to produce national estimates.

Results: Glyphosate weighted detection frequency was 75% (median (interquartile range): 0.392 (0.263-0.656) µg/L). Children aged 6-11 years had the highest geometric mean and 95th percentiles (both unadjusted and creatinine corrected); creatinine corrected geometric mean and 95th percentiles were greater in females than males. Asian persons had an increased proportion of non-detectable glyphosate concentrations. Participants who fasted for eight or more hours had lower glyphosate concentrations.

Conclusions: These NHANES nationally representative data suggest that there is widespread exposure to glyphosate and that food may be an important exposure source for the general U.S. population.

347 – Global river flow data developed from surface runoff based on the Curve Number Method to improve the assessment of consumer product disposal Down-The-Drain

Ryan Heisler[1], Kathleen Mcdough[2], Susan Csiszar[3], Raghu Vamshi[4], Amy Ritter[5]
[1]American Cleaning Institute, Senior Manager, Envrionmental & Scientific Affairs, Washington, USA. [2]P&G- Global Product Stewardship, Senior Director Research Fellow, Cincinnati, USA. [3]P&G- Global Product Stewardship, Senior Scientist,, Cincinnati, USA. [4]Waterborne Environmental LLC., Lead Scientist and Manager, Washington, USA. [5]Waterborne Environmental LLC, Principal Engineer, Washington, USA

Abstract
Despite advancements in exposure modeling and the growth of global and local data resources, environmental exposure assessment of chemicals disposed down-the-drain (DtD), including consumer product ingredients at the global scale within a consistent framework, has been a challenge over the years. Most assessments have been applied to specific geographies, using simplistic approaches to building a spatially resolved global assessment infrastructure. Challenges such as inconsistent or scarce data, particularly for countries with high assessment needs, have further complicated the evolution of spatially resolved global exposure assessment tools. Other limiting factors in developing detailed river flow datasets over large spatial scales have been paucity of detailed input spatial data and challenges in processing of these data as well. Reflecting on one existing and validated U.S based Down-the-Drain disposal model, iSTREEM®, we sought to apply the well-established USDA Curve Number (CN) method for spatially distributed hydrologic processing to estimate surface runoff. Publicly available global datasets for hydrologic soil groups, land cover, and precipitation were spatially processed by applying the CN equations to create a global mean annual surface runoff grid of 50 meters. Runoff was spatially combined with global hydrology of catchments and rivers from publicly available datasets to estimate daily mean annual flow (MAF) across the globe. Estimated daily MAF were compared with measured gauge flow at rivers in several countries which showed good correlation (R2 of 0.76 - 0.98). These flow estimates can be used for diverse applications at local watersheds to larger regions across the globe.

348 – Prenatal Exposures to Perfluoroalkyl and Polyfluoroalkyl Substances: Associations with Behavior in Childhood

Catherine M. Bulka[1], Lesliam Quiros-Alcala[2], Xiaoshuang Xun[3], T. Michael O'Shea[4], Kun Lu[5], Alison E. Hipwell[6], Joseph M. Braun[7], Vaia Lida Chatzi[8], Amy M. Padula[9], Dana Dabelea[10], Anne P. Starling[11], Anne Dunlop[12], Donghai Liang[12], Susan L. Schantz[13], Hyeong-Moo Shin[14], Jiwon Oh[15], Rebecca J. Schmidt[16], Tom O'Connor[17], Rebecca C. Fry[4]
[1]University of South Florida, Assistant Professor, Tampa, USA. [2]Johns Hopkins, Assistant Professor, Baltimore, USA. [3]Johns Hopkins, Senior Biostatistician, Baltimore, USA. [4]University of North Carolina, Professor, Chapel Hill, USA. [5]University of North Carolina, Associate Professor, Chapel Hill, USA. [6]University of Pittsburgh, Professor, Pittsburgh, USA. [7]Brown University, Associate Professor, Providence, USA. [8]Keck School

of Medicine of USC, Professor, Los Angeles, USA. [9]University of California, San Francisco, Associate Professor, San Francisco, USA. [10]Colorado School of Public Heatlh, Professor, Aurora, USA. [11]University of North Carolina Gillings School of Global Public Health, Assistant Professor, Chapel Hill, USA. [12]Emory University, Assistant Professor, Atlanta, USA. [13]University of Illinois Urbana-Champaign, Professor Emerita, Urbana-Champaign, USA. [14]Baylor University, Associate Professor, Waco, USA. [15]University of California, Davis, Postdoctoral Researcher, Davis, USA. [16]University of California, Davis, Associate Professor, Davis, USA. [17]University of Rochester, Professor, Rochester, USA

Abstract
Background: Animal studies suggest that prenatal exposures to per- and polyfluoroalkyl substances (PFAS) may adversely impact neurodevelopment; however, findings from epidemiologic studies remain inconclusive.

Methods: We pooled data from 10 cohorts in the ECHO program to investigate the relationship between 4 commonly detected PFAS (PFOS, PFOA, PFHxS, PFNA), measured in maternal serum during pregnancy, with parent-reported Child Behavior Check List (CBCL) scores ascertained when the children were 1.5-3 years (n=1212), 4-6 years (n=283), or 7-19 years (n=465). We fit 3 mixed-effects linear regression models (one for each age group), adjusting for potential confounders and incorporating cohort as a random intercept, to estimate differences in overall CBCL scores per 1-ng/mL increase in PFAS concentrations. Secondary analyses focused on CBCL score components measuring specific behavioral issues.

Results: PFAS were highly detected in maternal serum (>60%). After adjusting for covariates, a 1-ng/mL in maternal serum PFOA was associated with a 1.4-point (95% CI: 0.5-2.2) increase in overall CBCL score among children aged 7-19 years; secondary analyses revealed this relationship was driven by higher scores for internalizing problems. Associations for PFOS, PFHxS, and PFNA were all null, as were associations for PFOA with CBCL scores measured among younger children.

Conclusions: We found suggestive evidence that greater in utero exposures to PFOA could be associated with more internalizing behavioral problems in middle childhood through adolescence. Additional research is needed to determine if this relationship is causal and whether there are any critical windows of exposure during pregnancy.

349 – Exposure of Young Children to SVOCs in the Sleeping Micro-Environment

Sara Vaezafshar[1], Sylvia Wolk[2], Erica Phipps[3], Roxana Suehring[4], Victoria Arrandale[5], Liisa Jantunen[6], Miriam Diamond[5]
[1]University of Toronto, Ph.D. Candidate, Toronto, Canada. [2]University of Toronto, Student, Toronto, Canada. [3]Canadian Partnership for Children's Health and Environment, Executive Director, Ottawa, Canada. [4]Toronto Metropolitan University, Professor, Toronto, Canada. [5]University of Toronto, Professor, Toronto, Canada. [6]Environment and Climate Change Canada, Analytical Environmental Chemist - Manager of Organic Analysis Laboratory and Sinior Building Officer of the Centre for Atmospheric Research Experiments (CARE), Barrie, Canada

Abstract
Young children spend up to 15 hours per day sleeping. The sleeping micro-environment (SME) comprises the mattress, objects like stuffed animals, and air within the immediate crib environment. Children may be exposed to semi-volatile organic compounds (SVOCs) from one or all items in SME. Non-dietary exposure to elevated concentrations of some SVOCs has been associated with adverse effects, e.g., development of childhood asthma associated with exposure to di(2-ethylhexyl)phthalate (DEHP) and tris(2-butoxyethyl)phosphate (TBOEP).

SVOCs measured in SMEs of children aged 1-4 years in Greater Toronto Area using passive silicon rubber samplers. Participant parents (n=16) deployed 3 passive samplers in SMEs of their children: 1) on top of the mattress but under the bedsheet covered to air with aluminum foil (captured emissions from the mattress), 2) on top of the mattress cover but exposed to air, and 3) beside the bed to sample ambient bedroom air. Targeted SVOCs included phthalates (PAEs,n=8), organophosphate esters (OPEs,n=28), benzophenone (BPs,n=3) and salicylates (SALs,n=5). Phthalates and some OPEs are plasticizers, other OPEs are flame retardants, and the BPs and SALs are UV-filters/blocker/stabilizers.

Results from passive silicon rubber samplers deployed on top of mattress and covered indicated that at least one target SVOCs (i.e.PAEs, OPEs, BPs, and SALs) was in 100% of samples. The most abundant SVOCs were benzophenone-8 and 2-ethylhexyl salicylate. Dominant PAEs and OPEs were DEHP, diisononylphthalate (DiNP), tris(1,3-dichloro-2-propyl)phosphate (TDiCPP), TBOEP, and tris(chloropropyl)phosphate (TCPP). Findings highlight that mattresses and mattress covers are SVOC sources in SMEs

350 – High-resolution Profiling of Newborn Metabolome to Study the Effect of Tobacco Smoke Exposure During Pregnancy on Early Birth Outcomes in the Atlanta African American Maternal-Child Cohort

Youran Tan[1], Xiajie Lyu[2], Anne Dunlop[3], Dana Barr[4], Blake Rushing[5], Susan Mcritchie[6], Susan Sumner[7], Kaitlin Taibl[1], Stephanie Eick[8], Parinya Panuwet[8], Volha Yakimavets[9], Barry Ryan[10], Elizabeth Corwin[11], Dean Jones[12], Donghai Liang[8]
[1]Gangarosa Department of Environmental Health, Rollins School of Public Health, Emory University, PhD Student, Atlanta, USA. [2]Gangarosa Department of Environmental Health, Rollins School of Public Health, Emory University, MPH, Atlanta, USA. [3]Department of Gynecology and Obstetrics, School of Medicine, Emory University, Professor, Atlanta, USA. [4]Gangarosa Department of Environmental Health, Rollins School of Public Health, Emory University, Professor, Atlanta, USA. [5]University of North Carolina at Chapel Hill's Nutrition Research Institute, Assistant Professor, Chapel Hill, USA. [6]University of North Carolina at Chapel Hill's Nutrition Research Institute, Biostatistician, Chapel Hill, USA. [7]University of North Carolina at Chapel Hill's Nutrition Research Institute, Professor, Chapel Hill, USA. [8]Gangarosa Department of Environmental Health, Rollins School of Public Health, Emory University, Assistant Professor, Atlanta, USA. [9]Gangarosa Department of Environmental Health, Rollins School of Public Health, Emory University, Research Specialist, Atlanta, USA. [10]Gangarosa Department of Environmental Health, Rollins School of Public Health, Emory University, Research Professor, Atlanta, USA. [11]School of Nursing, Columbia University, Professor, New York City, USA. [12]Department of Medicine, School of Medicine, Emory University, Professor, Atlanta, USA

Abstract

Background: Environmental exposure to tobacco smoke during pregnancy is an established risk factor for early birth before full term (≥ 39 weeks), including preterm birth (PTB, <37 weeks) and early term birth (ETB, 37-38 weeks). Here, we aimed to characterize the molecular relationships between prenatal smoke exposure and early birth outcomes via the newborn metabolome.

Method: Participants were enrolled in the Atlanta African American Maternal-Child Cohort between 2016-2020 (N=269). Maternal urine samples were collected during 8-14 weeks gestation for cotinine and trans-3'-hydroxy-cotinine (3HC) measurement. Newborn dried blood spots (DBS) were collected at delivery for high-resolution metabolomics profiling and gestational age was ascertained from the medical record. Using an untargeted metabolomic workflow, newborn biological pathways and markers associated with exposure and outcome were estimated with multivariable regression, followed by pathway enrichment and chemical annotation, using the meet-in-the-middle approach.

Results: The geometric mean level of urinary maternal cotinine and 3HC was 7.44 μg/g creatinine and 14.9 μg/g creatinine, respectively. In total, 648 and 503 metabolomic signals were associated with maternal cotinine and 3HC levels, respectively ($p<0.05$). Seven pathways were enriched across metabolome-wide association studies, including tryptophan, leukotriene, and biopterin metabolism. Six metabolites associated with both tobacco biomarkers and early birth outcomes were confirmed with Level-1 evidence, including glutathione and 17-hydroxyprogesterone (all $p<0.05$).

Conclusion: The perturbations in newborns' metabolomics were closely involved in redox reactions and vasoconstriction. Understanding molecular mechanisms informed further health strategies to minimize the risk of prenatal smoke exposure on resultant birth effects.

351 – Saving collection efforts of mobile monitoring campaigns by transfer learning methods

Zhendong Yuan[1], Jules Kerckhoffs[2], Gerard Hoek[3], Roel Vermeulen[3]
[1]Institute for Risk Assessment Sciences, PhD student, Utrecht, Netherlands. [2]Institute for Risk Assessment Sciences, PostDoc, Utrecht, Netherlands. [3]Institute for Risk Assessment Sciences, Professor, Utrecht, Netherlands

Abstract

Generating hyperlocal air pollution maps for the whole Europe requires conducting numerous mobile monitoring campaigns in different cities. Previous studies illustrate that stepwise linear regression (SLR) can map long-term concentrations stably and accurately with only a few collection days. This work investigates whether transfer-learning-based land-use regression (LUR) models can further reduce collection durations without compromising accuracy. Our Amsterdam Google mobile campaign collected 165 days. SLR and two transfer-learning methods - CPH2AMS and Global2Local - were developed using these mobile measurements. Transfer-learning algorithms save training data by adapting external knowledge from another similar task. CPH2AMS does this by transferring mobile measurements from another city (Copenhagen), while Global2Local leverages long-term measurements from a larger geographic region (e.g., transfer stationary measurements from neighbouring countries). We plotted the performance (R2 and MAE) of all three models trained with the Amsterdam mobile measurements sliced into sequentially increasing continuous collection days (from 1 to 160) and resampled 20 times. Modelling accuracy was validated by independent long-term monitors (Palmes, n=82). The performance curves of CPH2AMS and Gloabl2Local increased faster than SLR with increasing collection days. With a small number of collection days (e.g., 1 to 20), the MAE of CPH2AMS was averagely only 62% of SLR'MAE which was 5.04 ug/m3 lower. With 10 collection days, CPH2AMS already achieved 70% accuracy of the SLR fitted with 160 collection days. Transferring mobile knowledge from another mobile-monitored city or long-term knowledge from larger geographic regions can shorten collection durations, which stimulates the progress of mobile monitoring covering entire Europe.

352 – Integrated Approach to Assess Exposure to Ultrafine Particles in Epidemiological Studies

Sabyne Audignon[1,2], Olivier Ramalho[3], Olivier Le Bihan[4], Fleur Delva[5,6], Aude Lacourt[7]
[1]INSERM BPH U1219, Epicene team, Univ. Bordeaux, Hygienist, Phd Student, Bordeaux, France. [2]Bordeaux University Hospital, Hygienist, Phd Student, Bordeaux, France. [3]Scientific and Technical Center for Building, Project manager, Marne-La-Vallée, France. [4]Air Breizh, Research project manager, Rennes, France. [5]Bordeaux University Hospital, Public health doctor, Bordeaux, France. [6]INSERM BPH U1219, Epicene team, Univ. Bordeaux, Public health doctor, Bordeaux, France. [7]INSERM BPH U1219, Epicene team, Univ. Bordeaux, Research manager, Bordeaux, France

Abstract
Ultrafine particles or UFPs (< 100 nm) are unintentionally released from many sources in all environments (e.g., engines, cooking and welding). Although the toxicological determinants of UFPs are fairly well understood, their effects on human health are poorly documented. Indeed, UFP exposure remains difficult to assess because their measurement involves new methods to characterize various parameters (e.g., surface area, number, diameter, shape and chemical composition).
In this context, we aim to provide an integrated approach to assess UFP exposure by developing indirect (non-instrumental) measurement methods.

First, an inventory of available exposure data was conducted using a literature review. Based on these data and expert judgment, two assessment tools were developed: a job-exposure matrix (MatPUF JEM) for occupational exposure and an activity-environment-exposure matrix (MA2E) for indoor exposure. They are presented like cross-tables. As rows, situations of exposure to UFPs are identified: jobs or domestic activities. As columns, the corresponding exposure is estimated according to semi-quantitative parameters (e.g., probability, intensity and frequency). Information on the nature of the particles is provided for each situation according to chemical families of UFPs (carbonaceous, metallic, mineral, wood, polymers and other organics). As a result, MatPUF JEM and MA2E provide a standardized estimate of UFP exposure in jobs and domestic activities reported by individuals.

Finally, by combining these tools with the modelling of outdoor exposure, an integrated approach to exposure assessment will be proposed. Its use in epidemiology to document individual exposure to UFPs in all settings will contribute to improving knowledge of their health effects.

353 – Spatio-temporal Trends of Air Quality in Kampala City, Uganda, 2020–2022

Mackline Ninsiima
Kampala Capital City Authority, Epidemiologist, Kampala, Uganda

Abstract
Background: Cities are more prone to poor air quality compared to non-urban areas. We assessed the spatio-temporal trends of air quality based on PM2.5 concentrations in Kampala City, Uganda.

Methods: PM2.5 concentrations generated by 25 Clarity Node Solar-Powered monitors from January 1, 2020–June 30, 2022 were abstracted from the Clarity dashboard. We computed 24-hour average PM2.5 concentration at city and division levels by combining data from all monitors in the respective areas. Average PM2.5 concentrations were compared by hour of the day to observe variations in air quality throughout the day. We used the Seasonal Mann-Kendall statistical test to assess trends in 24-hour average PM2.5.

Results: Overall, the 24-hour average PM2.5 from January 1, 2020–June 30, 2022 was 59 µg/m3 (range: 18–182 µg/m3) in Kampala City. PM2.5 concentration peaked from 10am–noon (74–73 µg/m3) and 8pm–9pm (73–77 µg/m3). There was a small decline in PM2.5 from January 2020 to June 2022 ($r = -0.27$, $p < 0.001$). PM2.5 increased during April–June [2020 ($r=0.56$, $p=0.006$), 2021 ($r=0.26$, $p=0.030$), and 2022 ($r=0.37$, $p=0.030$)] and declined during July-September, 2021 ($r=-0.43$, $p=0.008$) and January-March, 2022 ($r=-0.41$, $p=0.011$).

Conclusion: Unhealthy PM2.5 levels were observed even during times of less traffic and economic activity in Kampala City; specific, predictable annual patterns in high PM2.5 were seen. In 2022, Kampala Capital City Authority developed the Kampala City Clean Air Action Plan with interventions to be undertaken by multiple partners aimed at improving air quality.

354 – High Resolution and Spatiotemporal Place-Based Computable Exposures at Scale in the NIH ECHO Program

Erika Manning[1], Patrick Ryan[2,3], Jeff Blossom[4], Heike Luttmann-Gibson[5], Nathan Lothrop[6], Rima Habre[7], Diane R. Gold[8], Andrew Vancil[1], Joel Schwartz[8], James E. Gern[9], Cole Brokamp[2,3]
[1]Cincinnati Children's Hospital, Biostatistician/Epidemiologist, Cincinnati, OH, USA. [2]Cincinnati Children's Hospital, Professor, Cincinnati, OH, USA. [3]University of Cincinnati, Professor, Cincinnati, OH, USA. [4]Harvard University, GIS Services Manager, Cambridge, MA, USA. [5]Harvard University, Senior Research Scientist, Cambridge, MA, USA. [6]University of Arizona, Project Manager, Tucson, AZ, USA. [7]University of Southern California, Professor, Los Angeles, CA, USA. [8]Harvard University, Professor, Cambridge, MA, USA. [9]University of Wisconsin-Madison, Professor, Division Chief, Vice Chair of Research, Madison, WI, USA

Abstract
Place-based exposures, termed "geomarkers", are powerful determinants of health but are often understudied because of a lack of open data and integration tools. Existing DeGAUSS (Decentralized Geomarker Assessment for Multisite Studies) software has been successfully implemented in multisite studies, ensuring reproducibility and protection of health information. However, DeGAUSS relies on transporting geomarker data, which is not feasible for high-resolution spatiotemporal data too large to store locally or download over the internet. We expanded the DeGAUSS framework for high-resolution spatiotemporal geomarkers. Our approach stores data subsets based on coarsened location and year in an online repository, and appropriate subsets are downloaded to complete exposure assessment locally using exact date and location. We created two free and open-source DeGAUSS containers for estimation of high-resolution, daily ambient air pollutant exposures, transforming published exposure assessment models into computable exposures for geomarker assessment at scale. To validate the containers, we assigned daily PM2.5, O3, and NO2 exposure estimates for 762 children from the CCAAPS cohort, a participating ECHO cohort. Children had varying lengths of follow-up from 2001 to 2010, resulting in a total of 1,521 unique addresses and 1,691,985 days of assessed exposure. All daily exposure estimates for all three pollutants were equal to those obtained using existing exposure assessment methods within ECHO. The DeGAUSS approach was then implemented throughout ECHO, allowing us to estimate exposures for 17,587 participants across 53 cohorts, representing 1,590,931 person-months of follow up time, demonstrating the potential for reproducible and secure exposure assessment in large national and multisite cohort studies.

355 – Association of per-and polyfluoroalkyl substances (PFAS) and their mixture with inflammatory biomarkers in the Atlanta African American Maternal-Child Cohort

Youran Tan[1], Kaitlin Taibl[1], Anne Dunlop[2], Dana Barr[3], Parinya Panuwet[4], Volha Yakimavets[5], Kurunthachalam Kannan[6,7], Elizabeth Corwin[8], Barry Ryan[9], Jasmin Eatman[1], Donghai Liang[4], Stephanie Eick[4]
[1]Gangarosa Department of Environmental Health, Rollins School of Public Health, Emory University, PhD Student, Atlanta, USA. [2]Department of Gynecology and Obstetrics, School of Medicine, Emory University, Professor, Atlanta, USA. [3]Gangarosa Department of Environmental Health, Rollins School of Public Health, Emory University, Professor, Atlanta, USA. [4]Gangarosa Department of Environmental Health, Rollins School of Public Health, Emory University, Assistant Professor, Atlanta, USA. [5]Gangarosa Department of Environmental Health, Rollins School of Public Health, Emory University, Research Specialist, Atlanta, USA. [6]Department of Pediatrics, New York University School of Medicine, Professor, New York City, USA. [7]Department of Environmental Medicine, New York University School of Medicine,, Professor, New York City, USA. [8]School of Nursing, Columbia University, Professor, New York City, USA. [9]Gangarosa Department of Environmental Health, Rollins School of Public Health, Emory University, Research Professor, Atlanta, USA

Abstract
Background: Per- and polyfluoroalkyl substances (PFAS) have been identified previously as environmental contributors to adverse birth outcomes, although mechanisms remain unclear. One potential mechanism is the association between PFAS and inflammatory cytokine production. Here we examined associations between PFAS mixture and inflammatory biomarkers during both early and late pregnancy.

Method: Serum was collected from pregnant women enrolled in the Atlanta African American Maternal-Child Cohort (N=425). Serum concentrations of multiple PFAS including PFNA, PFOA, PFOS, and PFHxS were measured and detected in >90% samples at 8-14 weeks gestation. Serum concentrations of interferon-γ [IFN- γ], interleukin 6 [IL-6], IL-10, tumor necrosis factor-α [TNF-α]) and C-reactive protein [CRP]) were measured at 8-14 weeks and/or 24-30 weeks gestation. The mixture effect of PFAS on each inflammatory biomarker was examined with quantile g-computation, Bayesian kernel machine regression (BKMR) and Bayesian Weighted Sums (BWS).

Results: Across all models, PFAS mixture was associated with increased IFN-γ, IL-10 and TNF-α at both timepoints, with the strongest effects being observed in late pregnancy. Using quantile g-computation, increasing concentrations of PFAS mixture were associated with a 29% (95%CI:18.0%-40.7%) TNF-α increase during late pregnancy. Similarly, using BWS, summed PFAS concentrations were associated with increased TNF-α in late pregnancy (summed effect= 0.29, 95% highest posterior density [HDP] 0.17, 0.41 for late pregnancy). Results were consistent with BKMR and PFOS was the main driver of overall mixture effect in three methods.

Conclusion: An increase in prenatal PFAS exposure is associated with an increase in multiple pro-inflammatory cytokines, potentially contributing to PTB.

356 – POSTER: Methods for linking housing-unit-level characteristics with residential addresses in the electronic health record: A comparison of two approaches using messy, real-world addresses

Erika Manning[1], Qing Duan[1], Stuart Taylor[2], Alexandra M.S. Corley[3,4], Joseph Michael[5], John Egbo[6], Benjamin Foley[7], Chidiogo Anyigbo[3,4], Landon Krantz[8], Bijan Ketabchi[8], Ryan Gillette[9], Jeffrey Steller[10], Ndidi Unaka[3], David Hartley[3,4], Andrew F. Beck[3,4], Cole Brokamp[3,4]
[1]Cincinnati Children's Hospital, Biostatistician/Epidemiologist, Cincinnati, OH, USA. [2]Cincinnati Children's Hospital, Healthcare Economist, Cincinnati, OH, USA. [3]Cincinnati Children's Hospital, Professor, Cincinnati, OH, USA. [4]University of Cincinnati, Professor, Cincinnati, OH, USA. [5]Cincinnati Children's Hospital, Lead Analyst - Quality, Cincinnati, OH, USA. [6]Cincinnati Children's Hospital, Analyst - Quality, Cincinnati, OH, USA. [7]Cincinnati Children's Hospital, Clinical Research Coordinator, Cincinnati, OH, USA. [8]Cincinnati Children's Hospital, Clinical Fellow, Cincinnati, OH, USA. [9]Cincinnati Children's Hospital, Senior Specialist - Project Management, Cincinnati, OH, USA. [10]Cincinnati Children's Hospital, Program Manager, Cincinnati, OH, USA

Abstract
Housing characteristics can be predictive of adverse health outcomes in children, including for conditions such as asthma, depression, blood lead poisoning, and more. County auditors maintain databases of housing characteristics, conditions, and code violations, but address formatting inconsistencies present barriers when linking housing-unit-level data to patient-level data. We matched addresses using machine learning and token expansion and hashing methods. The python package csvdedupe parses addresses using extant natural language processing and creates rules for matching. As an alternative, we developed an R package called {parcel} that parses and expands addresses into all possible addresses based on abbreviations. Addresses are then hashed to create "hashdress" identifiers used to link "messy" addresses from different sources. We used both methods to match residential addresses from all hospitalizations at Cincinnati Children's from 2016 to 2021 to housing data available from the Hamilton County (OH) Auditor Database. The machine learning and token methods matched 58% and 57% of 32,885 unique addresses in Hamilton County, respectively; 52% were matched to the same parcel by both methods, 2% were matched to different parcels, 4% were matched by the machine learning method only, 3% were matched by the token method only, and 39% were not matched by either method. Using a combination of the two methods, 30% of the 124,173 hospitalizations had an address that matched a parcel in Hamilton County. Linking addresses to parcel-level information such as condition/value, tenure, and history of housing code violations could be useful for informing policy and identifying at-risk subpopulations.

357 – Application of alternative environmental and biomonitoring matrices for exposure evaluation in collaborative European human biological monitoring surveys on occupational exposure

Radu Duca[1,2], Emilie Hardy[3], Matteo Creta[3], Maria - Mirela Ani[4], An Van Nieuwenhuyse[5,6], Paul Scheepers[7], Kate Jones[8], Susana Viegas[9], Adrian Covaci[10], Simo Porras[11], Tiina Santonen[12]
[1]Nathional Health Laboratory (LNS), Head of Unit, Dudelange, Luxembourg. [2]University of Leuven (KU Leuven), Senior researcher, Leuven, Belgium. [3]Nathional Health Laboratory (LNS), Technical Responsible, Dudelange, Luxembourg. [4]Nathional Health Laboratory (LNS), MD, Dudelange, Luxembourg. [5]Nathional Health Laboratory (LNS), Head of Department, Dudelange, Luxembourg. [6]University of Leuven (KU Leuven), Prof, Leuven, Belgium. [7]Radboud University, Prof, Nijmegen, Netherlands. [8]Health and Safety Executive, Team lead, Stockport, United Kingdom. [9]NOVA National School of Public Health, Prof, Lisbon, Portugal. [10]University of Antwerp, Prof, Antwerpen, Belgium. [11]Finnish Institute of Occupational Health, Senior researcher, Helsinki, Finland. [12]Finnish Institute of Occupational Health, Prof, Helsinki, Finland

Abstract
Occupational biomonitoring studies performed under European Human Biomonitoring Initiative (HBM4EU) and Partnership for the Assessment of Risks from Chemicals (PARC) are intended to support European circular economy strategies and identify challenges related to occupational health and safety. In this context, apart from the more classical industrial hygiene (dust fraction, wipes) and biomonitoring (urine and blood) matrices, alternative matrices like silicon wristbands, settled dust (for industrial hygiene measurement) and like hair (for biomonitoring) were explored. The results obtained for the alternative environmental and biological matrices could distinguish the different groups of workers. For instance, in E-Waste survey higher levels of metals were obtained in the hair samples from battery worker (Cd: 0.10 – 3.10 µg/g; Cr: 2.85 - 141 µg/g; Pb: 0.69 – 751 µg/g) vs controls (Cd: 0.10 – 0.12 µg/g; Cr: 3.33 - 21.6 µg/g; Pb: 0.38 – 4.79 µg/g). Hair levels were further correlated with the settled dust metals levels. Moreover, positive correlation between settled dust data and urinary metabolite concentration was found for all phthalates, with the strongest correlation for DEHP (RSperman 0.689,

p<0.05). Also, the correlation between PBDE and PCB concentrations in serum, wristband, and settled dust was evaluated. Overall, alternative matrices are giving access to an extended window of exposure reflecting longer periods of exposure, which are not captured by the classical matrices.

358 – Characterizing Vulnerability, the Public Health Exposome, and Cumulative Impacts to Address Environmental Health Disparities

Andrew Geller[1], Charles Lee[2]
[1]EPA, Senior Science Advisor, RTP, USA. [2]EPA, Senior Policy Advisor, Washington, USA

Abstract
EPA has traditionally used its source-exposure-dose-effect paradigm for risk assessment of single chemicals and chemical mixtures. With an increased emphasis on equity and environmental justice, EPA and some states and municipalities have committed to incorporate the cumulative impacts of exposures to combinations of chemical and non-chemical stressors and their effects on health, well-being, and quality of life outcomes into assessments for permitting and other risk management actions. This presentation explores how the exposome and new approach methodologies can be brought to bear on the challenge of characterizing cumulative impacts and acting to reduce environmental health disparities. It examines how vulnerability can result from the literal biological embodiment of the material and social world in which we live. Exposure to stressors from the chemical, social, and built environment can perturb biological pathways resulting in early cellular changes. If these changes surpass tipping point thresholds, they can further trigger long-lasting altered biological function, resulting in vulnerability or morbidity. The vulnerable phenotype can be thought of as an allostatic state, mediated through weathering mechanisms such as epigenetic changes and reflecting the lifespan impacts of the exposome including social determinants of health and chemical toxicants. Using examples such as biomarkers of accelerated aging associated with both social and chemical stressors, asthma, and cardiovascular disease, we will discuss how the public health exposome can be linked back to actions to promote health and equity. The nexus of high impact exposomal factors + highly potentiating interactions + vulnerabilities can be used to identify public health interventions.

359 – PFAS and Differential miRNA Expression in Firefighters

Melissa Furlong[1], Shawn Beitel[2], Jefferey Burgess[3]
[1]University of Arizona, Assistant Professor, Tucson, USA. [2]University of Arizona, Research Scientist, Tucson, USA. [3]University of Arizona, Professor, Tucson, USA

Abstract
Firefighters are occupationally exposed to per-and poly-fluoryl alkyl substances (PFAS). Recently, firefighting was classified as a Group 1 carcinogen by IARC. Epigenetic regulation may be responsible for some of the observed relationship between firefighting, PFAS, and cancer, therefore, we investigated the associations of serum PFAS levels with miRNA expression in firefighters. Blood samples collected from 303 firefighters enrolled in the Firefighter Cancer Cohort Study from 6 sites across the USA were tested for 9 PFAS analytes along with miRNA expression. Linear regression was used to estimate associations between log PFAS and miRNA expression, with false discovery rate (FDR) adjustment. Gene set enrichment analysis (GSEA) pathway analysis was performed using miRTarBase's miRWalk pathways. We report pathways and diseases that were significantly enriched for at least three separate PFAS chemicals. Age, race/ethnicity, BMI, department, and gender were controlled for in all models. At FDR<0.05, n-PFOS was inversely associated with hsa-miR-128-1-5p expression (Beta = -0.146, 95% CI -0.216, -0.076). At a relaxed FDR of 0.20, we observed inverse associations for Sm-PFOS with 5 miRNAs (hsa-let-7d-5p, hsa-let-7a-5p, hsa-miR-423-5p, hsa-let-7b-5p, has-mir-629-5p). Five miRWalk pathways were separately enriched for ≥3 PFAS; these included IL 1 signaling, Selenium metabolism, P38 MAPK signaling, MAPK Cascade, and cytosolic DNA sensing. Several disease pathways were also enriched for multiple PFAS, including blood diseases (polycythemia, primary myelofibrosis, thrombocythemia, thrombocytosis), metaplasia, and inflammation. PFAS in firefighters were found to be associated with disruption of biological pathways with implications for blood disorders and cancers.

360 – A Novel Approach for Evaluation of Ketene Exposure from Vitamin E Acetate (VEA) Vaping

Ping Wang[1], Peyton Jacob[2], Zhong-Min Wang[3], Jefferson Fowles[4], Donal Oshea[5], Tracy Barreau[6], Kazukiyo Kumagai[7]
[1]California Department of Public Health, Research Scientist, Richmond, USA. [2]University of California, Research Chemist, San Francisco, USA. [3]California Department of Public Health, Research Scientist Supervisor, Richmond, USA. [4]California Department of Public Health, Staff

Toxicologist, Richmond, USA. [5]Royal College of Surgeons in Ireland, Department of Chemistry, Professor, Dublin, Ireland. [6]California Department of Public Health, Environmental Program Manager, Richmond, USA. [7]California Department of Public Health, Research Manager, Richmond, USA

Abstract
The outbreak of E-cigarettes Vaping Associated Lung Injury (EVALI) in the United States in 2019 led to a total of 2,807 hospitalizations with 68 deaths. While the exact causes for this vaping related lung illness are still being debated, laboratory analysis of products from victims of EVALI have shown that Vitamin E Acetate (VEA), an additive in some tetrahydrocannabinol (THC)-containing e-cigarette products, is strongly linked to the EVALI outbreak. Because of the similar appearance and viscosity with pure THC oil, VEA was used as a diluent agent in cannabis oils in illicit markets. A potential mechanism for EVALI may involve VEA's thermal decomposition product, ketene, a highly poisonous gas, being generated under vaping conditions. In this study, a novel approach was developed to evaluate ketene exposure from VEA and Vitamin E (VE) vaping. Ketene in generated aerosols was captured by two different chemicals and analyzed by GC/MS and LC/MS/MS. Our results confirmed the ketene formation when VEA was vaped. The amount of ketene increased as more VEA was consumed and showed a correlation to temperatures measured within e-devices. Device power strength, which affects heating temperature, played an important role in ketene formation. In addition to ketene, duroquinone (another pyrolysis product of VEA) was also observed when VEA was vaped. Ketene was not detected when VE was vaped under the same conditions. A density functional theory study indicates that other acetate moieties (e.g., THC-O-acetates and some flavor compounds) could also potentially produce ketene gas when vaped.

361 – Emerging Approaches to Advance Cumulative Impact Assessment and Environmental Justice

Charles Lee[1], Andrew Geller[2]
[1]EPA, Senior Policy Advisor, Washington, USA. [2]EPA, Senior Science Advisor, RTP, USA

Abstract
The issue of cumulative impact assessment is rapidly becoming an area of greater public awareness and urgency. More than a quarter of the states in the US currently have statutes or bills pending on cumulative impacts. The impetus for these developments comes from environmental justice advocacy on behalf of communities historically overburdened by a concentration of pollution and other stressors. In this context, important examples are emerging on approaches for characterizing cumulative impacts from the concentration of environmental burden in people of color, indigenous, and low-income communities, and means to reduce health disparities. The Minnesota Pollution Control Agency and the Massachusetts Department of Environmental Protection have developed similar protocols that use exposure indicators and health endpoints to identify, organize, and evaluate relevant stressors. The New Jersey Department of Environmental Protection developed protocols that rely on comparative analysis and seek to ensure that no community bears a disproportionate burden of adverse environmental exposures and public health impacts. And, the Chicago Department of Public Health utilized Health Impact Assessment to incorporate exposure, health, and compliance data to conduct a permit analysis of a proposed recycling facility. This presentation will draw lessons from these experiences for how to consider the totality of exposures to combinations of chemical and nonchemical stressors and their effects on health, well-being, and quality of life outcomes. Two areas of great interest are methods to i. account for the lived reality of overburdened communities and ii. characterize population vulnerability to consider cumulative impacts in environmental and/or regulatory decision-making.

362 – Establishing QA/QC Protocols and Tools to Support Non-targeted Analyses of Environmental Matrices

Nickolas Sayresmith[1], Laura Brunelle[2], Angela Batt[3], Alex Chao[4], Tyler Carr[5], Tommy Cathey[6], James McCord[7], Jeffrey Minucci[8], Thomas Purucker[9], Asif Rashid[10], Deron Smith[6], Antony Williams[11], Dave Alvarez[12], Edward Furlong[13], Susan Glassmeyer[3], Dana Kolpin[14], Marc Mills[15], Jon Sobus[16]
[1]USEPA, Office of Research and Development, Center for Computational Toxicology and Exposure, ORISE Postdoc, Research Triangle Park, NC, USA. [2]Oak Ridge Institute for Science and Education, ORISE Researcher, Oak Ridge, TN, USA. [3]USEPA, Office of Research and Development, Center for Environmental Solutions and Emergency Response, Research Chemist, Cincinnati, OH, USA. [4]USEPA, Office of Research and Development, Center for Computational Toxicology and Exposure, Research Chemist, Research Triangle Park, NC, USA. [5]USEPA, Office of Research and Development, Center for Computational Toxicology and Exposure, Researcher, Research Triangle Park, NC, USA. [6]USEPA, Office of Mission Support, Office of Information Security and Privacy, Researcher, Research Triangle Park, NC, USA. [7]USEPA, Office of Research & Development, Center for Environmental Measurement and Modeling, Research Chemist, Research Triangle Park, NC, USA. [8]USEPA, Office of Research and Development, Center for Public Health and Environmental Assessment, Physical Scientist, Research Triangle Park, NC, USA. [9]USEPA, Office of Research and Development, Center for Computational Toxicology and Exposure, Research Ecologist, Research Triangle Park, NC, USA. [10]USEPA, Office of Research and Development, Center for Computational Toxicology and

Exposure, EPA Innovation Fellow, Research Triangle Park, NC, USA. [11]USEPA, Office of Research and Development, Center for Computational Toxicology and Exposure, Computation Chemist, Research Triangle Park, NC, USA. [12]USGS, Columbia Environmental Research Center, Research Chemist, Columbia, MO, USA. [13]USGS, National Water Quality Laboratory, Research Chemist, Denver, CO, USA. [14]USGS, Iowa Water Science Center, Research Hydrologist, Iowa City, IA, USA. [15]USEPA, Office of Research and Development, Center for Environmental Solutions and Emergency Response, Environmental Engineer, Cincinnati, OH, USA. [16]USEPA, Office of Research and Development, Center for Computational Toxicology and Exposure, Physical Scientist, Research Triangle Park, NC, USA

Abstract
Non-targeted analysis (NTA) utilizing high-resolution mass spectrometry is an increasingly popular tool for environmental science given its ability to facilitate exploratory exposomic investigations and identify contaminants of emerging concern. Generalized reporting standards, like those promulgated in the NTA "Study Reporting Tool" (SRT), are now available for NTA practitioners. Yet, wide-spread standardization of NTA procedures is ongoing, and practical demonstrations of best-practices (as per the NTA SRT) remain lacking. Addressing this need, we report a proof-of-concept analysis in which vendor software and a prototype U.S. Environmental Protection Agency (USEPA) web application were used to generate critical QA/QC outputs for an environmental NTA study. NTA data used for this demonstration were from a study by the U.S. Geological Survey and USEPA to investigate the fate and transport of water contaminants throughout a watershed representative of typical de facto water reuse. We explored NTA QA/QC as it relates to: (1) Study Design (involving pooled samples, method blanks, tracer spikes, and system-suitability mixtures [SSMs]), (2) Data Acquisition (involving technical replicates and randomized run sequences), and (3) Data Outputs (including measures of accuracy, precision, sensitivity, and reproducibility). Physical outputs of the QA/QC workflows include summary tables of tracer and SSM data, run sequence plots demonstrating batch and matrix effects, and several visualizations (e.g., logic trees and heat maps) summarizing results of data filtering and flagging steps. After ensuring data quality, "clean" features are further examined to probe spatial and temporal trends, and direct prioritization efforts for compound annotation and follow-up confirmation.

363 – POSTER: Developing community-based actions to address air pollution concerns in Environmental Justice Neighborhoods: DIY Indoor Air Filter Workshops

Jennifer Amahuda[1], Dora Frietze-Armenta[2], Janelle Vidal[3], Dayane Duenas Barahona[4], Yoshira Ornelas Van Horne[5,6], Jill Johnston[7]
[1]University of Southern California, student, Los Angeles, USA. [2]Tacoma Beautiful, Lead Project Planner, Pacoima, USA. [3]University of Southern California, Project Specialist, Los Angeles, USA. [4]University of Southern California, Project Administrator, Los Angeles, USA. [5]University of Southern California, Adjunct Faculty, Los Angeles, USA. [6]Columbia University, Assistant Professor, New York, USA. [7]University of Southern California, Associate Professor, Los Angels, USA

Abstract
eople spend almost 90% of their time in indoor environments that can be up to 2 to 5 times more polluted than outdoor environments. The use of HEPA air purifiers is one way to reduce concentrations of particles indoors. However, most HEPA air purifiers tend to be inaccessible to communities of color who are disproportionately impacted by both indoor and outdoor pollution. Cheaper air purifiers often have a lower clean air delivery rates and can even generate ozone. DIY air filters are an efficient cost-effective alternative to portable air purifiers. In partnership with community organizations, we used a solutions-focused approach to conduct a series of workshops on indoor and outdoor air pollution. Together we developed bilingual lectures and materials so that community members could better share their air pollution stories. To address immediate concerns, community members built low-cost DIY air filters to use in their homes. These workshops serve as a tool for participants to learn about indoor air quality, identify pollutant sources in their own homes, and think of ways they can actively reduce their exposures. The workshops also encouraged participants to think of community-level solutions using the hierarchy of controls model. By building power with local community members, leaders, and youth, we can create community-owned solutions to reduce air pollution and build capacity to improve neighborhood air quality regulations.

364 – POSTER: Assessing Asbestos Exposure in Soil-Contaminated Areas of Sibaté, Colombia: An Agent-Based Modeling Approach

Françoise Duraffour[1], Juan P Ramos-Bonilla[2,3], Benjamin Lysaniuk[4]
[1]CNRS, UMR Prodig, Research Engineer, Aubervilliers, France. [2]Universidad de los Andes, Associate Professor, Bogotá, Colombia. [3]Collegium Ramazzini, Fellow, Bologna, Italy. [4]IRD (MàD by CNRS), UMR Prodig, Researcher, Aubervilliers, France

Abstract
Sibaté is a city in Colombia located 25 km southwest of Bogotá, the capital of the country. In this municipality, the first asbestos-cement factory of the country operated for nearly 80 years. Several concerning findings for this city have been reported in the scientific

literature, including the presence of a cluster of mesothelioma cases diagnosed at young ages (i.e., 38 to 58 y.o.) with no history of occupational exposure to asbestos, and the creation of asbestos-containing landfilled zones during the decades of 1980 and 1990. Residents report that children used these areas as playgrounds, even when residue materials contaminated with asbestos were being disposed. Agent-based modeling (ABM) was used to estimate the number of inhabitants who have been directly exposed to asbestos because of the interaction with two landfilled zones, integrating into the analysis both urban sprawling and demographic changes over time. Different scenarios were analyzed by modifying input values for variables such as the number of children playing in the landfills. The number of potentially exposed people varies greatly depending on the scenario studied, ranging from 31 to 3,800. A plausible scenario using conservative inputs suggests that more than 250 individuals could have been directly exposed to asbestos over 10 years. This could explain the high incidence of cases of mesothelioma diagnosed at young ages with no history of occupational exposure to asbestos.

366 – Use of remote sensing technologies to detect and locate corrugated asbestos-cement sheets

Pauline Gluski[1], Juan P Ramos-Bonilla[2,3], Benjamin Lysaniuk[4]
[1]IRD , UMR Prodig, Research Engineer, Aubervilliers, France. [2]Universidad de los Andes, Associate Professor, Bogotá, Colombia. [3]Collegium Rammazini, Fellow, Bologna, Italy. [4]IRD (MàD by CNRS), UMR Prodig, Researcher, Aubervilliers, France

Abstract
The application of machine learning tools has led to remarkable advancements, including spatial analysis problem solutions such as satellite image classification and remote object recognition. One of these applications is the identification of corrugated asbestos-cement sheets (CACS) using an object-oriented approach. The RoofClassify plugin, developed by an Italian team of scientists within an open-source geographic information system, uses multispectral satellite images to remotely identify CACS. In this study, the plugin was evaluated in Sibaté, Colombia, which has a more heterogeneous roof composition and less reliable input data than the Prato region where it was previously applied. The plugin was trained using 114 polygons that contained the four most common types of roof materials in the urban area of Sibaté, which were identified directly in the field. Then, the plugin was applied in an area where the roof materials were visually determined. The results showed a perfect match at the pixel scale between installed and detected CACS in 53.34% of the total roof area analyzed. Moreover, more than 93% of the polygons where the plugin at least partially recognized the presence of CACS were detected. RoofClassify can aid in identifying installed asbestos-containing products, especially in low- and middle-income countries, as it uses widely available satellite images, which have broader global coverage, and avoids costly aerial campaigns to obtain multispectral aerial photographs.

367 – POSTER: Assessment of community exposure to groundwater contaminated with TPH and PAH discharge from petroleum depot in Abia State, Nigeria

Joseph Ikechukwu Nwachukwu[1], Chinelo Lois Gilbert[2], Michael. A Nwachukwu[3], Chika Veronica Nwachukwu[4]
[1]Federal University of Technology, Lecturer, Owerri, Imo State, Nigeria. [2]Federal University of Technology, Research Student, Owerri, Imo State, Nigeria. [3]Federal University of Technology, Professor, Owerri, Imo State, Nigeria. [4]Federal University of Technology, Principal Laboratory Technologist, Owerri, Imo State, Nigeria

Abstract
The unrestrained discharge of waste water from the Nigerian National Petroleum Company (NNPC) into an unlined waste pit threatens the hygienic quality of Umuakpara community in Osisioma, Abia state of Nigeria. This study carried out a bi-seasonal (rainy and dry) sampling of 15 groundwater boreholes used by the community for drinking purposes. Samples collected under standard methods, were analyzed for Total TPH and PAH concentrations using the gas chromatography (GC-FID) instrument. Concentrations of both TPH and PAH in groundwater samples exceeded the WHO and EU limits for PAHs (0.001mg/l) and TPH (0.01mg/l) in both seasons. Anthracene showed the highest concentration among PAH components, with value range of between 0.0111 and 0.0624 (0.0303100.0063963) mg/l, while TPH fractions C15 and C17 had the highest means, and ranged from 0.22000 to 1.0000 (0.403 0.9577810) mg/l, and 0.22000 to 0.37000 (0.2910000 0.01433333) mg/l respectively. Significant variability of PAH existed across sampled boreholes whereas strong correlations were noted among the respective PAH components at $P < 0.01$. Contaminants' concentration reduced with distance from the unlined waste pit proving source of PAH in groundwater. Both the Hazard index and Target Hazard Quotient employed to evaluate the potential health risk of exposure by the community based on the result, revealed vulnerability trend of the different age categories. The study recommended urgent necessary measures that could be taken by relevant stakeholders to forestall outbreak of potential health hazard associated with the continuous operation of the depot and use of groundwater in the area for drinking purposes.

369 – POSTER: Size-selective Bioaerosol Sampler Combined with a Latex Agglutination Assay for Rapid Detection of Airborne Legionella pneumophila

Subin Han[1], Anthony W.G. Bovenschen[2], Li Liao[1], Jae Hong Park[3]
[1]Purdue University, Graduate student, West Lafayette, USA. [2]Purdue University, Undergraduate student, West Lafayette, USA. [3]Purdue University, Associate professor, West Lafayette, USA

Abstract
Bioaerosols are aerosols of biological origin that can cause adverse health effects. For example, airborne Legionella pneumophila (L. pneumophila) can develop Legionnaires' disease once inhaled. To assess the exposure to and protect people against airborne L. pneumophila, a rapid detection method is necessary. In conventional methods, airborne L. pneumophila is collected and then transported to the lab for analysis. However, these processes are time-consuming and not practical in the field. Therefore, an alternative method was developed in this study. The developed method is a combination of a lab-made inertial impactor and a latex agglutination test assay. The inertial impactor was designed to collect bioaerosols larger than 0.5 μm on the substrate used in the latex agglutination assay. To evaluate the collection efficiency and cut-off diameter of the inertial impactor, concentrations of test particles up and downstream of the impactor were measured using a scanning mobility particle sizer. The limit of detection (LOD) of the latex agglutination assay kit was also determined using a known concentration of L. pneumophila suspension. In the lab sampling test, L. pneumophila suspension was aerosolized in an exposure chamber. The sampler was located in the exposure chamber and particle number concentration was monitored using an optical particle sizer. The LOD of the test kit was 665,385 cells per test kit. In the results of the lab sampling test, L. pneumophila was detected when the number of deposited particles in the kit exceeded 250,000. The developed method showed its feasibility of use in the field.

372 – POSTER: Respiratory Effects of Co-exposure To Phthalates And Indoor Biological Agents

Luyao Zhang[1], Ruixue Dai[2], Wendy Lou[3], Piush Mandhane[4], Theo Moraes[5,3], Elinor Simons[6], Teresa To[3,7], Stuart Turvey[8], Padmaja Subbarao[3,9], Jeffrey Brook[3]
[1]University of Toronto, PhD candidate, Toronto, Canada. [2]Hospital for Sick Children, Senior Biostatistician, Toronto, Canada. [3]University of Toronto, Professor, Toronto, Canada. [4]University of Alberta, Professor, Edmonton, Canada. [5]Hospital for Sick Children, Division head, Respiratory Medicine, Toronto, Canada. [6]University of Manitoba, Professor, Winnipeg, Canada. [7]Hospital for Sick Children, Senior Scientist, Child Health Evaluative Sciences, Toronto, Canada. [8]The University of British Columbia, Professor, Vancouver, Canada. [9]Hospital for Sick Children, Associate Chief, Clinical Research, Toronto, Canada

Abstract
Children are co-exposed indoors to biological agents and chemicals, though the combined respiratory effects are unknown. We assessed whether exposure to indoor biological agents, namely pet allergens and endotoxin, modifies the association between di-(2-ethylhexyl) phthalate (DEHP) and respiratory outcomes in preschoolers. Using a Canadian birth cohort study (CHILD), we conducted two analyses. First, we combined pet ownership, a surrogate for pet allergens, with DEHP from house dust (N=726). Second, we focused on a subgroup of children with DEHP and biological agents in dust (N=261). Biological agent concentrations were dichotomized before interaction assessment. The association between DEHP and wheeze phenotypes and asthma at age 5/recurrent wheeze was investigated using multinomial and logistic regression, respectively. Interaction terms and stratified analyses were used to investigate the association between DEHP and respiratory outcomes by pet ownership and biological agent levels. Exposure to DEHP was significantly associated with higher risk of persistent and intermediate wheeze in children, and the association was modified by pet ownership and pet allergens. For persistent/recurrent wheeze, the association with DEHP became larger among children with dogs as well as based on higher dog allergens (P-interaction <0.1) while became smaller and insignificant when exposed to cats. Similarly, for asthma, the association with DEHP was larger among children with dogs (also higher dog allergens) and without cats (also lower cat allergens) at home, respectively. Endotoxin was not found to have significant interactions with DEHP. In conclusion, exposure to pets may modify the association between phthalates and respiratory outcomes in children.

373 – Characterizing Ultrafine Particle Mobile Monitoring Data for Epidemiology

Annie Doubleday[1], Magali Blanco[2], Elena Austin[3], Julian Marshall[4], Timothy Larson[5], Lianne Sheppard[4]
[1]University of Washington, Graduate student, Seattle, USA. [2]University of Washington, Postdoctoral Scholar, Seattle, USA. [3]University of Washington, Assistant Professor, Seattle, USA. [4]University of Washington, Professor, Seattle, USA. [5]University of Washington, Professor Emeritus, Seattle, USA

Abstract
Mobile monitoring is increasingly used to assess exposure to traffic-related air pollutants (TRAPs), including ultrafine particles (UFPs). Due to the rapid spatial decrease in the concentration of UFPs and other TRAPs with distance from roadways, mobile measurements may be non-representative of residential exposures, which are commonly used for epidemiologic studies. Our goal was to develop, apply, and test one possible approach for using mobile measurements in exposure assessment for epidemiology. We used an absolute principal component score (APCS) model to adjust the contribution of on-road sources in mobile measurements to provide exposure predictions representative of cohort locations. We then compared UFP predictions at residential locations from mobile on-road plume-adjusted versus stationary measurements to understand the contribution of mobile measurements and characterize their differences. We found predictions from mobile measurements are more representative of cohort locations after down-weighting the contribution of localized on-road plumes. Further, predictions at cohort locations derived from mobile measurements incorporate more spatial variation compared to those from short-term stationary data. Sensitivity analyses suggest this additional spatial information captures features in the exposure surface not identified from the stationary data alone. We recommend correction of mobile measurements to create exposure predictions representative of residential exposure for epidemiology.

374 – Flame Retardant Levels in Cabin Air and Car Seat Foam Collected from Personal Vehicles: Implications for Exposure

Rebecca Hoehn[1], Lydia Jahl[2], Nicholas Herkert[3], Kate Hoffman[4], Arlene Blum[5], Anna Soehl[6], Xuening Tang[7], Heather Stapleton[8]
[1]Duke University, PhD Student, Durham, USA. [2]Green Science Policy Institute, Science and Policy Associate, Berkley, USA. [3]Duke University, Research Project Manager, Durham, USA. [4]Duke University, Assistant Research Professor, Durham, USA. [5]Green Science Policy Institute, Executive Director, Berkley, USA. [6]Green Science Policy Institute, Science and Policy Consultant, Berkley, USA. [7]Duke University, Masters Student, Durham, USA. [8]Duke University, Professor, Durham, USA

Abstract
Silicone wristbands (WBs) are a popular passive sampling tool and were recently used in a study that found exposure to organophosphate ester (OPE) flame retardants (FRs) was significantly higher in people with longer commutes, suggesting vehicle interiors could be a significant exposure source. The present study sought to characterize the potential for human exposure to FRs in personal vehicles and to evaluate the influence of temperature.

Participants from across the US owning a vehicle of model year 2015 or newer were invited to participate. Participants hung a pre-cleaned WB on their rearview mirror for 7 days and collected a foam sample from a vehicle seat. WBs and foam were analyzed for 54 FRs using gas chromatography-mass spectrometry. 101 people participated in winter and 54 participated again in the summer. 51 participants submitted a foam sample.

OPEs, and particularly Tris(1-chloro-isopropyl)phosphate (TCPP), were detectable in all wristband samples. TCPP was detected at levels ranging from 0.2 to 11,624 ng/g WB. Concentrations were significantly correlated with average ambient temperature and were 2-4x higher in summer compared to winter. A 1-degree Celsius increase in temperature predicted a 12% increase in WB TCPP concentration. Analysis of the foam revealed that 45% contained TCPP. The presence of TCPP in foam resulted in ~4x higher median concentrations on WBs in winter, and ~9x higher concentrations in summer. These results suggest that flame retardant use in vehicle interiors, such as seat foam, can lead to OPE exposure and this exposure is exacerbated by warmer temperatures.

375 – POSTER: A Collaborative Partnership and Community-Based Low-Cost Air Monitoring to Reduce Disparities in Traffic-Related Air Pollution

Yonghong Luo[1], Shannah Nice[2], Gennaro Saliceto[3], Allison Yeh[4], Lizzie Ehrreich[5], Jason W. Beckstead[6], Dinorah Martinez Tyson[6], Amy L. Stuart[6,7]
[1]College of Public Health, University of South Florida, Doctoral Candidate, Tampa, USA. [2]Civil & Environmental Engineering, University of South Florida, Undergraduat Student, Tampa, USA. [3]College of Public Health, University of South Florida, Graduate Student, Tampa, USA. [4]Hillsborough Transportation Planning Organization, Assistant Executive Director, Tampa, USA. [5]Hillsborough Transportation Planning Organization, Planner, Tampa, USA. [6]College of Public Health, University of South Florida, Professor, Tampa, USA. [7]Civil & Environmental Engineering, University of South Florida, Professor, Tampa, USA

Abstract
Traffic-related air pollution (TRAP) disproportionately affects marginalized populations, but local-scale data are lacking. This study aimed to 1) choose cost-effective TRAP monitors for community use, 2) assess air quality at pilot sites in marginalized neighborhoods

near high-traffic roads in Hillsborough County, Florida, and 3) involve community members in efforts to decrease inequality. A partnership between university researchers, county government transportation planning and air monitoring agencies, and community organization leaders was established to facilitate the project. Methods included development of monitor and site selection criteria; field testing and evaluation of monitors; installation of community monitoring sites; and engagement with community participants through multiple activities. Seven low-cost monitors that measure particles, nitrogen dioxide, and/or volatile organic compounds were tested. Of these, the Clarity Node-S and PurpleAir PA-II were most suitable for community host installations due to their useability and performance. The Node-S was most useful for public facility installations (e.g., parks) where access to Wi-Fi and power was limited. Data calibration was required for adequate performance. Initial performance for 24-hr $PM_{2.5}$ measurement by both monitors was good (e.g., $R^2 \geq 0.8$), but changed substantially over time for the PA-II (R^2 of 0.36 during a summer period). Performance for NO_2 measurement was low. The Atmotube Pro personal monitor was found to be useful for engagement. Analysis of community site data showed maximum levels of daily $PM_{2.5}$ at some sites to be higher than at the near-road regulatory site. Results are being used to establish a sustainable community-engaged TRAP monitoring network for the county.

376 – The role of private institutions in advancing wastewater-based epidemiology to promote healthier communities

Kavya Boyina[1], Rochelle Holm[2], Himeshi Samarasinghe[3], Adam Gushgari[4], Petros Chigwechokha[5], Ted Smith[2], John Cosgrove[6]
[1]Eurofins Pandemic Prevention Services, Project Manager, West Sacramento, USA. [2]Christina Lee Brown Environment Institute, School of Medicine, University of Louisville, Associate Professor of Medicine, Louisville, USA. [3]Eurofins Pandemic Prevention Services, Principal Scientist - Epidemiology, West Sacramento, USA. [4]Eurofins Pandemic Prevention Services, National Business Development Manager, West Sacramento, USA. [5]Malawi University of Science and Technology, Senior Lecturer (Molecular Biology & Genetics), Limbe, Malawi. [6]Eurofins Pandemic Prevention Services, Senior Vice President, West Sacramento, USA

Abstract
Wastewater-based epidemiology (WBE) is an emerging method used for tracking pathogens and pharmaceutical compounds in populations as a way to monitor community-wide health. During the SARS-CoV-2 pandemic, various stakeholders rapidly established collaborative programs to provide near real-time COVID-19 surveillance with wastewater testing.

We conducted a systematic review to examine evidence from the peer-reviewed literature about the private sector role and their partnerships with government, academic, and other entities. We aimed to identify evidence of the scope and scale of collaborations, summarize our observations for private-sector opportunities, and identify areas of effective cooperation for future developments to inform and promote risk assessments, policy, and healthier communities.

We conducted a database search for articles published within the years 2000-2023 using relevant keywords. References were screened by title, abstract, and affiliation to determine relevance. Full text copies of manuscripts were evaluated based on the following criteria, focus on WBE, authorship of study, and available authorship credit. We analyzed the qualitative data and summarized the metrics.

A total of 55 articles were included. A limited number of public-private partnerships were identified in this field. The top categories private companies contributed to included writing-review & editing, followed by methodology and investigation. Private companies' roles were not limited to conducting laboratory analysis. Where laboratory analysis was conducted by the private sector, biological rather than chemical targets were commonly analyzed.

Evidence suggests there is a need for an integrative approach to expand public-private collaborations, while considering the roles private industry can play in advancing the field of WBE.

377 – POSTER: Novel approaches to infant formula safety screening: using non-target analysis and computational toxicological tools to assess organic contaminants in infant formula

Qinfan Lyu[1], Matthew Newmeyer[2], Sara Lupolt[3], Antony Williams[4], Carsten Prasse[5], Keeve Nachman[6,7]
[1]Department of Environmental Health & Engineering, Johns Hopkins Bloomberg School of Public Health, Graduate Student, Baltimore, USA. [2]Department of Environmental Health & Engineering, Johns Hopkins Bloomberg School of Public Health, Research Associate, Baltimore, USA. [3]Department of Environmental Health & Engineering, Johns Hopkins Bloomberg School of Public Health, Assistant Scientist, Baltimore, USA. [4]U.S. Environmental Protection Agency, Office of Research and Development, Center for Computational Toxicology and Exposure, Cheminformatician, Raleigh, USA. [5]Department of Environmental Health & Engineering, Johns Hopkins Bloomberg School of Public Health, Assistant Professor, Baltimore, USA. [6]Department of Environmental Health & Engineering, Johns Hopkins Bloomberg School of Public

Health, Associate Professor, Baltimore, USA. [7]Department of Health Policy and Management, Johns Hopkins Bloomberg School of Public Health, Associate Professor, Baltimore, USA

Abstract
Nearly a third of all US infants receive formula prior to six months of age. Although formula production is subject to stringent regulation and supervision, contamination can occur during the manufacturing process. Newborns may be highly vulnerable to contaminant exposures during critical developmental windows. While several studies have characterized elemental contaminants in infant formula, few have examined organic chemicals. We conducted a market basket study of infant formula products available in Baltimore, Maryland to identify and toxicologically assess organic contaminants present in infant formula. A Quick, Easy, Cheap, Effective, Rugged, and Safe (QuEChERS) method was employed to extract the samples, and analyze them by liquid chromatography–high-resolution mass spectrometry. Raw data were processed with Compound Discoverer to detect and identify compounds across samples. Tentatively identified organic chemicals present in at least one formula product were subjected to computational (in silico) toxicological screening tools (including the Hazard Comparison Cheminformatics Module and the EPA Generalized Read-Across model, for chemicals lacking formal in vivo or in vitro data) for developmental and neurological hazards to prioritize compounds for targeted confirmation. These data will help inform future studies to understand the broader occurrence of organic contaminants in infant formula. Our research may also be utilized to prioritize chemicals for regulation in infant formula to protect children's health. This abstract does not necessarily represent the views or policies of the U.S. Environmental Protection Agency.

378 – POSTER: The "perfect storm" of exposure and health inequality in an Indigenous population

Judith Zelikoff[1], Esther Erdei[2], shannon doherty-lyons[3], Jill Aquino[4], Tri Huynh[5]
[1]New York University School of Medicine, Professor, NY, USA. [2]University of New Mexico, Professor, Albuquerque, USA. [3]New York University School of Medicine, Research Scientist, NY, USA. [4]University of New Mexico, Community Coordinator, NY, USA. [5]New York University School of Medicine, Graduate Student, NY, USA

Abstract
Emerging from the "perfect storm" of injustice and contamination are Native Americans who face adverse health outcomes emerging from tainted soil that grows contaminated traditional food sources; degraded water and air quality; and a loss of infrastructure leading to reduced community strength. Native American (NA) families are 400% more likely to report food insecurity than all other Americans. Such is the case for the Ramapough Lunaape Turtle Clan. Such is the case for the Ramapough Lunaape Turtle Clan residing across northern NJ who have been exposed for over a decade of industrial dumping of thousands of gallons of paint sludge and other waste in their backyards. Measurements have shown elevated levels in the soil, ground- and surface water of toxic heavy metals, VOCs and PAHs, at concentrations well above federal and state standards. As a result, environmental contamination, poverty and food instability, good nutrition and food sovereignty are difficult to sustain for the Ramapough Nation. A major goal of this Research to Action (R2A) grant is to identify and intercede upon modifiable risk factors that help explain the observed associations between food sovereignty, and health disparities associated with disproportionate exposure to environmental chemicals. Our R2A grant, that begins with an extensive exposure assessment study builds on our 10-year Tribal-academic partnership with the Ramapough Nation of northern NJ to advance tradition-centered farming practice and management intervention strategies to support sustainable food systems to relieve local food insecurity, nutritional deficiency, prevent disease and promote health and well-being in this already vulnerable population.

379 – Perceptions and Experiences of PFAS Biomonitoring in Australian Communities Affected by Environmental Contamination

Kayla Smurthwaite[1], Nina Lazarevic[2], Sandra Nilsson[3], Susan Trevenar[4], Cathy Banwell[5], Martyn Kirk[6]
[1]Australian National University, Postdoctoral Research Fellow, Canberra, Australia. [2]Australian National University, Research Fellow, Canberra, Australia. [3]University of Queensland, Postdoctoral Research Fellow, Brisbane, Australia. [4]Australian National University, Research Manager, Canberra, Australia. [5]Australian National University, Emeritus Professor, Canberra, Australia. [6]Australian National University, Professor, Canberra, Australia

Abstract
Understanding human exposure to persistent environmental pollutants is critical to informing public health responses and addressing community concerns in contaminated areas. We aimed to investigate community perceptions and experiences of human biomonitoring for PFAS in Australian towns impacted by environmental contamination.

From 2016–2019, the Australian Government conducted the Voluntary Blood Testing Program for PFAS for residents and workers of three

Australian towns. From 2019–2020, we surveyed 812 people who provided a blood sample for the Program and used an inductive approach to thematically analyse responses on the usefulness of the blood testing for PFAS.
We observed thematic differences in community experiences and perceptions of the PFAS blood testing. Participants who reported that the blood testing was helpful communicated that the results confirmed their exposure. However, participants assumed that the test results could be used to make inferences to their health or that levels within reference values can be considered 'safe'. In contrast, participants who indicated that the test was not helpful reflected on the uncertainty in the relationships between exposure and potential health effects and had unanswered questions about their results.
Communities affected by local environmental contamination consistently request access to individual blood testing and results to understand their exposure. However, risk communication strategies related to the reporting of these results may not adequately consider the psychosocial implications. Globally, there is a need to evaluate and standardise approaches to reporting individual blood serum PFAS results, including the selection and reporting of reference values, in research and clinical settings.

380 – Associations Between Environmental Tobacco Smoke (ETS) and Ambient $PM_{2.5}$ Exposures and Urinary Monohydroxylated Polycyclic Aromatic Hydrocarbon Metabolites (OH-PAHs) in Three Pregnancy Cohorts: A Cross-sectional Study

Anne M. Riederer[1], Allison R. Sherris[2], Christopher D. Simpson[3], Adam A. Szpiro[4], Christine T. Loftus[5], Melissa Melough[6], Erin E. Masterson[7], Drew Day[8], Erin R. Wallace[5], Leonardo Trasande[9], Emily S. Barrett[10], Ruby H.N. Nguyen[11], Kurunthachalam Kannan[12], Morgan Robinson[13], Shanna H. Swan[14], W. Alex Mason[15], Nicole R. Bush[16], Sheela Sathyanarayana[3], Kaja Z. LeWinn[17], Catherine J. Karr[3]
[1]University of Washington, Clinical Associate Professor, Seattle, USA. [2]University of Washington, Postdoctoral Fellow, Seattle, USA. [3]University of Washington, Professor, Seattle, USA. [4]University of Washington, Associate Professor, Seattle, USA. [5]University of Washington, Epidemiologist, Seattle, USA. [6]University of Delaware, Assistant Professor, Newark, USA. [7]University of Washington, Clinical Assistant Professor, Seattle, USA. [8]Seattle Children's Research Institute, Epidemiologist, Seattle, USA. [9]New York University, Professor, New York, USA. [10]Rutgers University, Associate Professor, Piscataway, USA. [11]University of Minnesota, Associate Professor, Minneapolis, USA. [12]New York University, Research Professor, New York, USA. [13]New York University, Researcher, New York, USA. [14]Icahn School of Medicine at Mt. Sinai, Professor, New York, USA. [15]University of Tennessee, Professor, Memphis, USA. [16]University of California San Francisco, Professor, San Francisco, USA. [17]University of California San Francisco, Associate Professor, San Francisco, USA

Abstract
Prenatal PAH exposure is associated with adverse health outcomes in children, motivating research on how pregnant people are exposed. We examined associations between ETS and recent $PM_{2.5}$ exposure with 7 OH-PAHs in mid-pregnancy spot urine from 1,603 non-smokers in the ECHO PATHWAYS study (3 birth cohorts in 6 U.S. cities). Analyses were restricted to OH-PAHs with >60% detection in at least one cohort. ETS exposure was represented by urinary cotinine quartile, while $PM_{2.5}$ exposure was represented by same-day estimates of ambient $PM_{2.5}$ in residential census tracts from EPA's Downscaler Model. We fit separate models for log-transformed OH-PAH concentration associations with cotinine quartile or $PM_{2.5}$ adjusted for urinary specific gravity, site, batch, BMI, household income, education, employed/not, neighborhood deprivation index, season, and year. Tobit regression was used to account for left-censoring in the OH-PAH data, while multiple imputation by chained equations was used to impute missing covariate data. In adjusted models (all cohorts), highest ETS exposure quartile vs. lowest was associated with 43% (95% CI: 9%, 90%) higher urinary 1-hydroxynaphthalene, 36% (16%, 62%) higher 2-hydroxynaphthalene, 17% (0.2%, 36%) higher 2-hydroxyphenanthrene, 48% (27%, 70%) higher 3-hydroxyphenanthrene, 101% (49%, 172%) higher 1-hydroxypyrene, and 27% (-0.002%, 60%) higher 1/9-hydroxyphenanthrene. Second and third quartile cotinine were also associated with higher OH-PAHs although not consistently across analytes. Same-day ambient $PM_{2.5}$ was not significantly associated with any OH-PAH. ETS remains a major source of PAH exposure for pregnant people in the USA. Our findings underscore the importance of promoting ETS exposure reduction for pregnant individuals.

381 – When Analytical, Computational, Observational, and Statistical Efforts Meet: A Collaborative Approach to Estimate Children's Dust and Soil Ingestion Rates

Li Li[1], You Fu[2], Yeongkwon Son[3], Dingsheng Li[2], Amy Olsen[4], Patrick Hurbain[4], Andrey Khlystove[5]
[1]University of Nevada, Reno, Assistant Professor, Reno, USA. [2]University of Nevada, Reno, Associate Professor, Reno, USA. [3]Desert Research Institute, Research Associate Professor, Reno, USA. [4]University of Nevada, Reno, Graduate student, Reno, USA. [5]Desert Research Institute, Research Professor, Reno, USA

Abstract
Unintentional ingestion of soil and dust particles through touching and mouthing behaviors is a common pathway for children's exposure to chemical contaminants in the environment. Accurately estimating the rates of dust and soil ingestion is crucial for

researchers to understand the potential health risks associated with this exposure pathway. In this presentation, we introduce the development of a new approach that fuses analytical, computational, observational, and statistical efforts to estimate the rates of dust and soil ingestion. This approach builds on our recruitment of 100 participants (more to include, with 125 targeted) with diverse demographics, including sex, race/ethnicity, parent education, age, and anthropometrics. Two methods are used to estimate the ingestion rates of dust and soil particles. The first method involves observing the frequency of touching, mouthing, and handwashing in the children, as well as feeding these data into a computational model named PROTEX to back-calculate the ingestion rates for the participating children. The second method involves measuring concentrations of 32 metals and metalloids in urine, stool, dust, and soil samples collected from the children's living environments, and feeding these data into a toxicokinetic model to infer the best estimate of the ingestion rates. The collaboration of two methods demonstrates the use of data and knowledge from different subjects in facilitating a more robust estimation of the ingestion rates. Our preliminary results show that no statistically significant differences were found between dust and soil ingestion among children with different demographic factors.

382 – Adoption of Urban Greenspace Solutions as a Means for Public Health Improvement

Lyssa Losa[1], Garett Sansom[2], Galen Newman[3], Anna Glanzer[4]
[1]Texas A&M University School of Public Health, Research Assistant, College Station, USA. [2]Texas A&M University School of Public Health, Assistant Professor, College Station, USA. [3]Texas A&M University School of Architecture, Associate Professor, College Station, USA. [4]Texas A&M University School of Public Health, Graduate Research Assistant, College Station, USA

Abstract
Many neighborhoods located along the Houston Ship Channel, including Manchester, TX, have been documented as having excess risks of exposure to acute pollution, emergency chemical spills and incidents, and high-impact natural and manmade disaster events. Investigation of the environmental conditions and associated health risks within this community has revealed geographic concentrations of poor health outcomes associated with immediate proximity to industry and flooding risk. One proposed solution to environmental pollutants, and the subsequent negative public health outcomes, is the adoption of greenspace solutions. Despite this need, little research has quantified this potential adoption of new greenspace solutions within the same community. This presentation will (1) present reduction of heavy metals contamination of two greenspace solutions through urban landscape planes utilizing the Long-Term Hydrologic Impact Analysis (L-THIA) technique and (2) model future cancerous and non-cancerous chronic conditions changes based upon these changes. Our findings will demonstrate the importance of environmental sampling data and community engagement processes to inform planning and design. As a result of these processes, we can achieve the goals of reducing runoff, air and water pollutions, increasing carbon dioxide sequestration, facilitating groundwater replenishment, and mitigating environmental pollutants for improved public health outcomes.

383 – POSTER: Spatiotemporal Variance of Baltimore's Urban Heat Island: Disparate Exposures to Extreme Heat

Bianca Corpuz[1], Benjamin Zaitchik[2], Darryn Waugh[2]
[1]Johns Hopkins University, PhD Student, Baltimore, USA. [2]Johns Hopkins University, Professor, Baltimore, USA

Abstract
The combined effects of climate change and urban expansion will make cities more vulnerable to negative health effects caused by climate phenomena such as urban heat islands (UHIs). To reduce or prevent the harmful effects of various climate events, the development of mitigation and adaptation strategies are fundamental. However, for strategies to be implemented, understanding local-climate phenomena such as UHIs is required. UHIs are not uniform within a city and experience patterns of spatial variability with time and meteorological conditions. This study estimates the variability in air temperature within Baltimore to capture these spatial and temporal patterns. Using a ground-based sensor network of iButton thermochrons, a National Weather Service station, and satellites, we developed statistical models to predict air temperature anomalies throughout Baltimore. By understanding the impacts that local weather and physical properties have on UHI formation and variability throughout a city, we may be able to better monitor and predict their spatiotemporal patterns. Our time-varying temperature maps can be merged with socio-economic and demographic data which allow us to determine which regions are most vulnerable to extreme heat and when, and explore disparities in terms of spatiotemporal variation of air temperature in Baltimore City. These findings can be used by local government officials and community groups to implement more directed surveillance measures and heat mitigation efforts.

384 – Tracking a Chemical's Journey from Production Line to Body Burden using the PROduction-To-EXposure (PROTEX) Model

Li Li[1], Zhizhen Zhang[2], Dingsheng Li[3], Alessandro Sangion[4], Jon Arnot[4], Frank Wania[5]
[1]University of Nevada, Reno, Assistant Professor, Reno, USA. [2]University of Nevada, Reno, Graduate student, Reno, USA. [3]University of Nevada, Reno, Associate Professor, Reno, USA. [4]Arnot Research & Consulting, Senior Scientist, Toronto, Canada. [5]University of Toronto, Scarborough, Professor, Toronto, Canada

Abstract
Human and ecological exposures to synthetic chemicals involve chemical emissions from multiple lifecycle sources, multimedia environments at multiple scales, and multiple pathways of intake and elimination. These components are governed by the complicated interactions between chemical properties, environmental conditions, and human and ecological characteristics. Such multidimensionality and complexity warrant a systematic, mechanistic perspective in exposure modeling and necessitate a holistic approach to track a chemical's journey from the production line to body burden. In this presentation, we introduce a state-of-the-science model named PROduction-To-EXposure (PROTEX), which supports characterizing the entire continuum from chemical production, lifecycle multi-source emissions, multi-scale multimedia environmental fate and transport, and multi-route exposure, to the presence in humans and other organisms. PROTEX builds on theories of substance flow analysis, multimedia fate and transport, bioaccumulation, and exposure and toxicokinetics. It integrates mechanistic descriptions of various physical, chemical, biological, and socioeconomic processes. PROTEX's mechanistic nature allows it to be flexibly parameterized for a wide range of non-ionizable and ionizable chemicals (PCBs, PBDEs, QACs, etc.) in diverse regional environments (e.g., the subtropical U.S., the Canadian side of the Lake Ontario basin, central China, and the western Baltic drainage basin) for different populations (Whites, Blacks, Asians, etc.). When supplied with time-variant production or use data, PROTEX provides dynamic time-dependent estimates, allowing users to reconstruct the long-term history of chemical contamination. By considering interindividual variabilities in anthropometrics, behavior, and physiology, PROTEX additionally supports predicting population distribution of chemical exposure. Outputs from PROTEX support science-based decision-making in 21st-century chemical assessment and management.

385 – Cause-specific Excess Deaths in Korea during COVID-19 in 2020

Haemin Park[1], Kiyoung Lee[2,3], Domyung Paek[4,5,6]
[1]Institute of Health and Environment, Researcher, Seoul, Korea, Republic of. [2]Graduate School of Public Health, Seoul National University, Professor, Seoul, Korea, Republic of. [3]Institute of Health and Environment, Professor, Seoul, Korea, Republic of. [4]National Cancer Center, Visiting Professor, Gyeonggi-do, Korea, Republic of. [5]Wonjin Institute for Occupational and Environmental Health, Senior Scientist, Seoul, Korea, Republic of. [6]Graduate School of Public Health, Seoul National University, Professor Emeritus, Seoul, Korea, Republic of

Abstract
Coronavirus 19 (COVID-19) has changed our lives in various ways. To measure its overall impact on mortality, all-cause and cause-specific excess mortality were examined. The purpose of this study was to assess the impact of the waves based on weekly cause-specific excess deaths reflecting the spatio-temporal variation of COVID-19 in Korea.

Cause-specific deaths were estimated by week, region, cause, gender and age from 4th-47th week of 2020, with the quasi-Poisson generalized linear models by constructing time-series deaths data from 2015-2019. The prediction of cause-specific deaths was performed and compared with observed deaths by region, cause, and three separate periods.

In first period, Daegu-Gyeongbuk area, the epi-center of the first wave, had experienced the 5.7% of excess deaths with increase in deaths from infectious disease and genitourinary system, while deficit was observed in pneumonia deaths in all regions. During second period, 17.8% of excess deaths from septicemia in Capital-metropolitan area and 9.1% of circulatory system in Daegu-Gyeongbuk area were observed. As the second wave began to rise in third period, excess deaths from diabetes, infectious disease, and symptoms uncategorized in Capital-metropolitan area and ischemic heart disease in Daegu-Gyeongbuk were prominent.

While the direct impact on cause-specific excess deaths was evident in the epi-center, the rather indirect impacts could be suspected for the other parts of the country. Based on these findings, preparations for both direct and indirect impacts will be needed in the unfolding of next pandemic.

386 – Development of a passive sampler assessing time-weighted average exposure to isothiazolinones in indoor air

Wonsik Shin[1], Seon-Kyung Park[2], Pil-Gon Kim[3], Jung-Hwan Kwon[4]
[1]Division of Environmental Science and Ecological Engineering, Korea University, Master course student, Seoul, Korea, Republic of. [2]Wonjin Institute of Occupational Environmental Health, Ph.D, Seoul, Korea, Republic of. [3]Division of Environmental Science and Ecological Engineering, Korea University, Post doctor, Seoul, Korea, Republic of. [4]Division of Environmental Science and Ecological Engineering, Korea University, Professor, Seoul, Korea, Republic of

Abstract
Isothiazolinones, including 5-chloro-2-methyl-4-isothiazolin-3-one/2-methyl-4-isothiazolin-3-one (CMIT/MIT), benzisothiazolinone (BIT), and octylisothiazolinone (OIT), have been widely used as biocides in many products. Epidemiological investigations have shown that the use of CMIT/MIT as a humidifier disinfectant resulted in fatal lung injury in Korea. The N-S bond of the isothiazolone ring reacts with nucleophilic cells, making them active biocides. However, their high reactivity makes it difficult to quantify isothiazolinones in indoor air after disinfection uses. In this study, a passive sampler for measuring time-weighted average concentrations of isothiazolinones was developed. The passive sampler consists of an expanded polytetrafluoroethylene (ePTFE) tube and a mixture of glycerol and water solution containing glutathione in the ePTFE tube. The reaction products of isothiazolinones with glutathione were used to quantify isothiazolinones in indoor air by analyzing compound-specific adducts of glutathione. As a model isothiazolinone, CMIT was inserted into a 125 L acrylic chamber and the stable atmospheric concentration was attained after 2 hours, monitored using a proton transfer reaction time-of-flight mass spectrometer. Passive samplers were deployed for 24 hours and removed at predetermined time intervals. Time-dependent increase of the reaction product was used to obtain the sampling rate (0.11 L/h) of the passive sampler for CMIT. The passive sampler can be an alternative for quantifying reactive isothiazolinones in indoor air.

387 – Are air pollution and noise exposures associated with physical activity for older adults with different types of dementia?

Karl Brown[1], Ching-Hsuan Huang[2], Sarah Payne[1], Adrienne Jankowski[1], Katie Wilson[3], Yeilim Cho[4], Debby Tsuang[5,6], Edmund Seto[7]
[1]Geriatric Research Education and Clinical Center, VISN-20 VA Puget Sound, Researcher, Seattle, USA. [2]Department of Environmental & Occupational Health Sciences, University of Washington, Research Assistant, Seattle, USA. [3]Department of Biostatistics, University of Washington, Assistant Teaching Professor, Seattle, USA. [4]Department of Psychiatry and Behavioral Sciences, University of Washington, Affiliate Postdoctoral Scholar, Seattle, USA. [5]Geriatric Research Education and Clinical Center, VISN-20 VA Puget Sound, Professor, Seattle, USA. [6]Department of Psychiatry and Behavioral Sciences, University of Washington, Professor, Seattle, USA. [7]Department of Environmental & Occupational Health Sciences, University of Washington, Professor, Seattle, USA

Abstract
Environmental factors are associated with physical activity, however, little is known about the associations between noise and air pollution exposures and physical activity for those with dementia. Alzheimer's Disease (AD) and Dementia with Lewy Bodies (DLB) -- two types of dementia -- can involve movement-related symptoms. We assessed exposures to traffic-related noise and various air pollutants and correlations with physical activity measured by accelerometry for adults with AD (N=8) and DLB (N=12). Accelerometry metrics included activity counts, ENMO, and steps. Median daily accelerometry counts were higher for the AD group compared to DLB (1,558,931 vs 1,383,438 counts, $p = 0.13$), but step counts and ENMO were not different. Pollutant exposures were not significantly different between groups, however median exposures to CO and SO_2 were higher in the AD group compared to the DLB (CO 0.27 vs. 0.24 ppm, $p = 0.13$; SO_2 0.64 vs. 0.55 ppb, $p = 0.16$). The three accelerometry metrics were highly correlated (correlation $r \geq 0.83$). Weak correlations were observed between noise exposure and air pollutant exposures ($r \leq 0.36$). Among AD subjects, activity was moderately correlated with traffic-related pollution (ENMO to NO_2 $r = 0.51$, ENMO to PNC $r = 0.51$) and SO_2 (ENMO to SO_2 $r= 0.79$). Among DLB subjects, activity was also correlated with traffic-related pollution, but not with SO_2. We observed a different direction of correlation for particle exposures and activity between groups (negative for AD, positive for DLB), suggestive of lower activity for AD in neighborhoods with higher $PM_{2.5}$ and PM_{10}.

388 – POSTER: Electrochemical Detection of Manganese in Drinking Water: Validation with a Community-based Study

Elena Boselli[1], Zhizhen Wu[2], Alexa Friedman[3], Birgit Claus Henn[4], Ian Papautsky[5]
[1]University of Illinois Chicago, PhD student, Chicago, USA. [2]University of Illinois Chicago, Postdoctoral scientist, Chicago, USA. [3]Boston University School of Public Health, PhD student, Boston, USA. [4]Boston University School of Public Health, Faculty, Boston, USA. [5]University of Illinois Chicago, Faculty, Chicago, USA

Abstract
Manganese (Mn) is an essential trace element that plays a crucial role in enzymatic functions. However, excessive levels of Mn can lead to harmful effects[1]. Overexposure to Mn has been linked to neurodegenerative diseases in adults and neurodevelopmental impairments in children. While Mn is naturally present in the environment, human activities can elevate its concentration. Among the various sources of Mn intake in the general population, drinking water is a major one. The current drinking water guideline for Mn is 50 μg/L[2]. Conventional methods for Mn detection rely on centralized laboratory equipment and are associated with long turnaround times. Conversely, electrochemical methods offer a suitable alternative for point-of-use (POU) assessment of Mn. In this study, we validated an electrochemical approach using square wave stripping voltammetry (SWCSV) for Mn detection in environmental water samples from households in a community in Holliston, MA, which is at particular risk of Mn contamination[3]. The developed system showed a linear range of 0-100 μg/L and a limit of quantitation of 6 μg/L. We benchmarked the analysis of environmental water samples (n=78) against inductively coupled plasma mass spectrometry (ICP-MS) and achieved an accuracy of approximately 70%[4]. The median Mn concentration in tap water samples from participating households was 2.3 μg/L (range 0.3-5306 μg/L). All samples above 50 ug/L (n=11) were identified by the electrochemical system. Developing a POU sensor system will allow for more frequent Mn monitoring in drinking water, especially in rural communities or settings at greater risk of exposure.

389 – POSTER: Concentration distribution of bioaerosols and analysis of influencing factors according to the characteristics of Public-use Facilities

Bo Yeon Kwon[1], Sungchul Seo[2,3], Jihyeon Kim[1,4], Kyungmin Kim[1,4]
[1]Institute for Environment Health and Safety, Seokyeong University, Researcher, Seoul, Korea, Republic of. [2]Institute for Environment Health and Safety, Seokyeong University, professor, Seoul, Korea, Republic of. [3]Department of Nano Chemical and Biological Engineering, Seokyeong University, professor, Seoul, Korea, Republic of. [4]Department of Nano Chemical and Biological Engineering, Seokyeong University, Student, Seoul, Korea, Republic of

Abstract
Introduction: There have been conflicting conclusions from as many studies as there are numerous variables on the factors affecting the concentration and type of bioaerosols. We explored to identify contributing factors related to the level of total airborne bacteria (TAB) and airborne mold and to analysis their significant impact on the concentration in indoor air according to the characteristics of Public-use Facilities.

Methods: Air samplings for TAB and mold were performed from July 2021 to March 2023 in several public-use facilities classified into five categories(A~E). Other variables such as temperature, relative humidity (RH), floor (aboveground/underground), and the number of people occupied in indoor were measured. We analyzed the correlation between TAB/mold concentration and meteorological factors also, multiple regression analyses were performed for determining the statistical significance of contributing variables.

Results: Bioaerosol not only showed a significant correlation between temperature and humidity, but also showed the highest correlation between mold and humidity. Multiple regression analyses showed that A facility have higher TBA concentrations, and E facility have higher airborne mold concentrations than other facilities. These facilities were statistically significantly affected by temperature, humidity and floor. In addition, TBA were affected by number of occupants, whereas mold was not found to be affected.

Conclusion: In this study, it was confirmed that the effect on TBA and mold could be different depending on the environmental characteristics of Public-use Facilities. In order to manage indoor air quality in Public-use Facilities, customized measures may be required depending on vulnerable facilities.

390 – Nutritional Mitigation of the Effect of Mercury on Neurodevelopment in Surinamese Children

Firoz Abdoel Wahid[1], Anisma Gokoel[2], Lissa Soares[3], Wilco Zijlmans[4], Jeffrey Wickliffe[5], Arti Shankar[6], Maureen Lichtveld[7]
[1]University of Pittsburgh, Assistant Professor, Pittsburgh, USA. [2]Scientific Research Center Suriname, Researcher, Paramaribo, Suriname. [3]Stony Brook Medicine, Postdoc, New York, USA. [4]Anton De Kom University, Researcher, Paramaribo, Suriname. [5]University of Alabama at Birmingham, Professor, Birmingham, USA. [6]Tulane University, Professor, New Orleans, USA. [7]University of Pittsburgh, Professor, Pittsburgh, USA

Abstract
Background: The Caribbean Consortium for Research in Environmental and Occupational Health (CCREOH) cohort study showed that pregnant women living in Suriname have blood concentrations of Pb and Hg, exceeding USEPA and CDC action levels. Prenatal exposure

to these metals harms neurodevelopment, but Se found in fish and Brazil nuts may counteract neurotoxicity. We hypothesize that Se mitigates negative effects of Hg and Pb on neurodevelopment.

Methods: The Human Health Exposure Analysis Resource program analyzed CCREOH women's blood samples for Hg, Pb, and Se. CCREOH's children (n=783, aged 11-42 months) participated in neurodevelopmental testing using the Bayley Scales for Infant Development III (BSID III). Ordinary least square regression models were used to examine Se/Pb and Se/Hg interactions, with the BSID III composite scores as outcome variables, adjusted for socio-demographic factors.

Results: The mean prenatal levels of Se, Hg, and Pb were respectively 148.74 µg/L, 6.13 µg/L, and 3.40 µg/dL. The mean composite scores for the cognitive, social-emotional, language and motor scores were respectively 107, 93, 95 and 104. The regression models showed an interaction between Se/Hg, related to the cognitive domain. Specifically, cognitive scores were the highest at the lowest Hg concentrations at the top 20th percentile of Se levels. No significant interaction was shown for Se/Pb.
Conclusion: At low levels of Hg exposure, the highest levels of Se seem protective for cognitive function. No level of Se mitigates the effect of Pb on neurodevelopment.

391 – PFAS in Whole Blood Self-Collected using Volumetric Microsampling vs Traditional Serum Approach in a Community with Elevated Exposures

Courtney Carignan[1], Rachel Bauer[2], Andrew Patterson[3], Thep Phomsopha[4], Eric Redman[5], Heather Stapleton[6], Christopher Higgins[7]
[1]Michigan State University, Assistant Professor, East Lansing, USA. [2]Michigan State University, Doctoral Student, East Lansing, USA. [3]Eurofins Environment Testing America, Technical Director, West Sacramento, USA. [4]Eurofins Environment Testing America, Scientist, West Sacramento, USA. [5]Eurofins Environment Testing America, Director of Technical Services, West Sacramento, USA. [6]Duke University, Professor, Durham, USA. [7]Colorado School of Mines, Professor, Golden, USA

Abstract
A remote sampling approach was developed at Eurofins for quantifying PFASs in whole blood samples collected using volumetric absorptive micro-samplers (VAMS), which allow for self-collection of blood using a finger prick. This study compares PFAS exposure measured by self-collection of blood using VAMS to the standard venous serum approach. Blood samples were collected from participants (n=53) in a community with prior PFAS drinking water contamination using a venous blood draw as well as participant self-collection using VAMS. Whole blood from the venous tubes were also loaded onto VAMS to compare differences in capillary vs venous whole blood PFAS levels. Samples were quantified for PFASs using liquid chromatography tandem mass spectrometry and online solid phase extraction. PFAS levels in serum were highly correlated with measurements in capillary VAMS ($r \geq 0.91$, $p<0.05$). Serum PFAS levels were generally two-fold higher than whole blood, reflecting expected differences in their composition. Of interest, FOSA was detected in whole blood (both venous and capillary VAMS) but not in serum. Overall, these findings indicate that VAMS are useful self-collection tools for assessing elevated human exposure to PFASs.

393 – Prediction of ambient ozone-attributable mortality risk modified by future temperature under climate change scenarios in China

Chen Chen[1], Menghan Wang[2], Tiantian Li[3], Xiaoming Shi[3]
[1]China CDC Key Laboratory of Environment and Population Health, National Institute of Environmental Health, Chinese Center for Disease Control and Prevention, Associate Professor, Beijing, China. [2]China CDC Key Laboratory of Environment and Population Health, National Institute of Environmental Health, Chinese Center for Disease Control and Prevention, Student, Beijing, China. [3]China CDC Key Laboratory of Environment and Population Health, National Institute of Environmental Health, Chinese Center for Disease Control and Prevention, Professor, Beijing, China

Abstract
Background: Given the hazards of the co-exposure of ozone and temperature under climate change, much work remains on exploring and analyzing the population health burden of ozone modified by temperature.

Methods: Based on a national population death surveillance network from 2013-2018, we adopt a two-stage statistical approach to construct the exposure-response (E-R) relationships between independent exposure to ozone and mortality, and the relationships under the modification of temperature. We use the carbon-neutral emissions scenarios along with importing the E-R relationships of independent exposure to ozone and temperature modification to estimate the ozone-related mortality burden in China.

Result: A 10-μg/m^3 increase in lag 0-1 day ozone is associated with a 0.12% increase in mortality. The effect of ozone is enhanced by modification of extreme heat and cold. We estimate around 16,970 deaths nationwide attributable to independent exposure to ozone during the baseline period (2015 to 2019). Under the SSP1-2.6 emissions scenario, with concentrations of ozone pollution substantially decreasing, we find the number of deaths drops to 274 in the mid-century (2056-2060). When temperature modification is included, we find the number of deaths increases to 103,538 during the baseline period, and decreases to 2,956 during the mid-century. The mortality burden of ozone would be further exacerbated under rapid population development.

Conclusions: Under climate change, extreme heat and cold could aggravate the mortality burden of ozone. Although ozone pollution has decreased, the irreversible effects of global warming would increase the burden of ozone-related deaths.

394 – Air Pollution Exposure and the Maternal Gut Microbiome in Early Pregnancy: Implications for Risk of Preterm and Early-term Births among African Americans

Ziyin Tang[1], Anne L. Dunlop[2], Yi-Juan Hu[3], Natalie Shen[1], Elizabeth A. Holzhausen[4], Bridget Chalifour[4], Youran Tan[1], Jeremy A. Sarnat[5], Tanya L. Alderete[6], Donghai Liang[7]
[1]Gangarosa Department of Environmental Health, Rollins School of Public Health, Emory University, Graduate Student, Atlanta, USA. [2]Department of Gynecology and Obstetrics, School of Medicine, Emory University, Professor, Atlanta, USA. [3]Department of Biostatistics and Bioinformatics, Rollins School of Public Health, Emory University, Associate Professor, Atlanta, USA. [4]Department of Integrative Physiology, University of Colorado Boulder, Postdoctoral Associate, Boulder, USA. [5]Gangarosa Department of Environmental Health, Rollins School of Public Health, Emory University, Associate Professor, Atlanta, USA. [6]Department of Integrative Physiology, University of Colorado Boulder, Assistant Professor, Boulder, USA. [7]Gangarosa Department of Environmental Health, Rollins School of Public Health, Emory University, Assistant Professor, Atlanta, USA

Abstract
Introduction: Air pollution exposure has been associated with increased risks of preterm birth (PTB) and early-term birth (ETB), as well as the composition and functional potential of the adult gut microbiome. However, the complex mechanisms underlying these associations remain largely unknown.

Methods: 264 African American women were recruited and provided rectal swabs in early pregnancy for gut microbiome profiling via 16S rRNA amplicon sequencing. We used spatiotemporally-resolved models to estimate prior week, one-, three-, and six-month average exposure to particulate matter ($PM_{2.5}$), nitrogen dioxide, and daily 8-h maximum ozone (O_3). LDM (Linear decomposition model) was used to examine the associations between gut bacteria (relative abundance, presence/absence) with exposures and early birth. Additionally, we tested the potential mediation effects of microbes on the associations between air pollution and early birth using LDM. A 20% false discovery rate was deemed statistically significant.

Results: Higher three- and six-month average O_3 exposures were associated with the presence of gut bacteria of genus Fastidiosipila. Compared to participants with full-term birth, mothers with ETB were less likely to have gut bacteria of phylum Actinobacteriota, while those with PTB were more likely to have presence and higher relative abundance of the genus Dialister. The presence of Actinobacteriota may be a potential mediator of the association between prior month $PM_{2.5}$ exposure and ETB.

Conclusions: This study suggested that maternal air pollution exposure may alter the composition of the maternal gut microbial community in early pregnancy, which may be associated with an increased risk of early birth.

395 – A Hybrid National Level Model for Formaldehyde, the Most Important Air Toxic

Sachraa G Borjigin[1], Md Kamrul Islam[2], Zhiyi Yang[3], Tian Xia[1], Stuart Batterman[4]
[1]Department of Environmental Health Sciences, University of Michigan, Postdoctoral Research Fellow, Ann Arbor, USA. [2]Department of Environmental Health Sciences, University of Michigan, Research Analyst, Ann Arbor, USA. [3]Department of Biostatistics, University of Michigan, MS Student, Ann Arbor, USA. [4]Department of Environmental Health Sciences, University of Michigan, Professor, Ann Arbor, USA

Abstract
Formaldehyde (HCHO) is ubiquitous in the troposphere, the most common aldehyde in the environment, the most important urban air toxic from a health perspective, and a major precursor in the formation of ambient ozone. This simple aldehyde is a unique air toxic due to the significance of contributions from biogenic and anthropogenic sources and secondary atmospheric formation; it is also a key indoor air pollutant due to off-gassing of many products used in buildings. This study presents the development and evaluation of a

hybrid model to predict annual and seasonal levels of HCHO across the USA. The model incorporates a wide range of predictor variables, including land use variables (e.g., water coverage, impervious surface area, tree canopy coverage, population density, highway length, vehicle traffic, normalized difference vegetation index) derived for buffers spanning 0.1 to 50 km distance; climatic data including temperature and precipitation, chemical transport model (CTM) predictions of primary and secondary HCHO, and satellite data (HCHO, NO_2 and O_3 from TROPOMI). Model variables are selected using LASSO and other approaches that best fitted cleaned HCHO measurements collected at 275 monitoring sites across the USA from 2002 to 2021. The hybrid model significantly improved the agreement to measurements (r^2>0.66) compared to any single predictor model (e.g., satellite and CTM estimates had r^2<0.22). Results from our final model can be used to estimate population exposures and health risks, important given temporal and spatial variation of HCHO, the sparse monitoring network, and the magnitude of concentrations that can frequently exceed guideline levels.

396 – POSTER: Associations of triazine herbicides exposure with blood lipid parameters in general Chinese population: The mediating effect of systemic inflammation

Mengyi Wang[1], Min Zhou[2], Qiyou Tan[1], Ruyi Liang[1], Yongfang Zhang[1], Weihong Chen[3], Tao Jing[3]
[1]School of Public Health, Tongji Medical College, Huazhong University of Science and Technology, PhD candidate, Wuhan, China. [2]School of Public Health, Tongji Medical College, Huazhong University of Science and Technology, PhD, Wuhan, China. [3]School of Public Health, Tongji Medical College, Huazhong University of Science and Technology, Prof. PhD, Wuhan, China

Abstract
Objectives: The adverse health effects of triazine herbicides (THs) exposure have attracted broad concern. While the effects of THs on blood lipid metabolism remain unclear. We aimed to assess the associations between THs and blood lipid parameters and explore the underlying mechanism.

Methods: We enrolled 4191 urban adults without previous use of antihyperlipidemic drugs from the Wuhan-Zhuhai cohort, and measured their concentrations of serum THs (atrazine, cyanazine, and simazine), blood lipid parameters (total cholesterol [TC], triglycerides [TG], high-density lipoprotein cholesterol [HDL-C]), and plasma C-reactive protein (CRP). Generalized linear models, bayesian kernel machine regression models, and mediation analyses were used to explore the relationships between THs, CRP, and blood lipid parameters.

Results: The continuous and categorical analyses revealed significant and positive relationships of atrazine with TC, HDL-C, cyanazine with TC, TG, and simazine with TC, and TG. Besides, the increase of THs mixture levels was significantly and monotonously associated with an increase in TC, TG, and a decrease in HDL-C. Furthermore, elevated THs were significantly associated with increased CRP, which was positively associated with TC and TG, and negatively related to HDL-C. Mediation analyses showed that CRP partly mediated the positive associations of atrazine, and cyanazine with TC and TG, with a mediated proportion of 1.61%-3.20%.

Conclusions: Our results suggested THs exposure was positively associated with abnormal blood lipids, and systemic inflammation may play a part of the mediation role in these associations.

397 – Preconception and Prenatal Exposure to Ambient Air Pollution and Depression 1-3 Years Postpartum in the MADRES Pregnancy Cohort

Yuhong Hu[1], Tingyu Yang[2], Xinci Chen[2], Mario Vigil[2], Zhongzheng Niu[3], Sandrah P. Eckel[4], Nathan Pavlovic[5], Fred Lurmann[6], Shohreh F. Farzan[7], Rima Habre[4], Carrie Breton[8], Theresa M. Bastain[9]
[1]Department of Population and Public Health Sciences, Keck School of Medicine, University of Southern California, PhD student, Los Angeles, USA. [2]Department of Population and Public Health Sciences, Keck School of Medicine, University of Southern California, Statistician, Los Angeles, USA. [3]Department of Population and Public Health Sciences, Keck School of Medicine, University of Southern California, Postdoctoral Scholars, Los Angeles, USA. [4]Department of Population and Public Health Sciences, Keck School of Medicine, University of Southern California, Associate Professor of Population and Public Health Sciences, Los Angeles, USA. [5]Sonoma Technology, Inc., Data Scientist, Petaluma, USA. [6]Sonoma Technology, Inc., Manager of Exposure Assessment Studies, Petaluma, USA. [7]Department of Population and Public Health Sciences, Keck School of Medicine, University of Southern California, Assistant Professor of Population and Public Health Sciences, Los Angeles, USA. [8]Department of Population and Public Health Sciences, Keck School of Medicine, University of Southern California, Professor of Population and Public Health Sciences, Los Angeles, USA. [9]Department of Population and Public Health Sciences, Keck School of Medicine, University of Southern California, Associate Professor of Clinical Population and Public Health Sciences, Los Angeles, USA

Abstract
BACKGROUND/AIMS: Air pollution during pregnancy has been linked with postpartum depression. However, few studies have investigated critical windows of exposure. This study aimed to evaluate preconception and prenatal ambient air pollution exposure and risk of postpartum depression up to 3 years after childbirth and to identify windows of susceptibility.

METHODS: This study followed 354 predominantly low-income Hispanic/Latina participants with full-term pregnancies in the MADRES cohort 1-3 years postpartum. We estimated daily residential $PM_{2.5}$, PM_{10}, NO_2, and O_3 concentrations from 12 weeks prior to conception to 37 gestational weeks using inverse-distance squared spatial interpolation from ambient monitoring data and calculated weekly averaged levels. Depression was assessed by the 20-item Center for Epidemiologic Studies-Depression (CES-D) scale at 12, 24, and 36 months postpartum. Poisson log-linear distributed lag models via generalized estimating equations (GEE) were performed to estimate the adjusted risk ratio (RR) and 95% CI of weekly air pollution for postpartum depression.

RESULTS: Depression was observed in 17.9%, 16.5%, and 13.5% of participants at 12, 24, and 36 months, respectively. We found increased risk of depression up to 3 years postpartum with prenatal exposure to PM_{10} and NO_2 during the significant sensitive windows of gestational weeks 14-27 and 11-27, respectively. The strongest associations (PM_{10} RR = 1.05, 95% CI: 1.02-1.08; NO_2 RR = 1.05, 95% CI: 1.03-1.07) were both observed at gestational week 21.

CONCLUSION: Mid-pregnancy PM_{10} and NO_2 exposures were associated with an increased risk of persistent postpartum depression, which has critical implications for prevention of mental health outcomes.

398 – Poly- and Perfluoroalkyl Substances (PFASs) in Farm Raised and Farmers Market Foods near a Midwestern PFAS Impacted Site

Ankita Bhattacharya[1], Rachel Bauer[1], Sarah Choyke[2], Juliane Brown[3], Christopher Higgins[4], Heather Stapleton[5], Jason Conder[6], Courtney Carignan[7]
[1]Michigan State University, Doctoral Student, East Lansing, USA. [2]Colorado School of Mines, Post Doctoral Fellow, Golden, USA. [3]Colorado School of Mines, Doctoral Student, Golden, USA. [4]Colorado School of Mines, Professor, Golden, USA. [5]Duke University, Professor, Durham, USA. [6]Geosyntec consultants, Principal Scientist, Costa Mesa, USA. [7]Michigan State University, Assistant Professor, East Lansing, USA

Abstract
Food safety concerns have been raised regarding agricultural fields that have received biosolids containing elevated levels of poly- and perfluoroalkyl substances (PFASs) that have been traced back to PFAS-using industries (e.g., metal plating, paper mills, textiles, etc.). Migration from point sources to farmland can occur via water, dry deposition, and application of PFAS- contaminated biosolids. When taken up by crops, this has the potential to contribute to dietary PFAS exposure. Dietary exposure from local foods is a concern among a Midwestern community with a point source of PFAS contamination. Therefore, we collected 166 samples of 10 different food types from farms within a 15-mile radius of a midwestern community with PFAS contamination as well as from background locations from a neighboring county. Several samples were collected from 'site' and 'background' farms during the growing seasons of 2020/21 and combined by food type into 28 composite pairs. Food samples were tested for 48 PFASs using LC-MS/MS (animal products) and LC-QTOF-MS (produce) with improved methods designed to minimize bias during sample extraction. PFAS concentrations in eggs, spinach, and butternut squash (Sum PFAS ranged up to 6 ng/g) collected from markets near the site were modestly elevated compared to the background, whereas other leafy greens and squashes were low or modestly elevated among both site and background samples. In addition to sharing details of these findings, we will report exposure estimates for the local foods using site-specific PFAS concentration and consumption frequency data.

399 – Principles and Elements for Creating and Sustaining Successful Public-Private Partnerships (PPPs) for Community Environmental Monitoring Programs: Preliminary Results

Ana M. Rule[1], Christel J. Tajouoh[2], Fernando A. Wagner[3], Judy S. LaKind[4,5]
[1]Johns Hopkins Bloomberg School of Public Health, Assistant Professor, Baltimore, USA. [2]School of Social Work, University of Maryland, Graduate Student, Baltimore, USA. [3]School of Social Work, University of Maryland, Professor, Baltimore, USA. [4]LaKind Associates, LLC, President, Baltimore, USA. [5]Department of Epidemiology and Public Health, University of Maryland School of Medicine, Adjunct Professor,

Baltimore, USA

Abstract
Background: Public-private partnerships (PPP) (a cooperative investment of resources that results in joint goals, risk-taking, sharing of authority, and mutual benefits for all partners) can play a critical role in advancing our understanding of environmental exposures by maximizing cross-disciplinary expertise and resource sharing among communities, government, academic, and industry researchers. PPPs associated with community environmental monitoring have not been well-explored in the published literature. The goal of this project is to understand the elements that distinguish successful PPPs from those with limitations, and to better develop guidance needed to ensure successful future PPPs.

Methods: We conducted a scoping review of the literature on existing PPPs associated with community environmental monitoring. We also interviewed community members, industry representatives, regulators and academics who had been involved in PPPs; thematic qualitative analysis allowed us to identify critical elements for successful PPPs.

Results: Our initial literature search yielded over 15,000 potentially relevant publications. Focusing on environmental monitoring PPPs reduced that number to a couple hundred relevant papers. Using stakeholders' input we refined and expanded on the literature-based set of principles and elements. We identified the key elements and principles necessary for creating and sustaining successful PPPs for community environmental monitoring.

Conclusions: The resulting comprehensive and current set of elements and principles can be used by stakeholders to enable parties to trust each other and the findings from PPP research. We anticipate that results will serve as the foundation for future PPPs and for improving our understanding of community exposures and health.

400 – Biomonitoring Equivalents for in vitro Toxicity Studies using PBPK Modeling: Case study of As, Cd and Hg

Mi-yeon Shin[1], Haesong Sher[2], Joohyeon Park[2], Woozong Jang[2], Sungkyoon Kim[3,4]
[1]Graduate School of Public Health, Seoul National University, PhD researcher, Seoul, Korea, Republic of. [2]Graduate School of Public Health, Seoul National University, Master's student, Seoul, Korea, Republic of. [3]Graduate School of Public Health, Seoul National University, Professor, Seoul, Korea, Republic of. [4]Institute of Health and Environment, Professor, Seoul, Korea, Republic of

Abstract
It has been challenging to determine appropriate levels of testing in in vitro toxicology studies. This study aimed to refine Biomonitoring Equivalents (BEs) for heavy metals using physiologically based pharmacokinetic (PBPK) modeling to improve in vitro toxicity testing.

Forward dosimetry with published PBPK models was utilized to estimate exposure levels of arsenic (As), cadmium (Cd), and mercury (Hg) in liver, blood, and urine corresponding to exposure guidance values. Oral exposure was assumed to occur three times daily until steady state. The estimated BEs were compared to existing ones.

Liver concentrations of As, Cd, and Hg were 8 nM (inorganic), 14.1 nM (organic), 10 nM, 0.9 nM (inorganic Hg), and 2.1 nM (organic Hg), respectively. Urinary total As (Σ all species) and Cd, and blood methyl Hg concentrations were 0.39 μg/L, 1.0 μg/L, and 0.65 μg/L, respectively. The estimated BEs were 1.5 (total As), 31 (Cd), and 17 (methyl Hg) times lower than existing ones.

The study's findings highlight the potential of using PBPK modeling to refine BEs for heavy metals, contributing to more accurate in vitro toxicity testing. The estimated liver concentrations provide essential information for determining appropriate exposure ranges for liver toxicity tests. The results indicate a need for re-evaluating the current BEs, possibly leading to more conservative values.

401 – Network analysis to examine the association of persistent organic pollutants with diabetes risk in Great Lakes sport fish consumers using Gaussian graphical model

Chak Kwong (Tommy) Cheng[1], Mary Turyk[2], Hua Yun Chen[3]
[1]University of Illinois at Chicago, PhD Student in Biostatistics, Chicago, USA. [2]University of Illinois at Chicago, Professor of Epidemiology, Chicago, USA. [3]University of Illinois at Chicago, Professor of Biostatistics, Chicago, USA

Abstract
Background: Many studies indicate that exposure to persistent organic pollutants (POPs) is associated with increased diabetes risk. However, due to strong correlations among POPs it is difficult to identify toxic agents within the mixture. Our objective was to explore the use of Gaussian graphical models in examining the association of a mixture of POPs with diabetes risk.

Methods: We conducted network analyses using Gaussian graphical models to examine the conditional dependence associations among POPs [dichlorodiphenyldichloroethene (DDE), polychlorinated biphenyls (PCBs), polybrominated diphenyl ethers (PBDEs), and polyfluoroalkyl substances (PFAS)]; sport and commercial fish consumption rates; serum biomarkers associated with diabetes risk (adiponectin, C-reactive protein, gamma glutamyl transferase, triglycerides, and cholesterol); and age, sex and body mass index in 513 participants of the Great Lakes Sport Fish Consumption Study (GLSFC). We considered both incident diabetes and hemoglobin A1c (adjusting for diabetes medication use) as measures of diabetic risk.

Results: Our analysis indicated that DDE was associated with diabetes risk after adjusting for all the exposures, biomarkers, and covariates simultaneously. There was no evidence that biomarkers of diabetes risk were potential intermediates between POPs and health outcomes.

Conclusions: The result from the network analysis confirmed the known association between DDE and diabetes, which has been shown in previous analyses of this cohort that used traditional single-exposure regression models, and did not identify any additional POPs impacting diabetes. Our methodology was a straightforward approach to examine the relationship of multiple chemical exposures and intermediate health outcomes in comparison to traditional regression analyses.

402 – POSTER: Exploring Candidate Metabolites of Alternative Plasticizers ATBC and ATEC in Human Urine Using LC-qTOF-MS: A Preliminary Study

Eu-Kyung Jo[1], Jinhyun Kwon[1], Younglim Kho[2], Junho Jeon[3], Mi-Yeon Shin[4], Sungkyoon Kim[5]
[1]Graduate School of Public Health, Seoul National University, Student, Seoul, Korea, Republic of. [2]Department of Health, Environment and Safety, Eulji University, Professor, Seongnam, Korea, Republic of. [3]Department of Environmental Engineering, Changwon National University, Associate Professor, Gyeongsangnam-do, Korea, Republic of. [4]Graduate School of Public Health, Seoul National University, Research Professor, Seoul, Korea, Republic of. [5]Graduate School of Public Health, Seoul National University, Professor, Seoul, Korea, Republic of

Abstract
The use of alternative plasticizers, such as acetyl tributyl citrate (ATBC) and acetyl triethyl citrate (ATEC), is growing due to concerns over the toxicity of conventional plasticizers. Despite their widespread use, little is known about the human exposure biomarkers of these compounds. This study aims to identify candidate metabolites of ATBC and ATEC in human urine using liquid chromatography coupled to quadrupole-time-of-flight mass spectrometry (LC-qTOF-MS) according to suspect screening workflow.

One healthy male and one healthy female volunteer were orally administered with deuterium-labeled ATBC and non-labeled ATEC, respectively. Urine samples were collected at various time points within 48 hours following administration. Pretreatment of the urine samples was carried out in two ways (liquid-liquid extraction, LLE; solid-phase extraction, SPE). Samples were analyzed using LC-qTOF-MS and the data were processed with Waters UNIFI software for suspect screening of metabolites.
From the SPE samples, six candidate metabolites were identified for each compound including metabolites undergone one or more reactions of hydroxylation, hydrolysis, and deacetylation. Results will be compared to determine the suitable extraction method after further analysis of LLE samples.

Though this study has limitation of small sample size, it demonstrates a novel approach for detecting candidate metabolites of alternative plasticizers ATBC and ATEC in human urine using LC-qTOF-MS. The findings provide a starting point for understanding the potential human exposure biomarkers of the compounds, contributing to future exposure assessment and toxicological studies.

403 – POSTER: Estimation of Exposure Amounts of Bisphenols in the South Korean General Population with a Physiologically Based kinetic (PBK) Model

Mi-yeon Shin[1], Jinhyun Kwon[2], Eu-Kyung Jo[3], Haesong Sher[3], Joohyeon Pak[3], Woozong Jang[3], Sungkyoon Kim[4,5]
[1]Graduate School of Public Health, Seoul National University, Seoul, Seoul, Korea, Republic of. [2]Graduate School of Public Health, Seoul National University, PhD student, Seoul, Korea, Republic of. [3]Graduate School of Public Health, Seoul National University, Master's student,

Seoul, Korea, Republic of. [4]Graduate School of Public Health, Seoul National University, Professor, Seoul, Korea, Republic of. [5]Institute of Health and Environment, Seoul National University, Professor, Seoul, Korea, Republic of

Abstract
Biomonitoring data have demonstrated that exposure of general population to bisphenols is widespread. The purpose of this study was to estimate daily intake (EDIs) of the chemicals for general population using PBK model and Fractional urinary excretion (F_{ue}).

We applied published PBK models to calculated the EDIs from urinary BPA and BPS collected from Korean National Environmental Health Survey (KoNEHS 2015-2017); the estimates were compared with the ones from the published F_{ue}.

The median urinary concentrations of BPA and BPS were 1.52 μg/L and 0.025 μg/L for male adults, 1.14 μg/L and 0.034 μg/L for female adults, respectively. The median EDIs (95%tile) of BPA were 16 (91) ng/kg/day for male adults (n = 1,647) and 12 (76) ng/kg/day for female adult (n = 2,133). For BPS, the EDIs were 0.63 (6.9) ng/kg/day for male adults and 0.84 (9.9) ng/kg/day for female adult, the EDIs were 14 ~ 25 times lower compared to those in BPA. Generally, the EDIs from PBK model were 0.5 times lower than those from F_{ue} in BPA and BPS.

The PBPK models were useful in estimating the exposure amounts from urinary BPA and BPS. The EDIs of BPA in adult were exceeded temporary tolerable daily intake (t-TDI, 0.004 μg/kg/day). We need to continuously monitor exposure levels of BPA and the alternatives.

404 – Tracking daily 1 km $PM_{2.5}$ chemical composition in China since 2000 from space via deep learning

Jing Wei[1], Zhanqing Li[2], Xi Chen[3], Yuewei Liu[4]
[1]University of Maryland, Research Associate, College Park, USA. [2]University of Maryland, Professor, College Park, USA. [3]Chinese Center for Disease Control and Prevention, Professor, Beijing, China. [4]Sun Yat-sen University, Associate Professor, Guangzhou, China

Abstract
Fine particulate matter ($PM_{2.5}$) chemical composition has strong and diverse impacts on the planetary environment, climate, and health. We developed a four-dimensional spatiotemporal deep forest (4D-STDF) model to estimate daily $PM_{2.5}$ chemical composition at a spatial resolution of 1 km in China since 2000 by integrating measurements from a high-density observation network, satellite $PM_{2.5}$ retrievals, atmospheric reanalyses, and model simulations. Cross-validation results illustrate the reliability of sulfate (SO_4^{2-}), nitrate (NO_3^-), ammonium (NH_4^+), and chloride (Cl^-) estimates, with high coefficients of determination (CV-R^2) with ground-based observations of 0.74, 0.75, 0.71, and 0.66, and average root-mean-square errors (RMSE) of 6.0, 6.6, 4.3, and 2.3 μg/m^3, respectively. The three components of secondary inorganic aerosols (SIA) account for 21% (SO_4^{2-}), 20% (NO_3^-), and 14% (NH_4^+) of the total $PM_{2.5}$ mass in eastern China. We observed significant reductions in the mass of inorganic components by 40–43% between 2013 and 2020, slowing down since 2018. Comparatively, the ratio of SIA to $PM_{2.5}$ increased by 7% across eastern China except in Beijing and nearby areas, accelerating in recent years. SO_4^{2-} has been the dominant SIA component in eastern China, although it was surpassed by NO_3^- in some areas, e.g., Beijing-Tianjin-Hebei region since 2016. SIA, accounting for nearly half (~46%) of the $PM_{2.5}$ mass, drived the explosive formation of winter haze episodes in the North China Plain. A sharp decline in SIA concentrations and an increase in SIA-to-$PM_{2.5}$ ratios during the COVID-19 lockdown were also revealed, reflecting the enhanced atmospheric oxidation capacity and formation of secondary particles.

405 – POSTER: Tetrachloroethylene Exposure and Neurobehavioral Performance among Children Living Near Multiple Contamination Sites

Marwan Alajlouni[1], Jung Hyun Lee[1], Alaina Bryant[2], Brandon Tso[3], Abigail Valdez[3], Ian Kurz[3], Thomas Wallace[4], Steven Koch[5], Mary Turyk[6], Brandon Boor[7], Antonios Tasoglou[8], Sa Liu[9]
[1]School of Health Sciences, Purdue University, PhD Graduate Student, West Lafayette, USA. [2]School of Health Sciences, Purdue University, Master Graduate Student, West Lafayette, USA. [3]School of Health Sciences, Purdue University, Undergraduate Student, West Lafayette, USA. [4]Martinsville Indiana Superfund Site Association, Founder, Martinsville, USA. [5]School of Medicine, Indiana University, Associate Professor of Clinical Pediatrics, Indianaplois, USA. [6]School of Public Health, University of Illinois, Professor, Chicago, USA. [7]Lyles School of Civil Engineering, Purdue University, Associate Professor of Civil Engineering, West Lafayette, USA. [8]Montrose Environmental, Denver , CO, USA, Emerging Technology Project Manager, Denver, USA. [9]School of Health Sciences, Purdue University, Assistant Professor, West Lafayette, USA

Abstract
BACKGROUND: Tetrachloroethylene (PCE) is a neurotoxicant with data mainly from high levels of occupational exposure. The impact of community exposure on children's neurobehavioral performance is poorly understood.

AIM: This study evaluated the association between PCE exposure and neurobehavioral performance in 6-11 year old children living in a community with multiple PCE contamination sites.

METHODS: Exposure was quantified by measuring PCE in exhaled breath. Neurobehavioral performance was assessed using the Wide Range Achievement Test Fifth Edition (WRAT-5), the Behavior Assessment System for Children Third Edition (BASC-3), and the NIH Toolbox cognitive measures. A questionnaire was administered to collect demographic and covariate data. Associations of outcome measures with PCE tertiles were assessed using linear regression models adjusted for significant covariates.

RESULTS: Eighty-nine participants were enrolled. Seventy-three of them completed data collection. The average scores for fluid composite, total composite, early childhood composite, behavioral index, and math were significantly lower than the standard scores, with 30% of participants having at least one score clinically significantly worse than the standard scores. Fluid composite ($p = .009$), total composite ($p = .016$), early childhood composite ($p = .013$), and academic performance in math ($p = .002$) and reading ($p = .006$) were negatively associated with PCE exposure when comparing the highest exposure tertile to the lowest.

CONCLUSIONS: The study suggests PCE exposure may impact children's cognitive function, problem-solving ability, adaptability, and academic performance among those living in this community. Further investigations are needed to confirm these findings.

406 – Functional Exposomics: Concept and Its Utility for Systematic Discovery and Mapping of Biologically Significant Exposures in Biomedical Research

Ming Kei (Jake) Chung[1], Vy Kim Nguyen[1,2,3], Thomas P van der Meer[4], Chirag Patel[5]
[1]Department of Biomedical Informatics, Harvard Medical School, Postdoc, Boston, USA. [2]Department of Environmental Health Sciences, School of Public Health, University of Michigan, Ann Arbor, Postdoc, Michigan, USA. [3]Department of Computational Medicine and Bioinformatics, Medical School, University of Michigan, Ann Arbor, Postdoc, Michigan, USA. [4]Department of Endocrinology, University of Groningen, PhD candidate, Groningen, Netherlands. [5]Department of Biomedical Informatics, Harvard Medical School, Associate professor, Boston, USA

Abstract
High-resolution mass spectrometry (HRMS) serves as the gold standard for conducting non-targeted exposome analyses. Although tens of thousands of signals can be detected, only a fraction can be putatively labeled, with even fewer being confirmed. There is a considerable challenge in converting raw signals into actionable knowledge.

We recently introduced the concept of the functional exposome as a solution to overcome critical bottlenecks in exposome research. In this presentation, our objectives are to: 1) provide a brief overview of the functional exposome concept; 2) explain how the functional exposome can be measured; 3) showcase recent studies that utilized affinity-based methods to uncover exposures within the functional exposome; and 4) present preliminary results from the analysis of antibody complexes from a healthy individual.

Building on the idea that the human adaptive immune system is highly responsive to environmental factors and the ability of B cells to secrete antibodies that bind to immunological exposures (i.e., circulating immune complexes, CICs), we employed an immunoprecipitation technique to capture CICs and performed a non-targeted analysis of the recovered molecules using HRMS. In comparison with the control, we found over 100 differentiating signals, with at least 20 exhibiting a signal fold change greater than 100 000. The putatively annotated features included five pharmaceuticals (remikiren, loperamide, a quinidine barbiturate, and metergoline), a natural product derived from fruit (1,5-dibutylmethylhydroxycitrate), and a potential microbial metabolite (all-trans-hexaprenyl diphosphate).

Upon further development, the method could be one of the high-throughput technologies propelling the next generation exposomics in public health and biomedicine.

407 – Primary School Extracurricular Activities: Fostering Citizen Science Education for the Next Generation

David Kocman[1], Rok Novak[2], Cirila Jeraj[3]
[1]Jožef Stefan Institute, Senior Researcher, Ljubljana, Slovenia. [2]Jožef Stefan Institute, Research Assistant, Ljubljana, Slovenia. [3]Vodice Elementary School, Teacher, Vodice, Slovenia

Abstract
Active involvement of individuals and communities in Citizen Science (CS)-based research, concerning the presence and exposure to various stressors, requires certain skills and competences. By developing these skills at a young age, individuals can be better prepared to be part of a CS project and related activities in the future. To this end, we developed an extracurricular activity at the elementary school Vodice in Slovenia, entitled "Citizen Scientist", targeting pupils aged 7-9. Participating pupils had two distinct roles: (1) a student, learning about the scientific approach and the concepts of Citizen Science, and (2) a citizen scientist, performing real measurements, collecting data, analyzing it and interpreting the results. The primary topics were noise, air quality, heat stress and wellbeing near an elementary school in a semi-rural area in the subalpine basin of Ljubljana. Low-cost noise, air quality and temperature sensors were used to collect the data. The pupils were involved in research design (where and why to perform the measurements), implementation (logging the data), analysis (interpreting trends and correlations) and dissemination (presenting the results to other pupils, parents, teachers and other stakeholders). In this contribution, we evaluate the activities conducted on a monthly basis throughout one school year, and report on the lessons learned.

408 – In vitro exposure to e-cigarette constituents and condensates are cytotoxic to cultured trophoblasts

Viktoriia Goriainova[1], Judith Zelikoff[2], Maria Katsigeorgis[1], John Adragna[1], Cassandra Meakin[3], Lauren Alekskunes[4]
[1]New York University School of Medicine, Graduate Student, NY, USA. [2]New York University School of Medicine, Professor, NY, USA. [3]Rutgers University, Post-doctoral, New Brunswick, USA. [4]Rutgers University, Professor, New Brunswick, USA

Abstract
As of 2018, 41 million people addicted to nicotine were using electronic cigarettes (e-cigs), including about 150 thousand women during pregnancy. A recent study finds that about 4% of pregnant American women are vaping, and the rate of e-cig use is actually higher among pregnant women than women who aren't pregnant. Extravillous trophoblasts are critical for establishment of uteroplacental exchange and proper placentation, and therefore a successful pregnancy outcome. The purpose of this research was to determine how the constituents of e-cigarettes and their heated condensate affects viability, proliferation and migration of extravillous trophoblasts, as well as signal transduction pathways associated with placental injury and adverse birth consequences. The findings from this study suggest that e-cig constituents and heated aerosols are toxic to extravillous trophoblasts critical for establishment of uteroplacental exchange and proper placentation. Moreover, that the over-migration stimulated by e-cig condensate suggests a potential increased risk of placenta accreta – an anomaly in which placental tissue penetrates myometrium.

409 – Residents' Experiences During a Hydrogen Sulfide Emergency: A Qualitative Analysis

Arbor Quist[1], April Hovav[2], Bhavna Shamasunder[3], Jill Johnston[4]
[1]University of Southern California, Postdoctoral Fellow, Los Angeles, USA. [2]Occidental College, Postdoctoral Scholar, Los Angeles, USA. [3]Occidental College, Associate Professor, Los Angeles, USA. [4]University of Southern California, Associate Professor, Los Angeles, USA

Abstract
In October 2021, thousands of residents in Carson, California began complaining of noxious odor, headaches, and nausea. Hydrogen sulfide (H_2S), a toxic odorous gas, was measured at concentrations up to 7000 parts per billion (ppb) and remained above California's acute air quality standard of 30 ppb for 4 weeks, with episodic elevations continuing for 3 months. Prior research has yielded conflicting results on the health effects of the H2S exposure at levels that were experienced in Carson during this event. Furthermore, there is a gap in understanding how people perceive and experience environmental emergencies like this H2S event. Expanding on a rapid health survey conducted in the aftermath of the event, we conducted 6 focus groups with a total of 36 participants to understand the complex ways in which this incident affected residents. Conversations focused on the inadequacy of government response, the health effects of H2S exposure, and issues of environmental injustice. Many participants expressed feeling gaslit and distrustful, as local leaders communicated that the levels of H2S were not harmful, but residents were personally experiencing adverse health effects. Participants also expressed difficulty obtaining clear information on H2S, stress regarding the unknown effects of the odors, and increased concern about environmental contaminants and future disasters. This study demonstrates the necessity of clear and prompt risk communication during disasters and the importance of engaging with community members to broaden our understanding of the physical and mental health impacts of odors and chemical disasters.

410 – Phthalate and alternative plasticizer exposure and risk assessment in patients with chronic kidney disease

Inae Lee[1], Jeonghwan Lee[2], Yaerim Kim[3], Younglim Kho[4], Sunhey Jung[5], Gowoon Lee[1], Bareum Kwon[6], Jung Pyo Lee[2], Kyungho Choi[7]
[1]Seoul National University, Research associate professor, Seoul, Korea, Republic of. [2]SMG-SNU Boramae Medical Center, Professor, Seoul, Korea, Republic of. [3]Keimyung University School of Medicine, Professor, Daegu, Korea, Republic of. [4]Eulji University, Professor, Seongnam, Korea, Republic of. [5]Eulji University, Master student, Seongnam, Korea, Republic of. [6]Seoul National University, postdoc researcher, Seoul, Korea, Republic of. [7]Seoul National University, Professor, Seoul, Korea, Republic of

Abstract
Phthalates have been suggested as a risk factor for chronic kidney disease, but research on the level of exposure among the chronic kidney disease (CKD) patients is limited. In addition, the concentration of phthalate metabolites in urine is highly variable due to their short half-lives and complex exposure characteristics. A panel of CKD patients (Study on Kidney Disease and Environmental Chemicals, SKETCH) was developed. The SKETCH panel was recruited among the patients who regularly visit 5 university hospitals in Korea (n=307). From June 2020 to December 2021, urine samples were collected at 5 time points at approximately 3-month intervals, and the urinary levels of phthalates and alternative plasticizers were analyzed by HPLC/MS/MS. Exposure and risk assessments were performed for chemicals with high detection frequencies: DEP, DiBP, DnBP, BBzP, DEHP, DEHTP, DINCH, and DINP. Intraclass correlation coefficient (ICC) of the chemicals ranged from 0.2 to 0.6, and most of chemicals showed moderated ICC values. When the 50th and 75th percentiles of exposure were compared, DEHP, DnBP, and DEHTP were ranked the highest in levels of respective urinary metabolites. At the 95th percentile, DEP, DEHP, and DnBP were observed to be the highest. The proportion of HQs of that exceeded 1 was low (DnBP (1.3%), DEHP (0.3%)), and the proportion of HI that exceeded 1 was 6.5%. Among the alternative plasticizers, oxidized metabolites of DEHTP and an ATEC metabolite were most often detected.

Acknowledgement: This work was supported by grants from National Research Foundation of Korea (NRF-2020R1A2C3011269 and NRF-2022R1C1C2006982).

411 – The AirPen: A Wearable Monitor for Characterizing Exposures to Particulate Matter and Volatile Organic Compounds

Jessica Tryner[1], Casey Quinn[1], Emilio Molina Rueda[2], Marie J. Andales[3], Christian L'Orange[4], John Mehaffy[1], Ellison Carter[4], John Volckens[5]
[1]Colorado State University, Research Scientist, Fort Collins, USA. [2]Colorado State University, Research Assistant, Fort Collins, USA. [3]Colorado State University, Research Associate, Fort Collins, USA. [4]Colorado State University, Assistant Professor, Fort Collins, USA. [5]Colorado State University, Professor, Fort Collins, USA

Abstract
Exposure to air pollution is a leading risk factor for disease and premature death, but technologies to assess personal exposure to particulate and gaseous air pollutants, including the timing, source(s) and location of such exposures, are limited. We developed a small (200 g), quiet, wearable monitor, called the AirPen, to quantify personal exposures to fine particulate matter ($PM_{2.5}$) and volatile organic compounds (VOCs) as a function of time, location, and activity. The AirPen combines physical sample collection (PM onto a filter and VOCs onto a sorbent tube) with a suite of low-cost sensors (for PM, VOCs, temperature, humidity, light intensity, location, and motion). We validated the AirPen against conventional personal sampling equipment in the laboratory and then conducted a field demonstration to measure at-work and away-from-work exposures to $PM_{2.5}$ and VOCs among employees at an agricultural facility. Time-integrated $PM_{2.5}$ and benzene concentrations collected using AirPens were strongly correlated with (Pearson's $r > 0.94$), but slightly lower than, concentrations derived from conventional personal samplers. Integration of multiple data streams from the AirPen indicated that personal exposures to benzene, toluene, ethylbenzene, and xylenes were dominated by a specific workplace location. Personal exposures to $PM_{2.5}$ and BTEX were typically lower when participants were away from work. Our results illustrate how the AirPen can be used to advance our understanding of the air exposome by deriving time, location, source, and activity, even in the absence of detailed activity diary data.

412 – HBM4EU e-waste study: Occupational Exposure of Electronic Waste Workers to Phthalates and Alternative Plasticizers in Europe

Paulien Cleys[1], Emilie Hardy[2], Radu Duca[2], Yu Ait Bamai[1], Adam Cseresznye[1], Adrian Covaci[1], Giulia Poma[1], Malarvannan Govindan[1], Tiina Santonen[3], Simo Porras[3], Paul T.J. Scheepers[4], Susana Viegas[5], Carla Martins[5], Lode Godderis[6], Jelle Verdonck[6], Katrien Poels[6], Maria João Silva[5], Henriqueta Louro[5], Inese Martinsone[7], Lāsma Akūlova[7], An van Nieuwenhuyse[2], Martien Graumans[4], Selma Mahiout[3]

[1]University of Antwerp, Antwerp, Belgium. [2]Laboratoire National de Santé, Dudelange, Luxembourg. [3]Finnish Institute of Occupational Health, Helsinki, Finland. [4]Radboud University, Nijmegen, Netherlands. [5]NOVA University Lisbon, Lisbon, Portugal. [6]KU Leuven, Leuven, Belgium. [7]Rīga Stradiņš University, Riga, Latvia

Abstract
Workers involved in the processing of electronic waste (e-waste) are potentially exposed to toxic chemicals, such as phthalates (PHs) and alternative plasticizers (APs). These compounds are highly present in electronic devices, but have a high tendency to leach into the environment. Dismantling and shredding of e-waste will therefore lead to the production of dusts that contain these plasticizers, to which workers are potentially exposed. The aim of this study was to investigate the exposure to PHs and APs in e-waste workers (n=106) from ten European companies by (i) analysing urine samples from e-waste workers collected in the beginning and at the end of the workweek, (ii) comparing them with urine samples from a control population, not occupationally exposed (n=63), (iii) analysing settled floor dust collected in e-waste premises. Significantly higher urinary concentrations of seven out of thirteen metabolites were found in e-waste workers compared to the control population. However, no significant difference was found between pre- and post- shift concentrations in the e-waste workers. Moreover, concentrations of DBP, DEHP and DiDP in dust were positively correlated with their corresponding urinary metabolites concentrations of the e-waste workers (Spearman's ρ= 0.4, 0.3 and 0.2 respectively). Additionally, significantly lower urinary concentrations of nine metabolites were found in e-waste workers using respiratory protective equipment (RPE) during their work activities, reflecting that proper use of RPE can protect occupational exposure to plasticizers. This study on occupational exposure in e-waste processing is part of the European Human Biomonitoring Initiative (HBM4EU).

413 – Assessing the Risks and Impacts of Stubble Burning on Health and Environment

Vignesh Prabhu[1], Udhaya Kumar[1], Pratima Singh[2]
[1]Center for Study of Science, Technology & Policy, Bengaluru, 560094, India, Senior Associate, Bengaluru, India. [2]Center for Study of Science, Technology & Policy, Bengaluru, 560094, India, Senior Research Scientist, Bengaluru, India

Abstract
Agriculture is one of the major industries in India and contributes around one-third of its GDP. Due to the lack of incentives and funding opportunities for the farmers, unscientific agricultural practice like stubble burning is practised in several states of India. This activity has been associated with a range of negative externalities including air pollution-related health risks, economic impact, etc., This study presents an overview of the risks and impacts associated with stubble burning on health and the environment in Punjab, based on ground-based measurement and evidence from the literature. Monitoring using low-volume samplers was carried out to measure and quantify PM2.5 and the chemical markers during the Kharif (Apr-May) and rabi crop burning season (Oct-Nov) in five representative districts of Punjab. Emissions from stubble burning and other agriculture activities like the operation of tractors and motor pumps were quantified using secondary data. Further, BenMAP analysis was carried out to estimate the exposure of the farmers and the economic benefits owing to PM2.5 reduction. The results were then extrapolated to a wider region using modelling and statistical tools. The daily average (range) PM2.5 during the Kharif and rabi crop burning season was observed to be 107.4 µg m-3 (51.5 – 191.5 µg m-3) and 83.7 µg m-3 (52.1 – 138.7 µg m-3), respectively. Furthermore, high levels of OC1 and K+ signify the extent of contribution from stubble-burning emissions. This study reveals that we require concrete measures to address the issue of stubble burning.

414 – How Could Earlier Phase-out of PCB Production Have Reduced Cancer and Diabetes Cases in the United States?

Nevaeh Ledesma[1], Amy Olsen[2], Dingsheng Li[3], Li Li[4]
[1]University of Nevada, Reno, Research assistant, Reno, USA. [2]University of Nevada, Reno, Graduate student, Reno, USA. [3]University of Nevada, Reno, Associate Professor, Reno, USA. [4]University of Nevada, Reno, Assistant Professor, Reno, USA

Abstract
Polychlorinated biphenyls (PCBs) were widely produced in the U.S. from 1930 to 1979 as heat transfer fluids in electrical equipment and as plasticizers in paints and sealants. Despite scientific evidence since the 1940s suggesting their adverse health impacts, manufacturers continued production for other decades, prioritizing commercial profits over public health. In this presentation, we report our model-based estimation of the health impacts that could have been avoided in three "parallel universes" if manufacturers had stopped PCB production immediately after learning about their toxicity, decades earlier than reality. Using a comprehensive exposure evaluation model named PROTEX, we reconstructed the time-variant environmental releases of PCBs in the U.S. and the life-course trajectories of serum PCB concentrations in different generations of Americans from 1855 to 2015, with the real PCB production history and in three alternate PCB production scenarios. Model predictions agree well with NHANES biomonitoring data. Our estimation reveals that the real

PCB production history is associated with over 1,000 cancer cases each year nationwide for generations born between the 1950s to 1980s and 0.5 to 2 million adult diabetes cases nationwide from the 1980s to 2020s. By using alternate production histories, we estimated the number of avoidable cancer and diabetes cases and monetized these cases as the economic burden to the U.S. healthcare system if PCBs had been phased out decades before reality. Our findings demonstrate the importance of considering the potential health risks associated with chemical production and everyday use to prevent avoidable human suffering and economic costs.

415 – POSTER: Development of alternative evaluation methods for fragrance allergens in consumer chemical products

SeonYeong Lee[1], JiYeon Yang[2], Sungwon Chun[1], Jinyong Kim[1], YeShin Kim[3], JinBeom Cho[4]
[1]Korea Environmental Industry&Technology Institute, researcher, Seoul, Korea, Republic of. [2]Yonsei university, research professor, Seoul, Korea, Republic of. [3]Riskcom, phD, Seoul, Korea, Republic of. [4]Korea Conformity Laboratories, researcher, Seoul, Korea, Republic of

Abstract
Consumer chemical products such as air fresheners, deodorants, and cleaners contain various fragrance substances, which can be inhaled and cause contact dermatitis, respiratory allergies, and other health issues. Nevertheless, there is currently no standardized risk assessment for the inhalation of fragrances, so alternative evaluation methods should be developed for risk management.

An Inflammatory response test was applied using BEAS-2B cell lines, human bronchial epithelial cells. In this study, we have selected three representative substances(Cinnamaldehyde, Isoeugenol, Citral) that have been reported to have allergenic potential in consumer chemical products, along with three newly identified allergenic substances(sage oil, spearmint oil, eucalyptus oil). A representative respiratory allergen, 2,4-TDI, was selected as the positive control. The BEAS-2B cells were treated with fragrance substances at concentration of 0, 200, 500 ug/mL, and the expression levels of the allergen indicators IL33 and TSLP were evaluated using the ELISA method.

All substances showed significantly higher expression of IL-33 and TSLP, as compared to the negative control group. Cinnamaldehyde, isoeugenol, and sage oil exhibited expression levels similar to the positive control 2,4-TDI, while spearmint oil showed significantly higher expression levels. The level of IL-33 and TSLP expression for fragrance allergens exceeded the positive reaction criteria, indicating their potential to cause respiratory allergies.

This study aimed to develop an evaluation method for potential respiratory allergenic fragrance substances using an in-vitro screening method. Our results could provide fundamental data for the management of respiratory allergenic substances related to fragrance substances, which are widely used in consumer products.

416 – POSTER: Sustainability of artisanal mining in DR Congo as reflected by assessing exposure to metals using human biomonitoring

Radu Duca[1,2], Maria-Mirela Ani[3], Matteo Creta[4], An Van Nieuwenhuyse[5,6], Paul Musa Obadia[7,8], Joseph Pyana Kitenge[9,10], Tony Kayembe-Kitenge[11,12], Trésor Carsi Kuhangana[2,13], Paterne Safari Mudekereza[14], Célestin Banza Lubaba Nkulu[15], Benoit Nemery[6]
[1]National Health Laboratory (LNS), Head of Unit, Dudelange, Luxembourg. [2]University of Leuven (KU Leuven), Senior researcher, Leuven, Belgium. [3]National Health Laboratory (LNS), MD, Dudelange, Luxembourg. [4]National Health Laboratory (LNS), Technical Responsible, Dudelange, Luxembourg. [5]National Health Laboratory (LNS), Head of Department, Dudelange, Luxembourg. [6]University of Leuven (KU Leuven), Prof, Leuven, Belgium. [7]University of Leuven (KU Leuven), PhD student, Leuven, Belgium. [8]University of Lubumbashi, PhD student, Lubumbashi, Congo, the Democratic Republic of the. [9]University of Leuven (KU Leuven), MD, Leuven, Belgium. [10]University of Lubumbashi, MD, Lubumbashi, Congo, the Democratic Republic of the. [11]University of Lubumbashi, Senior researcher, Lubumbashi, Congo, the Democratic Republic of the. [12]Institut Supérieur des Techniques Médicales, Senior researcher, Lubumbashi, Congo, the Democratic Republic of the. [13]School of Public Health, University of Kolwezi, Prof, Kolwezi, Congo, the Democratic Republic of the. [14]Université Catholique de Bukavu, Prof, Bukavu,, Congo, the Democratic Republic of the. [15]University of Lubumbashi, Prof, Lubumbashi, Congo, the Democratic Republic of the

Abstract
Critical materials are essential for numerous modern applications, like rechargeable batteries for smartphones, laptop computers and electric vehicles. In this study, biomonitoring of trace elements was applied for workers of different artisanal mines in DR Congo, to determine the levels of internal exposure. Workers from diamond (n=83), copper-cobalt (n=166) and gold mines (n=49) were recruited (by convenience sampling at their workplace) between 2017 and 2021. Spot samples of urine (n=277) and venous blood (n=255) were

collected to quantify trace metal(loid)s by ICP-MS. The highest concentrations [median (IQR)] of blood Pb were observed in gold miners [450 (299–703) μg/L], and the highest urinary concentrations (μg/g creatinine) of Co were observed in copper-cobalt mineworkers [25.1 (13.5–60.5)], both median values exceeding the ACGIH Biological Exposure Index (BEI) for occupational exposure. High urinary Ni concentrations were observed in gold miners [6.3 (4.6–9.2)] and diamond miners [5.9 (4.0–8.6)]. Infrequently measured elements, such as Ga, Ge, Nb, Ta, W and U were also quantified, with higher blood concentrations of Ge being observed among copper-cobalt mineworkers [19.0 (11.2–20.8) μg/L]. These data indicate that artisanal extraction of strategic metal commodities may lead to high exposures to toxic hazards.

417 – Dietary Exposure to Persistent and Current-Use Pesticides: A Look into Pesticide Mixtures in Duplicate Diets

Daniel Figueiredo[1], Neus González[2], Hans Mol[3], Paul Scheepers[4]
[1]IRAS, Utrecht University, Post-doc researcher, Utrecht, Netherlands. [2]Ciències Mèdiques Bàsiques, Universitat Rovira i Virgili, Researcher, Tarragona, Spain. [3]Wageningen Food Safety Research, Senior Researcher, Wageningen, Netherlands. [4]Radboud Institute for Biological and Environmental Sciences, Radboud University, Professor (Associate), Nijmegen, Netherlands

Abstract
Pesticides are still indispensable in agricultural production. However, they have been linked to increased risk for certain diseases. Dietary intake is the most important exposure route, and one can be continuously exposed to a cocktail of pesticides via food intake.

We collected duplicate diets (food and drinks) for 24-hours from 43 different participants from Europe and Argentina. We quantitatively measured 202 active ingredients (174 parent compounds and 28 metabolites) in the samples, by multi-residue methods based on targeted liquid and gas chromatography-mass spectrometry.

Results showed that 112 pesticides were found in at least one sample, with metalaxyl-M, pirimiphos-methyl and tebuconazole observed in more than 70% of them. Together with trifloxystrobin, these pesticides were detected in more than 50% of the samples. The median concentrations ranged from 0.03 and 65 μg/kg in the homogenized duplicate diet material, with a maximum value of 102.33 μg/kg for pyrimethanil. Overall, the correlations between pesticides was poor (Pearson's rho ≈ 0.1), although for few cases, such as imazalil and thiabendazole, correlation was strong (Pearson's rho ≈ 0.95). For none of the pesticides the daily intake (μg/kg BW/day) exceeded the acceptable daily intake (ADI), indicating no health risk.

We measured exposure to pesticide mixtures through food intake. Although the level of exposure was below the ADI for the individual pesticides, the effects of simultaneous exposures to multiple pesticides can be of concern. Further studies are necessary to understand the long-term effects of repeated exposure to low levels of pesticide mixtures.

418 – Kidney Function in Relation to Pesticide Exposure and Heat Stress Among Migrant Farmworkers

Nicolas Lopez-Galvez[1], Rietta Wagoner[2], Robert Canales[3], Isaiah Casarez[4], Paloma Beamer[5]
[1]San Diego State University, Assistant Professor, San Diego, USA. [2]University of Arizona, Doctoral Student, Tucson, USA. [3]Gorge Washington, Associate Professor, DC, USA. [4]San Diego State University, Student, San Diego, USA. [5]University of Arizona, Professor, Tucson, USA

Abstract
The risk factors of Chronic Kidney Disease of Undetermined Etiology (CKDu) are not fully understood, but pesticides and heat exposure, have been proposed to explain the pattern of this disease. We recruited 72 male migrant farmworkers in a large-scale grape farm. To determine kidney functioning during the pre-harvest and late-harvest season, urine and serum were collected and analyzed for several biomarkers. The estimated glomerular filtration rate (eGFR) was calculated from serum creatinine. The association between heat stress was measured using the Physiological Strain Index (PSI), which was calculated from participants' ear temperature and heart rate at rest and during their working activities. Several urinary pesticide biomarkers including those for organophosphates and pyrethroids, were analyzed to evaluate a possible association between pesticides and kidney function. Linear mixed model regressions were built with eGFR as the dependent variable, while PSI and pesticide biomarkers were the main predictors after adjusting for demographic-occupational factors

419 – POSTER: Development of an analytical method for indoor polycyclic aromatic hydrocarbons and their halogenated derivatives by using thermal separation probe coupled to gas chromatography-tandem mass spectrometry

Tetsuya Takikawa[1], Wang Qi[2], Kazushi Noro[2], Yuichi Miyake[3], Takashi Amagai[1]
[1]University of Shizuoka, student, Shizuoka, Japan. [2]University of Shizuoka, Assistant Professor, Shizuoka, Japan. [3]Yokohama National University, Assistant Professor, Yokohama, Japan

Abstract
Polycyclic aromatic hydrocarbons (PAHs) and their halogenated derivatives (XPAHs) have been a concern owing to their high toxicity. Monitoring of indoor PAHs and XPAHs concentrations should be important for risk assessment because human beings typically spend more than 90% of time indoors. However, background level of indoor PAHs and XPAHs concentrations are unknown because of low sensitivity of conventional analytical methods. In the present study, we developed a high-sensitive analytical method using thermal separation probe coupled to gas chromatograph with triple quadrupole mass spectrometer (TSP method) for 26 PAHs and 40 XPAHs. The method quantification limit (MQL) values of the TSP method were 1.1 (3,8-dichlorofluoranthene)–906 (dibenzo[a,e]pyrene) times lower than those of conventional method. The regression line comparing the TSP method to the conventional method was $y=(0.944\pm0.04)x$ with good correspondence, revealing that TSP method can be applicable to indoor air analysis. The total concentrations of PAHs and XPAHs were 944 and 73.5 pg m^{-3} for the house, and 735 and 0.924 pg m^{-3} for the office, respectively. Thirteen types of the PAHs and XPAHs could not be detected by the conventional methods due to their high MQL values. The composition of total toxicity equivalency quantity values in house were dominated by dibenzo[a,i]pyrene (DBaiP: 43.2%) and dibenzo[a,h]pyrene (DBahP: 27.1%), although DBaiP and DBahP could not be detected by the conventional method. Therefore, the TSP method contributes to improve the risk assessment for indoor PAHs and XPAHs.

421 – POSTER: Global Comparison of Human Breastmilk Consumption and Patterns in Infants and Toddlers

Joyanne Terry[1], Nicolle Tulve[2], Lindsay Stanek[3], Lisa Melnyk[4]
[1]EPA, Literature Review, Durham, USA. [2]EPA, Research Physical Scientist, Durham, USA. [3]EPA, Physical Scientist, Durham, USA. [4]EPA, Research Chemist, Durham, USA

Abstract
Breastmilk consumption, lipid content, and feeding frequency and duration are oftentimes used as data inputs for modeling analyses, exposure, and risk assessments. Our objective was to systematically review the literature to report the latest data on breastmilk consumption, lipid content, and feeding frequency and duration. We conducted a systematic review of peer-reviewed literature (2012-2020) on partially and exclusively breastfed infants using Web of Science and PubMed. Example search terms included "human," "milk," "intake," and "consumption" and relevant variations. Initial screening evaluated titles and abstracts against the inclusion criteria. For papers that passed this screening, full text screening was completed to ensure relevant extractable data which included cohort demographics, sample size, milk and lipid intakes, feeding frequency and duration, and supporting environmental, socioeconomic, and health information. Fifty-five studies met the criteria and were included in the analysis. Milk intake was the greatest for infants aged 1 to <3 months (134+24 mL/kg*day) and least for infants aged 1 to 2 years (38+42 mL/kg*day). Maximum lipid intake occurred when infants were 0 to <1 month old (6+2 g/kg*day), while the minimum lipid intake occurred when infants were 6m to <1 year old (3+2 g/kg*day). Feeding frequency was greatest for infants aged 0 to <1 month (11 feeds/day). Maximum average feeding duration was 87+48 min/feed at 3 months. Our preliminary analysis showed some fluctuations in average milk intake by age. In general, human milk intake is comparable across the world for each age range.

422 – Association between paraben exposure and early renal impairment among the general Taiwanese

Po-Chin Huang[1], Jung-Wei Chang[2], Yen-Hsuan Huang[3], Wan-Ting Chang[4]
[1]National Institute of Environmental Health Sciences, National Health Research Institutes, Associate Research Fellow, Miaoli, Taiwan. [2]Institute of Environmental and Occupational Health Sciences, National Yang Ming Chiao Tung University, Associate Professor, Taipei, Taiwan. [3]Institute of Environmental and Occupational Health Sciences, National Yang Ming Chiao Tung University, graduate student, Taipei, Taiwan. [4]National Institute of Environmental Health Sciences, National Health Research Institutes, Research assistant, Miaoli, Taiwan

Abstract
Exposure to consumer chemicals has been associated with chronic kidney disease among humans, but their associations with

estimated glomerular filtration rate (eGFR) are inconsistent. We aimed to assess the relationship between paraben exposure and renal function from a population-based study. We enrolled 361 participants (≥18 y, N = 269; <18 y, N = 92) who provided questionnaire information as well as blood and urine samples. Urinary parabens including methylparaben (MeP), ethylparaben (EtP), propylparaben (PrP), and butylparaben (BuP) are measured by LC/MS-MS. From the renal function index, we measured the serum level of blood urea nitrogen (BUN), and the urinary levels of microalbumin, albumin, protein, and creatinine. We used multiple regression models to evaluate the relationship between paraben exposure and renal function in our participants.

After adjusting for the same confounding factors in adults, the positive association of ACR with urinary MeP (β: 0.251; 95%CI=0.004 to 0.498), the positive association of BUN and the negative association of eGFR with urinary EtP (β: 0.061; 95% CI =0.004 to 0.119; β: -0.049; 95% CI = -0.093 to -0.005) were confirmed, which means ACR increased by 0.251×10^{-2} mg/g cr. with a 1% increase in urinary MeP level, BUN increased by 0.061×10^{-2} mg/dL, and eGFR decreased by 0.049×10^{-2} mL/min/1.73 m^2 with a 1% increase in urinary EtP level, respectively. Our findings suggested that daily exposure to EtP were significantly positively associated with an increased risk of higher BUN and lower eGFR in Taiwanese ≥18 y.

423 – Operator safety - assessing operator exposure during drone application of pesticides

Maxie Kohler[1], Thomas Anft[2], Lirong Chen[3], Christian Timmermann[2], Georg Hamacher[4], Edgars Felkers[2], Christian Kuester[5]
[1]Bayer AG, ORE Team Lead, Monheim, Germany. [2]Bayer AG, ORE Expert, Monheim, Germany. [3]Bayer AG, ORE Expert APAC, Singapore, Singapore. [4]AGREXIS AG, ORE Expert, Basel, Switzerland. [5]Bayer AG, Regulatory Scientific Affairs Expert, Monheim, Germany

Abstract
Drones gain increasing relevance in pesticide applications. Regulations and registrations to allow drone uses are developing at a fast pace, especially in Asia. The use of drones for application of pesticides is expected to improve operator safety significantly.

Bayer has developed and published operator safety standards [1] that are applied to our portfolio to make sure the products we develop are safe for the operators in countries without specific regulatory requirements for operator safety in the registration process.

An operator exposure study was performed in accordance with OECD guidance No. 9 [2] to allow for drone-specific exposure assessment. The drones (DJI, 4 nozzles, 10 L spray tank) were operated by one person and re-filled by a second person. The spray solution contained a tracer (food colorant, Brilliant Blue) as analyte. Samples of the extracts of inner and outer dosimeters as well as gloves, face and neck wipes and hand washes (12 replicates) were analyzed by a validated LC-MS/MS method.

The results show that the majority of exposure occurs during mixing and loading while exposure of the pilot is significantly lower. For mixing and loading, exposure was reduced by >98% wearing one layer of clothes and gloves. Comparison of the drone study data of the pilot with the handheld application exposure models showed that the exposure of the applicator is reduced by >89% when spraying by drones. These results and additional study data will serve as a basis for model development regarding operator exposure from drone applications around the world.

424 – POSTER: Assessment of nicotine personal exposure of non- or passive smokers by using passive sampler

Miyu Moriya[1], Kazushi Noro[2], Ayana Komatsu[1], Yoko Kai[3], Wang Qi[2], Takashi Amagai[4]
[1]University of Shizuoka, Student, Shizuoka, Japan. [2]University of Shizuoka, Assistant professor, Shizuoka, Japan. [3]University of Shizuoka, Technician, Shizuoka, Japan. [4]University of Shizuoka, Professor, Shizuoka, Japan

Abstract
Nicotine has been used as a marker for environmental tobacco smoke. This is because the amount of nicotine sources other than cigarettes are almost negligible. We have developed a passive sampler that can measure trace amounts of nicotine exposure. Nicotine personal exposure concentrations of non- or passive smokers were measured in metropolitan areas in Japan in October to December 2022. We asked 5,390 participants to answer a questionnaire on passive smoking status on the web, and 258 of them (130 males and 128 females) were asked to spend 24 hours with a passive sampler. All 258 participants were non-smokers, and 80 of them answered that they were not exposed to passive smoking in their daily lives. The arithmetic mean and standard deviation of nicotine personal exposure concentrations were 203 ng m^{-3} and 570 ng m^{-3}, respectively. The median value was 63 ng m^{-3}. The distribution of nicotine exposure concentration followed a logarithmic normal distribution. There was no significant difference in exposure levels between male and female. There were no significant differences in exposure levels by age or occupation. There were also no significant differences in nicotine exposure levels among groups exposed to passive smoking outdoors, indoors, both, and none during the study. In addition, the

median nicotine concentration (63 ng m^{-3}) was higher than those of 'acceptable (10^{-6}) lung cancer mortality risk' (7.5 ng m^{-3}). These results indicated that lung cancer mortality risk of the passive or non-smoker in this study were between 10^{-5} and 10^{-6}.

425 – Sensor Data Enrichment for Better Interpretation to Target Preventive Measures

Maaike le Feber[1], Bas Henzing[2], Anjoeka Pronk[3]
[1]TNO, Senior intergrator and scientist, Utrecht, Netherlands. [2]TNO, Business developer and scientist, Utrecht, Netherlands. [3]TNO, Principal Scientist, Utrecht, Netherlands

Abstract
Low cost PM sensors are increasingly used by citizens to monitor air quality. These sensor data come with a relatively high uncertainty, but this is balanced by their costs of use; a much denser network can be created for the same budget.

Comparable PM sensors are being explored in occupational settings to assess personal exposures. And despite the relatively high uncertainty of the data, their high time resolution offers a wealth of information.

With the aim of prevention, further interpretation of this high resolution sensor data can be achieved by combining the sensor data with other (high resolution) data sources. Two examples will be presented.

To monitor outdoor air quality, dynamic concentration maps are composed based on national measurements stations and air pollution models. We demonstrated that these concentration maps can be improved by using data of sensor networks, including citizen science initiatives. Areas with relatively high pollutant concentrations will cause health effects if individuals are in this area at that time. We combined dynamic concentration maps with population dynamics data to prioritize areas of high concern.

In the occupational setting we measured with a low cost PM sensor personal exposure of a construction worker to respirable dust. Instead of observing this worker to log activities and circumstances to understand the exposure, we used context sensors to collect this information. The exposure and context data were synchronized and visualized to assist interpretation of the exposure profile into information on when and under which (working) conditions exposure peaks occur.

426 – POSTER: Heavy Metal Content in Cosmetics from Seven Asian Countries and Potential Health Risks to Consumers

Jae-Hyun Kim[1], Aram Lee[2], Young-Jun Jeon[1], Kang-Hyun Lee[1], Won Kim[3], In-ja Choi[3], Jeongim Park[4]
[1]Soonchunhyang University, researcher, Asan, Korea, Republic of. [2]Soonchunhyang University, postdoctral researcher, Asan, Korea, Republic of. [3]Wonjin Institute for Occupational and Environmental Health, researcher, Seoul, Korea, Republic of. [4]Soonchunhyang University, professor, Asan, Korea, Republic of

Abstract
Cosmetics are products that are commonly used, but they can contain heavy metals that pose health risks to consumers. The purpose of this study was to analyze the heavy metal content in cosmetics and assess the risk of dermal absorption. We purchased a total of 230 facial and eye cosmetic products from seven Asian countries (Bangladesh, India, Indonesia, South Korea, Malaysia, Philippines, and Vietnam). We used XRF to prescreen products with Hg over 1 ppm, and for products with Hg below 1 ppm, we used a mercury analyzer for further analysis. We also analyzed six other heavy metals using ICP-OES. The detection limits for Hg, As, Cd, Cr, Ni, Pb, and Sb were 0.003, 10.6, 1.4, 2.4, 1.5, 9.1, and 12.8 ppm, respectively. In facial cosmetics (n=169), the detection rates of Hg, As, Cd, Cr, Ni, Pb, and Sb were 58%, 20%, 17%, 100%, 94%, 21%, and 9%, respectively. For eye cosmetics (n=61), the detection rates were 36%, 77%, 6%, 94%, 100%, 6%, and 27%, respectively. 160 (70%) samples exceeded cosmetic safety standards for metals set by the EU. Additionally, 33% of facial cosmetics (55 samples) showed HQ values exceeding 1 with Hg from dermal absorption. These findings suggest that one third of facial cosmetics in those Asian markets may pose a risk of Hg exposure. Further risk assessments are necessary to evaluate the cumulative exposure to heavy metals from cosmetic use and their potential hazards.

427 – Advancing Local Urban Heat Action: Qualitative Interviews with Decision-makers and Evaluation of a Photovoice Project

Leila Heidari[1], Alina McIntyre[1], Michael Hagen[2], Bianca Bowman[3], Roseann Bongiovanni[4], Jonathan I. Levy[5], M. Patricia Fabian[6], Madeleine K. Scammell[6]
[1]Boston University, Doctoral Candidate, Boston, USA. [2]Boston University, Research Assistant, Boston, USA. [3]GreenRoots, Inc., Climate Justice Coordinator, Chelsea, USA. [4]GreenRoots, Inc., Executive Director, Chelsea, USA. [5]Boston University, Professor, Boston, USA. [6]Boston University, Associate Professor, Boston, USA

Abstract
Photovoice is participatory action research that aims to engage policy makers. Through a Photovoice campaign in two environmental justice communities in Massachusetts, we recruited residents to document social, economic, and environmental barriers and facilitators to coping with heat. They publicly shared their findings, including calls for action related to tree equity, water, heat vulnerability, and keeping cool.

We conducted interviews with nineteen (n=19) representatives of local government, and public health and policy organizations in Chelsea and East Boston. We invited each to view the exhibit and report. Interview domains included perceptions of heat exposure and vulnerability, their role in heat adaption, and their responses to the Photovoice.

Overall reactions to Photovoice were positive, with decision-makers calling it "compelling," as well as "grounding" regarding the power of local residents sharing their lived experiences, especially in the context of advocacy. One-third of interviewees named that the stories resonated with their own personal experiences. While participants across sectors valued hearing and uplifting residents' voices, both elected city council members emphasized this: "it's our constituents, our people," and "the personal should always inform policy."

In responding to specific calls to action, participants described their limits of authority, scope or jurisdiction in their role and/or organization, while also talking about the importance of collaborators. For many of the interviewees, the calls to action were congruent with current and planned projects.

Participatory evaluation, such as post-Photovoice interviews, presents a useful tool to engage decision-makers and assess facilitators, barriers, and opportunities towards action.

428 – POSTER: Investigation of Biocide Content in Biocidal Consumer Products in South Korea

Sohyun Kang[1], Kiyoung Lee[2], Miyoung Lim[3]
[1]Seoul National University, Ph.D. Student, Seoul, Korea, Republic of. [2]Seoul National University, Professor, Seoul, Korea, Republic of. [3]Seoul National University, Research Professor, Seoul, Korea, Republic of

Abstract
The use of biocidal consumer products increased since the Coronavirus disease-2019 as the demand for personal hygiene increased. However, unintentional adverse health effects could occur without proper safety management. Since 2019, consumer chemical products and biocides have been under regulation by the Korean Ministry of Environment and the contents approved were to be mandatorily disclosed to consumers through public web pages. This study assessed the usage status and characteristics of biocides included in Korean biocidal consumer products. Information of 1,605 biocidal consumer products were collected using crawling on the Ecolife web page. The names of the products, the active ingredients, their concentrations, and additives were examined. Of the 1,605 biocidal products listed in Ecolife web page, 1,228 items were identified their ingredients and concentrations. Depending on the purpose of use, products were classified into insecticide (n=1,076), disinfectant (n=106), and rodenticide (n=46). Pyrethroids (such as phenothrin, deltamethrin, tetramethrin) were most frequent active ingredient in insecticides, and their concentration ranged from 0.003 to 25 wt%. Quaternary ammonium was most frequent active ingredients in disinfectants, and its concentration ranged from 0.0096 to 20 wt%. Flocoumafen was most frequent active ingredients in rodenticides, and its concentration ranged from 0.003 to 10 wt%. The registered additives were used as solvent, surfactant, fragrance, aerosol propellent, etc. This study examined the biocide in biocidal consumer products using regulatory ingredient disclosure database. This study provides data essential for consumer risk assessment and could aid in reducing consumer's potential exposure harmful substances in biocidal consumer products.

429 – Comparison of low birth weight and preterm birth as adverse pregnancy outcomes in rural and urban women of Islamabad due to polycyclic aromatic hydrocarbons exposure from wood smoke

Munnaza Khalid[1], Abid Ali Mughal[2], Humaira Bilqis[3], Habib Ullah[4], Asfandyar Shahab[5], Qaisar Mahmood[6], Audil Rashid[7]

[1]Department of Obstetrics and Gynaecology (Unit-I) Holy Family Hospital Rawalpindi Medical University, Resident Medical Officer, Rawalpindi, Pakistan. [2]PMAS Arid Agriculture University, PhD Scholar, Rawalpindi, Pakistan. [3]Department of Obstetrics and Gynaecology (Unit-I) Holy Family Hospital Rawalpindi Medical University, Assistant Professor, Rawalpindi, Pakistan. [4]Zhejiang University, Post-Doc Fellow, Hangzhou, China. [5]Guilin University of Technology, Assistant Professor, Guilin, China. [6]COMSATS University Islamabad, Professor, Abbottabad, Pakistan. [7]EcoHealth Lab, PMAS Arid Agriculture University, Professor, Rawalpindi, Pakistan

Abstract
Exposure to Polycyclic aromatic hydrocarbons (PAHs) among women inside houses is inevitable during wood and other biomass which is linked to adverse pregnancy outcomes, such as low birth weight, preterm birth, and congenital anomalies. Rural women may be severely exposed to PAHs from wood smoke than their urban counterparts. In Pakistan, association between exposure to PAHs from indoor and outdoor sources and the risk of low birth weight and preterm birth has been investigated other cities but not from Islamabad. The present study aimed to investigate the relationship between exposure to PAHs from wood smoke and adverse pregnancy outcomes such as low birth weight, preterm birth in both rural and urban populations of Islamabad, the federal capital of Pakistan. We found that the rural population in Islamabad showed exposure to wood smoke during pregnancy was associated with an increased risk of preterm birth (OR=2.3, 95% CI: 1.2-4.5) and a higher risk of low birth weight (OR=1.81, 95% CI: 1.24-6.31) than urban population in Islamabad. Overall, while both rural and urban women may be at risk of adverse pregnancy outcomes due to exposure to PAHs from wood smoke, the evidence suggests that rural women may be at a higher risk. This may be due to differences in heating and cooking practices, as well as other environmental and social factors. It is important for policymakers to prioritize interventions to reduce exposure to PAHs from wood smoke in both rural and urban areas to improve maternal and child health outcomes.

430 – Indoor Levels and Health Risk Assessment of Volatile Organic Compounds in Portuguese Homes

Klara Slezakova[1], Carla Costa[2,3,4], Carla Valongo[5], João Paulo Teixeira[2,3,4], Joana Madureira[2,3,4]
[1]LEPABE-ALiCE, Faculty of Engineering, University of Porto, Junior Researcher, Porto, Portugal. [2]Environmental Health Department, National Institute of Health, Auxiliary Researcher, Porto, Portugal. [3]EPIUnit-Institute of Public Health, University of Porto, Auxiliary Researcher, Porto, Portugal. [4]Laboratory for Integrative and Translational Research in Population Health (ITR), Auxiliary Researcher, Porto, Portugal. [5]Environmental Health Department, National Institute of Health, Senior Laboratory Technician, Porto, Portugal

Abstract
This study aimed to investigate concentrations and seasonal variations of volatile organic compounds (VOCs) in Portuguese households and to assess the respective health risks. VOCs were collected at 64 Portuguese homes. Samples were collected for 8 h in various indoor spaces and outdoors simultaneously, using sorbent tubes. Automated thermo-desorption/GC-FID method was used to quantify total VOCs (TVOCs) and seven specific compounds. Carcinogenic (CR) and non-carcinogenic (NCR) health risks were then assessed according to U.S. EPA methodology. The results showed that indoor TVOCs (mean: 217 $\mu g/m^3$) were five times higher than outdoor ones; pinene and limonene were the most abundant indoor VOCs. Concerning the seasonal variations, the highest level of indoor TVOCs was observed in winter (mean: 444 $\mu g/m^3$), being three times higher than in spring. For the individual VOCs, pinene and limonene in winter were identified at the highest levels compared to those measured in other seasons. The risk analysis indicated that the CR and NCR were higher than the negligible risk levels. Mitigation strategies are needed to minimise health risks; emission source control is the most effective manner to reduce VOCs levels in private dwellings which requires, among others, further "knowledge literacy" of the respective occupants.

431 – POSTER: Effect of weekdays and weekends on ambient PM2.5 levels in five Indian megacities

Darpa Jyethi
Indian Statistical Institute, North East Centre, Assistant Professor, Tezpur, India

Abstract
Urban air quality is an urgent environmental issue, more so in the Indian cities which mostly register high particulate matter levels (>100 microgram/m3) in the atmosphere. Chronic and acute exposure to poor air quality is detrimental to primarily the respiratory health of the human body. One of the strategies to cope with increasing outdoor air pollution in cities is to limit personal exposure to pollutants. In this context, air quality during weekends have been perceived to be relatively cleaner due to reduction in vehicular traffic on roads. Various studies have reported differences in urban air quality with respect to weekday-weekend, especially with respect to tropospheric ozone, a secondary, photochemical pollutant. The present study investigates the existence of any effect of weekdays and weekends on ambient PM2.5 levels in major Indian megacities with population of more than10 million). Given the variable effect reported by studies in other parts of the world and the severe problem of particulate pollution in many Indian cities, the study sheds light on the perception of

cleaner weekends in cities due to lower on-road vehicular traffic. The study uses hourly PM2.5 data from five cities (Delhi, Kolkata, Chennai, Mumbai, and Hyderabad) during 2016-22, and employs Wilcoxon signed-rank test to determine the weekday-weekend effect.

432 – POSTER: WAR IN UKRAINE: AIR POLLUTION AND CLIMATE CHANGE

Arina Petrosian[1], Olena Turos[2], Tetiana Maremukha[3], Hanna Davydenko[3], Polina Yaryfa[4]
[1]State Institution "O.M. Marzieiev Institute for Public Health of the National Academy of Medical Sciences of Ukraine", Leading Researcher of Air Quality Laboratory, Kyiv, Ukraine. [2]State Institution "O.M. Marzieiev Institute for Public Health of the National Academy of Medical Sciences of Ukraine", Head of Air Quality Laboratory, Kyiv, Ukraine. [3]State Institution "O.M. Marzieiev Institute for Public Health of the National Academy of Medical Sciences of Ukraine", Senior Researcher of Air Quality Laboratory, Kyiv, Ukraine. [4]School 306, Student, Kyiv, Ukraine

Abstract
Background: The war in Ukraine is a global problem today, the result of which is the emission of a large number of pollutants into the air. The aims of this work was to assess the level of air pollution and analyze the danger to the health of the population in the conditions of wartime and climate change.
Methods: data of monitoring PM10, PM2.5, O3, NO2, CO, SO2 by gas analyzers HORIBA; statistical methods (StatSoft STATISTICA 10.0 portable).

Results: Uncharacteristically high daily average values of concentrations exceeding the limit values AQS were established for (min-max; M±m): PM10 (3.8-103.3 μg/m3; 18.5±0.95 μg/m3); PM2.5 (3.2-91.2 μg/m3; 12.6±0.66 μg/m3); O3 (21.0-90.0 μg/m3; 60.0±0,92 μg/m3 in "peak season"); NO2 (2.0-66.0 μg/m3; 21.6±0.81 μg/m3); concentrations of CO, SO2 were within the normative values of AQS. The obtained results are correlated and were primarily due to the conduct of active war actions and their consequences (fires, rocket attacks) and intensified in the spring-summer period adverse weather conditions.

Conclusions: The conducted studies showed that during the period of active hostilities, an excess was observed for: PM2.5 in 1.6-2.8 times; O3 in 1.5 and NO2 in 1.1 times, in accordance to the values of the pre-war period. A strong negative correlation ($p<0,0001$) was found between O3 concentrations and relative air humidity ($rS=-0.70$), an moderate positive correlation with air temperature ($rS=0.55$); as well as the mean force feedback between PM and air speed ($rS = -0.50$ and $rS = -0.51$ with PM10 та PM2.5, respectively).

433 – POSTER: How different livestock farm types affect spatial exposure patterns of particulate matter and ammonia

Ceder Raben[1], Hans Erbrink[2], Serigne Lô[1], Gerard Hoek[3], Dick Heederik[4], Wietske Dohmen[5]
[1]Utrecht University, PhD candidate, Utrecht, Netherlands. [2]Erbrink Stacks Consult, Modeler, owner, Soest, Netherlands. [3]Utrecht University, Associate Professor, Utrecht, Netherlands. [4]Utrecht University, Professor, Utrecht, Netherlands. [5]Utrecht University, Assistant Professor, Utrecht, Netherlands

Abstract
Livestock farms housing different production animal species emit different amounts of ambient particulate matter (PM) and ammonia (NH3); PM and NH3 affect health and biodiversity. This study aims to show how different species (veal calves, dairy cows, laying hens, broilers, other poultry, sows, fattening pigs, goats) affect spatial patterns of ambient PM and NH3. The study area (40 x 48 km) is situated in the Foodvalley+ region of the Netherlands (in Dutch: 'Geldersche Vallei'), an intensive livestock production area with 1975 livestock farms. PM and NH3 emissions were calculated for each stable source. Laying hens emitted the greatest amount of PM: 79.2% of total emitted PM in the study area. Veal calves emitted the greatest amount of NH3: 31.1% of total emitted NH3 in the study area. Based on stable sources, a dispersion model that accounts for deposition will be used to calculate ambient concentrations of PM and NH3 for a total of 192885 receptor points on a 100 x 100 m grid that covers the study area. The resulting spatial pattern of ambient PM and NH3, and how this pattern is affected by different species, will be presented and also visualized in contour plots. For NH3, 71 modelled concentrations are compared to field measurements from corresponding locations in the study area. The research presented was supported by the Regio Deal Foodvalley.

434 – POSTER: Source Apportionment and Risk Assessment of Elemental Load in Rural Household Dust in Northeast India

Darpa Jyethi
Indian Statistical Institute, North East Centre, Assistant Professor, Tezpur, India

Abstract
House dust samples from rural region in the River Brahmaputra flood plain in Northeast India were analyzed for elemental composition. Four trace elements (Cu, Zn, As, and Pb) and Cl were found to be significantly enriched. Principal component analysis (PCA) and positive matrix factorization (PMF) indicated three major sources of elements in house dust— traffic+ re-suspension of road dust (35.8%), soil dust (22.2%), and river sediment deposit (16.4%). Two minor sources—biomass burning (13.3%), and construction activities (12.3%) were also identified. Based on health risk assessment (HRA), both children and adult were found to be susceptible to non-carcinogenic and carcinogenic risks.

435 – Towards systematic adoption of (low cost) respirable aerosol sensors following an industrial hygiene framework

Emanuele Cauda, Justin Patts, Cody Wolfe, Milan Yekich
NIOSH, Researcher, Pittsburgh, USA

Abstract
The exposure of workers to respirable dust in occupational environments can be monitored using established methodologies which entail the collection of a representative sample on a media and consequent analysis in a laboratory. While these methodologies are very accurate, they only provide information on the average exposure during a period of time, generally a work shift. Respirable aerosol sensors can fill this gap and the availability of relatively inexpensive or low-cost sensors can allow researchers and practitioners to generate data and ultimate information with unprecedented spatial and temporal granularity. Many factors have prevented so far, a systematic adoption of respirable dust sensors for exposure monitoring: concern for their analytical performance, complexity of use, and lack of understanding their value are some factors. This contribution will propose using the industrial hygiene framework [anticipation, recognition, evaluation, control, and confirmation] to analyze the possible systematic adoption of different types of respirable aerosol sensors. Every task of the framework benefits from the use of different methodologies present in the toolbox since every task is characterized by different specific questions. Different data analytical techniques can and should be adopted to answer different questions and different is the engagement of users and workers with the technologies and the information generated. Not following the framework can lead to the selection of the wrong sensor technology, an increased mistrust in sensors, and an overall dissatisfaction from the researchers and users. Specific examples will be presented on the adoption of (low cost) respirable aerosol sensors as a complementary tool.

436 – Tiered approach to aggregate exposure assessment

Wouter Fransman[1], Femke Affourtit[2], Christiaan Delmaar[2], Henk Goede[3], Stefan Hahn[4], Calvin Ge[3]
[1]TNO, Senior Scientist, Utrecht, Netherlands. [2]RIVM, Scientist, Bilthoven, Netherlands. [3]TNO, Scientist, Utrecht, Netherlands. [4]Frauenhofer ITEM, Scientist, Hannover, Germany

Abstract
In the context of a chemical safety assessment, typically tiered approaches are recommended to lower the burden of having to provide very accurate exposure estimations in cases when the exposure is well below the limit value. In a tiered approach, the risk characterization is conducted in an iterative fashion: it starts with a simple estimation procedure with low input requirements, but with high uncertainty. Remaining uncertainty is covered by making reasonable worst-case assumptions on unknown aspects to ensure that the estimated risk potential represents an over-estimation of the real risk. When this low-tier assessment does not show safe use, in next iterations the assessment of exposure is refined by including more detailed, accurate information until the assessment is sufficiently accurate to show safe use.

In the RISKHUNT3R project, we propose a tiered approach for a stepwise refinement of the assessment of aggregate exposure of (members of) a population to a single chemical substance from multiple sources, via multiple pathways and routes. This approach has been tested in various case studies by applying measured data and predictive models for consumer and occupational exposure scenarios through different routes of exposure (inhalation, dermal, oral). The overall objective in RISKHUNT3R is that health-based limit values will be delivered for internal exposures (e.g., absorbed doses) that are converted from external exposure for each pathway and source. Evaluating the aggregate risk for various sources and routes in combination would be done by adding individual Risk Characterisation Ratio's and requiring that their sum be less than 1.

437 – A Summer Research Academy on Sensor-Based Environmental Health Monitoring for Middle School Students

Madeline L. Jones[1], Charlie Blake[2,3], Georgia Bracey[4], Carol E. Colaninno[5], Andreia F. Dexheimer[6], Candice Johnson[7], Sharon M. Locke[8,9], Ben K. Greenfield[10]
[1]University of Southern Maine, Graduate Assistant, Portland, ME, USA. [2]Center for STEM Research, Education, & Outreach, Southern Illinois University Edwardsville, Research Fellow, Edwardsville, IL, USA. [3]Environmental Sciences Department, Southern Illinois University Edwardsville, Lecturer, Edwardsville, IL, USA. [4]Center for STEM Research, Education, & Outreach, Southern Illinois University Edwardsville, Research Assistant Professor, Edwardsville, IL, USA. [5]Center for STEM Research, Education, & Outreach, Southern Illinois University Edwardsville, Research Associate Professor, Edwardsville, IL, USA. [6]Center for STEM Research, Education, & Outreach, Southern Illinois University Edwardsville, Post-Doctoral Fellow, Edwardsville, IL, USA. [7]Center for STEM Research, Education, & Outreach, Southern Illinois University Edwardsville, Resource Center Manager & Program Coordinator, Edwardsville, IL, USA. [8]Center for STEM Research, Education, & Outreach, Southern Illinois University Edwardsville, Director, Edwardsville, IL, USA. [9]Environmental Sciences Department, Southern Illinois University Edwardsville, Professor, Edwardsville, IL, USA. [10]Public Health Program, Muskie School of Public Service, University of Southern Maine, Associate Professor, Portland, ME, USA

Abstract
Environmental Health Investigators (EHI) Summer Academy is a sensor-based environmental health monitoring program for middle school students in the Metro-East Saint Louis Area, Illinois, USA. The program is held in the summers of 2021 to 2023, with a recruitment and inclusion strategy intended to engage low-income students and students of color. Inquiry-based learning and science interest development conceptually underlie the program, which is funded by an NIH Science Education Partnership Award. EHI students are trained on and use data logging sound monitors and portable airborne particulate matter monitors (including the Dylos DC1700, TSI SidePak AM520, and PurpleAir PA-II). They also participate in monitoring soil contamination with a handheld X-ray fluorescence spectrometer. Student participants use these tools and training to explore local variations in air, soil, and noise environments, and potential human and wildlife health hazards. Participants then work in teams to develop their own place-based research questions, collect data, answer their questions, and present their findings. The curriculum is aligned to disciplinary content in the Earth and Space Science and Physical Science domains of the Next Generation Science Standards (NGSS) for K-12 education. Challenges encountered during this program include adaptations to the COVID-19 pandemic, attendance, community partnerships, and adherence to the principles of informal education. We describe the EHI program and share the curriculum as a novel integration of exposure science with K-12 STEM education, which employs multimedia field sensors to foster inquiry-based learning.

440 – The Impact of Temperature and Relative Humidity on Ventricular Arrhythmias in Patients with Implanted Cardiac Devices: A Case Time Series Analysis in North Carolina, 2010-2021

Stephanie Cleland[1,2], Lindsey Rosman[3], K. Lloyd Hill[4], Anthony Mazzella[5], Cavin Ward-Caviness[6], Ana Rappold[7]
[1]ORISE at US Environmental Protection Agency, Center for Public Health and Environmental Assessment, ORISE Research Fellow, Chapel Hill, USA. [2]University of North Carolina at Chapel Hill, Department of Environmental Sciences & Engineering, ORISE Research Fellow, Chapel Hill, USA. [3]University of North Carolina School of Medicine, Division of Cardiology, Assistant Professor of Medicine, Chapel Hill, USA. [4]ORAU at US Environmental Protection Agency, Center for Public Health and Environmental Assessment, Research Associate, Chapel Hill, USA. [5]University of North Carolina School of Medicine, Division of Cardiology, Cardiology Fellow, Chapel Hill, USA. [6]US Environmental Protection Agency, Center for Public Health and Environmental Assessment, Senior Scientist, Chapel Hill, USA. [7]US Environmental Protection Agency, Center for Public Health and Environmental Assessment, Statistician, Chapel Hill, USA

Abstract
Medical professionals are interested in using remote monitoring to identify at-risk populations, reduce outcome disparities, and provide timely interventions to vulnerable patients. We examined the effect of warm season weather on the risk of ventricular tachycardia/ ventricular fibrillation (VTVF) in patients with implanted cardiac devices. We used daily remote monitoring data on daily VTVF occurrence in 526 patients enrolled in University of North Carolina's Cardiovascular Device Surveillance Registry between 2010-2021. Local temperature and relative humidity (RH) exposure during mid-April to mid-October was estimated using weather station data. Associations between meteorology and VTVF occurrence were estimated using a case time-series design with distributed lag non-linear models. Models were adjusted for temporal trends and physical activity. Temperature was not associated with VTVF, nor did it confound the association with RH. Conversely, high RH increased VTVF risk in 1-2 days following exposure. When RH increased from the median (73.8%) to the 95th percentile (90.0%), odds of VTVF occurring the next day increased by 11.7% (95% CI: 3.4-20.7%). Patients in neighborhoods with low urbanicity and high deprivation, as well as male, older, and white patients, had an elevated risk of VTVF when exposed to high RH. Exposure to humid conditions during the warm season increased risk of VTVF in patients with implanted cardiac devices, with key differences by individual- and area-level characteristics. Additional research in larger populations is needed to

determine generalizability of these findings. Remote monitoring data may provide opportunities to increase resilience and reduce inequalities in patient outcomes.

441 – Conducting a watershed-scale simulation and risk assessment of tetracycline exposure

Wen Jiang[1], Yaobin Meng[2]
[1]Beijing Normal University, Graduate student, Beijing, China. [2]Beijing Normal University, Associate Professor, Beijing, China

Abstract
China is the world's leading country in both the production and use of chemicals. However, the abundance of new organic pollutants such as antibiotics can potentially pose severe threats to both human health and the ecological environment. Despite this, the research and management of these pollutants are still in their early stages. A case study on tetracycline was conducted to investigate the risks of exposure to new pollutants at a watershed level. By conducting a thorough literature review and analysis, a simplified estimation method for human and animal tetracycline emissions was proposed. The results of this study can serve as a reference for the effective management and control of new organic pollutant residues in China's industrial and agricultural production activities. Based on the SWAT model, we constructed a SWAT-TC model by integrating environmental behaviors such as adsorption analysis, diffusion and migration of tetracycline in both water and soil phases, to achieve spatiotemporal differentiation simulation of tetracycline exposure concentration on a daily scale. The team utilized the 75th percentile simulated concentration as the predicted exposure concentration (PEC) and combined it with the predicted ineffective stress concentration (NPEC) to obtain the ecological risk assessment results. They also incorporated climate model data such as CMIP6 to calculate the ecological risk of tetracycline under multiple meteorological scenarios. Based on this framework, the researchers explored a comprehensive heavy metal simulation framework and the underlying mechanisms, quantifying the combined risk of tetracycline and heavy metals. Ultimately, this study will provide valuable insights for regional pollution control.

442 – Using Semantic Technologies to Support Exposome Research: Insights from the Human Health Exposure Analysis Resource (HHEAR) Data Center

Jeanette Stingone[1], Sofia Bengoa[2], James Masters[3], Patricia Kovatch[4], Carolina Valle[2], Nancy Mervish[5], Susan Teitelbaum[6]
[1]Columbia University, Assistant Professor, New York, USA. [2]Icahn School of Medicine at Mount Sinai, Research Coordinator, New York, USA. [3]Icahn School of Medicine at Mount Sinai, Director of Semantic Technology, New York, USA. [4]Icahn School of Medicine at Mount Sinai, Professor and Dean of Scientific Computing and Data Science, New York, USA. [5]Icahn School of Medicine at Mount Sinai, Senior Scientist, New York, USA. [6]Icahn School of Medicine at Mount Sinai, Professor, New York, USA

Abstract
Implementation of the exposome paradigm is a critical aspect of the upcoming generation of environmental health research. Addressing complex research questions on the exposome often requires data that extends beyond a single study or data resource. The HHEAR Data Center, funded by NIH, establishes the foundation for this research by hosting a public data repository to facilitate pooling and sharing of data generated by the HHEAR program. The latest semantic technologies are used to efficiently conduct data harmonization across studies and promote data sharing by aligning the repository with the FAIR (Findable, Accessible, Interoperable, Reusable) data principles. The HHEAR Data Center's semantic strategy is made up of three key components: 1) the development of an ontology to bring together standardized concepts and terms to represent the diverse data within environmental health studies; 2) the application of a semantic data dictionary approach to construct and query the HHEAR knowledge graph; and 3) the development of structured, metadata templates to capture essential study details needed to align with FAIR. This semantic pipeline facilitates pooling, visualization, and analysis of exposure and health outcomes data across studies, including comparison of exposure patterns within research-based studies to national population-based biomonitoring data and generation of SHINY dashboards. Here we will demonstrate how these semantic tools can be adopted by other researchers and initiatives to advance data sharing and promote interoperability across data repositories.

443 – Optimization and Validation of a New Custom-made Activated Charcoal Cloth Patch in Real Work Conditions

Rani Claus[1], Dahlia Van De Velde[2], Frauke E. I Lemaire[1], Steven Ronsmans[3], Nora Marain[4], Peter H. M. Hoet[5], Jeroen A. J. Vanoirbeek[5]
[1]Environment and Health, KU Leuven, PhD Student, Leuven, Belgium. [2]Environment and Health, KU Leuven, Master Student, Leuven, Belgium.

[3]Environment and Health, KU Leuven, Post-Doc, Leuven, Belgium. [4]Laboratory of Respiratory Diseases and Thoracic Surgery, KU Leuven, PhD Student, Leuven, Belgium. [5]Environment and Health, KU Leuven, Professor, Leuven, Belgium

Abstract
Assessing dermal exposure to volatile organic compounds (VOCs) is becoming increasingly important in industrial settings. However, no standardized methods exist to quantify dermal exposure. In this study, a new custom-made activated charcoal cloth (ACC) patch was designed, validated, and tested in a workplace environment to evaluate its suitability in real work conditions. Initially, we evaluated the sample stability of thirteen VOCs during storage (at 20°C, 4°C and -20°C), the desorption efficiency (DE) of 188 VOCs and the cleaning procedure (baking the ACC for 2h at 120°C). Subsequently, we used the patches to assess dermal exposure in eighteen workers. Each worker wore three patches (thumb, index finger and neck). Results of the sample stability tests demonstrated a maximal decay of 10% over fourteen days at 4°C and -20°C. Among the 188 VOCs tested, 33 showed DE concentration dependency (as described by a dubinin-radushkevich isotherm) within a 0.0005–1mg/mL range. After the cleaning procedure, only minimal residual concentrations (0.77–6.13µg/patch) were measured. The patches allowed us to differentiate between several tasks by comparing the VOC concentration found on the fingers with the VOC concentration found on the neck patch. The relative concentration (fingers/neck) of paint mixers and painters (open or confined areas) were all above 75%, where concentrations of drivers of the cherry pickers were between 9–30%. The concentrations of non-painters were found below LOQ. These findings demonstrate the ability of the custom-made ACC patches to measure dermal VOC exposure in real work conditions.

444 – Evaluating Long-Term Occupational Exposures Among Structural Firefighters Using Silicone Wristbands

Samantha Samon[1], Viviana Geron[2], Nicholas Herkert[3], Duncan Hay[4], Sharon Zhang[3], Daniel Erhlich[2], Kate Hoffman[5], Heather Stapleton[6]
[1]Duke University, Postdoctoral Associate, Durham, USA. [2]Duke University, Student, Durham, USA. [3]Duke University, Research Project Manager, Durham, USA. [4]Duke University, Research Technician II, Durham, USA. [5]Duke University, Research Assistant Professor, Durham, USA. [6]Duke University, Professor, Durham, USA

Abstract
Firefighting has been associated with numerous adverse health consequences. However, past studies are limited by cross-sectional designs and poor hazard characterization, making identifying specific occupational exposures associated with disease risks challenging. Herein, we evaluate the utility of silicone wristbands as a measure of cumulative, longer-term exposure experienced by firefighters.

Firefighters (n=29) wore silicone wristbands over a four to five-month period. A new wristband was worn during each shift, which consisted of five on-duty and four off-duty days (~eight bands per firefighter). Wristbands were analyzed for 147 chemicals, including PAHs, PFAS, plasticizers, pesticides and flame retardants. Bands were analyzed individually, and composite samples were also generated for each participant, which contained segments of all eight bands worn by a participant.

In preliminary analyses (n=6 participants; 48 individual and 6 composite samples), 54 compounds were detected in >70% of samples. Phthalates had the highest median concentrations (e.g., DEHT median=167,150 ng/g), along with several organophosphate esters (e.g., TDCIPP median=625 ng/g). Individual exposure levels were highly variable over time, suggesting firefighter occupational exposures are not consistent over multiple shifts. However, data suggest composite samples are strongly correlated with total exposure (sum concentration of individual bands). Chemical concentrations were associated with activities performed while firefighting (e.g., attending a fire).

Evaluating chemical exposure among firefighters is challenging because their work environment is constantly changing, and exposures are highly variable. Data suggest that silicone wristbands provide an efficient tool for capturing information about firefighters' cumulative exposure to a wide range of compounds over time.

445 – Using EPA's ChemExpo Tool to Build Updated Quantitative Structure-Use Relationship (QSUR) Models

Katherine Phillips[1], Victoria Hull[2], Sakshi Handa[3], Kristin Isaacs[4]
[1]U.S. EPA, Physical Scientist, Research Triangle Park, USA. [2]Oak Ridge Associated Universities, Student Services Contractor, Research Triangle Park, USA. [3]U.S. EPA, Biologist, Research Triangle Park, USA. [4]U.S. EPA, Research Physical Scientist, Research Triangle Park, USA

Abstract
There are thousands of chemical substances in need of prioritization for risk evaluation by the U.S. Environmental Protection Agency (EPA). While high-throughput toxicology methods can be utilized to obtain screening level hazard values for most of these chemicals,

there are still gaps in data available to support analogous methods for developing screening level exposure values. Recently, the EPA's Office of Research and Development has developed a suite of tools to aid in advancing high-throughput exposure estimates. One tool in this suite is the Chemicals and Products Database (CPDat) which contains information on composition and use of chemicals in consumer and industrial settings. In addition to CPDat, EPA has developed an internal data management and curation tool, Factotum, that facilitates rapid collection, curation, and categorization of new sources into CPDat. Here, we demonstrate how CPDat (and the underlying Factotum) can aid in constructing new Quantitative Structure-Use Relationship (QSUR) models. These models have proven to be helpful in parameterizing high-throughput exposure models (including estimating key emissions to indoor and ambient environments), bolstering tentative identifications from non-targeted analyses, and identifying candidate chemical alternatives. Information required to build and validate these models can quickly be identified and downloaded via EPA's new public-facing Chemical Exposure Knowledgebase (ChemExpo) application for exploring CPDat data. QSUR models for 44 different functional uses were constructed and validated with a training set of more than 10k chemical substances. Finally, we evaluate these models against public regulatory substance inventories.

This abstract does not necessarily reflect U.S. EPA policy.

446 – POSTER: The Critical Review of Exposure Assessment in Human Cancer Studies: Procedures and Lessons Learned in the IARC Monographs Programme

Roland Wedekind[1], Daniel R. S. Middleton[2], Eero Suonio[3], Nathan L. DeBono[3], Mary K. Schubauer-Berigan[1]
[1]International Agency for Research on Cancer, Scientist, Lyon, France. [2]Queen's University Belfast, Fellow, Belfast, United Kingdom. [3]International Agency for Research on Cancer, Consultant, Lyon, France

Abstract
Introduction: Epidemiological studies provide critical evidence for the identification of carcinogenic hazards and evaluating the quality of exposure assessment in these studies is essential to judging their informativeness. Since the update of the Preamble in 2019, the IARC Monographs on the Identification of Carcinogenic Hazards to Humans contain a formal critical appraisal of the exposure assessment quality in human cancer and mechanistic studies that are key to a given evaluation of carcinogenicity. This work describes the rationale and workflow of this new procedure with application to specific examples.

Methods: The steps of the exposure assessment critique as outlined by the IARC Monographs Preamble were reviewed. Former Monographs evaluations were used as case-studies to identify the exposure assessment domains that were most informative for judging study quality.

Results: A common approach was used to evaluate the exposure assessment quality in the Monographs on night shift work, aniline, and 1,1,1-trichloroethane but some domains were adjusted for different agents, such as information on exposure routes and the assessment of co-exposures. Agent-specific knowledge is required concerning exposure assessment methods and exposure duration and intensity and potential co-exposures to identify the strengths and limitations of the exposure assessments performed in human cancer and mechanistic studies.

Conclusion: Since the new Preamble, an increased effort was made in the IARC Monographs programme to evaluate the quality of exposure assessment in human studies. This structured process can improve transparency and consistency in the application of evidence synthesis frameworks for future IARC working groups and other scientific review panels.

447 – NC HHEAR Hub Metabolomics and Exposomics Workflow: Data Capture, Quality Control, and Data Processing via ADAP

Aleksandr Smirnov[1], Blake Rushing[2], Yuanyuan Li[3], Susan McRitchie[4], Susan Sumner[5], Xiuxia Du[6]
[1]University of North Carolina at Charlotte, Research Assistant Professor, Charlotte, USA. [2]University of North Carolina at Chapel Hill, Assistant Professor, Chapel Hill, USA. [3]Biotage AB, Manager, Senior Application Scientist, Charlotte, USA. [4]University of North Carolina at Chapel Hill, Program Manager, Chapel Hill, USA. [5]University of North Carolina at Chapel Hill, Professor, Chapel Hill, USA. [6]University of North Carolina at Charlotte, Professor, Charlotte, USA

Abstract
Processing untargeted metabolomics and exposomics big data acquired from liquid chromatography-tandem mass spectrometry (LC-MS/MS) presents many challenges including handling big data (up to a few dozens of batches with ~100 samples in each batch),

aligning data from multiple batches, and performing batch-effect correction. The NC HHEAR Hub uses the Automated Data Analysis Pipeline (ADAP) to overcome these challenges. ADAP consists of ADAP-BIG, ADAP-KDB, and other components. ADAP-BIG is a desktop software for pre-processing raw LC-MS/MS data, that performs peak detection, peak filtering, alignment, normalization, batch-effect correction, etc. It was specifically designed to handle data from large-scale studies, includes statistical analysis capabilities for quality control, and has an easy-to-use graphical interface. ADAP-KDB is a cloud application for matching mass spectra against public and/or in-house libraries. Together, ADAP-BIG and ADAP-KDB form the pipeline used for processing raw untargeted data by the NC HHEAR Hub. This pipeline has been compared to Progenesis and Compound Discoverer. This comparison was based on the percentage of common peaks detected by the three software tools, overall sample positions on PCA plots, and overlap of high-confidence annotations produced by matching the mass, retention time, and MS/MS. The comparison showed a comparable or better performance in terms of quality control PCA plots and much faster execution time by the ADAP pipeline. The latter allowed the NC HHEAR Hub to considerably reduce the time needed for processing HHEAR data.

448 – Exploring the associations of exposures to toxic substances and the devastating disease, Amyotrophic Lateral Sclerosis (ALS)

Gregory Zarus[1], Paul Mehta[2], Jaime Raymond[3], Theodore Larsen[4], Stephan Brenner[5], D. Kevin Horton[6]
[1]ATSDR, Director Office of Innovation and Analytics, Atlanta, USA. [2]ATSDR, Principle Investigator ALS, Atlanta, USA. [3]ATSDR, Epidemiologist, Atlanta, USA. [4]ATSDR, Senior Scientist, Atlanta, USA. [5]ATSDR, Medical Economist, Atlanta, USA. [6]ATSDR, Section Chief Registries & Surveillance, Atlanta, USA

Abstract
ALS is a devastating disease that qualifies one for expedited Medicare coverage. There is no known cause or cure, and current therapeutics add on average only months to the patients' lives. Since 2013, the Agency for Toxic Substances and Disease Registry (ATSDR) and partners have studied potential associations between developing ALS and exposure to toxic substances, with the goal of identifying risk factors and preventing future disease. This presentation will summarize ATSDR research findings related to the etiologic role of environmental exposures and ALS.

To date, various studies have shown associations between the risk of developing ALS and

- Exposure to metals including lead and mercury
- Exposure to pesticides and polychlorinated biphenyl congeners
- Hobbies involving lead
- Past occupations with potential for environmental exposures
- Stronger associations with earlier life exposure vs late life exposure
- Living or recreating near particular water bodies

We will also present additional research findings suggesting possible environmental effects on ALS risk, including studies of estrogen-related factors and ALS onset and studies of metabolism alterations associated with both ALS and substance exposures. And close with recently initiated research to study environmental justice issues related to ALS and proposed studies investigating associations with exposure to emerging contaminants such as per- and polyfluoroalkyl substances and microplastics and to study impacts.

This body of research works towards the goal of increasing understanding of the etiologic role of environmental exposures and ALS. It is hoped that eventually this research will support development of mitigation strategies to limit or reduce risk in the future.

449 – Estimation of Infiltration level of outdoor air into residential biuldings with air change per hour (ACH) measured in urban, rural, and industrial areas over four seasons in South Korea :

Birhan Gurmu[1], Mingi Kim[2], Chiyu Song[1], Sangoon Lee[1], Sungroul Kim[3]
[1]Soonchunhyang Univ., Ph.D student, Asan, Korea, Republic of. [2]Soonchunhyang Univ., Researcher, Asan, Korea, Republic of.
[3]Soonchunhyang Univ., Professor, Asan, Korea, Republic of

Abstract
Indoor air quality is crucial for human health, especially for vulnerable groups such as the elderly, who spend most of their time indoors. Understanding the air exchange per hour (ACH) is of paramount importance in the assessment of indoor air quality. However, information on ACH in different areas and seasons in South Korea remains limited. In this study, we measured the natural ACH through four seasons, i.e., fall, winter, spring, and summer (2 weeks/season, between 1:00–5:00 a.m., 5 minute interval) in 81 residential buildings

located in urban, industrial, and rural areas from 2021 to 2022. To estimate ACH, indoor sensors were installed in 81 residential buildings. We compared their seasonal and spatiotemporal variations using descriptive statistics and evaluated the association of ACH with indoor air quality. Our multivariate regression modeling results showed that the ACH value increased with larger CO2 concentration difference after adjusting for seasonality (summer versus others) and region (rural,city, and industrial area). We also found a significant difference in ACH between seasons (especially summer versus others) and regions (cities versus others) ($P<0.05$). Our study utilized indoor CO2 data to determine the ACH. In conclusion, using real-time CO2 data to determine the ACH in residential buildings is a cost-effective and easy-to-use method that may prove beneficial in estimating the infiltration of outdoor air into buildings and the degree of contribution of outdoor air to indoor air quality.

451 – POSTER: Individual carbonyl reductase 3 genetic susceptibility to atopic dermatitis in adolescents under chronic particulate matter exposure

Sara Kress[1], Ashtyn Areal[2], Michael Lau[3], Claudia Wigmann[1], Frederick Hartung[2], Thomas Haarmann-Stemmann[4], Tamara Schikowski[4]
[1]IUF– Leibniz Research Institute for Environmental Medicine, PostDoc, Düsseldorf, Germany. [2]IUF– Leibniz Research Institute for Environmental Medicine, PhD student, Düsseldorf, Germany. [3]Mathematical Institute, Heinrich Heine University, PhD student, Düsseldorf, Germany. [4]IUF– Leibniz Research Institute for Environmental Medicine, Head of working group, Düsseldorf, Germany

Abstract
The pathogenesis of atopic dermatitis (AD) involves skin barrier impairments by an interplay of genetic and environmental factors. Carbonyl reductase 3 (CBR3) is correlated with inflammatory responses and is inducible by particulate matter (PM) in keratinocytes. Thus, we investigated the effect of a CBR3 genetic risk score (GRS) on AD under chronic PM exposure. We used data between birth and age 15 from 428 adolescents enrolled in the GINIplus/LISA North study. Individual chronic exposures to PM with median aerodynamic diameters of ≤2.5, ≤10, 2.5–10 µm ($PM_{2.5}$, PM_{10}, PM_{coarse}), and the reflectance of $PM_{2.5}$ filters ($PM_{absorbance}$) were modeled within the ESCAPE project and estimated using the means of exposures over 15 years. Out of 52 CBR3 single nucleotide polymorphisms (SNPs), we constructed and evaluated a GRS utilizing bagging and random forests. The effect of the GRS on AD (at age 15, defined as ever-diagnosed) was investigated by logistic regression adjusted for the exposure to each PM separately and potential confounders. Our results showed significant effects of CBR3 GRS on AD under adjustment for chronic exposures to each PM (AD diagnoses: 34.8%; $PM_{2.5}$, PM_{10}, $PM_{absorbance}$: p-value=0.009; PM_{coarse}: p-value=0.008). Furthermore, we identified 8 SNPs presenting the highest protective (rs881712, rs879892, rs12626189, rs9305578, rs1971504, rs2242803, and rs10470174) or risk (rs45622637) trends of CBR3 SNPs. In conclusion, the individual susceptibility to AD in adolescents given the observed chronic PM exposure is likely affected by CBR3 variations. To gain further insights into the underlying individual pathophysiology, combinations of epidemiological and mechanistic studies are required.

452 – Applications of Non-Targeted Analysis for Rapid and Community Response

John Sloop[1], Alex Chao[2], Jennifer Gundersen[3], Allison Phillips[4], Jon Sobus[5], Elin Ulrich[6], Antony Williams[7], Stephen Jackson[2], Seth Newton[5]
[1]US EPA, Postdoctoral Researcher, RTP, USA. [2]US EPA, Chemist, RTP, USA. [3]US EPA, Chemist, Narragansett, USA. [4]US EPA, Toxicologist, Cincinnati, USA. [5]US EPA, Physical Scientist, RTP, USA. [6]US EPA, Research Chemist, RTP, USA. [7]US EPA, Cheminformatician, RTP, USA

Abstract
Several thousand intentional and unintentional chemical releases occur annually in the U.S., with the contents of almost 30% being of unknown composition. When targeted methods are unable to identify the chemical(s) present, additional approaches like non-targeted analysis (NTA) can be used to identify unknown analytes. With recent advances in cheminformatics tools and data processing workflows, it has become possible to achieve confident chemical identifications via NTA in a timescale useful for rapid response. To evaluate the usefulness of NTA in rapid response situations, three mock scenarios were designed that reflect real-world releases of unknown substances, including surrogate chemicals mimicking a chemical warfare agent attack, the contamination of a home with illicit drugs, and an accidental industrial spill. Chemicals of interest were correctly identified by a blinded analyst in each of these scenarios with LC-HRMS and NTA approaches utilizing both existing and new data processing and analysis methods. Importantly, positive identifications occurred within 72 hours of sample receipt, suggesting that NTA could be a useful complement to targeted methods traditionally used in rapid response. Further utility of the NTA approach was demonstrated by analyzing, via GC-HRMS, air samples that were collected in response to a community's concerns of a pervasive odor, providing a case study of a real-world application of NTA to address environmental and human health concerns. The results showcase the usefulness of NTA in rapid and community response scenarios to assist in identifying unknown chemical stressors. This abstract does not necessarily reflect Agency policy.

454 – Applied Exposomics: A Population-centered Approach to Integrate Exposomics with Addressing Public Health Concerns

Jeanette Stingone[1], Andrew Geller[2], Darryl Hood[3], Konstantinos Makris[4], Charles Mouton[5], J. Christopher States[6], Susan Sumner[7], Lily Wu[8], Arcot Rajasekar[7,9]
[1]Columbia University, Assistant Professor, New York, USA. [2]US Environmental Protection Agency, Senior Science Advisor, RTP, USA. [3]The Ohio State University, Professor, Columbus, USA. [4]Cyprus International Institute for Environmental and Public Health, Cyprus University of Technology,, Associate Professor, Limassol, Cyprus. [5]University of Texas Medical Branch, Professor, Galveston, USA. [6]University of Louisville, Professor, Louisville, USA. [7]University of North Carolina, Chapel Hill, Professor, Chapel Hill, USA. [8]California Environmental Protection Agency, Acting Chief, Safer Alternatives Asessment and Biomonitoring Section, Sacramento, USA. [9]Renaissance Computing Institute, Chief Scientist, Chapel Hill, USA

Abstract
Environmental factors affecting health and vulnerability far outweigh genetics in accounting for disparities in health status and longevity in US communities. The concept of the exposome, the totality of exposure from conception onwards, has provided a paradigm for researchers to investigate the complex role of the environment on the health of individuals. We propose a complementary framework, applied exposomics, for exposomal assessment at the population-level. The goal of this framework is to bring the discovery of the exposome paradigm to research and practice on the health of populations, defined by various axes including geographic, social and occupational. This framework includes the integration of community-level measures of the built and social environments, environmental pollution derived from conventional and community science approaches, internal markers of exposure that can be measured at the population-level and early responses associated with health status that can be tracked using population-based monitoring. Primary challenges to the implementation of the proposed framework include needed advancements in population-level measurement, lack of existing models with the capability to produce interpretable and actionable evidence and the ethical considerations of labeling geographically-bound populations by exposomic profiles. To address these challenges, we propose a set of recommendations that begin with greater engagement with and empowerment of affected communities and targeted investment in community-based solutions. Applications to urban and rural settings and disaster epidemiology are discussed as examples for implementation.

455 – Time Patterns of Internal Exposure Data for Bisphenols, Phthalates, Organophosphate Flame Retardants, Cadmium and Polyaromatic Hydrocarbons in Europe.

Laura Rodriguez Martin[1], Liese Gilles[1], Agneta Åkesson[2], Jonas Tägt[2], Emilie Helte[2], Ovnair Sepai[3], Lorraine Stewart[3], Marike Kolossa-Gehring[4], Marta Esteban López[5], Argelia Castano[5], Jos Bessems[1], Greet Schoeters[1], Eva Govarts[1]
[1]VITO Health, Flemish Institute for Technological Research (VITO), 2400 Mol, R&D, Mol, Belgium. [2]Institute of Environmental Medicine, Karolinska Institutet, 171 77 Stockholm, Sweden, Researcher, Stockholm, Sweden. [3]UKHSA UK Health Security Agency, Harwell Science Park, Chilton OX11 0RQ, United Kingdom, Researcher, Chilton, United Kingdom. [4]German Environment Agency (UBA), Department of Toxicology, Health-related Environmental Monitoring, 1419506844 Berlin, Germany, Researcher, Berlin, Germany. [5]National Centre for Environmental Health, Instituto de Salud Carlos III, 28220 Majadahonda, Spain, Researcher, Majadahonda, Spain

Abstract
In the European Human Biomonitoring Initiative (HBM4EU), Human biomonitoring (HBM) data were used to assess internal exposure levels within the European population over time.

HBM data from three time periods were used. Comparable HBM data were generated and aggregated within the HBM4EU Aligned Studies activity from recent samples 2014-2021) and from biobanked samples of the DEMOCOPHES studies (2011-2012). For the period 2000-2010, aggregated data were collected in a harmonized way across available HBM studies, complemented with literature data. Time patterns in internal exposure were evaluated visually and statistically using the 50th and 90th percentiles (P50/P90) for phthalates/DINCH and organophosphorus flame retardants (OPFRs) in children (6-12 years), and cadmium, bisphenols, and polycyclic aromatic hydrocarbons (PAHs) in women (20-39 years).

Decreasing patterns have been identified in children for various phthalate (metabolites). The phthalate substitute DINCH shows an increasing pattern, though not statistically significant. OPFRs show variable patterns for the average exposure values (P50) as for the higher exposed population (P90), although no trends were statistically significant. For women, there is a clear decreasing pattern for BPA, while substitutes BPF and BPS show an increasing pattern which coincides with the BPA restrictions introduced in Europe. No clear pattern is observed for PAHs or cadmium.

Although the causal relations were not studied as such, exposure levels for chemicals for which restrictions at EU level are established visually decrease, levels of some of their substitutes (seem to) increase. The results support the policy-efficacy monitoring and policy-supportive role that human biomonitoring may play.

456 – Harmonized Human Biomonitoring in Europe: EU-wide Exposure Data of 11 Chemical Substance Groups from the HBM4EU Aligned Studies (2014-2021)

Eva Govarts[1], Liese Gilles[2], Laura Rodriguez Martin[2], Tiina Santonen[3], Petra Apel[4], Holger Koch[5], Marike Kolossa-Gehring[4], Marta Esteban López[6], Argelia Castaño[6], Lorraine Stewart[7], Ovnair Sepai[7], Greet Schoeters[2]
[1]VITO, Researcher - Biostatistician, Mol, Belgium. [2]VITO, Researcher, Mol, Belgium. [3]FIOH, Researcher, Työterveyslaitos, Finland. [4]UBA, Researcher, Dessau-Roßlau, Germany. [5]IPA, Researcher, Bochum, Germany. [6]ISCIII, Researcher, Madrid, Spain. [7]UKHSA, Researcher, Chilton, United Kingdom

Abstract
As one of the core elements of HBM4EU, a human biomonitoring (HBM) survey was conducted in 23 countries to generate EU-wide comparable HBM data. The HBM4EU Aligned Studies included a total of 10,795 participants of three age groups: children (6-12y), teenagers (12-18y) and adults (20-39y), recruited between 2014 and 2021. Depending on the age group, internal exposure to phthalates and substitute DINCH, halogenated and organophosphorus flame retardants, per- and polyfluoroalkyl substances (PFASs), cadmium, bisphenols, polycyclic aromatic hydrocarbons (PAHs), arsenic species, acrylamide, mycotoxins (deoxynivalenol (total DON)), benzophenones and selected pesticides was assessed. For substance groups analyzed in different age groups higher average exposure levels were observed in the youngest age group. Many biomarkers in teenagers and adults varied significantly according to educational attainment, with higher exposure levels of bisphenols, phthalates, benzophenones, PAHs and acrylamide in participants (from households) with lower educational attainment, while teenagers from households with higher educational attainment have higher exposure levels for PFASs and arsenic. Geographical variations were seen for all exposure biomarkers. For 15 biomarkers, the available health-based HBM guidance values were exceeded with highest exceedance rates for toxicologically relevant arsenic (40%), the pyrethroid metabolite 3-PBA (36%), and between 11-14% for total DON, Σ (PFOA + PFNA + PFHxS + PFOS), bisphenol S and cadmium. The generated data serve as a reference for comparison at the global level, provide a baseline to compare the efficacy of the European Commission's chemical strategy for sustainability and will give leverage to national policy makers for the implementation of targeted measures.

457 – Toxic Substances with Evidence of Ototoxicity

Gregory Zarus[1], Patricia Ruiz[2], Rae Benedict[3], Krystin Carlson[4], Thias Morata[5]
[1]ATSDR, Director, Office of Innovation and Analytics, Atlanta, USA. [2]ATSDR, Simulation Science Sections Chief, Atlanta, USA. [3]ATSDR, Toxicologist, Atlanta, USA. [4]NIOSH, Toxicologist, Cincinatti, USA. [5]NIOSH, SENIOR SERVICE FELLOW, Cincinatti, USA

Abstract
Ototoxicity is an important public health concern. Few studies involve non-pharmacological substances. Auditory problems in workers exposed to toxic non-pharmacological substances have been frequently documented. However, hearing is seldom measured in case control exposure studies involving these substances.

The Agency for Toxic Substances and Disease Registry (ATSDR) develops hundreds of Toxicological Profiles on substances frequently found in the environment, including volatile organic substances, pesticides, herbicides, metals, inorganic substances, and radionuclides. Each peer-reviewed profile reflects an extensive evaluation, summary, and interpretation of the available toxicological and epidemiological information. Although not presented as "ototoxicity" in the profiles, there are several signs and symptoms associated with "ototoxicity", such as hearing loss, tonal alteration, and effects to the hair cells of the Organ of Corti. Similarly, association of exposure to chemical substances with vestibular effects such as dizziness, vertigo, visual problems, and headaches are nonspecific making them hard to discern from other neurological causes. Study data on tinnitus are rarer still. While literature searches document tinnitus among the frequent complaints of populations exposed to toxic substances, dose-response studies are lacking for this reported affect.

Here we systematically present the evidence of ototoxicity associations of more than fifty hazardous substances that people are exposed to, both in the environmental and in an occupational setting. These analyses will alert health care providers to potential toxic substance exposures and ototoxicity and facilitate more informed management decisions. Furthermore, identification of ototoxicity assessment data needs will be discussed, thus encouraging further research in this field.

458 – Poly- and Perfluoroalkyl Substances in Firefighter Turnout Gear, Skin Wipes and Serum

Amulya Vankayalapati[1], Ankita Bhattacharya[2], Rachel Bauer[2], Heather Stapleton[3], Graham Peaslee[4], Courtney Carignan[5]
[1]Michigan State University, Research Assistant, East Lansing, USA. [2]Michigan State University, Graduate Research Assistant, East Lansing, USA. [3]Duke University, Professor, Durham, USA. [4]University of Notre Dame, Professor, Notre Dame, USA. [5]Michigan State University, Assistant Professor, East Lansing, USA

Abstract
Due to their unique properties, per- and polyfluoroalkyl substances (PFASs) are widely used in firefighting foams and gear. While studies have been conducted to better understand exposure to PFAS during firefighting, there are currently no investigations on the transfer of PFAS from the gear to the skin or investigations of PFAS in the latest generation of turnout gear (PFZero®). Therefore, we conducted a biomonitoring study of firefighters (n=18) who provided forearm wipes before and after a training event, a sample or wipe of their turnout gear, a blood sample, and completed a brief exposure questionnaire. Samples were analyzed for >30 PFASs using GC-MS/MS and LC-MS/MS for volatile and non-volatile PFASs, respectively. Concentrations were higher in forearm wipes collected after vs. before the training event. PFOA was detected in the majority of gear samples and after-training forearm wipes from participants with legacy (pre-2010) or next generation (post-2010) gear but not those with PFZero. Next generation turnout gear contained primarily 6:2 FTMAC with concentrations in the outer layer an order of magnitude higher than total PFAS in the legacy gear. PFZero turnout gear was found to contain small amounts of 6:2 and 8:2 FTMAC in wipes of the moisture barrier. These trends were reflected in the after-training forearm wipes. Overall, these findings reflect changes in PFAS mixtures used in firefighter turnout gear and indicate the potential for dermal exposure.

459 – Geospatial Infrastructure and Methods for Enabling Place-Based Research in the NIH ECHO Program

James Cajka[1], William Wheaton[2], Janet Moore[3], Timothy Shields[4], Amii Kress[5], Emily Knapp[5], Rima Habre[6], Cole Brokamp[7], Allan Just[8], Peter James[9], Diane Catellier[10]
[1]Research Triangle Institute, Geographic Information Systems Analyst, Research Triangle Park, NC, USA. [2]Research Triangle Institute, Retired, Research Triangle Park, NC, USA. [3]Research Triangle Institute, Statistician, Research Triangle Park, NC, USA. [4]Johns Hopkins University, Senior Scientist, Baltimore, MD, USA. [5]Johns Hopkins University, Assistant Scientist, Baltimore, MD, USA. [6]University of Southern California, Associate Professor, Los Angeles, CA, USA. [7]University of Cincinnati, Associate Professor, Cincinnati, OH, USA. [8]Mount Sinai School of Medicine, Associate Professor, New York, NY, USA. [9]Harvard TH Chan School of Public Health, Associate Professor, Boston, MA, USA. [10]Research Triangle Institute, Senior Statistician, Research Triangle Park, NC, USA

Abstract
The NIH ECHO program aims to understand the effects of a range of early life environmental influences on child health and development over the life course, broadly defined as environmental, social, and biological factors. The national ECHO Cohort consists of 50,000+ participants from diverse populations and geographies across the United States. This talk will present the multi-pronged approaches and systems developed by the Geospatial Working Group which includes investigators from the ECHO Data Analysis Center and participating cohorts to meet the unique geospatial challenges and needs of the program. These include but are not limited to securely standardizing address data, geocoding participant locations, constructing accurate lifetime residential histories, linking spatiotemporally resolved environmental exposure and social determinants of health data in space and time to participants to examine key research questions and influences on health outcomes (in five major domains), enabling researcher access to data and analysis platforms in a secure and centralized fashion, ingesting and harmonizing existing data, establishing metadata standards for a wide range of geospatial domains, and developing data collection tools for consistently capturing and updating residential locations over time in a child's life. The geospatial infrastructure and data pipelines developed provide an important contribution to the field and establish approaches and standards for future, nationwide geospatial data integration efforts into human health studies.

460 – Toxic Substances with Evidence of Ototoxicity

Gregory Zarus[1], Patricia Ruiz[2], Rae Benedict[3], Krystin Carlson[4], Thais Morata[5]
[1]ATSDR, Director, Office of Innovation and Analytics, ATSDR, USA. [2]ATSDR, Chief, Simulation Science, ATL, USA. [3]ATSDR, Toxicologist, ATL, USA.

[4]NIOSH, Associate Service Fellow, Cin, USA. [5]NIOSH, Senior Service Fellow, Cin, USA

Abstract
Ototoxicity is an important public health concern. Few studies involve non-pharmacological substances. Auditory problems in workers exposed to toxic non-pharmacological substances have been frequently documented. However, hearing is seldom measured in case control exposure studies involving these substances.

The Agency for Toxic Substances and Disease Registry (ATSDR) develops hundreds of Toxicological Profiles on substances frequently found in the environment, including volatile organic substances, pesticides, herbicides, metals, inorganic substances, and radionuclides. Each peer-reviewed profile reflects an extensive evaluation, summary, and interpretation of the available toxicological and epidemiological information. Although not presented as "ototoxicity" in the profiles, there are several signs and symptoms associated with "ototoxicity", such as hearing loss, tonal alteration, and effects to the hair cells of the Organ of Corti. Similarly, association of exposure to chemical substances with vestibular effects such as dizziness, vertigo, visual problems, and headaches are nonspecific making them hard to discern from other neurological causes. Study data on tinnitus are rarer still. While literature searches document tinnitus among the frequent complaints of populations exposed to toxic substances, dose-response studies are lacking for this reported affect.

Here we systematically present the evidence of ototoxicity associations of more than fifty hazardous substances that people are exposed to, both in the environmental and in an occupational setting. These analyses will alert health care providers to potential toxic substance exposures and ototoxicity and facilitate more informed management decisions. Furthermore, identification of ototoxicity assessment data needs will be discussed, thus encouraging further research in this field.

461 – Towards advancing precision environmental health: Developing a customized exposure burden score to PFAS mixtures to enable equitable comparisons across population subgroups, using mixture item response theory

Shelley Liu[1], Leah Feuerstahler[2], Yitong Chen[3], Joseph Braun[4], Jessie Buckley[5]
[1]Icahn School of Medicine at Mount Sinai, Associate Professor, New York, USA. [2]Fordham University, Assistant Professor, New York, USA. [3]Icahn School of Medicine at Mount Sinai, Biostatistician, New York, USA. [4]Brown University, Associate Professor, Providence, USA. [5]Johns Hopkins University, Associate Professor, Baltimore, USA

Abstract
Quantifying a person's cumulative exposure burden to PFAS mixtures is important for risk assessment and biomonitoring. A summary metric would also be beneficial for report-back, enabling people to understand how their total exposure burden compares to others. However, different people may be exposed to different sets of PFAS due to complex heterogeneity in exposure sources and patterns. We used a data science approach (mixture item response theory) to estimate a person's total exposure burden to PFAS, accounting for the fact that different people have different diets and behaviors that expose them to different sets of PFAS chemicals. This ensures that PFAS burden scores can be equitably compared across population subgroups, rather than using a one-size-fits-all approach to quantifying PFAS burden for everyone (e.g. using summed PFAS concentrations as the summary metric), which assumes that all people are exposed to the same PFAS chemicals. We applied our methods to targeted PFAS biomonitoring data from the United States National Health and Nutrition Survey (2013-2018). Using our data science approach, we found that Asian Americans have significantly higher PFAS burden compared with non-Hispanic Whites and all other race/ethnicity groups, even when controlling for socio-economic status. However, this disparity was sometimes hidden if we used summed PFAS concentrations as the summary metric. Our work suggests that biomonitoring and risk assessment may want to consider a summary PFAS exposure metric that accounts for complex exposure heterogeneity, so that the summary metric is fair and informative for all people.

462 – Monitoring Human Exposure to Parabens and Triclosan: Comparing Silicone Wristbands to Spot Urine Samples as Predictors of Internal Dose

Jessica Levasseur[1], Kate Hoffman[2], Sharon Zhang[3], Ellen Cooper[4], Heather Stapleton[5]
[1]Duke University Nicholas School of the Environment, PhD Candidate, Durham, USA. [2]Duke University Nicholas School of the Environment, Research Professor, Durham, USA. [3]Duke University Nicholas School of the Environment, Research Project Manager, Durham, USA. [4]Duke University Nicholas School of the Environment, Research Scientist, Durham, USA. [5]Duke University Nicholas School of the Environment, Professor, Principal Investigator, Durham, USA

Abstract
Americans use an average of nine personal care products (PCPs) per day, with an estimated exposure to 126 chemicals daily. Silicone wristbands (WBs) are exposure tools that capture integrated dermal and inhalation exposure routes while worn. WBs eliminate the burden of traditional biomonitoring methods, but understanding how WB measurements compare to measurements of internal dose is needed. This study compared spot urine samples and WBs to predict the total mass excreted for 4 parabens and triclosan, chemicals commonly used in PCPs. Participants began with 5 WBs on one arm, removing one each day for the 5-day study period. Daily, random spot urine samples and all other urine excreted were collected. Pooled urine was used to determine total mass excreted or internal dose. WBs had 100% detection frequency for all compounds by day 3. Hydrophobic compounds such as butylparaben and triclosan demonstrated linear uptake on WBs over 5 days (3.5 and 3.6 ng/g/day, respectively). Detection frequencies for both spot and pooled urine were >85%, except for butylparaben (<10%). Spearman correlations between methylparaben, ethylparaben, and triclosan were significantly ($p<0.05$) and positively correlated between day 5 WBs and total mass excreted (0.62, 0.73, and 0.64). Spearman correlations between random spot urine samples as compared to total mass excreted for methylparaben, ethylparaben, propylparaben, and triclosan were also significantly and positively correlated (0.75, 0.65, 0.81, and 0.78, respectively). Overall, WBs performed similarly to spot urine samples in predicting total exposure to parabens and triclosan, compounds typically associated with PCP use.

463 – POSTER: Urinary phthalate metabolite concentrations and lipid traits among Chinese adults: a cross-sectional study from China National Human Biomonitoring (CNHBM)

Saisai Ji[1,2], Yingli Qu[3], Zheng Li[1], Xiaoming Shi[4]
[1]National Institute Of Environmental Health, China Cdc, Assistant Researcher, Beijing, China. [2]School of Public Health, Peking University, Student, Beijing, China. [3]National Institute Of Environmental Health, China Cdc, Associate Researcher, Beijing, China. [4]National Institute Of Environmental Health, China Cdc, Professor, Beijing, China

Abstract
Background: Increasing evidence suggests that phthalate exposure may contribute to increased risks of metabolic disorders. The effect of phthalate exposure on lipid metabolism in adults remains unexplained.

Objectives: This study aims to elucidate the associations between phthalate exposure and lipid metabolism indicators in Chinese adults.

Methods: This cross-sectional study included 896 participants aged 18 to 79 years enrolled in China National Human Biomonitoring (2017-2018). We quantified 12 urinary phthalate metabolite concentrations and assessed 6 lipid metabolism indicators in fasting serum, including triglycerides, total cholesterol, high density lipoprotein cholesterol (HDL-C), low-density lipoprotein cholesterol (LDL-C), apoA-I and apoB. Generalized linear mixed models were used to assess the association between urinary phthalates and lipid biomarkers. Weighted quantile sum (WQS) regression was used to estimate the influence of the phthalate mixture.

Results: Among the 12 phthalate metabolites, the median concentrations of mono-n-butyl phthalate (MnBP) was the highest (431.72 nmol/L), followed by mono-iso-butyl phthalate (MiBP) (113.17 nmol/L), and mono-ethyl phthalate (MEP) (48.15 nmol/L). In the single phthalate models, the mono-(2-ethyl-5-oxyhexyl) phthalate (MEOHP) exhibited a significant negative association with the lipid metabolite indicator of HDL-C (β: -0.019, 95% CI: -0.037, <0.000) and apoA-I (β: -0.021, 95% CI: -0.036, -0.006) in the fully adjusted models. MEHHP was associated with lower apoA-I with β (95%CI) of -0.018 (-0.033, -0.002). WQS regression analysis showed that MEOHP was the largest contributor for decreased HDL-C (35.86%) and apoA-I (35.38%).

Conclusions: The metabolites of di-(2-ethylhexyl) phthalate (DEHP) were associated with decreased HDL-C and apoA-I in Chinese adults.

464 – Systematic Evidence Mapping for Dietary and Product Use Correlates with Human Exposure to PFAS

Chris Holder[1], Elaine A. Cohen Hubal[2], Jeanne Luh[3], Madison G. Lee[4], Lisa Jo Melnyk[5], Kent Thomas[6]
[1]ICF, Science Director in Air Quality Modeling, Durham, USA. [2]U.S. EPA, Office of Research and Development, Center for Public Health and Environmental Assessment, Senior Science Advisor, Research Triangle Park, USA. [3]ICF, Lead Environmental Scientist, Durham, USA. [4]ICF (formerly), Health Scientist, Durham, USA. [5]U.S. EPA, Office of Research and Development, Center for Public Health and Environmental Assessment, Research Chemist, Cincinnati, USA. [6]U.S. EPA, Office of Research and Development, Center for Public Health and Environmental Assessment, Research Physical Scientist, Research Triangle Park, USA
Abstract

Per- and poly-fluoroalkyl substances (PFAS) often are found in environmental media, foods, indoor environments, and consumer products/articles. Understanding relationships between PFAS exposure sources and concentrations in human biomatrices can inform source-contribution assessments and provide targets for exposure reduction. A process of systematic evidence mapping (SEM) was implemented to collate information and develop a database for correlations between 16 PFAS measured in biomatrices with food consumption and product use. The SEM yielded 84 studies reporting correlates of PFAS exposures; 58 were published after 2015. Diet and products/articles were assessed in 77 and 47 studies, respectively. PFOA and PFOS were assessed in most studies; PFNA, PFHxS, and PFDA were included in many. Significant correlations were reported for foods in 62 studies for one or more PFAS; most commonly for seafood, meats/eggs, and combined foods/overall diet categories (>67% of the studies investigated these food categories). Product/article categories of smoking/tobacco, carpet/flooring/furniture/housing, and cosmetics/toiletries categories were referenced most often, with significant findings for one or more PFAS in at least 56% of the studies that investigated these product categories. Seven of 11 product/article categories included five or fewer studies. Substantial evidence was found for correlations between dietary intake and biomatrix levels for several PFAS. Studies examining product use were relatively sparse, except for smoking/tobacco, and would benefit from additional research. The resulting database allows further assessment of dietary and product use exposures and can inform new research to better understand PFAS source-to-exposure relationships.

465 – The Urban Exposome and Social Determinants in CARDIA

Kai Zhang[1], Laura Jones[2], Jie Yao[2], Brian Joyce[3], Yinan Zheng[4], Kyeezu Kim[5], Ravi Shah[6], Venkatesh Murthy[7], Aimin Chen[8], Donald Lloyd-Jones[9], Lifang Hou[9]
[1]University at Albany, State University of New York, Empire Innovation Associate Professor, Rensselaer, USA. [2]University at Albany, State University of New York, Doctoral Student, Rensselaer, USA. [3]Northwestern University, Research Assistant Professor, Chicago, USA. [4]Northwestern University, Assistant Professor, Chicago, USA. [5]Northwestern University, Research Fellow, Chicago, USA. [6]Vanderbilt University, Associate Professor, Nashville, USA. [7]University of Michigan, Professor, Ann Arbor, USA. [8]University of Pennsylvania, Professor, Philadelphia, USA. [9]Northwestern University, Professor, Chicago, USA

Abstract
Understanding the interplay between the urban exposome and social determinants is crucial for promoting health equity and intervention. In this study, we explored these relationships using data from the Coronary Artery Risk Development in Young Adults (CARDIA) study, a longitudinal study of over 5,000 Black and White men and women in four US cities (Chicago, Minneapolis, Oakland, and Birmingham). We constructed urban exposome profiles for participants from Exam Year(Y) 0, 7, 10, 15, 20 to Y25, comprising air pollution, green space, built and food environment, and then sought associations between each exposome component s and SEP variables by city, year, sex, and race. We ran generalized estimating equations (GEE) models to account for multiple visits, regressing exposome covariates for air pollution variables (e.g., PM2.5, PM10, ozone), built environment covariates such as distance to roads, waters, and green space, and food/social environment covariates such as distances to grocery shops and schools; against SEP variables such as education, employment, income, and financial stress by city. Exposome values that were significant at α = 0.05 for SEP covariates included PM2.5, sulfate, nitrate, black carbon, NO_2, SO_2, CO, road area, distance to water and green space, and counts of food shops at 1km and 3km from participant residences. Our analysis revealed that lower exposures to adverse environmental factors were generally associated with better social position. Our findings shed light on the interplay between the urban exposome and social determinants and highlight need for further research to promote health equity in urban areas.

466 – Characterizing Pathways of Exposure for Risk-Based Chemical Prioritization

Caroline Ring[1], Paul Kruse[2,3], John Wambaugh[4], Kristin Isaacs[5]
[1]US EPA, Computational Exposure Data Scientist, Research Triangle Park, USA. [2]ORISE, Postdoctoral Fellow, Oak Ridge, USA. [3]US EPA, Postdoctoral Fellow, Research Triangle Park, USA. [4]US EPA, Physical Scientist, Research Triangle Park, USA. [5]US EPA, Research Physical Scientist, Research Triangle Park, USA

Abstract
The magnitude of daily intake for a chemical is roughly correlated with the type and number of pathways of exposure to that chemical. We categorized more than 60,000 chemical substances by broad pathways of exposure: consumer, industrial, pesticide, food, pharmaceuticals, and other pathways. These pathway categorizations can be used as a training set for automated read-across of chemical exposure pathways to better inform rapid exposure modeling as part of risk-based chemical prioritization.

Pathway categorizations integrated multiple heterogeneous data streams that reported chemical use and occurrence. Data streams included large public inventories of chemicals—for example, lists of consumer product ingredients; lists of substances added to food;

lists of pharmaceutical active ingredients; and government inventories of chemicals with reported use in industrial and consumer sectors. Chemical use and occurrence information from each data stream was organized into 19 harmonized, hierarchical categories for pathways of exposure and 3 aggregated pathways. Each chemical was categorized for each pathway as positive, negative, or unknown (no data), depending on the use/occurrence data available.

Using these pathway categorizations, a machine-learning model was trained to predict pathways based on chemical descriptors (such as structure and/or properties). Information about pathways of exposure has been shown to improve the predictive performance of consensus models of human exposure to environmental chemicals and can be applied to inform rapid decision-making about chemical prioritization.

The views expressed in this presentation are those of the authors and do not necessarily reflect the views or policies of the U.S. EPA.

467 – Introducing the Public-Facing ChemExpo Knowledgebase Web Application Supporting Delivery of Chemical Use and Consumer Product Data to Inform Exposure Assessment

Sakshi Handa[1], Kenta Baron-Furuyama[2], Katherine Phillips[3], Taylor Wall[1], Mary Horton[4], Christian Lutz[4], Allison Larger[5], Kristin Isaacs[6], Colleen Elonen[7]
[1]US Environmental Protection Agency, Biologist, Durham, USA. [2]US Environmental Protection Agency, Software Developer, Durham, USA. [3]US Environmental Protection Agency, Physical Scientist, Durham, USA. [4]Oak Ridge Associated Universities, Data Curator, Durham, USA. [5]General Dynamics Information Technology, Data Curator, Durham, USA. [6]US Environmental Protection Agency, Research Physical Scientist, Durham, USA. [7]US Environmental Protection Agency, Supervisory Biologist, Duluth, USA

Abstract
To support chemical decision-making, EPA's Office of Research and Development (ORD) must identify and characterize exposure pathways. How a chemical is used (e.g., in a consumer, occupational, or industrial context) is critical to determining exposure pathways. The ORD data management and curation application, Factotum, facilitates the rapid collection and distribution of high-quality chemical use and exposure-related data from publicly available documents to inform chemical exposure and risk assessments. Factotum has been used to manage and curate data from over 560,000 documents, representing 4 million individual chemical datapoints and 36,000 unique chemical substances. To disseminate data curated within Factotum, ORD has developed a new publicly available data search and visualization tool, called Chemical Exposure Knowledgebase (ChemExpo). The initial release of ChemExpo will allow for search, exploration, and download of chemical use data, including specific chemical substances, chemical composition of consumer products, functional role of chemicals within products and processes, and presence of chemicals on reported specific or general use lists. ChemExpo contains more detailed exposure data than currently available via EPA's CompTox Chemicals Dashboard, including additional document metadata, product-level chemical ingredient information, and product category, function category, and chemical summaries. Application functionality and data terminology are well-documented via a comprehensive user guide, including use cases to aid user exploration. By making this application publicly available, EPA ORD aims to increase transparency and accessibility of exposure relevant data to researchers, regulators, and consumers to support exposure assessments and other risk-based chemical evaluations.
This abstract does not necessarily reflect U.S. EPA policy.

468 – Intraurban Environmental Inequities in the Spatial Distribution of Polycyclic Aromatic Hydrocarbons and their Corresponding Inhalation Cancer Risk

Elysia G. Fuller-Thomson[1], Jasmin K. Schuster[2], Jacob Mastin[2], Tom Harner[3], Jack Cheng[1], Scarlett Rakowska[1], Sarah Faisal[1], Charity Delos Reyes[1], Gabrielle Olmedo[1], Matthew D. Adams[4]
[1]University of Toronto Mississauga, Student, Mississauga, Canada. [2]Environment and Climate Change Canada, Physical Scientist, Toronto, Canada. [3]Environment and Climate Change Canada, Senior Research Scientist, Toronto, Canada. [4]University of Toronto Mississauga, Professor, Mississauga, Canada

Abstract
Environmental inequities have long been documented in air quality exposure; however, most previous research has concentrated on a narrow range of the whole air mixture. Polycyclic aromatic hydrocarbons (PAHs) and their associated inhalation cancer risk have rarely been assessed at a fine urban spatial resolution, and research on inequities in PAH exposure distribution seldom goes beyond the anecdotal. Our research aimed to determine both the intraurban distribution of PAHs and investigate the association of PAHs with socioeconomic inequalities. Targeting a study site with more elevated pollutant exposures than surrounding urban centres and a wide

socioeconomic spread, Hamilton (43°15′24″N 79°52′09″W) and the adjacent western Burlington, Canada were selected to capture ambient U.S. EPA priority pollutant PAH concentrations. PAHs were measured by passive air samplers equipped with polyurethane foam (PUF-PAS) and sampled between late July to late September 2022 at 29 sites throughout Hamilton and West Burlington. Spatial modelling, using both kriging and land use regression, were used to combine the network of PAH inhalation cancer risk measurements into a predicted continuous pollution surface of the city, which was then assessed for correlations with socioeconomic inequality using the four metrics of the Ontario Marginalization Index. A clear pattern emerged of marginalization in close association to elevated PAH exposure which was distinctive to more transport-related pollutants' distribution. Two major conclusions are thus derived: intraurban pollution varies widely within an urban city, and for a true picture of environmental justice in air quality, a multipollutant lens is needed.

469 – Constructing Lifetime Residential Histories and Understanding Residential Mobility in the NIH ECHO Cohort

Rima Habre[1], Janet Moore[2], James Cajka[3], William Wheaton[4], Roxana Khalili[5], Heike Gibson[6], Cole Brokamp[7], Patrick Ryan[8], Timothy Shields[9], Emily Knapp[10], Amii Kress[10], Diane Catellier[11]
[1]University of Southern California, Associate Professor, Los Angeles, CA, USA. [2]Research Triangle Institute, Statistician, Research Triangle Park, NC, USA. [3]Research Triangle Institute, Geographic Information Systems Analyst, Research Triangle Park, NC, USA. [4]Research Triangle Institute, Retired, Research Triangle Park, NC, USA. [5]University of Southern California, Postdoctoral Research Associate, Los Angeles, CA, USA. [6]Harvard TH Chan School of Public Health, Senior Research Scientist, Boston, MA, USA. [7]University of Cincinnati, Associate Professor, Cincinnati, OH, USA. [8]University of Cincinnati, Professor, Cincinnati, OH, USA. [9]Johns Hopkins University, Senior Scientist, Baltimore, MD, USA. [10]Johns Hopkins University, Assistant Scientist, Baltimore, MD, USA. [11]Research Triangle Institute, Senior Statistician, Research Triangle Park, NC, USA

Abstract
Exposure measurement error is introduced in large population health studies of pregnancy or life course epidemiology by ignoring residential mobility (i.e., residential address changes in time). Given timing, location (to/from) and reasons for moving are not random, and changes in residential neighborhood exposures and opportunity factors can have important influences on subsequent child health and development, it is important to accurately capture lifetime residential location, understand when mobility occurs, and how it influences exposures. The NIH ECHO Geospatial Working Group developed methods to harmonize residential address information and construct complete, spatiotemporally resolved residential timelines for the national ECHO Cohort. These timelines enable accurate and spatiotemporally variable exposure assessment to social, environmental, and other exposures and risk factors and understanding of mobility over the life course. We hypothesize that the peri-natal period will be a time of increased residential mobility, and that there will be differences across individual level factors (e.g., income, family size, etc.), geography, and time in the frequency and timing of moving and its resultant effect on exposures (e.g., upward vs downward mobility as defined by neighborhood socioeconomic, environmental and opportunity indicators). The goal of this talk is 1) To describe methods developed to harmonize address and date (time) information in ECHO and construct complete residential history timelines for all participants, 2) To analyze patterns of residential mobility (timing, frequency, distance, etc.) in ECHO by life stage, and 3) To understand how residential mobility relates to changes in neighborhood level opportunity factors and environmental exposures.

471 – Towards Harmonized Human Biomonitoring (HBM) in Europe - What Have We Learned and How to Improve in PARC and Beyond

Liese Gilles[1], Eva Govarts[1], Marta Esteban López[2], Susana Pedraza Diaz[2], Loïc Rambaud[3], Philipp Zimmermann[4], Lorraine Stewart[5], Ovnair Sepai[6], Argelia Castano[2], Greet Schoeters[7], Sebastien Denys[8], Marike Kolossa-Gehring[9]
[1]VITO, R&D, Mol, Belgium. [2]ISCIII, R&D, Madrid, Spain. [3]SPF, Epidemiologist, Saint Maurice, France. [4]UBA, R&D, Berlin, Germany. [5]UKHSE, R&D, London, United Kingdom. [6]UKHSE, Program lead general toxicology and Biomonitoring, London, United Kingdom. [7]University Antwerp, Professor, Antwerp, Belgium. [8]SPF, scientific director, Saint Maurice, France. [9]UBA, Head of section toxicology, Berlin, Germany

Abstract
HBM provides information on real-life human exposures and can inform the development of environment and health policies such as the European Green Deal. To make optimal use of HBM in Chemicals Risk assessment, a sustainable HBM program representative for Europe is desirable. Examples of such programmes already exist in several countries outside the EU. However, in Europe generation of HBM data is fragmented, unstable, and limited in time due to the absence of legal requirements and lack of structural funding.

Joint European initiatives launched since 2005 have nevertheless increased alignment and capacity building at EU level. In 2005 the Expert Team to Support BIOmonitoring in Europe (ESBIO) was established until 2007. This was followed by the (DEMO)COPHES project

(2010-2012), which collected comparable HBM data of 1844 mothers and their children (5-11 years) in 17 European countries. After that, as part of the European Human Biomonitoring Initiative (HBM4EU), the HBM4EU Aligned Studies were organized (2014-2021), aligning national or regional HBM studies in Europe. The study focused on children (6-11 years), teenagers (12-19 years) and young adults (20-39 years), included 23 countries and a total of 10,795 participants. Currently, as part of the 7-year PARC programme (The European Partnership for the Assessment of Risks from Chemicals), PARC Aligned Studies will be performed between 2023-2026, targeting children (6-11 years), teenagers (12-17 years) and young adults (18-39 years).

We present what we have learned from these joint European efforts and how we can improve towards the PARC Aligned Studies and beyond.

472 – Community-Engaged Research with Social Media Influencer Networks to Increase the Environmental Health Literacy of Black Women

Elissia Franklin[1], Robin Dodson[2], Aleyana Momplaisir[3], Lilly Marcelin[3]
[1]Silent Spring Institution, Postdoctoral Research Fellow, Newton, USA. [2]Silent Spring Institution, Associate Director of Research Operations, Newton, USA. [3]Resilient Sisterhood Project, Program Coordinator, Boston, USA

Abstract
Black women are disproportionately exposed to many endocrine-disrupting chemicals (EDCs), which may contribute to health inequities such as more aggressive forms of breast cancer, infertility, uterine fibroids, and metabolic disorders. Use of consumer products is an important and modifiable source of EDC exposure. Access to reliable health information that promotes EDC reduction through knowledge about toxic chemicals in consumer products is limited for Black women due to historical systems of oppression. Our goal is to provide accurate and relevant educational material delivered in a high-impact way to support reducing EDC exposures that contribute to health risks experienced by Black women. We designed an educational campaign to reach Black women on social media, a virtual setting they already frequent. We worked with social media influencers to share information on EDC exposures using their voice and platforms. Of the 100,000 followers reached, many reported that the content increased their concern about chemicals in products and that they will start taking steps to limit their exposures. Our work demonstrated that by tapping into online communities with culturally sensitive and educationally appropriate messaging we can start conversations to address health issues with audiences who are most vulnerable. To our knowledge, this is the first environmental health project that leverages the power and reach of social media influencers in spreading health-related information with a focus on Black women's chemical exposures. We are now launching a follow-up program to assess the effectiveness of this innovative approach at supporting behavioral change and increasing environmental health literacy.

473 – Role and development of a multi-class urine assay for assessing human exposure to environmental chemical mixtures

Syam Andra[1], Ravikumar Jagani[2], Divya Pulivarthi[2], Dhavalkumar Patel[2,3], Rosalind Wright[1], Robert Wright[1], Manish Arora[1], Mary Wolff[1]
[1]Institute for Exposomic Research, Department of Environmental Medicine and Public Health, Icahn School of Medicine at Mount Sinai, Professor, New York City, USA. [2]Institute for Exposomic Research, Department of Environmental Medicine and Public Health, Icahn School of Medicine at Mount Sinai, Senior Associate Researcher, New York City, USA. [3]Texas Tech University Health Sciences Center, Mass Spec Core Facility Manager, Amarillo, USA

Abstract
Environmental exposures are complex chemical mixtures, with diverse properties and differences in metabolism. Often only trace level residues are found in humans. Thus, existing approaches are primarily single chemical panel assays that cover 3–30 biomarkers per assay, require a large biospecimen volume to assay, are expensive, and do not account for mixtures of multiple exposures. Epidemiological studies often have modest biospecimen volumes, making exposomic analysis difficult. To overcome these obstacles, multi-class techniques are gaining popularity. These methods optimize cleanup procedures for multiple classes of chemicals combined with instrumental setup and sensitivity for trace-level detection. In a single low volume urine aliquot (0.2 mL), we developed a multi-class assay for quantifying biomarkers of exposure to a mix of phenolic chemicals from consumer products, polycyclic aromatic hydrocarbons, organophosphate flame retardants, current use pesticides, volatile organic compounds, tobacco alkaloids, illicit drugs, and pharmaceuticals. This multi-class assay was validated in international proficiency testing programs. Method expansion was demonstrated by incorporating endogenous biomarkers of nutrients, as well as oxidative and psychosocial stress biomarkers. We now quantify >100 biomarkers from ten chemical classes in three to four panels, including nearly all NHANES urinary biomarkers, while reducing cost by one-third and sample volume by one-fifth. This exposome-scale multi-class assay is being used for population

studies, saving assay costs and biospecimen use while providing quantitative results and even uncovering unexpected exposures. Its application will improve our understanding of chemical mixture interactions and their role in human health.

474 – POSTER: A review of pediatric serum values for six per-and polyfluoroalkyl substances

Lee Blum
NMS Labs, Toxicologist, Horsham, USA

Abstract
Per-and polyfluoroalkyl substances (PFAS) are a class of synthetic chemicals widely used in industry and consumer products that have been associated with several adverse health effects including those affecting children. Over the past 5 years, approximately 415 pediatric serum samples from children up to 19 years of age with unknown exposure to PFAS were tested for six PFAS contaminants including perfluorobutanesulfonic acid (PFBS), perfluoroheptanoic acid (PFHpA), perfluorohexanesulfonic acid (PFHxS), perfluorononanoic acid (PFNA), linear perfluorooctanoic acid (n-PFOA), and linear perfluorooctancsulfonic acid (n-PFOS). These samples were assayed using HPLC separation with negative-ion electrospray tandem mass spectrometry (LC-MS/MS). The median and 95th percentile values, respectively, for PFBS were <0.050 and 0.055 ng/mL; for PFHpA <0.050 and 0.210 ng/mL; for PFHxS 0.750 and 2.68 ng/mL; for PFNA 0.260 and 0.967 ng/mL; for n-PFOA 0.820 and 2.60 ng/mL; and for n-PFOS 1.20 and 6.84 ng/mL. The samples were further grouped by reported ages as <2 yr (n=13), 3-5 yr (n=16), 6-11 yr (n=29) and 12-19 yr (n=357) then compared to serum values of age-related data from the National Health and Nutrition Examination Survey (NHANES). In the 3-5 yr group, the median and 95th percentile serum concentrations for PFHpA, PFHxS and n-PFOA exceeded the NHANES values while only higher 95th percentile concentrations were observed for PFNA and n-PFOS. The 6-11 yr group exhibited higher 95th percentile values for PFHpA, PFHxS, and n-PFOS. The findings demonstrate these samples have greater PFAS exposure than the general U.S. pediatric population.

475 – POSTER: Exposure pathway forensics for chemical source identification in non-targeted analysis

Kristin Isaacs[1], Jon Sobus[2], Seth Newton[2], John Sloop[3], Katherine Philiips[1]
[1]US EPA, Research Physical Scientist, Research Triangle Park, USA. [2]US EPA, Physical Scientist, Research Triangle Park, USA. [3]US EPA, Postdoctoral Fellow, Research Triangle Park, USA

Abstract
Non-targeted analysis (NTA) technologies and supporting data analysis platforms hold great potential for identifying thousands of chemicals in environmental and biological samples. However, challenges in linking identified chemical substances with upstream exogenous exposure sources must be addressed. Here, a method was developed for exploring exposure pathways that integrated available chemical use and monitoring data with mining of chemical co-occurrence patterns within NTA samples. A fingerprinting approach was developed that encoded available exposure pathway information from U.S. Environmental Protection Agency (EPA) Office of Research and Development chemical use and monitoring databases as hierarchical binary descriptor sets. These descriptors can be queried for enrichment for subsets of chemicals co-occurring in biological or environmental samples using NTA to provide evidence of exposure source. Descriptors represented occurrence of chemicals in commercial, industrial, consumer, and functional categories, as well as known occurrence in the environment. In a proof-of-concept case study, an enrichment analysis was applied to a set of chemicals tentatively identified using NTA in 46 geospatially diverse drinking water samples collected in California. A clustering approach was used to identify chemical groups co-occurring across subsets of multiple samples. The approach identified unique chemical clusters reflecting pharmaceutical, manufacturing, and consumer uses. These use patterns may be indicative of unique sources impacting specific sampling locations. However, use information was missing for 75% of the identified chemicals, highlighting the need for improved use data and annotation of degradants of commercial chemicals. These approaches have the potential to inform exposure source identification and mitigation.

476 – Analysis of Head and Beard Hair Reveals Chemical Exposome Trajectory over a 10-Year Period

Brice M.R. Appenzeller[1], Achilleas Pexaras[2], Paul Palazzi[3], Alba Iglesias Gonzalez[4], FengJiao Peng[4], Maria Ruiz Castell[1]
[1]Luxembourg Institute of Health, Head of Unit, Strassen, Luxembourg. [2]Luxembourg Institute of Health, Technician, Strassen, Luxembourg. [3]Luxembourg Institute of Health, Engineer, Strassen, Luxembourg. [4]Luxembourg Institute of Health, PostDoc, Strassen, Luxembourg

Abstract

Hair gives access to extended time windows and is well adapted to chronic exposure assessment. However, in most studies, the windows usually do not exceed 12 months, and no study investigated longer periods. Moreover, no hair from other body part (e.g. beard) has been used to analyze pollutants. This study investigated the chemical exposome trajectory of a subject over 10 years by means of repeated head and beard hair samples.

Head hair samples were collected from a single subject (male, age 36 y at t0) on 48 different occasions (1-7 per year) over 10 years. In parallel, 19 beard hair samples were collected for matched comparison. Samples were analyzed for 152 different pollutants/metabolites. The variability of the subject's exposure over time was compared to a cohort of 450 adults living in the same region.

The number of chemicals detected in head and beard hair was 42 and 39 respectively. Chemicals concentration was highly reproducible over time: in the single subject, the intra-individual variability observed for pentachlorophenol, 3-phenoxybenzoic acid, and PCB 180 was 111%, 65%, 48% whereas inter-individual variability in the cohort was 739%, 462%, 440% for the same compounds. Most chemicals presented significant association between head and beard hair (e.g. RPearson=0.83, $p<0.001$ for PCB 180).

The study suggests that information obtained from hair analysis is representative of chronic exposure over several years. Hair would thus be suitable to investigate health outcomes associated with past exposure. In the absence of hair, beard hair might represent an interesting surrogate.

477 – Challenges and Successes Conducting New Jersey's First Statewide Biomonitoring Population Study

Elisabeth Cook[1], Emily Lax[2,3], Colleen Donohoe[4,5], Chang Ho Yu[6], Eric Bind[7], Zhihua (Tina) Fan[8]
[1]New Jersey Department of Health, Research Scientist, Ewing, USA. [2]New Jersey Department of Health, CDC PHAP Associate, Ewing, USA. [3]Centers for Disease Control and Prvention, PHAP Associate, Ewing, USA. [4]New Jersey Department of Health, APHL-CDC Fellow, Ewing, USA. [5]APHL, Fellow, Ewing, USA. [6]New Jersey Department of Health, Program Manager, Data Management & Analysis, Ewing, USA. [7]New Jersey Department of Health, Program Manager, Biomonitoring, Ewing, USA. [8]New Jersey Department of Health, Director, Environmental and Chemical Laboratory Services, Ewing, USA

Abstract
The NJ Health and Nutrition Examination Survey (NJHANES) is the State's first population-based biomonitoring study. NJHANES aims to collect exposure data for ~120 environmental contaminants combined with health and nutritional data for NJ residents using questionnaires and specimen collection to identify factors that may impact health and exposure. NJHANES initially randomly selected 4,800 NJ households through a two-stage cluster sampling frame using mailing recruitment to target 500 residents. REDCap software, MSTeams, and encrypted email are utilized for questionnaires, consent calls, and delivering participant results to address social distant challenges during the COVID-19 pandemic. Successes include accommodations for non-English speaking respondents, accessibility for those without technology, and implementing a socially distant enrollment process.

However, recruitment challenges included a low response rate, respondents skewed towards highly educated and high-income households, and many returned mailings. To address the gaps, recruitment strategies were modified as follows: a reminder postcard was introduced after the initial recruitment mailing, pre-verifying mailing addresses, added text to the recruitment envelope, added a second round of household selection (n=4,800), and doubled the incentive to $100.
Since these changes were implemented, responses per month increased approximately 4X (5 to 20) and respondent education level shifted closer to U.S. Census data for NJ, while respondent income level remained unchanged.

Future plans include further validating addresses before mailings, strengthening community outreach with local health departments, adding a post-sampling survey to examine factors that impact response rate, and increasing public awareness for NJHANES and biomonitoring. Presentation will address other planning and implementation challenges.

478 – Trimester-Specific Associations Between Gestational Exposure to Low Level Air Pollution and HbA1c

Yan Lin[1], Ruoxue Chen[2], Yihui Ge[2], Philip Hopke[3], Timothy Stevens[3], Sally Thurston[3], Thomas O'Connor[3], David Rich[3], Junfeng Zhang[4]
[1]Duke University, Postdoc, Durham, USA. [2]Duke University, PhD student, Durham, USA. [3]University of Rochester, Professor, Rochester, USA. [4]Duke University, Professor, Durham, USA

Abstract
Emerging evidence suggests that air pollution exposure may increase the risk of gestational diabetes. This study aimed to determine to

what extent exposure to low level air pollution increased HbA1c during pregnancy and identify the sensitive time window. We measured HbA1c in 608 blood samples collected from 237 pregnant women in Rochester, NY (≤3 samples per woman) and calculated residential exposure to PM2.5 and NO2 based on high-resolution spatiotemporal models. We used a generalized additive model to examine the associations between HbA1c and air pollution exposure, adjusting for non-linear effects of post-conceptional age. The median (IQR) 9-week concentrations of PM2.5 and NO2 were 6.87 (5.92 – 7.65) μg/m3 and 10.7 (6.1 – 14.0) μg/m3, respectively. The median (IQR) Hbalc level was 5.18 (4.86 – 5.54) % with a U-shaped trajectory during pregnancy. Compared with the first trimester, HbA1c decreased by 0.14% (95%CI: 0.06% to 0.19%) in the second trimester and increased by 0.31% (95%CI: 0.22% to 0.40%) in the third trimester. One IQR increase in 9-week PM2.5 concentration during the first trimester was associated with 0.077% (95%CI: 0.002% to 0.153%) increases in Hbalc. One IQR increase in 9-week NO2 concentration during the third trimester was associated with 0.156% (95%CI: 0.057% to 0.256%) increases in Hbalc. These results indicated that gestational exposure to PM2.5 and NO2 even at low levels was associated with impaired glucose metabolism in pregnant women. The first trimester was the sensitive window for PM2.5 exposure while the third trimester was the sensitive window for NO2 exposure.

479 – Creating environmental education curriculum for elementary, middle, and high school students in Atlanta with a focus on air quality and soil lead contamination

Eri Saikawa[1], Alexander Avramov[2]
[1]Emory University, Associate Professor, Atlanta, USA. [2]Emory University, Associate Scientist, Atlanta, USA
Abstract
This project aims to build on and expand our current project to educate the elementary, middle and high school students, and the community members, in Georgia by providing unique active-learning curriculum on topics related to the environment. For air quality management, we have validated low-cost air quality sensors and created an air quality monitoring network by placing them on a dozen elementary, middle, and high school campuses in Atlanta and surrounding areas in Georgia. The sensors were validated by co-locating them with regulatory monitors. All the data are available to students and teachers via semi-annual reports that include both raw and bias-corrected values. For assessing potential soil lead contamination, we have conducted soil lead screenings, called community science soil screening, health, outreach and partnership (soilSHOPs) and developed a volunteer soil lead monitoring network among high school students to educate participants about the impacts of soil contamination on human health. We also hosted educational garden classes for young children at community gardens and worked on promoting reducing food waste and increasing composting in schools and in homes. We will also share the experience of hosting a virtual Hackathon event for middle school and high school students to work on environmental health issues.

480 – Water? Food? Shelter? What about air? The San Joaquin Valley Center for Air Injustice Reduction (SJV-CAIR)

Kate DeMarsh[1], Kimberly Valle[2], Tim Tyner[3], Jermaine Reece[4], Estrella Herrera[5], Matt Holmes[6], Alec Chan-Golston[7], Sandie Ha[7], Sidra Goldman-Mellor[8], Trevor P. Hirst[9], Asa Bradman[10]
[1]University of California, Merced, Graduate Student, Merced, USA. [2]University of California, Merced, PhD Candidate, Merced, USA. [3]Central California Asthma Collaborative, Co-Director, Fresno, USA. [4]Little Manila Rising, Operations and Air Quality Manager, Stockton, USA. [5]University of California, Merced, Postdoctoral Scholar, Merced, USA. [6]Little Manila Rising, Environmental Justice Outreach Coordinator, Stockton, USA. [7]University of California, Merced, Assistant Professor, Merced, USA. [8]University of California, Merced, Associate Professor, Merced, USA. [9]University of California, Merced, Executive Director, Health Science Research Institute, Merced, USA. [10]University of California, Merced, Professor, Merced, USA

Abstract
The San Joaquin Valley (SJV) is a broad, inland California valley that extends south from Stockton to Bakersfield, CA and is bounded by the Sierra Nevada mountains to the east and coastal ranges to the west. This topography, combined with local and regional pollution sources, creates an environment for air pollution formation and retention. Air quality in the SJV often fails to meet Environmental Protection (EPA) Standards for Particulate Matter (PM) 2.5. PM sources in the SJV include agricultural production and transportation, power generation, industry, and traffic-related air pollutants (TRAP) along the Interstate 5 and CA-99 corridors.

The San Joaquin Valley Center for Air Injustice Reduction (SJV-CAIR) at the University of California, Merced is partnering with community-based organizations to better understand air pollution in the SJV. Specifically, we are deploying a large network of air quality monitors in Fresno, Stockton, and neighboring areas to increase citizen-science based air monitoring to better characterize air pollution trends and spatial variability. We are currently examining the relationship between regulatory and citizen-science monitors and also evaluating the association between PM exposure and hospital emergency department visits for respiratory and other health outcomes. To date, we have demonstrated improved characterization of spatial variability in PM exposure using PM 2.5 interpolation maps using (i) only regulatory monitors, (ii) existing citizen science monitors, and (iii) the new SJV-CAIR monitoring network. Our ultimate goal is to better assess TRAP exposures and health impacts to inform air pollution mitigation policies in some of CA's most disadvantaged communities.

481 – Association between Prenatal Exposure to Neurotoxic Chemicals and Neurodevelopment of Children: Application of Multi-target Prediction Methods

Janja Snoj Tratnik[1,2], Martin Breskvar[1], Darja Mazej[1], Milena Horvat[3,4], David Neubauer[5], Zdravko Špirić[6], Igor Prpić[7], Sašo Džeroski[8,4]
[1]Jožef Stefan Institute, Research Associate, Ljubljana, Slovenia. [2]Jožef Stefan International Postgraduate School, Research Associate, Ljubljana, Slovenia. [3]Jožef Stefan Institute, Head of Department of Environmental Sciences, Ljubljana, Slovenia. [4]Jožef Stefan International Postgraduate School, Professor, Ljubljana, Slovenia. [5]University Medical Centre Ljubljana, Professor, Ljubljana, Slovenia. [6]Green Infrastructure Ltd., Professor, Zagreb, Croatia. [7]Clinical Hospital Centre Rijeka, Professor, Rijeka, Croatia. [8]Jožef Stefan Institute, Head of Department of Knowledge Technologies, Ljubljana, Slovenia

Abstract
We have applied machine learning algorithms with an aim to explore association(s) between potentially neurotoxic pollutants (mercury, lead, arsenic, manganese, organophosphorous pesticides) micronutrients, socio-economic, health- or life-style related features in prenatal, early life and childhood and neuropsychological development. The following data has been used from the existing Mediterranean longitudinal birth cohort study including mother-child pairs from Slovenia and Croatia: life-style; concentration of elements in biological samples; Bayley-III scores (18 months) and WISC-IV scores (7-8 years of age). We have applied single- and multi-target (semi-) supervised machine learning methods to build predictive models for the neurodevelopmental scores. Additionally, we have used feature ranking methods to estimate the importance of individual attributes on the predicted scores. Overall, the trees, rules and ranking produced by the machine learning approaches demonstrate a complex relationship between environmental chemicals, nutrients, SES, and life-style. In addition to previously established negative associations between mercury and fine motor development, some additional features have been uncovered that seem to condition the response to environmental chemical stress. Among those, the most evident are Zn with beneficial effect at higher levels and Cu with the opposite effect, certain socio-economic and body weight related attributes, and (to a lesser extent) also co-exposure to Pb, Cd and As, and status of micronutrient Se. The use of machine learning algorithms is a novelty in this research field. Although certain relationships that were observed remain to be explained, the overall observation of small effects of chemicals seem to be masked or compensated by various factors.

482 – Improving Pesticide Handler Safety by Integrating Risk Assessment, Training, and Stewardship

Steve McEuen[1], Mark Best[2], Udo Blaschke[3], Philip Fisher[4], Carrie Fleming[5], Suzy Gao[6], Marcela Giachini[7], Nevena Hristozova[8], Daniele Lautenschalaeger[9], Wibke Meyer[10], Tharacad Ramanarayanan[11], Natalie Shevchuv[12], Jennifer Thomasen[13], Andrew Ward[14], Christiane Wiemann[15]
[1]FMC Corporation, Assoc. Global R&D Fellow, Human Safety, Newark, Delaware, USA. [2]Syngenta Crop Protection, Senior Technical Expert / Operator Safety Assessment EAME - Team Leader, Bracknell, United Kingdom. [3]BASF SE, Agricultural Solutions, Team Leader Global OPEX, Ludwigshafen, Germany. [4]Bayer SAS, Crop Science Division, Occupational and Residential Exposure Expert, Sophia Antipolis, France. [5]Corveta, Senior Toxicologist and Risk Assessor, Indianapolis, USA. [6]BASF China Company Ltd, Agricultural Solutions, APAC Regional Operator Exposure Scientist, Beijing, China. [7]Corteva, Global Risk Assessment SME, Sao Paulo, Brazil. [8]CropLife International, Regulatory Affairs Coordinator, Brussels, Belgium. [9]Bayer Crop Science do Brasil Ltda, Human Safety Manager, Sao Paulo, Brazil. [10]CropLife International, Director, Regulator Affairs, Brussels, Belgium. [11]Syngenta Crop Protection, LLC, Syngenta Fellow, Greensboro, USA. [12]FMC Corporation, Global Regulatory Affairs Director, Philadelphia, USA. [13]Bayer, Crop Science Division, Occupational and Residential Exposure Lead, North America, Chesterfield, USA. [14]CropLife International, Director Of Stewardship, Brussels, Belgium. [15]BASF Oesterreich GmbH, Agricultural Solutions, Senior Toxicologist and Senior Human Risk Assessor, Vienna, Austria

Abstract
CropLife International (CLI) and its member companies have been promoting risk assessment, product stewardship, and training to protect the health of agricultural workers and pesticide handlers. Potential exposure through dermal and inhalation routes, determined from statistical models developed from empirical worker exposure data is an important aspect of risk assessment. CropLife International has developed an Operator Exposure (OPEX) tool based on established methodologies as a part of the Principles of Regulation project. This tool, designed to be used by countries and regions without the necessary resources to develop the data and tools for exposure assessment, was deployed in 2014. Since then, the FAO has been recommending the CropLife International OPEX and European Food Safety Authority tools for countries that have limited resources to develop data and models. Now CropLife International is engaging with

ICPPE-LMIC, an initiative coordinated by the International Center for Personal Protective Equipment to develop risk assessment and mitigation measures for operators in Low- and Middle-Income Countries, to incorporate more application scenarios and newer, more comprehensive data to build upon the success of the CLI-OPEX tool. CropLife International has also invested significantly in Stewardship programs and Training to raise awareness on relevant factors with the aim to reduce exposure. Without these measures on good agricultural practices for handling pesticides, sophisticated regulations and risk assessment may result in less effective pesticide handler health protection. This presentation will outline recent and on-going CropLife International initiatives and collaborations in these areas of agricultural worker health protection.

483 – A Community Study of Drinking Water Supply and Gastroenteritis in Puerto Rico: Application of a Salivary Immunoassay for Detection of Waterborne Infection

Shannon Griffin[1], Melitza Crespo-Medina[2,3], Cristina Maldonado[4], Timothy Wade[5], Andrey Egorov[5], Graciela Ramirez-Toro[6,7]
[1]US EPA, Biologist, Cincinnati, USA. [2]Inter American University of Puerto Rico, Associate Professor, San German, Puerto Rico. [3]Center for Environmental Education, Conservation and Research, Researcher, San German, Puerto Rico. [4]US EPA, Environmental Scientist, San Juan, Puerto Rico. [5]US EPA, Health Scientist, Chapel Hill, USA. [6]Center for Environmental Education, Conservation and Research, Institutional Director, San German, Puerto Rico. [7]Inter American University of Puerto Rico, Professor, San German, Puerto Rico

Abstract
Waterborne illness presents public health and economic challenges to communities with limited resources and impaired infrastructure. In Puerto Rico (PR), approximately one third of the 240 drinking water systems not served by the PR Aqueduct and Sewer Authority (PRASA) offer little to no treatment. An ongoing prospective intervention study in three non-PRASA communities aims to evaluate public health benefits of low-cost, sustainable water treatment technologies by comparing rates of infections before and after introduction of filtration and reliable disinfection. This study involves the detection of pathogens in stool samples obtained from users of the small public water supply systems, using the Luminex Gastrointestinal Pathogen Panel kit, and analysis of saliva samples for antibody responses to waterborne pathogens, including noroviruses and Cryptosporidium, using an in-house multiplex Luminex immunoassay. Salivary antibody testing is a non-invasive, easy-to-collect, and inexpensive alternative to serological testing. In the pre-intervention phase, 97 stool and 90 saliva samples were collected from community members along with information regarding self-report of gastroenteritis symptoms. Preliminary results show that nine (9.3%) participants shed pathogens in stool, including enterotoxigenic E. coli, Salmonella, Shigella, adenoviruses, noroviruses, Cryptosporidium, Giardia, and Entamoeba histolytica. Additionally, there was a significant association between self-reported diarrhea and strong (>90th percentile) IgG antibody responses to GII.9 noroviruses (OR: 6.54, 95%CI: 1.55 – 27.67, p=0.01). The post-intervention phase is expected to begin in summer 2023. Results of this work help to prioritize risks and inform decisions regarding water supply and treatment that may improve the health and welfare of non-PRASA communities in PR.

484 – POSTER: Occupational Exposure to Particulate Matter Among Metal Sculptors in Chicago

Marcos Gonzalez[1], Hamed Hamza[1], Yuan Shao[2], Linda Forst[3]
[1]University of Illinois at Chicago, Graduate Student, Chicago, USA. [2]University of Illinois at Chicago, Assistant Professor, Chicago, USA. [3]University of Illinois at Chicago, Professor, Chicago, USA

Abstract
Metal sculptors are those who shape, form, and expose the hidden nature of metals: an inanimate object is brought to life using tools of creation such as forging, grinding, and welding. The art piece is subject to all forms of manipulation to reach its achieved state; metal processes include grinding for polish, chemical exposure and etching for patinas, smashing and milling for shapes, abrading for texture, and heating to conjoin or deconstruct pieces. These techniques and processes generate metal dust, fine particles, and fumes detrimental to human health. Little research has been conducted on artists' exposure to such environments. To address this gap, we conducted indoor air monitoring in three Chicagoland metal sculptor studios. Indoor air quality (IAQ) and particulate matter ($PM_{2.5}$, PM_4, and PM_{10}) concentrations were monitored real-time; particle diameter and size were also characterized. Our IAQ data showed a range of daily averages of 0 ppm, 563 – 776 ppm, 13.0 – 15.7°C, and 22.3 – 31.3% for carbon monoxide (CO), carbon dioxide (CO_2), temperature (C°) and relative humidity (%) respectively. PMs had a daily mean exposure range from 1,200 – 6,500 μg/m³ for $PM_{2.5}$, 1,300 – 7,300 μg/m³ for PM_4, and 1,700 – 11,380 μg/m³ for PM_{10}. 76-96% of the particle sizes were bound in the range 0.3 – 0.579 μm. Artist studios are

unregulated and understudied, often not considered workplaces with occupational/environmental health hazards. Further studies of private metal sculpting studios are needed to determine which practices pose the greatest risk.

485 – Advances in Sample Preparation and Analyses when using Wristband Personal Passive Samplers to Support Research on the Exposome

Nicholas Herkert[1], Duncan Hay[2], Taylor Hoxie[3], Heather Stapleton[4]
[1]Duke University, Research Project Manager, Durham, USA. [2]Duke University, Research Technician II, Durham, USA. [3]Duke University, PhD Candidate, Durham, USA. [4]Duke University, Ronie-Richele Garcia-Johnson Distinguished Professor, Durham, USA

Abstract
Silicone wristbands are increasingly being used as an exposure tool for monitoring an individual's personal exposure to chemicals in the ambient environment. Wristbands have been shown to overcome some of the limitations on relying upon biological samples to evaluate exposures, and previous research by our group has shown that wristband measurements are correlated with internal dose, sample a wide range of chemical contaminants, and are stable for long periods of time after sampling. However, the analysis can be challenging due to the complexity of the sampler and advances are needed for greater throughput and data quality.

To address limitations, we developed a new heat-based pretreatment of wristbands utilizing an Entech 3700 Thermal Conditioner. The heat pretreatment replaces solvent cleaning and saves on cost and time while producing wristbands of the same high quality. Secondly, we have developed a tin-based storage system, replacing a foil wrapped storage system, that cuts down on preparation time and participant biases.

To improve sample processing throughput, we have developed a dispersed Florisil cleanup protocol that replaces a large Florisil packed glass column approach and allows us to perform both targeted analyses on 120 different chemicals and suspect screening analyses in the same run. Additionally, we have developed new protocols for quantification of 14 volatile per- and polyfluorinated alkyl substances (PFAS) via positive chemical ionization mode on a Thermo Fisher Orbitrap GC-HRMS. GC Method optimization reduced the detection limits for PFAS to ~20 pg/g wristband, a 2 order of magnitude reduction for some analytes.

486 – Disparities in urban greenness and epigenetic aging

Kyeezu Kim[1], Brian Joyce[2], Drew Nannini[1], Yinan Zheng[3], Penny Gordon-Larsen[4], James Shikany[5], Donald Lloyd-Jones[6], Ming Hu[7], Mark Nieuwenhuijsen[8], Douglas Vaughan[1], Kai Zhang[9], Lifang Hou[10]
[1]Northwestern University Feinberg School of Medicine, Postdoctoral Fellow, Chicago, USA. [2]Northwestern University Feinberg School of Medicine, Research Assistant Professor, Chicago, USA. [3]Northwestern University Feinberg School of Medicine, Assistant Professor, Chicago, USA. [4]University of North Carolina, Professor, Chaple Hill, USA. [5]University of Alabama, Professor, Birmingham, USA. [6]Northwestern University Feinberg School of Medicine, Chair, Department of Preventive Medicine, Chicago, USA. [7]University of Maryland, Associate Professor, College Park, USA. [8]ISGlobal, Director, Urban Planning, Environment and Health Initiative, Barcelona, Spain. [9]University of Albany, Empire Innovation Associate Professor, School of Public Health, Albany, USA. [10]Northwestern University Feinberg School of Medicine, Director, Robert J. Havey, MD Institute for Global Health, Chicago, USA

Abstract
Background: Emerging evidence links epigenetic age, a novel biological aging biomarker, with environmental/social determinants and aging health. However, the longitudinal associations between greenness and epigenetic age have not been well-studied, particularly among minority groups.

Methods: We included 924 participants from a biracial (Black and White races) cohort in four US cities, the Coronary Artery Risk Development in Young Adults (CARDIA) study. We estimated surrounding greenness with the Normalized Difference Vegetation Index (NDVI) around participants' residential addresses from baseline (1985-1986) to exam year (Y) 20 (2005-2006). We derived DNA methylation-based epigenetic age acceleration (EAA) estimates at Y20. Associations of long-term exposures to greenness with EAA were assessed using generalized estimating equation regression models adjusted for individual and neighborhood socioeconomic status (SES). The role of self-reported race, sex, and neighborhood SES was evaluated through stratification analysis.

Results: Longitudinally, each 10%-point increment of NDVI within 5 km (Y0-Y20) was associated with 2.64-year [β: -2.64, 95% confidence interval [CI]: -5.04, -0.24] lower EAA. Black participants had a lower amount of greenness and fewer advantages of greenness in association with lower EAA [β: -0.80, 95% CI: -4.75, 3.13 vs. β: -3.03, 95% CI: -5.63, -0.43 for White participants]. Women exhibited stronger associations than men. Participants with lower neighborhood SES showed stronger associations compared with those with higher neighborhood SES.

Conclusions: Greater exposure to greenness was associated with slower epigenetic aging. Our results support the hypothesis that expanding green infrastructure may promote healthy biologic aging and reduce health disparities among disadvantaged communities.

487 – POSTER: Portable Sensor Array for Monitoring of Exposure to Wildfire Smoke

Tamara Russ[1], Joseph Stetter[2], David Peaslee[3], Mel Findlay[4], Vinay Patel[5], Bennett Meulendyk[6]
[1]Interlink Electronics Inc., Sensor Scientist, Newark, USA. [2]Interlink Electronics Inc., Technical Fellow, Newark, USA. [3]Interlink Electronics Inc., Senior Physicist, Newark, USA. [4]Interlink Electronics Inc., Senior Sensor Scientist, Newark, USA. [5]Interlink Electronics Inc., Senior Sensor Engineer, Newark, USA. [6]Interlink Electronics Inc., Senior Electrical Engineer, Newark, USA

Abstract
The human health impact of smoke on air quality (AQ) is growing each year. Studies have found that annually approximately 400,000 people die due to poor AQ caused by wildfire smoke exposure. Wildfire smoke is a complex mixture of thousands of different components including particulate matter, carbon monoxide, sulfur dioxide, nitrogen dioxide, ozone and VOCs. We present here a low-cost, low-power and portable sensor array of four electrochemical sensors, a low-cost particulate matter sensor for particulates smaller than 2.5 μm, and a temperate, pressure and humidity sensor. The sensor array data was compared to EPA AQ stations data in the SF Bay area for smoke pollutants over months of monitoring to assess the potential for widespread quantification of smoke exposure. This low-cost AQ array can be used to provide accurate AQ data in regional gaps where current AQ systems are lacking precision (satellite-based systems) or are simply absent (in between EPA air quality monitoring stations). The AQ sensor arrays are >100 times less expensive than current EPA-type AQ stations and therefore suitable for deployment in widespread networks to monitor AQ in smoke impacted areas. Wildfire smoke travels hundreds of kilometers from the actual burn site and impacts populations that often are not even aware of the airborne dangers nor near EPA AQ stations for assessment. The data from the low-cost sensor array is evaluated quantitatively as part of a distributed portable AQ monitoring system that can quantitate the impact of wildfire smoke as well as potentially improve existing exposure models.

488 – Can biomarkers improve COPD diagnosis?

Romain FREUND[1], Jean-Jacques Sauvain[1], Guillaume Suarez[1], Pascal Wild[1], Thomas Charreau[1], Amelie Debatisse[2], Kirushanthi Sakthithasan[2], Valérie Jouannique[2], Jacques A Pralong[3,4,5], Irina Guseva Canu[1]
[1]Center from primary care and public health (Unisanté), University of Lausanne, Lausanne, Switzerland. [2]Autonomous Paris Transport Authority (RATP), Paris, France. [3]Faculty of Medicine, University of Geneva, Geneva, Switzerland. [4]SwissMedPro Health Services, Geneva, Switzerland. [5]Hôpital de la Tour, Geneva, Switzerland

Abstract
Background: Chronic obstructive pulmonary disease (COPD) affecting 334 million people in the world remains a major cause of morbidity and mortality. Proper diagnosis of COPD is still a challenge and largely solely based on spirometric criteria. We aimed to investigate the potential of nitrosative/oxidative stress and related metabolic biomarkers in exhaled breath condensate (EBC) to discriminate COPD patients.

Methods: 303 participants were randomly selected from a 15,000-transit worker cohort within the Respiratory disease Occupational Biomonitoring Collaborative Project (ROBoCoP). COPD was defined using the Global Initiative for Chronic Obstructive Lung Disease (GOLD) criteria as post-bronchodilator ratio of Forced Expiratory Volume in 1st second to Forced Vital Capacity <0.7 in spirometry validated by an experienced pulmonologist. Discriminative power of biomarkers profiles in EBC was analyzed using linear discriminant analyses.

Results: Amongst 300 participants with validated spirometry, 50.3% were male, 52.3 years old in average, 36.0% were current smokers, 12.7% ex-smokers with mean tobacco exposure of 5.3 pack-years. Twenty-one participants (7.0%) were diagnosed as COPD, including 19 new diagnoses, 12 of which with a mild COPD stage (GOLD 1). Amongst 8 biomarkers measured in EBC, combination of 2 biomarkers, Lactate and MDA significantly discriminated COPD subjects from non-COPD, with a 71%-accuracy, area under the receiver curve of 0.78 (p-value<0.001), and a negative predictive value of 96%.

Conclusion: These findings support the ability of biomarkers in EBC to discriminate COPD patients even at an early stage. These biomarkers from EBC, a non-invasive and drugless technique can improve early CODP diagnosis in the future.

489 – Species differences in paralytic shellfish poisoning (PSP) toxin levels across Southeast Alaska

Matthew Gribble[1], Baylin Bennett[2], Christopher Whitehead[3], Esther Kennedy[4], Shannon Cellan[5], Kari Lanphier[6], Hugh Roland[7], Jeffrey Wickliffe[8], Tod Leighfield[9], Jeff Feldpausch[10], John Harley[11]
[1]University of Alabama at Birmingham, Associate Professor, Birmingham, USA. [2]University of Alabama at Birmingham, Research Assistant, Birmingham, USA. [3]Sitka Tribe of Alaska, Environmental Program Manager, Sitka, USA. [4]University of California, Davis, Graduate Student, Davis, USA. [5]Sitka Tribe of Alaska, Environmental Lab Manager, Sitka, USA. [6]Sitka Tribe of Alaska, Environmental Project Manager and Tribal Consortium Coordinator, Sitka, USA. [7]University of Alabama at Birmingham, Postdoctoral Trainee, Birmingham, USA. [8]University of Alabama at Birmingham, Professor/Chair, Birmingham, USA. [9]National Oceanic and Atmospheric Association, Chemist, Charleston, USA. [10]Sitka Tribe of Alaska, Resource Protection Director, Sitka, USA. [11]University of Alaska Southeast, Research Assistant Professor, Juneau, USA

Abstract
Paralytic shellfish poisoning (PSP) toxin levels drastically differ among shellfish species in Southeast Alaska as well as across the year. This study used data from routine shellfish monitoring samples and ad hoc community-provided food samples collected 2015 – 2022. The samples were analyzed for PSP toxin (saxitoxin equivalents) using a receptor-binding assay by the Sitka Tribe of Alaska Environmental Research Lab. Using conditional logistic regression for matched comparisons of shellfish collected from the same sites, we estimated the conditional odds ratio of shellfish toxins exceeding the food safety standard (80 micrograms saxitoxin equivalents per 100 g shellfish tissue) comparing butter clams (Saxidomus gigantea) to littleneck clams (Leukoma staminea) was 125.86 (95 confidence interval [CI]: 50.63, 312.85); for cockles (Clinocardium nuttalli) vs. littleneck clams was 7.18 (95% CI: 2.02, 25.57] and for mussels (Mytilus edulis) vs. littleneck clams was 3.72 (95% CI: 1.72, 8.05). The geometric mean ratio (GMR) of continuous toxin levels for butter clams vs. littleneck clams estimated from a multilevel Tobit regression of log-transformed toxin levels, accounting for a left-censoring limit of ln(4) saxitoxin equivalents, and conditional on-site random effects, was 28.05 (95% CI 19.65, 40.05). The conditional GMR for mussels vs. littleneck clams was 3.64 (95% CI 2.29, 5.79) and for cockles vs. littleneck clams was 3.68 (95% CI 2.22, 6.10). Thus, littleneck clams may pose substantially lower toxin exposure risks than three other shellfish types in Southeast Alaska, and butter claims may confer the highest risks of PSP toxin exposure among these four species.

490 – Neighborhood-Scale Air Quality and Health Impacts Associated with Electrifying Heavy-Duty Vehicles in the U.S. Midwest

Sara F. Camilleri[1], Anastasia Montgomery[2], Maxime A. Visa[3], Jordan L. Schnell[1,4], Zachariah E. Adelman[5], Mark Janssen[6], Emily A. Grubert[7], Susan C. Anenberg[8], Maria H. Harris[9], Daniel E. Horton[10]
[1]Northwestern University, Postdoctoral Scholar, Evanston, USA. [2]Northwestern University, PhD Candidate, Evanston, USA. [3]Northwestern University, Undergraduate Researcher, Evanston, USA. [4]NOAA/Global Systems Laboratory, Research Associate, Boulder, USA. [5]LADCO, Executive Director, Chicago, USA. [6]LADCO, Emissions Director, Chicago, USA. [7]University of Notre Dame, Associate Professor, Notre Dame, USA. [8]George Washington University, Associate Professor, Washington, USA. [9]Environmental Defense Fund, Senior Scientist, San Francisco, USA. [10]Northwestern University, Assistant Professor, Evanston, USA
Abstract
Heavy duty vehicles (HDVs) constitute ~6% of total on-road vehicles in the U.S. but are responsible for a much larger proportion of greenhouse gases such as CO2 as well as primary and secondary pollutants such as PM2.5, NOX and O3, with 43% of traffic-related PM2.5 and O3 attributable mortality linked to on-road diesel pollution. Air pollution from HDVs also disproportionately impacts racial and ethnic minority populations living in close proximity to truck routes. Shifting to electric HDVs (eHDVs) is an ongoing and accelerating climate mitigation action, however, the associated air quality and health implications are rarely modeled at equity-relevant geographic scales. In this work we use the two-way coupled Weather Research Forecast and Community Multiscale Air Quality (WRF-CMAQ) chemical transport model at 1.3 km to determine changes in simulated pollutant concentrations and the associated census-tract level health impacts in two electrification scenarios over a central Midwestern domain: (i) a decarbonization mid-transition scenario wherein we instantaneously electrify 30% of on-road HDVs and (ii) a best case scenario of 100% eHDVs with additional electricity demand met by emission-free electricity generating units (EGUs). Electrifying 30% HDVs decreases on-road NO2 and PM2.5 concentrations, outweighing increases at EGUs but increases daily maximum 8-hr running mean O3 (MDA8O3), especially in urban settings. Results suggest overall health benefits largely driven by NO2 reductions. In addition, we note relatively larger air quality and health co-benefits within communities with higher fractions of Black and Hispanic/Latino residents, thus highlighting the potential for eHDVs to help address long-standing environmental injustices.

491 – Description of children's chemical exposome and possible determinants of exposure in the region of Luxembourg based on hair analysis

Alba Iglesias-González[1], Mylene Schweitzer[2], Paul Palazzi[3], Fengjiao Peng[1], Serge Haan[4], Elisabeth Letellier[5], Brice M.R. Appenzeller[6]
[1]Human Biomonitoring Research Unit, Department of Precision Health, Luxembourg Institute of Health, Postdoc, Strassen, Luxembourg. [2]Human Biomonitoring Research Unit, Department of Precision Health, Luxembourg Institute of Health, Student, Strassen, Luxembourg. [3]Human Biomonitoring Research Unit, Department of Precision Health, Luxembourg Institute of Health, Research Engineer, Strassen, Luxembourg. [4]Molecular Disease Mechanisms Group, Department of Life Sciences and Medicine, Faculty of Science, Technology and Medicine, University of Luxembourg, Professor, Belvaux, Luxembourg. [5]Molecular Disease Mechanisms Group, Department of Life Sciences and Medicine, Faculty of Science, Technology and Medicine, University of Luxembourg, Assistant professor, Belvaux, Luxembourg. [6]Human Biomonitoring Research Unit, Department of Precision Health, Luxembourg Institute of Health, Head of unit, Strassen, Luxembourg

Abstract
Chemical exposure during childhood is associated with different pathologies, and a quarter of all deaths of children below 5y/o are caused by exposure to unhealthy environment. Children's specific physiology and behaviour, which differ from adults, make them particularly vulnerable to chemical exposure. Therefore, children exposure cannot be extrapolated from data obtained in adults and specific studies must be conducted on children. In this study, hair samples were collected from 264 children below 13y/o inhabitants of Luxembourg to investigate their chemical exposome and determinants of exposure.

Each sample was analysed with a multi-residue method targeting 153 biomarkers (140 pesticides, 4 PCBs, 7 BDEs, 2 bisphenols). Information on diet, residence area and presence of pets at home were collected through questionnaires.
Twenty-nine to 88 compounds (median = 61) were detected per sample. Twenty-three biomarkers were detected in ≥95% of the samples, including 13 in all samples. Results showed higher exposure in younger children and higher exposure to non-persistent pesticides in boys. Results showed that diet (organic vs conventional), residence area, and living with pets at home are determinants of exposure.

The present study demonstrates that children are exposed simultaneously to multiple pollutants from different chemical classes, and confirms that hair is well adapted to investigate children exposure. The presence of persistent organic pollutants (POPs) in most children, despite their young age, strongly suggests that exposure to POPs is still ongoing. The results obtained here set the basis for further investigations to better understand the determinants of exposure to pollutants in children.

492 – Integrating Participant Feedback and Concerns to Improve Community and Individual Level Chemical Exposure Assessment Reports

Samantha Samon[1], Micheal Barton[2], Kim Anderson[3], Diana Rohlman[4]
[1]Duke University, Postdoc, Durham, USA. [2]Oregon State University, Research Administration, Corvallis, USA. [3]Oregon State University, Professor, Corvallis, USA. [4]Oregon State University, Associate Professor, Corvallis, USA

Abstract
As exposure assessment has shifted towards personal exposure assessment and community-engaged research, there has been an increased trend toward reporting results to participants. Reports aim to increase environmental health literacy, but this can be challenging due to the many unknowns regarding chemical exposure and human health effects. The Houston Hurricane Harvey Health (Houston-3H) project evaluated chemical exposure using silicone wristbands with the goal of giving participants their results. Three focus groups were conducted in separate neighborhoods in the Houston metropolitan area to evaluate responses to prototype community and individual level report-backs. Participants (n=31) evaluated prototype drafts (cover letter, description of study, sample chemical results pages). Focus groups were audio-recorded, and transcripts were analyzed using a qualitative data analysis program for common themes, and quantitative data were statistically analyzed. Four main themes emerged from the analysis of the transcripts: (1) views on the report layout, including the appropriate use of images, colors, and language level; (2) expression of concern over how chemicals might impact their individual or community health, as well as concern over their communities' level of chemical exposure; (3) participants emotional response towards the researchers; and (4) participants ability to comprehend and evaluate environmental health information. Differences were seen across the focus groups held regarding items of importance for inclusion in the report and key concerns. However, there was agreement amongst the focus groups about the desire to obtain personal exposure results despite the uncertainty of what the participant results meant.

493 – Characterization of Potential Exposure and Inhalation Risks During Firsthand and Secondhand Vaping

Jennifer Jeon[1], Qian Zhang[2], Joseph Hess[3], Xiaojia He[4], Lillie Barnett[4], Jonathan Shannahan[5], Marilyn Black[6], Christa Wright*[7]
[1]Chemical Insights Research Institute, UL Research Institutes, Research Scientist I, Marietta, USA. [2]Chemical Insights Research Institute, UL Research Institutes, Research Scientist IV, Marietta, USA. [3]Chemical Insights Research Institute, UL Research Institutes, Data Scientist, Marietta, USA. [4]Chemical Insights Research Institute, UL Research Institutes, Research Scientist II, Marietta, USA. [5]Purdue University, Associate Professor, West Lafayette, USA. [6]Chemical Insights Research Institute, UL Research Institutes, VP & Senior Technical Advisor, Marietta, USA. [7]Chemical Insights Research Institute, UL Research Institutes, Research Director, Marietta, USA

Abstract
Along with the growing evidence of electronic nicotine delivery systems (ENDS) related health risks, secondhand vaping exposure in indoor environments is an emerging public health concern. However, the composition and exposure risk of secondhand vaping have not been fully characterized. In this study, we utilized environmental exposure chambers to characterize firsthand and secondhand vaping emissions and estimated inhalation risks. Exhaled emissions from ENDS users (secondhand) and non-ENDS users (baseline) were collected, while firsthand emissions were generated using an automated ENDS aerosol generation system programmed to simulate puffing topography profiles collected from the ENDS users. Particle concentrations and size distributions along with volatile organic compounds (VOCs) were characterized using aerosol monitoring, sampling, and analytical chemical instrumentation. Our results showed that secondhand exposure scenarios produced significantly higher total particle number concentrations ($p<0.05$) compared to the baseline, although significantly lower than firsthand exposures. Particle deposition modeling revealed that while secondhand emissions had smaller deposited mass, total and pulmonary particle deposition fractions were higher than firsthand deposition levels, possibly due to smaller secondhand emission particle diameters (approximately 50 nm). Furthermore, 64 unique VOCs such as benzaldehyde were only observed in the secondhand exposure scenario. Acute secondhand inhalation risks for nonanal and hexanal were 4 and 8 times greater, respectively, for youth and adults compared to firsthand exposure scenarios. This study highlights that secondhand exposure to ENDS aerosols results in high level of exposure to ultrafine particles and VOCs, which can pose a serious threat to respiratory health especially in vulnerable populations.

494 – Making air pollution data actionable: learning from Breathe London community-led initiatives

Kayla Schulte[1], Andrew Grieve[2], Mohammad Mead[3], Benjamin Barratt[4]
[1]Imperial College London, Postdoctoral research associate, London, United Kingdom. [2]Imperial College London, Senior Air Quality Analyst, London, United Kingdom. [3]Imperial College London, Head of Air Quality Measurement, London, United Kingdom. [4]Imperial College London, Reader in Environmental Exposures & Public Health, London, United Kingdom

Abstract
Citizen or community-led air quality sensing has become a worldwide phenomenon following the proliferation of lower-cost air pollution sensors. Efforts in this space have supported policy shifts aimed at reducing air pollution exposures on multiple scales. While much of the literature about community-led sensing initiatives has focused on questions of data quality and validity, less explores the barriers that arise after communities gain access to, and become involved in the generation of calibrated, institutionally backed air pollution data. The Breathe London project is the world's first real-time calibrated air sensor network delivering high quality, freely available PM2.5 and NO2 data at over 400 locations across London, UK. The network is managed by the London Mayor's office and maintained by Imperial College London's Environmental Research Group. This study analyzes interviews with members from 40 community groups across London with local Breathe London nodes to identify persisting barriers when leveraging air pollution data. The results indicate communities spend substantial time and effort working to interpret and communicate the data. Individuals from this study consistently cite having "limited bandwidth" or time capacity, along with the core challenge of how to render the data "meaningful". Communities are having to navigate how to make air pollution data actionable towards their respective exposure reducing policy or practical aims. These findings highlight a need for co-designed resources and tools that move beyond traditional modes of presenting and communicating air pollution data, placing community-based knowledge and needs at their core.

495 – Chrono-differentiated metabolism of xenobiotics: evidence from human studies

Konstantinos C. Makris
Cyprus University of Technology, Associate Professor of Environmental Health, Limassol, Cyprus
Abstract

Tailoring medical models to the right person or risk subgroups delivered at the right time is important in personalized medicine/ prevention initiatives. A paradigm shift in the human chronotoxicity of xenobiotics would involve two-sided desynchronized phenomena of interfacial interactions between cyclic or periodic environmental insults and the endogenous response and recovery profile. These systems-based networks are under the influence of well-synchronized biological clocks and their metabolic regulators. This presentation will present evidence from human studies that depict the concept of synchronization involving critical life windows of susceptibility, or circadian rhythms, or 24-hour (periodic) diurnal rhythms and answering whether these disruptions in synchronization would affect biological response and recovery. Desynchronized interfaces during critical life windows or in time-repeated exposure events of xenobiotics would likely lead to initiating a cascade of adverse health effects associated with differentiated disease phenotypes.

496 – Characterization of Extreme Air Pollution in Uttar Pradesh, India: Insights from a Dense Low-Cost Sensor Network

Mark Campmier[1], Saumya Singh[2], Sofiya Rao[3], Adithi Upadhya[4], Harsh Raj Mishra[5], Ravi Kant Pathak[6], Sagnik Dey[7], Joshua Apte[8]
[1]University of California, Berkeley, Graduate Student Researcher, Berkeley, USA. [2]University of California, Berkeley, Research Scientist, Berkeley, USA. [3]Indian Institute of Technology, Delhi, Principal Project Scientist, Delhi, India. [4]ILK Labs, Researcher, Bangalore, India. [5]University of Gothenburg, Research Assistant, Gothenburg, Sweden. [6]University of Gothenburg, Associate Professor, Gothenburg, Sweden. [7]Indian Institute of Technology, Delhi, Professor, Delhi, India. [8]University of California, Berkeley, Associate Professor, Berkeley, USA

Abstract
India faces the highest mortality burden attributable to fine particulate matter in the world, with over 1 million deaths attributable to $PM_{2.5}$ annually. Within India itself, both the highest mass concentrations and population are in the northern state of Uttar Pradesh, with a population-weighted annual $PM_{2.5}$ mass concentration of 76 μg/m³. Although reference-grade monitoring has increased recently, there are still critical gaps within cities, and the large distributed rural population completely lacks monitoring. We have previously demonstrated the low-cost PurpleAir PA-II PM2.5 monitor can be robustly calibrated with multilinear regression on an hourly basis (NRMSE < 25%, $R^2 > 0.8$). Furthermore, here we demonstrate that error can be additionally constrained with hierarchical modeling, limiting NRMSE below 22% across a dynamic range of hourly resolved mass concentrations (10 – 500 μg/m³) and meteorological conditions. Since September 2021, we have deployed over 100 PA-II across the urban-rural interface in Uttar Pradesh and the surrounding region. Overall, most sites are well correlated with all other sites, indicating a large regional contribution. Observing diurnal trends across the urban-rural gradient reveals archetypal patterns. Broadly, these patterns are associated with site-specific activities including industrial facilities, traffic conditions, and rural biomass burning. Our work provides additional evidence demonstrating that extreme air pollution in India is not confined to megacities, but also poses challenges to smaller cities and rural communities which face burdens equal to or greater than the urban $PM_{2.5}$ burden.

497 – POSTER: Challenges in the Integration of Tools for Real-time Assessment of Residential Air Quality and Asthma Symptoms

Luz Huntington Moskos[1], Kamal Eldeirawi[2], Sharmilee Nyenhuis[3], Emily Cramer[4], Kathryn Krueger[5], Olivia Bimbi[6], Barbara Polivka[7]
[1]University of Louisville School of Nursing, Associate Professor, Louisville, USA. [2]University of Illinois Chicago, Associate Professor, Chicago, USA. [3]University of Chicago, Associate Professor, Chicago, USA. [4]Children's Mercy Hospital, Associate Professor, Kansas City, USA. [5]University of Kansas Medical Center, Research Manager, Kansas City, USA. [6]University of Illinois Chicago, PhD Student, Chicago, USA. [7]University of Kansas Medical Center, Professor and Associate Dean for Research, Kansas City, USA

Abstract
Purpose: This presentation will discuss challenges during integration of an air quality monitor, home spirometer, and ecological momentary assessment (EMA) to assess residential air quality and asthma symptoms among adults with asthma.
Methods: Three software platforms (Awair Omni air quality monitor, home spirometry, EMA application) with diverse dashboards and technological requirements were used. Additional time for novel integration between Awair Omni and EMA platforms plus testing of integration was required. A cloud-based spreadsheet catalogued issues encountered and resolutions implemented. A thematic analysis was completed.

Results: Over 30 problems/issues were documented and led to the development of creative solutions. Examples include the extensive iterative process required to develop study instructional materials to minimize burden on study participants and the management of software updates from platforms during data collection. Securing feedback from an asthma Community Advisory Board led to additional modifications including revising instructional materials, reducing EMA notifications, limiting data collection to 14 days, and adding a 3-day run-in period. A CONSORT diagram was developed and updated weekly to monitor all aspects of study progress. Adjustments were completed to allow for unblinding of the air quality monitor and data harmonization during data analysis.

Conclusions: This presentation shares insight into the integration of diverse technologies to assess real-time indoor air quality events with real-time asthma symptoms from a diverse group of participants across the U.S. The lessons presented will guide development of future studies using multiple data collection measures in real-time. Despite these challenges, these devices were successfully integrated.

498 – Harmonized multi-national occupational surveys to support the policy making in Europe

Radu Duca[1,2], Henriqueta Louro[3], Beatrice Bocca[4], Radia Bousoumah[5], Aleksandra Fucic[6], Karen Galea[7], Lode Godderis[8,9], Thomas Göen[10], Ivo Iavicoli[11], Beata Janasik[12], Kate Jones[13], Elizabeth Leese[14], Veruscka Leso[15], Sophie Ndaw[16], Katrien Poels[2], Simo Porras[17], Flavia Ruggieri[4], Maria Silva[18], An Van Nieuwenhuyse[19,20], Jelle Verdonck[21], Wojciech Wasowicz[22], Ana Tavares[3], Ovnair Sepai[23], Paul Scheepers[24], Susana Viegas[3], Tiina Santonen[25]

[1]National Health Laboratoey (LNS), Head of Unit, Dudelange, Luxembourg. [2]University of Leuven (KU Leuven), Senior Researcher, Leuven, Belgium. [3]NOVA University, Senior Researcher, Lisbon, Portugal. [4]Istituto Superiore di Sanità, Senior Researcher, Rome, Italy. [5]INRS, Senior Researcher, Nancy, France. [6]Institute for Medical Research and Occupational Health, Senior Researcher, Zagreb, Croatia. [7]Institute of Occupational Medicine (IOM), Head of Workplace Exposure Science, Edinburgh, United Kingdom. [8]University of Leuven (KU Leuven), Prof, Leuven, Belgium. [9]IDEWE, CEO, Leuven, Belgium. [10]Friedrich-Alexander-Universität Erlangen-Nürnberg, Prof, Erlangen, Germany. [11]University of Naples Federico II, Prof, Naples, Italy. [12]Nofer Institute of Occupational Medicine, Senior Researcher, Łódź, Poland. [13]Health and Safety Executive, Team lead, Stockport, United Kingdom. [14]Health and Safety Executive, Senior Researcher, Stockport, United Kingdom. [15]University of Naples Federico II, Senior Researcher, Naples, Italy. [16]INRS, Head of Department, Nancy, France. [17]Finnish Institute of Occupational Health, Senior Researcher, Helsinki, Finland. [18]National Institute of Health. Doutor Ricardo Jorge, Prof, Lisbon, Portugal. [19]National Health Laboratoey (LNS), Head of Department, Dudelange, Luxembourg. [20]University of Leuven (KU Leuven), Prof, Dudelange, Belgium. [21]University of Leuven (KU Leuven), PhD student, Leuven, Belgium. [22]Nofer Institute of Occupational Medicine, Prof, Łódź, Poland. [23]UK Health Security Agency, Lead General Toxicology Group, London, United Kingdom. [24]Radboud University, Prof, Nijmegen, Netherlands. [25]Finnish Institute of Occupational Health, Prof, Helsinki, Finland

Abstract

Providing policy relevant data on chemical exposures was a major aim of the EU biomonitoring initiative, HBM4EU. Similarly, the recently launched "Partnership for the risk assessment of chemicals" (PARC) project emphasizes the relevance of the research findings to regulatory decision making.

Occupational biomonitoring studies performed under HBM4EU focused on HBM4EU priority substances with specific regulatory relevance. Chromates and diisocyanate studies were developed since these substances are high concerns for workers and to support regulatory measures under EU chemicals regulation (REACH) and occupational safety and health (OSH) legislation. E-waste study was targeted to support European circular economy strategies and identify issues relevant for workers' protection under OSH legislation. In PARC, two other occupational surveys are currently prepared. Focusing on circular economy in particular the waste management sector, and on the the health care sector in particular the hazardous medicinal drugs, anaesthetic gases and cleaning products. In addition, innovative techniques will be evaluated for feasibility and added value.

Already, based on the results of the chromates study we drew several policy relevant conclusions, which support the future updating of occupational exposure limits and the development of monitoring and risk management practices in industry. A survey to policy makers on the usefulness of the study results gave us confidence. We emphasize the importance of early and continued engagement of policy makers to ensure the usability of the results. Early communication with regulators of the anticipated benefits and formatting outputs appropriately are important and have been considered and improved within the PARC project.

499 – Non-Chemical Determinants of Serum PFAS Concentrations in Wisconsin and their Associations with Obesity

Rachel Pomazal[1], Amy Schultz[2], Noel Stanton[3], Meshel Lange[4], Roy Irving[5], Jonathan Meiman[6], Brandon Shelton[7], Kristen Malecki[8]

[1]University of Wisconsin-Madison, Graduate Student, Madison, USA. [2]University of Wisconsin-Madison, Associate Director, Survey of the Health of Wisconsin, Madison, USA. [3]Wisconsin State Laboratory of Hygiene, Chemical Response Coordinator, Madison, USA. [4]Wisconsin State Laboratory of Hygiene, Chemical Emergency Response Program Supervisor, Madison, USA. [5]Wisconsin Department of Health Services, Hazard Assessment Section Chief, Madison, USA. [6]Wisconsin Department of Health Services, State Epidemiologist and Chief

Medical Officer, Madison, USA. [7]Wisconsin State Laboratory of Hygiene, Advanced Chemist, Madison, USA. [8]University of Illinois at Chicago, Professor, Director of the Division of Environmental And Occupational Health Sciences, Chicago, USA

Abstract

Per- and polyfluoroalkyl substances (PFAS) present ongoing population health risks because they bioaccumulate and are linked to numerous systemic health impacts. Geography is an important driver of population-level PFAS exposure, but few statewide population-based exposure assessments exist. No such study has been conducted in Wisconsin to-date, despite the presence of known PFAS contamination sites. This study aims to use data from the Survey of the Health of Wisconsin to identify determinants of serum PFAS in a statewide sample, and to examine associations between BMI and serum PFAS. Survey data, serum samples, and exam-based physical measurements were collected from 605 adults. Serum PFAS for 38 congeners were analyzed by the Wisconsin State Laboratory of Hygiene using high-speed liquid chromatography and mass spectrometry. Geometric means (GM), linear spline and quantile regression were used to examine associations for each PFAS individually and in total. PFOS (GM=4.51 95% CI 4.18,4.87), PFOA (GM=1.20 95% CI 1.12,1.27), PFNA (GM=0.45 95% CI 0.43,0.48), PFHxS (GM=1.14 95% CI 1.04,1.24), PFDA (GM=0.14 95% CI 0.13,0.15), and PFHpS (GM=0.17 95% CI 0.16,0.18) were detected in nearly all samples. Older individuals ($p<0.0001$), males ($p<0.0001$), and whites ($p<0.0001$), had higher serum PFAS. [MK1] Eating any sport caught fish and fruit, vegetable, and legume intake above three cups per day [MK2] were also associated with higher serum PFAS. PFNA was positively associated with BMI at the 90th percentile, but no other PFAS were associated with BMI. Future research should further explore sport caught fish and diet as areas for intervention for PFAS exposure.

500 – Titanium Dioxide in Personal Care Products: Consumer Risks and Regulatory Considerations

Frank Pagone[1], Jacob Persky[2]
[1]RHP Risk Management Inc., Senior Manager, Chicago, USA. [2]RHP Risk Management Inc., Principal, Chicago, USA

Abstract

Titanium dioxide (TiO_2) is an insoluble white powder that is used in commercial products such as plastics, food, and cosmetics as an anticaking or whitening agent. Listed in 2011, the WHO IARC classified TiO_2 as Group 2B "Possible Human Carcinogen". This was followed by the listing of TiO_2 (airborne and unbound particles of respirable size) under California Prop 65 regulations. Currently, the USFDA lists TiO_2 as an allowed color additive in consumer products, such as cosmetics and personal care products, if it meets certain specifications, while regulations continue to evolve in the United Kingdom and the European Union.

To determine the human health risk from exposure to TiO_2 during application of cosmetics, a study was designed to evaluate the exposure and risk during full-face application of white pigmented makeup containing TiO_2. Task duration personal breathing zone samples were collected for titanium and TiO_2 during full-face airbrush application of make-up by a professional make-up artist within an ISO Class-8 Cleanroom isolation chamber. Time weighted average (8-hr) results ranged from 0.01 to 0.03 mg/m^3 which were far below the OSHA PEL and ACGIH TLV exposure limits. Fine and ultrafine Particle Size Fraction (PSF) results were also below NIOSH REL and derived No Significant Risk Levels (NSRL). The calculated cancer risk ranged from 1 to 3 in a million cancer risk for a daily (full-face) use pattern over a 70-year lifetime. Results indicate that "worst-case scenario" use of airbrush applied make-up containing TiO_2 does not exceed current regulatory and non-regulatory limits.

501 – POSTER: Biological Responses of Human Lung Cells Following in vitro Exposure to Aerosolized PET Nanoplastics

Ryan Chartier[1], Archana Krovi[1], Jean Kim[2], Erin Piel[3], Leah Johnson[4]
[1]RTI International, Research Chemist, RTP, USA. [2]RTI International, Research Microbiologist, RTP, USA. [3]RTI International, Engineer, RTP, USA. [4]RTI International, Senior Director, RTP, USA

Abstract

Annual global production of plastics has increased exponentially from 1.5 million metric tons in the 1950s to almost 400 million metric tons in 2021. Degradation and fragmentation of ubiquitous plastics into microplastics (MPs, 1 µm to 5 mm) and nanoplastics (NPs, defined here as <1 µm) has resulted in the permeation and accumulation of plastic waste across all ecosystems, and recently, detection in human blood, tissues, and organs. MPs are pervasive in the atmosphere and have been found in even the most remote parts of the planet. However, the potential health impacts of human exposure to MPs and NPs remains poorly understood. Here, we attempt to begin to address this emerging concern using an vitro model.

Polyethylene terephthalate (PET) NPs were generated in the laboratory using RTI's standard protocol. We exposed two human lung cell lines (alveolar A549, bronchial 16HBE) to aerosolized PET NPs at the air-liquid interface (ALI) for approximately 2 hours (total dose ~5.8 ng/

cm2), in triplicate, and analyzed the cells 4 hours post-exposure. Both PET-exposed cell lines displayed a very significant increase in cell death when compared to cells exposed to HEPA-filtered air ($p<0.001$). Genetic markers of inflammation (IL-6, IL-8) and oxidative stress (HMOX) were significantly elevated ($p<0.05$) in the A549 cells when compared to cells exposed to HEPA-filtered air, but there was no difference in inflammatory or oxidative stress markers for the 16HBE cells. These preliminary data indicate that human inhalation of NPs may elicit negative toxicological and biological responses following exposure.

502 – Visualizing state biomonitoring data on the Tracking Network

Gonza Namulanda[1], Kristin Dortch[2]
[1]Centers for Disease Control and Prevention, Health Scientist, Atlanta, USA. [2]Centers for Disease Control and Prevention, Public Health Analyst, Atlanta, USA

Abstract
The Centers for Disease Control and Prevention's (CDC) National Environmental Public Health Tracking Program (Tracking Program) leads the effort to build the Environmental Public Health Tracking Network (Tracking Network), which brings together standardized data on environmental hazards (e.g., contaminants in drinking water), exposure to environmental hazards (e.g., lead in blood), and potentially related health outcomes (e.g., birth defects) along with risk factor data such as socioeconomic information and lifestyle behaviors. Data on the Tracking Network can be visualized in tables, charts, and maps, with the capability to view different topics side-by-side.

The Tracking Network currently displays national-level biomonitoring data on a select number of chemical groups from CDC's National Report on Human Exposure to Environmental Chemicals. A goal of the Tracking Program is to also display state-level biomonitoring data of the same chemical groups as the national-level on the Tracking Network. The Tracking Program has been collaborating with CDC's Division of Laboratory Sciences Biomonitoring Program and funding recipients from both programs to pilot submitting state-level biomonitoring data and visualizing it on the Tracking Network. The team worked on standards to structure the data to be submitted, describe the data and biomonitoring projects comprehensively, and visualize the data on a test data portal of the Tracking Network. Next, we will use these standards to visualize data from ongoing state biomonitoring projects when they are completed.

We present the process, challenges and lessons learned from this pilot to submit to and visualize state-level biomonitoring data on the Tracking Network.

503 – Characterizing air quality and health impacts of heavy-duty vehicle idling in warehouse settings using neighborhood-scale WRF-CMAQ simulations

Victoria Lang[1], Sara Camilleri[2], Anastasia Montgomery[3], Zac Adelman[4], Mark Janssen[5], Maria Harris[6], Daniel Horton[7]
[1]Department of Earth and Planetary Sciences, Northwestern University, Graduate Student, Evanston, IL, USA. [2]Department of Earth and Planetary Sciences, Northwestern University, Postdoctoral Researcher, Evanston, IL, USA. [3]Department of Earth and Planetary Sciences, Northwestern University, Ph.D. Candidate, Evanston, IL, USA. [4]Lake Michigan Air Directors Consortium, Executive Director, Rosemont, IL, USA. [5]Lake Michigan Air Directors Consortium, Emissions Director, Rosemont, IL, USA. [6]Environmental Defense Fund, Senior Scientist, San Francisco, CA, USA. [7]Department of Earth and Planetary Sciences, Northwestern University, Assistant Professor, Evanston, IL, USA

Abstract
Warehouses are essential to modern economies, but their day-to-day operations contribute to local air pollution due to emissions from heating and cooling systems, temperature-controlled storage, forklifts, and the transportation of goods. Heavy-duty, primarily diesel, vehicles are commonly used to transport goods to and within warehouse locations. In the United States, diesel vehicles are responsible for 43% of annual mortality attributable to transportation related O_3 and $PM_{2.5}$ emissions. Observations have shown that heavy-duty vehicles (HDVs) spend as much as 30-50% of their workday idling at locations such as warehouses and that the magnitude of idling emissions can be much greater than on-road emissions, with warehouses disproportionately located in minoritized and low-income communities. Current regulatory-grade emissions inventories attempt to capture HDV idling emissions using land surface development intensities from the National Land Cover Database. However, such approaches fail to capture the point source emission intensity of idling vehicles at warehouse locales, which may bias estimates of pollutant exposure and health impacts. In this study, we augment the U.S Environmental Protection Agency's spatial allocation of HDV idling emissions to include warehouse point sources within a Chicago-centric midwestern domain. Our augmented emissions dataset is then used in high spatial resolution (~1 km) two-way coupled Weather Research Forecast and Community Multiscale Air Quality chemistry transport model simulations to better constrain the impacts of warehouse HDV idling. Our results assess O_3, NO_x, and $PM_{2.5}$ pollutant concentration changes associated with idling HDVs and quantify associated environmental, health, and equity disparities at the census tract level.

504 – Something, Somewhere, Sometime - not '100 Years of Solitude': Retrospective Exposure Assessment for your Parents, Grandparents and Great Grandparents

Stuart Batterman
Univ. Michigan, Professor, Ann Arbor, USA

Abstract
The tools and datasets that can be used to estimate internal and external exposures have blossomed remarkable over the last decades, and good estimates for certain environmental pollutants like PM2.5 are now available globally at quite high spatial and temporal resolution. But what exposure data do we have prior to the availability of satellite data (around 2000), "modern" measurements at regulatory air quality monitoring sites (1970s), and some localized smoke and dustfall measurements in a few urban cities decades earlier? Or for decedents and others for which we have no biological specimens but only residential and possibly work history? Considering the etiology of many diseases that may begin in utero, childhood, or adulthood, and the extensive health records and specialized datasets and biorepositories available since 1940, there is a need to understand exposures over the last century or even earlier. Is this possible? This talk will define some needs and explore possible approaches for determining long term retrospective exposure estimates that can be used in epidemiological studies. Focusing on air quality, we discuss the availability and utility of records for visibility, fuel use, technological innovation, and other factors to estimate air pollution levels that can be harmonized with modern estimates. Such estimates would complement current data and strengthen our understanding of the role of environmental exposures in causing chronic disease, thus complementing the longer records that are available for many sociological and genetic determinants of health and disease.

505 – POSTER: Factors Influencing Dust Loading in Homes across Three United States Regions

Olusola Ogunseye[1], Jenna Honan[1], Jacqueline Larson[2], Paloma Beamer[3], Cristina Martinez[4], Maribeth Gidley[5], Helena Solo-Gabriele[6], Foluke Adelabu[7], Neema Olagbemiro[7], Emmanuel Obeng-Gyasi[8], Alesia Ferguson[8]
[1]Mel and Enid Zuckerman College of Public Health, University of Arizona, PhD Student, Tucson, USA. [2]Mel and Enid Zuckerman College of Public Health, University of Arizona, Student, Tucson, USA. [3]Mel and Enid Zuckerman College of Public Health, University of Arizona, Faculty, Tucson, USA. [4]University of Miami, PhD Student, Coral Gables, USA. [5]University of Miami, Researcher, Coral Gables, USA. [6]University of Miami, Faculty, Coral Gables, USA. [7]North Carolina Agricultural and Technical State University, PhD Student, Greensboro, USA. [8]North Carolina Agricultural and Technical State University, Faculty, Greensboro, USA

Abstract
Exposure to metals and other contaminants found in indoor dust are of increasing concern to public health. Sociodemographic characteristics and geographical location may influence dust loading in homes in the United States. For example, regional weather patterns, structure and design of homes may contribute to the infiltration of outdoor soil into indoor environments. The objective of this study was to determine the potential factors associated with dust loading in homes across three US regions: the metropolitan areas of Greensboro, North Carolina; Miami, Florida and Tucson, Arizona. Dust loading (g/m2) on floors were collected by vacuuming a measured area of the floor for 5 minutes. Survey questionnaire was administered to collect sociodemographic characteristics, children's play activities, housing dynamics, and housing layout of participants across the three regions. We used classification and regression tree (CART) analysis to assess the factors that are most strongly associated with dust loading in the homes. Among the variables in the model, initial associations have been found for dust loading with location and educational status. Stronger associations will be explored in the expanded model. Efforts directed towards awareness and tailored public health promotion messages may contribute to reduction in dust loading in US homes and potentially reduce the risk of exposures to contaminants in indoor dust.

506 – Non-target analysis of indoor dust and children's urine to identify candidate organic tracers to estimate children's dust ingestion rates

Matthew Newmeyer[1], Qinfan Lyu[2], Sara Lupolt[3], Keeve Nachman[4], Carsten Prasse[5]
[1]Department of Environmental Health and Engineering, Johns Hopkins Bloomberg School of Public Health, Research Associate, Baltimore, USA. [2]Department of Environmental Health and Engineering, Johns Hopkins Bloomberg School of Public Health, Graduate Student, Baltimore, USA. [3]Department of Environmental Health and Engineering, Johns Hopkins Bloomberg School of Public Health, Assistant Scientist, Baltimore, USA. [4]Department of Environmental Health and Engineering, Johns Hopkins Bloomberg School of Public Health, Associate Professor, Baltimore, USA. [5]Department of Environmental Health and Engineering, Johns Hopkins Bloomberg School of Public Health, Assistant Professor, Baltimore, USA

Abstract
Incidental ingestion of indoor dust is a poorly quantified pathway of exposure to chemical contaminants. Young children spend considerable time in direct contact with the floor and may have higher dust exposure rates than older lifestages. Elemental tracers proposed for estimating dust exposures do not perform well due to their ubiquity in multiple environmental media. The goal of this study is to identify novel candidate organic tracers present in urine that will facilitate the estimation of children's dust ingestion rates. Caregivers of infants between 6–12 months old living in the Baltimore metropolitan area were trained to collect their child's urine over ~24 h using study-provided disposable diapers containing cotton pads. Upon return to the laboratory, urine was extracted from cotton pads via centrifugation, pooled, and stored at -80 °C until analysis. Indoor dust was collected via vacuuming in the caregiver's home from the room where their child spent most of their waking hours. Dust was sieved with a 106 µm mesh and stored at -20 °C until analysis. Urine extracts were prepared by dilution and protein precipitation by centrifugation, and dust extracts were prepared by sonication into solvent. Extracts were analyzed by liquid chromatography–high-resolution mass spectrometry. Raw data were processed with a suite of computational tools to detect and identify compounds, and identities were confirmed, when possible, with reference standards. These data are part of a larger effort, incorporating multiple environmental and biological media, to identify candidate organic tracers to improve estimates of children's dust ingestion.

507 – Development of a Community Health Survey to Support Public Health Research and Advocacy Efforts in Fenceline Communities in Southeastern Pennsylvania: A Community-Based Participatory Research Approach

Andrea A. Chiger[1], Echo Alford[2,3], Kearni N. Warren[4], Eve S. Miari[5,3], Lora Snyder[3], Thom Nixon[3], Mike Ewall[6,4], Mary A. Fox[7], Ryan D. Kennedy[8], Peter F. DeCarlo[8], Keeve E. Nachman[8], Sara N. Lupolt[9]
[1]Johns Hopkins University, PhD Candidate, Baltimore, MD, USA. [2]Clean Air Council, Delaware County Advocacy Coordinator, Philadelphia, PA, USA. [3]Marcus Hook Area Neighbors for Public Health, Community Organizer, Marcus Hook, PA, USA. [4]Delco Environmental Justice, Community Organizer, Chester, PA, USA. [5]Clean Air Council, Delaware County Programs Director, Philadelphia, PA, USA. [6]Energy Justice Network, Executive Director, Philadelphia, PA, USA. [7]Johns Hopkins University, Assistant Professor, Baltimore, MD, USA. [8]Johns Hopkins University, Associate Professor, Baltimore, MD, USA. [9]Johns Hopkins University, Assistant Scientist, Baltimore, MD, USA

Abstract
Community organizers in Southern Delaware County, PA expressed a desire to collect comprehensive data on environmental, health, and social conditions in their neighborhoods, with the goal of informing research and advocacy efforts to protect public health. To address this need, our team of academic and community co-investigators developed an online community health survey to characterize the health concerns of residents and the strengths, burdens, and needs of fenceline communities in Southern Delaware County. We included questions on chemical exposures, sources of pollution, financial stressors, healthcare, medical conditions, and priorities for policymakers. Wherever possible, we used validated survey questions to enable comparison to national or other local statistics. The 10-15 minute survey includes multiple-choice, Likert, and open-ended questions. We will present our preliminary results using quantitative (e.g., descriptive statistics, ANOVA to test for potential differences by factors such as race/ethnicity and neighborhood of residence) and qualitative (e.g., inductive and deductive analytical approaches to coding) methods. Our approach to data analysis, interpretation and presentation includes training opportunities and active contributions from both academic and community co-investigators. We will follow the survey with focus groups to further explore community concerns and preferred solutions. In addition to empowering future community research efforts, our study provides detailed local data on chemical and non-chemical stressors to support both community advocacy and cumulative impact assessment. Our work demonstrates the importance of community-based participatory research approaches for effectively responding to local environmental justice issues.

508 – Fugitive and settled dust is not just windblown soil - an analysis in an environmental justice community

Jeff Gearhart[1], Sun-Hee Seo[2], Melissa Sargent[1], Natalie Sampson[3], Jacob Napieralski[3], Stuart Batterman[4]
[1]Ecology Center, Scientist, Ann Arbor, USA. [2]Univ. Michigan, Postdoctoral Scholar, Ann Arbor, USA. [3]Univ. Michigan, Prof., Dearborn, USA. [4]Univ. Michigan, Prof., Ann Arbor, USA

Abstract
Fugitive dust is emitted from undefined or open sources. It is a nuisance and potential health threat that has received little attention, possibly because impacts tend to be localized and its contribution to PM2.5 is limited. Fugitive dust derived from ongoing or past industry may contain toxic substances, and exposure is particularly likely for environmental justice communities, specifically residential neighborhoods adjacent to industry. This community-initiated study examines the composition of fugitive dust surrounding a scrap metal processor in Detroit, Michigan, an industrialized community with a large minority population. We collected 57 deposition samples at 27 residential locations, and 20 composite samples of road-dust and surface soils. All samples were analyzed for over 40 elements in triplicate using XRF, and soil and road dust samples were analyzed for semivolatile organic compounds by GC/MS-MS. Samples

collected near the industrial facility had higher abundances of metals associated with iron and steel processing, e.g., Cu, Fe, Ni and Zn levels were 4 to 14 times greater than background levels; Cr and Mn also were elevated, although uncertainties are larger. Concentrations were lower, but remained elevated, in residential areas 1 to 3 blocks from the metal processor. In addition, polychlorinated biphenyls levels near the facility were elevated by 10-fold, and some concentrations approached or exceeded residential screening criteria. We discuss how low-cost screening methods can help identify likely sources, how soil and deposition sampling complements other monitoring, and the need for mitigation, follow-up assessments, and policy solutions to address environmental injustice.

510 – The Effect of Pesticide Spray Season and Residential Proximity to Agriculture on Glyphosate Exposure in Pregnant Women

Cynthia Curl[1], Carly Hyland[2], Meredith Spivak[3], Lianne Sheppard[4], Bruce Lanphear[5], Michael Antoniou[6]
[1]Boise State University, Associate Professor, Boise, USA. [2]Boise State University, Post-Doctoral Fellow, Boise, USA. [3]Boise State University, Epidemiologist, Boise, USA. [4]University of Washington, Professor, Seattle, USA. [5]Simon Fraser University, Professor, Vancouver, Canada. [6]Kings College London, Professor, London, United Kingdom

Abstract
Glyphosate is among the most heavily used pesticides in the world, but little is known about sources of glyphosate exposure in pregnant people living in agricultural regions. We evaluated glyphosate exposure during pregnancy in relation to residential proximity to agriculture and agricultural spray season. We quantified glyphosate concentrations in 453 urine samples collected biweekly from 40 pregnant women from February through December 2021. We estimated each participant's glyphosate exposure as the geometric mean (GM) of glyphosate concentrations measured in all samples (average n=11 samples/participant), as well as the GM of samples collected during the pesticide "spray season" (defined as those collected May 1 – Aug 15; average n=5 samples/participant) and the "non-spray season" (defined as those collected before May 1 or after Aug 15; average n=6 samples/participant). We defined participants residing <0.5 km from an actively cultivated agriculture field to live "near fields" and those residing >0.5 km from an agricultural field to live "far from fields" (n=22 and 18, respectively). Among near filed participants, urinary glyphosate was detected more frequently and at significantly higher GM concentrations during the spray season compared to the non-spray season (81% vs 55%; 0.24 µg/L vs. 0.15 µg/L, $p<0.001$). In contrast, among participants living far from fields, neither glyphosate detection frequency nor GMs differed in the spray vs non-spray season (66% vs. 64%; 0.15 µg/L vs. 0.17 µg/L, p=0.45). These findings suggest that agricultural spray is a source of glyphosate exposure for people living near fields.

511 – Building Institutional Collaboration at the US Environmental Protection Agency to Address Contaminants of Emerging Concern

Guiseppi-Elie Annette[1], Kathie Dionisio[2], José Zambrana, Jr.[3]
[1]US EPA Office of Research and Development, National Research Program Director for Chemical Safety for Sustainability, RTP, USA. [2]US EPA Office of Research and Development, Principal Associate National Research Program Director for Chemical Safety for Sustainability, RTP, USA. [3]US EPA Office of Research and Development, Senior Science Advisor in the Center for Environmental Measurement and Modeling, Los Angeles, USA

Abstract
Contaminants of Emerging Concern (CECs) are challenging local, regional, tribal, and federal governments in the US and abroad with ways to systematically identify, monitor and manage any potential risks. CECs can cover a wide range of agents (e.g., chemical, biological, radiological), environmental media and chemistries. The U.S. Environmental Protection Agency (EPA) has several ways in which its varying offices address CECs, depending on the CEC use and relevant environmental media. However, in order to help expedite the agency's response to potential CECs, EPA has developed a framework for addressing CECs that includes a process called Screening Risk of Emerging Contaminants (SIREN) to help identify situations for wider agency mobilization on addressing any particular CEC. The framework includes ways to identify potential CECs for consideration; and a technical approach that accounts for key factors such as potential hazards, exposure, persistence, bioaccumulation, cross-media impacts, and the degree to which the potential CEC is already being addressed by one or more offices. The output of the SIREN process is a plan of action for intra-agency coordination on addressing a potential CEC when needed. The presentation will provide an overview of SIREN including the technical approach, and how it may be used and adapted by external decision makers.

512 – Building Institutional Collaboration Across the US Federal Government to Implement a National Emerging Contaminants Research Initiative

Melanie Buser[1], Annette Guiseppi-Elie[2], David Balshaw[3]
[1]Executive Office of the President, Asst. Director for Environmental Health, Office of Science and Technology Policy, Washington, USA. [2]US EPA Office of Research and Development, National Program Director for Chemical Safety for Sustainability, RTP, USA. [3]US NIH National Institute of Environmental Health Sciences, Acting Director, Division of Extramural Research and Training, RTP, USA

Abstract
The Fiscal Year 2020 United States National Defense Authorization Act directed the White House Office of Science and Technology Policy (OSTP), in coordination with other federal agencies, to create a National Emerging Contaminants Research Initiative (NECRI) to improve the identification, analysis, monitoring, and treatment methods of Contaminants of Emerging Concern (CECs in drinking water (DW). In the NECRI, CECs are "are newly identified or reemerging manufactured or naturally occurring physical, chemical, biological, radiological, or nuclear materials that may cause adverse effects to human health or the environment and do not currently have a national primary DW regulation". While DW is the medium of focus for the NECRI, it is recognized that CECs exist in multiple media that may be relevant for addressing public and environmental health needs in varied settings. The capabilities and approaches developed under the NECRI are expected to lead to a holistic treatment of CECs. The presentation will provide examples of critical research and coordination priorities around the five (5) NECRI goals:

1. Decrease the time from DW CEC identification to risk mitigation;
2. Promote technological innovation in tools to discover, track, understand, and mitigate DW CECs;
3. Develop and deploy tools and approaches for DW CEC decision making;
4. Coordinate transdisciplinary DW CEC research activities among Federal and non-Federal partners; and
5. Foster transparency and public trust when communicating about DW CECs.

513 – Documenting improvements in indoor air quality in schools in environmental justice areas

Tian Xia[1], Julia Raneses[2], Raquel Garcia[3], Md Kamrul Islam[4], Stuart Batterman[5]
[1]University of Michigan, Postdoc, Ann Arbor, USA. [2]University of Michigan, Researcher, Ann Arbor, USA. [3]Southwest Detroit Environmental Vision, Executive Director, Detroit, USA. [4]University of Michigan, Research Analyst, Ann Arbor, USA. [5]University of Michigan, Professor, Ann Arbor, USA

Abstract
Poor indoor air quality (IAQ) in schools affects student and staff performance, contributes to asthma and other diseases, and promotes transmission of airborne infectious diseases including COVID-19. This particularly applies to environmental justice (EJ) communities where schools are located in polluted areas, and more broadly as most schools have rudimentary air filtration and inadequate ventilation, and many have old or improperly maintained HVAC systems. Our objectives in this community-based participatory research project were to assess indoor and outdoor air quality in schools located in EJ communities, evaluate impacts of an intervention using enhanced particle filtration, and recommend further IAQ improvements. We conducted pre- and post-intervention measurements and IAQ surveys in four school buildings. Measured IAQ parameters included black carbon (BC), particulate matter ($PM_{2.5}$), carbon dioxide (CO_2), and air change rates (ACRs). Prior to the intervention, most classrooms had inadequate ACRs (1.2 – 1.4 h^{-1}), and many had high CO_2 (up to >2000 ppm), moderate levels of $PM_{2.5}$ (2.9 – 9.9 μg/m³), elevated BC (0.5 - 1.6 μg/m³), and similar indoor and outdoor levels of PM2.5 and BC, showing sizable contributions from diesel-powered vehicles; additionally, PM_{10} levels were elevated indoors. After upgrading filters to MERV-13 and adding portable purifiers in classrooms without mechanical ventilation, levels of $PM_{2.5}$ and BC, and indoor/outdoor ratios decreased by approximately half in most classrooms. The study documents the improvement in IAQ in classrooms achievable using enhanced filtration, the need to increase ventilation, and the teacher and staff engagement needed to support and maintain IAQ programs.

514 – EU Human Biomonitoring Initiative (HBM4EU) Chromates Study: PFAS exposure in electroplaters and bystanders

Thomas Göen[1], Annalisa Abballe[2], Radia Bousoumah[3], Radu Duca[4,5], Lode Godderis[6,7], Ivo Iavicoli[8], Anna Maria Ingelido[2], Veruscka Leso[9], Johannes Müller[1], Sophie Ndaw[10], Simo Porras[11], Jelle Verdonck[12], Tiina Santonen[13]
[1]Friedrich-Alexander-Universität Erlangen-Nürnberg, Senior Researcher, Erlangen, Germany. [2]Istituto Superiore di Sanità, Senior Researcher, Rome, Italy. [3]French National Research and Safety Institute (INRS), Senior Researcher, Vandœuvre-Lès-Nancy, France.

[4]National Health Laboratory (LNS), Head of Unit, Dudelange, Luxembourg. [5]University of Leuven, Senior Researcher, Leuven, Belgium. [6]University of Leuven, Prof, Leuven, Belgium. [7]IDEWE, CEO, Leuven, Belgium. [8]University of Naples Federico II, Prof, Naples, Italy. [9]University of Naples Federico II, Senior Researcher, Naples, Italy. [10]French National Research and Safety Institute (INRS), Head of Department, Vandœuvre-Lès-Nancy, France. [11]Finnish Institute of Occupational Health (FIOH), Senior Researcher, Helsinki, Finland. [12]University of Leuven, PhD student, Leuven, Belgium. [13]Finnish Institute of Occupational Health (FIOH), Prof, Helsinki, Finland

Abstract

In the frame of the EU Human Biomonitoring Initiative, HBM4EU, a multi-national study was conducted across Europe to characterize occupational exposure to hexavalent chromium [Cr(VI)]. The use of Cr(VI) compounds (chromates and chromium trioxide) are authorized under the European regulation (EC 1907/2006) REACH and no alternative to the Cr(VI) use in some industrial activities has been found, so far. Workers exposure occurs mainly by inhalation while performing activities such as welding, Cr(VI) electroplating, and other surface treatment activities. Apart of the actual exposure to Cr(VI), we also evaluated the exposure to perfluoroalkyl substances (PFAS) in different industry sectors. PFAS exposure of 172 individuals from four countries was assessed by the determination of 8 perfluoroalkyl acids and 4 perfluoroalkylsulfonic acids in plasma samples. Participants were 52 chrome-plating workers, 43 welders, 3 surface treating workers and 74 workers without any occupational chrome exposure as controls. Significant differences between workers with chromium exposure and controls were found for the perfluoroalkyl sulfonic acids, particularly for perfluorooctylsulfonic acid (PFOS). Median and maximum levels were 4.83 and 789 µg/L for chrome plating workers, 4.97 and 1512 µg/L for welders, and 3.01 and 12.2 µg/L for controls, respectively. The very high maximum levels of PFOS for chromium platers and for welders can be explained by the application of PFOS as mist suppressants in electroplating baths, which resulted in an exposure of the operators but also welders, who performed maintenance and repair service at these workplaces.

515 – POSTER: Exposure to Phthalates from Electronic Cigarette Use: A Pilot Study

Payal R. Kumar[1], Angela Aherrera[2], Mina W. Tehrani[2], Yingyue Ni[3], Joyce Lin[4], Lesliam Quiros-Alcala[5], Ana M. Rule[6]
[1]Johns Hopkins Bloomberg School of Public Health, MPH Student, Baltimore, USA. [2]Johns Hopkins Bloomberg School of Public Health, Postdoctoral Fellow, Environmental Health and Engineering, Baltimore, USA. [3]Johns Hopkins Bloomberg School of Public Health, Masters Student,Environmental Health and Engineering, Baltimore, USA. [4]Johns Hopkins Bloomberg School of Public Health, Doctoral Student, Environmental Health and Engineering, Baltimore, USA. [5]Johns Hopkins Bloomberg School of Public Health, Assistant Professor, Environmental Health and Engineering, JHBSPH, Baltimore, USA. [6]Johns Hopkins Bloomberg School of Public Health, Assistant Professor, Environmental Health and Engineering, Baltimore, USA

Abstract

Introduction: The US has seen a significant increase in e-cigarette (e-cig) use, especially among youth. Electronic cigarettes rapidly evolved from "MOD" devices mostly used by smokers as a risk reduction strategy to "POD" devices with disposable cartridges. There are emerging concerns that the aerosol generated is a potential source of phthalates which are endocrine disrupting compounds (EDCs). Still, exposures to EDCs in e-cig users have not been characterized.

Objective: We aimed to characterize exposure to phthalates among e-cig users and controls.

Methods: We measured and compared concentrations of 16 phthalate biomarkers in urine samples from 30 e-cig users (15 POD, 15 MOD) and 24 non-user controls, ages 18-52 years. Demographics, e-cig use behaviors and other potential sources of phthalates were captured via investigator-administered questionnaire. We conducted non-parametric analyses to assess differences in biomarker concentrations by user type. Exploratory analysis was done to visualize the trends.

Results: Overall, detection frequency of phthalates was >90%: >91% in MOD users, >93% in POD users, and >91% in non-users. In general, we observed higher median biomarker concentrations among e-cigarette users compared to controls. For example, concentrations of mono-2-ethyl-5-carboxypentyl terephthalate (MEHHP) and mono-carboxy-octyl phthalate (mCOP) were lower in controls than in e-cig users. POD users had higher concentrations than MOD users except for mono ethyl hexyl phthalate (mEHP) ($p < 0.05$).

Conclusions: This study suggests differences in phthalate exposure between e-cig users and controls, as measured with urine metabolites. Further, there are indications that POD users have higher exposure to most phthalates than MOD users.

516 – Evaluating an Environmental Health Community-Academic Research Partnership Through Traditional and Non-Traditional Methods

Seana Cleary[1], Robert B. Hood[2], Emily Olack[1], Martha Scott Tomlinson[3], M. Elizabeth Marder[4], Hillary Barton[5], Metrecia Terrell[6], Michele Marcus[7], Melanie Pearson[8]
[1]Emory University Rollins School of Public Health, Graduate Research Assistant, Atlanta, USA. [2]Emory University Rollins School of Public Health, Post Doctoral Fellow, Atlanta, USA. [3]Emory University Rollins School of Public Health, Community Engagement Coordinator, Atlanta, USA. [4]California Environmental Protection Agency, Senior Environmental Scientist, Sacramento, USA. [5]Emory University Rollins School of Public Health, Associate Director of Research Projects, Atlanta, USA. [6]Emory University Rollins School of Public Health, Biostatistician & Data Manager, Atlanta, USA. [7]Emory University Rollins School of Public Health, Lead Scientist, Atlanta, USA. [8]Emory University Rollins School of Public Health, Director of Community Engagement, Atlanta, USA

Abstract
Background: Community-academic research partnerships can improve science and its translation into benefits for communities impacted by environmental health issues. Here, we describe an evaluation of our partnership with a community affected by one of the largest agricultural exposure incidents in US history, in which millions of Michiganders were accidentally exposed to polybrominated biphenyl (PBB) through the food supply to an extent that many still have measurable serum levels. A registry was formed by the Michigan Department of Health and Human Services (MDHHS) to assess exposure and health effects.

Methods: Since Emory University took over PBB Registry management in 2011, we have worked with community-based partners to guide research questions based on community concerns and feedback. To assess our community-academic partnership, we conducted two community partner questionnaires (2013 and 2020). We assessed several partnership dimensions, including motivation for board membership, bidirectional communication, engagement, trust, accessibility, and openness. We augmented these questionnaires using the NIEHS Partnerships for Environmental Public Health (PEPH) Evaluation Metrics Manual to thematically code our monthly meeting notes with the leadership board.

Results: Questionnaire responses generally indicated strong trust and good communication within the partnership. Respondents also identified challenges related to funding and bureaucratic roadblocks. Coding the monthly notes using the PEPH manual revealed strengths in Partnership themes of partner and community involvement, partner communication, and partner commitment.

Conclusion: The traditional questionnaire method revealed several strengths and areas for improvement. The novel thematic coding method added context to these results and will enhance our ability to improve our partnership.

520 – Application of Metamodeling Approaches to Improving Cohort-Specific High-Throughput Exposure Predictions

Mary Jacketti[1,2], Zachary Stanfield[2], Caroline Ring[3], John Wambaugh[4], Kristin Isaacs[4]
[1]Oak Ridge Institute for Science and Education (ORISE), Postdoctoral Fellow, Research Triangle Park, USA. [2]US EPA, Postdoctoral Fellow, Research Triangle Park, USA. [3]US EPA, Computational Exposure Scientist, Research Triangle Park, USA. [4]US EPA, Research Physical Scientist, Research Triangle Park, USA

Abstract
The U.S. Environmental Protection Agency's (EPA) Exposure Forecasting (ExpoCast) project has developed high-throughput (HT) models for estimating exposures to thousands of chemicals for use in chemical prioritization. Within the ExpoCast Systematic Empirical Evaluation of Models (SEEM) framework, predictions from multiple process-based models are integrated with biomonitoring data to produce calibrated consensus predictions for thousands of chemicals. However, for chemicals with no specific predictions from any of the component models, SEEM relies on the use of poorly resolved component model mean values when estimating consensus exposure. In addition, the current version of the SEEM framework does not consider stratification by age or gender cohort. Recently, machine learning-based metamodeling approaches have increasingly been used to model the response of complex exposure models, reducing computational time and parameterization burden thus allowing for the efficient expansion of predictions to new chemicals or model conditions. Here a metamodeling framework was applied to a process-based consumer exposure model, (the HT Stochastic Human Exposure and Dose Simulation model, SHEDS-HT) to expand SHEDS-HT predictions to thousands of additional chemicals based on chemical properties and potential occurrence in specific product use categories (PUCs). The framework comprised a multilabel classification model that utilized chemical structure descriptors to predict PUC occurrence and a support vector regression model of SHEDS-HT response. Predictions from this framework are being incorporated into a new cohort-specific consensus model for median population exposure. The multilabel classification model and SHEDS-HT metamodel may also be useful in identifying and evaluating reasonably foreseen consumer uses for emerging commercial chemicals.

521 – POSTER: Associations of urban blue (water bodies) and green spaces with coronary artery calcification among adults of self-reported Black race

Kyeezu Kim[1], Brian Joyce[2], Yinan Zheng[3], Drew Nannini[1], Penny Gordon-Larsen[4], James Shikany[5], Donald Lloyd-Jones[6], Mercedes Carnethon[7], Ming Hu[8], Ankeet Bhatt[9], Kelley Gabriel[10], Kai Zhang[11], Lifang Hou[12]
[1]Northwestern University Feinberg School of Medicine, Postdoctoral Fellow, Chicago, USA. [2]Northwestern University Feinberg School of Medicine, Research Assistant Professor, Chicago, USA. [3]Northwestern University Feinberg School of Medicine, Assistant Professor, Chicago, USA. [4]University of North Carolina, CARLA SMITH CHAMBLEE DISTINGUISHED PROFESSOR OF GLOBAL NUTRITION, Chapel Hill, USA. [5]University of Alabama, Professor, Birmingham, USA. [6]Northwestern University Feinberg School of Medicine, Chair, Department of Preventive Medicine, Chicago, USA. [7]Northwestern University Feinberg School of Medicine, Vice Chair, Department of Preventive Medicine, Chicago, USA. [8]University of Maryland, Associate Professor, College Park, USA. [9]Kaiser Permanente, Cardiologist, Kaiser Permanente San Francisco Medical Center, San Francisco, USA. [10]University of Alabama Birmingham, Professor, Birmingham, USA. [11]University of Albany, Empire Innovation Associate Professor School of Public Health, Albany, USA. [12]Northwestern University Feinberg School of Medicine, Chief of Cancer Epidemiology and Prevention in the Department of Preventive Medicine, Chicago, USA

Abstract
Background: Blue and green spaces have been associated with higher cardiovascular health; however, few studies have examined the role of race and other social determinants of health in this association. Here, we assessed the association of urban blue and green spaces with coronary artery calcification (CAC), among Black- and White adults in midlife.

Methods: We included 2920 participants from the Coronary Artery Risk Development in Young Adults (CARDIA) study, a longitudinal study with balanced race (Black and White) and sex (men and women) at inception. We used generalized estimating equation regression models to assess the associations of blue and green spaces exposures (1985-1986 to 2010-2011), represented by distance to rivers and land cover (Normalized Difference Vegetation Index [NDVI]), with presence of CAC (total Agatston score >0; 2010-2011). Models were adjusted for demographics, socioeconomic status (SES), and health-related behaviors and morbidity. We performed stratified analyses by race.

Results: Shorter distance to rivers and greater NDVI were associated with lower odds of CAC in Black participants, but not in White participants. Black participants living closer to rivers (1st tertile) had 32% lower odds of CAC compared to participants living farthest (3rd tertile) from rivers (OR=0.68, 95% CI=0.55-0.86 versus OR=0.88, 95% CI=0.74-1.04 among White participants). Among Black participants, a 10%-point increase of NDVI was associated with 15% lower odds of CAC (OR=0.85, 95% CI=0.75-0.95 versus OR=0.98, 95% CI=0.90-1.07 among White participants).

Conclusions: Our findings suggest greater exposure to blue and green spaces may contribute to higher cardiovascular health, especially among Black individuals.

522 – Partitioning of Neutral Per- and Polyfluoroalkyl Substances (PFAS) From Indoor Air to Cloth in North Carolina Homes During the Indoor PFAS Assessment (IPA) Campaign

Clara Eichler[1], Naomi Chang[1], Jason Surratt[2], Glenn Morrison[2], Barbara Turpin[3]
[1]University of North Carolina at Chapel Hill, Graduate Student, Chapel Hill, NC, USA. [2]University of North Carolina at Chapel Hill, Professor, Chapel Hill, NC, USA. [3]University of North Carolina at Chapel Hill, Professor and Chair, Chapel Hill, NC, USA

Abstract
The indoor environment can be an important contributor to population exposure for per- and polyfluoroalkyl substances (PFAS) present in consumer products. PFAS emitted into indoor air can partition to indoor surfaces and materials, including clothing, which may prolong the residence time of PFAS indoors. Among other things, we measured concentrations of nine neutral PFAS in air and cotton cloth, which had been deployed in 10 homes in North Carolina, USA, for up to 9 months during the Indoor PFAS Assessment (IPA) Campaign. We found that in indoor air, fluorotelomer alcohols (6:2 FTOH, 8:2 FTOH, and 10:2 FTOH) are the dominant gas-phase species, with concentrations in the range of 1.8-49 ng/m^3, 1.2-53 ng/m^3, and 0.21-5.7 ng/m^3, respectively. In cloth, perfluorooctane sulfonamidoethanols (MeFOSE and EtFOSE) accumulated most significantly over time, reaching concentrations of up to 0.26 ng/cm^2 and 0.24 ng/cm^2, respectively. From paired measurements of neutral PFAS in air and suspended cloth, we derived cloth-air partition coefficients (K_{ca}) for 6:2 FTOH, 8:2 FTOH, and 10:2 FTOH, ethylperfluorooctane sulfonamide (EtFOSA), MeFOSE, and EtFOSE. Mean log(K_{ca}) values range from 4.7 to 6.6, and results correlate well with the octanol-air partition coefficient (K_{oa}). We further investigated the effect of the cloth storage method on PFAS accumulation and found that cloth kept in dresser drawers accumulates more FTOHs than suspended cloth. This work adds valuable information to PFAS exposure models that can provide population-based exposure predictions and informs policy design for PFAS exposure mitigation.

523 – POSTER: Effective guidance for seafood consumption during pregnancy: Lessons learned from the harmonized HBM4EU-MOM study in five European countries

Andromachi Katsonouri[1,2], Catherine Gabriel[3], Marta Esteban López[4], Sonia Namorado[5,6,7], Thorhallur I.Halldorsson[8], Ricardo Assunção[5,9], Noelia Domínguez-Morueco[4], Elena Anastasi[10], Janja Snoj Tratnik[11], Laura Rodriguez Martin[12], Achilleas Karakoltzidis[3], Dimosthenis Sarigiannis[13,14,15]

[1]State General Laboratory, Ministry of Health, Senior Chemist, Nicosia, Cyprus. [2]The European Human Biomonitoring Initiative, HBM4EU, Chemical Group Leader for Mercury and HBM4EU-MOM coordinator, Nicosia, Cyprus. [3]HERACLES Research Center – KEDEK, Aristotle University of Thessaloniki, Researcher, Thessaloniki, Greece. [4]Environmental Toxicology Unit, National Centre for Environmental Health. Instituto de Salud Carlos III (CNSA-ISCIII), Researcher, Madrid, Spain. [5]Department of Epidemiology, National Institute of Health Dr. Ricardo Jorge (INSA), Researcher, Lisbon, Portugal. [6]NOVA National School of Public Health, Public Health Research Centre,UniversidadeNOVA de Lisboa, Researcher, Lisbon, Portugal. [7]Comprehensive Health Research Center (CHRC), Researcher, Lisbon, Portugal. [8]University of Iceland, Professor, Reykjavík, Iceland. [9]Current Affiliation: Egas Moniz School of Health & Science, Associate Professor, Lisbon, Portugal. [10]State General Laboratory, Ministry of Health, Researcher, Nicosia, Cyprus. [11]Jožef Stefan Institute, Department of Environmental Sciences, Researcher, Ljubljana, Slovenia. [12]VITO Health, Flemish Institute for Technological Research, Researcher, Mol, Belgium. [13]HERACLES Research Center – KEDEK, Aristotle University of Thessaloniki, Director, Thessaloniki, Greece. [14]Environmental Engineering Laboratory, Department of Chemical Engineering, Aristotle University of Thessaloniki, Professor, Thessaloniki, Greece. [15]Environmental Health Engineering, School for Advanced Study IUSS, Professor, Pavia, Italy

Abstract

Maternal seafood consumption may cause fetal exposure to methylmercury (MeHg), a potent developmental neurotoxicant. At the same time, seafood provides vital nutrients for optimal fetal development and maternal wellbeing and is culturally/economically important for many European countries. Suitable advice for seafood consumption to pregnant women must carefully balance the health benefits with potential risks from MeHg.

The HBM4EU-MOM study (2020-22) harmonized, communicated and evaluated advice for seafood consumption during pregnancy in five European countries (Portugal, Spain, Greece, Cyprus, Iceland), representing distinct seafood consumption patterns and traditions. In all, 654 pregnant women participated in the study.

The advice was based on the central theme 'fish is good, eat good fish'. It highlighted the health benefits of seafood consumption, suggested choosing a variety of preferable fresh seafood high in omega-3 fatty acids and low in levels of MeHg, explained what mercury is, why it may contaminate fish and what harm it may cause. The 'traffic light system' was used to communicate species 'to be avoided': 'red'; 'for occasional consumption': 'orange'; 'recommended': 'green' and the classification of locally relevant seafood species in these categories was harmonized. The advice was adapted for national use based on locally consumed species and was communicated to pregnant women participants. Maternal mercury biomonitoring and information collected during assisted interviews with the participants showed that it is possible to promote safe seafood consumption for prenatal MeHg control. Educated healthcare providers, armed with suitable tools, are important for effective communication, which should commence before conception.

524 – Optimizing Use of Novel Chemicals to Assess Children's Dust Ingestion Rates

Deborah Bennett

UC Davis, Professor, Davis, USA

Abstract

Dust and soil ingestion can be important exposure pathways to quantify for young children, particularly for chemicals that tend to partition to solids. Tracer elements found in soil have been used effectively to estimate ingestion rates of soil, but there is not a defined tracer for dust ingestion. The goal of this project is to develop an innovative method for determining dust ingestion rates using compounds found primarily in indoor dust rather than outdoor soil. In doing so, we will also determine the concentrations of a broad range of target and non-target compounds in house dust, skin wipes, and food to improve understanding of indoor chemical exposures. Over 80 chemicals will be measured in these environmental media. Additionally, a novel chemical panel will be run on the urine of the children. Summing metabolized forms (mol/d) and dividing by concentrations in dust (mol/g) yields a chemical- and child-specific dust ingestion rate (g/d) estimate. A variety of approaches to will be used to assess non-dust contributions (inhalation, food, product use) to chemical exposures to select optimal tracers for each child using both the Limiting Tracer Method and the Best Tracer Method used in determining soil ingestion rates. We will present the results of our analysis of all the environmental media for 20 children aged 2-3.

525 – The Role of Metabolic Perturbations in Mediating the Effects of Ambient Air Pollution on Lung Cancer in the Cancer Prevention Studies

Sabrina S. Chow[1], Ying Wang[2], Jeremy A. Sarnat[3], W. Ryan Diver[4], Emily L. Deubler[5], Douglas I. Walker[3], Ziyin Tang[1], C. Arden Pope III[6], Michael Jerrett[7], Michelle C. Turner[8], Donghai Liang[9]
[1]Rollins School of Public Health, Emory University, Graduate Student, Atlanta, USA. [2]American Cancer Society, Senior Principal Scientist, Kennesaw, USA. [3]Rollins School of Public Health, Emory University, Associate Professor, Atlanta, USA. [4]American Cancer Society, Senior Director, Kennesaw, USA. [5]American Cancer Society, Scientist, Kennesaw, USA. [6]Brigham Young University, Professor, Provo, USA. [7]Fielding School of Public Health, University of California, Los Angeles, Professor, Los Angeles, USA. [8]Barcelona Institute for Global Health, Associate Research Professor, Barcelona, Spain. [9]Rollins School of Public Health, Emory University, Assistant Professor, Atlanta, USA

Abstract
Background: Exposure to air pollution is an established risk factor for lung cancer. Investigation on mechanisms of air pollution carcinogenicity remains challenging. We aimed to address gaps by applying untargeted high-resolution metabolomics to identify metabolic perturbations associated with both ambient air pollution exposure and lung cancer incidence.

Methods: We included 680 case-control pairs within the CPS-II and -3 cohorts with data on medical condition and lifestyle. The plasma metabolome was profiled with ultrahigh-performance liquid chromatography-tandem mass spectrometry. Exposure assessment on carbon monoxide (CO), nitrogen dioxide (NO2), particulate matter (PM10), fine particulate matter, sulfur dioxide, and ozone was conducted using land use regression models based on residential address at blood draw. We conducted metabolome-wide association studies with multivariate linear regression models to assess associations of air pollution and lung cancer incidence using meet-in-the-middle (MITM) and high-dimensional mediation analysis (HDMA) approaches.

Results: Among 1,232 extracted metabolic features, MITM identified seven features significantly associated with air pollution and lung cancer (false discovery rate, q-value<0.2). Six features were significant through HDMA. All confirmed metabolites are prominent within peptide, lipid, and amino acid pathways. Gamma-glutamylglutamine and gamma-glutamylmethionine was significantly associated with CO, NO2, and PM10 exposure and lung cancer incidence in both MITM and HDMA.

Conclusion: This is the largest prospective metabolomics study examining biological perturbations associated with air pollution and lung cancer incidence. The findings indicate a potential central role of peptide metabolism in mediating cancer associations with air pollution. Findings from this study support future validations of biological mechanisms underlying air pollution carcinogenicity.

526 – Harmonization of human biomonitoring (HBM) in Europe as a driver for public health protection through policy support and citizen engagement

Andromachi Katsonouri[1], Joana Lobo Vicente[2], Argelia Castaño[3], Greet Schoeters[4], Marike Kolossa-Gehring[5]
[1]State General Laboratory, Ministry of Health, Senior Chemist, HBM4EU Chemical Group Leader for Mercury, Nicosia, Cyprus. [2]European Environment Agency, Researcher, Copenhagen, Denmark. [3]National Centre for Environmental Health. Instituto de Salud Carlos III (CNSA-ISCIII), HBM4EU Pillar Leader, Madrid, Spain. [4]Department of Biomedical Sciences, University of Antwerp, HBM4EU co-coordinator, Antwerp, Belgium. [5]German Environment Agency, HBM4EU coordinator, Berlin, Germany

Abstract
Responding to public concern and the need to protect citizens from the health impacts of toxic chemicals, the European Union has established strategic objectives, regulatory actions and programs to control their adverse effects based on sound scientific information and strengthened cooperation of stakeholders (including public authorities in Member States, European institutions and civil society). Using the example of mercury, a global pollutant requiring international cooperation for effective control, we demonstrate how harmonized European HBM drives, under this frame, public health protection through citizen engagement and policy support.

Under the 'European HBM Initiative' (HBM4EU), citizens contributed to the prioritization of mercury via a Stakeholders Forum; raised concern about the risk posed by mercury in the environment in focus-group discussions and online surveys; learned about mercury through targeted tools (video, infographic, factsheet); co-created validated guidelines for safe fish consumption in pregnancy through their participation in the HBM4EU-MOM study and learned personalized measures for mercury control through receiving the guidelines and their personal results.

Policy needs were addressed through a comprehensive framework for determining the impact of mercury exposure on European citizens (e.g. development of harmonized exposure assessment methods and HBM-based guidance values, identification of effect-biomarkers, exploring genetic susceptibility to mercury toxicity, measuring pregnant women's mercury exposure and its determinants, developing effective fish-consumption guidelines for prenatal mercury control). Input was provided to the United Nations Minamata Convention on Mercury, the European Food Safety Authority and national governments.

A sustainable European HBM system is essential to sustain this drive and ensure public health protection.

527 – Assessing wildland fire smoke exposure estimates and surrogates

Nathan Pavlovic[1], Fred Lurmann[2], Crystal McClure[3], Shih Ying Chang[4], Daniel King[5]
[1]Sonoma Technology, Inc, Lead Geospatial Data Scientist, Washington, DC, USA. [2]Sonoma Technology, Inc., Manager of Exposure Assessment Studies, Petaluma, CA, USA. [3]Sonoma Technology, Inc., Atmospheric Scientist, Petaluma, CA, USA. [4]Sonoma Technology, Inc., Air Quality Scientist / Project Manager, Petaluma, CA, USA. [5]Sonoma Technology, Inc., Geospatial Data Scientist, Petaluma, CA, USA

Abstract
Wildland fire (WLF) smoke is increasingly a dominant contributor to air pollution across regions. In addition to enhancing concentrations of criteria pollutants, WLF smoke contains a unique and variable mixture of pollutants that may render the smoke especially hazardous to human health. It is therefore important to provide accurate assessments of WLF smoke-specific exposures to support health analysis. A wide range of pollutant-specific smoke exposure estimates and general exposure surrogates have been used in health research, including chemical transport model (CTM) results, proximity to active WLF, or satellite estimates of smoke density in the atmospheric column. However, selection of an appropriate exposure metric is complicated by the difficulty of quantifying the accuracy of any given metric due to the lack of directly measurements of WLF smoke.

In this presentation, we will present the evaluation results for several WLF smoke exposure estimates and surrogates for PM2.5,, such as CTM and dispersion model results, proximity to fire, or satellite data, for use in health exposure assessment. Given the challenges related to direct measurement of WLF smoke, we use several independent WLF smoke indicators to assess the performance of these smoke results, including PM2.5 enhancement and PM2.5 chemical composition. Given uncertainties in exposure assignments for WLF smoke, it remains beneficial to consider several indicators of exposure in health analyses. The results of these assessments can be used to support prioritization of potential exposure estimates or surrogates in health exposure research.

528 – Production and Use of Human Biomonitoring Data to Steer European Public Policies Towards a Better Human Health Protection - The Case of Bisphenols Exposure Data Valorisation within the HBM4EU Project

Loïc RAMBAUD[1], Romuald TAGNE-FOTSO[2], Margaux RIOU[2], Abdelkrim ZEGHNOUN[3], Abdessatar SAOUDI[3], Liese GILLES[4], Eva GOVARTS[4], Marike KOLOSSA-GEHRING[5], Sebastien DENYS[6], Clemence FILLOL[7]
[1]Santé publique France, Epidemiologist, Saint-Maurice, France. [2]Santé publique France, Study Manager, Saint-Maurice, France. [3]Santé publique France, Statistician, Saint-Maurice, France. [4]Vito, Project Manager, Mol, Belgium. [5]UBA, Director, Berlin, Germany. [6]Santé publique France, Director, Saint-Maurice, France. [7]Santé publique France, Head of unit, Saint-Maurice, France

Abstract
Human biomonitoring (HBM) data are helpful to guide policy makers establishing regulations in order to protect human health by reducing exposure to chemicals. HBM programs aim to characterise human population exposures to a large panel of chemicals and provide relevant data at national scale. The HBM4EU project (2017-2022) built a transnational framework to produce HBM data for some chemicals of health concern at European scale. Among them, Bisphenols, for which several studies have shown their harmful effects on human health.

During the project, we initially collected existing data on Bisphenols exposure in Europe from national HBM surveys and research studies. Lack of availability and heterogeneity of the collected data narrowed us to an assessment of the exposure to Bisphenol A in European Women and did not permit the research of exposure determinants. In addition, the use of harmonised data generated by the HBM4EU Aligned Studies allowed us to characterize European adult's exposure to Bisphenols A, F and S, to describe exposure levels depending on age-, sex- or geographical factors and to identify a limited number of exposure determinants.

This work highlights the importance of the pre-harmonisation of HBM survey protocols to maximise the potential for producing standardised high quality data. A common framework to put national HBM programs into a European perspective seems crucial to better exploit HBM data. This in order to provide policy makers with relevant tools to improve current chemicals' regulations.

529 – Taking Stock of Toxic Chemical Variability in Products Used by Black Women and Latinas in South Los Angeles

Robin Dodson[1], Elissia Franklin[2], Astrid Williams[3], Sandy Navarro[4], René LaPointe Jameson[5], Bhavna Shamasunder[6]
[1]Silent Spring Institute, Research Scientist, Newton, USA. [2]Silent Spring Institute, Postdoctoral Research Fellow, Newton, USA. [3]Black Women for Wellness, Manager, Environmental Justice, Los Angeles, USA. [4]LA Grit Media, Founder, Social Documentarian, and Community Health Promoter, Los Angeles, USA. [5]Silent Spring Institute, Research Assistant, Newton, USA. [6]Occidental College, Professor, Los Angeles, USA

Abstract
Women of color are more highly exposed to consumer product chemicals, some of which are associated with hormone-mediated conditions including uterine fibroids and breast cancer. Comprehensive product use data among women of color are limited. In the Taking Stock Study, we sought to document product use by 70 Black women and Latinas in South Los Angeles. We used a new smartphone application to collect detailed product information, including brand, ingredients, and times of use, surpassing traditional survey approaches focused on product types. Participants logged product use over five days. We used this information to create an ingredient database. Participants logged >1,400 products. We screened these products' >8,000 listed ingredients for >30 endocrine-active and asthma-associated chemicals. Some participants had at least one target chemical in every product, while for others as few as 40% of products contained a target chemical. Every participant used at least one product with fragrance, nearly 80% used product(s) with parabens, and 60% used product(s) with formaldehyde releasers. We investigated whether self-reporting of product type could lead to misclassification of exposures to specific chemicals. Previous studies found an association between lotion use and paraben urine concentrations; however, in this study, only 30% of lotions listed parabens as an ingredient. Common methods for estimating exposure by product category may lead to exposure misclassification because not all products within a category contain the same ingredients of health concern. Detailed product data collected using new technologies, including apps, can improve exposure source identification, particularly in biomonitoring studies.

530 – The Impact of Phthalate Exposure on the Newborn Metabolome in the Atlanta African American Maternal Child Cohort

Susan Hoffman[1], Ziyin Tang[2], Anne Dunlop[3], Dana Barr[4], Blake Rushing[5], Susan McRitchie[6], Susan Sumner[7], Thompson Huynh[8], Parinya Panuwet[9], Stephanie Eick[10], Kaitlin Taibl[2], Grace Lee[11], Elizabeth Corwin[12], P Barry Ryan[13], Dean Jones[13], Donghai Liang[10]
[1]Emory University, Ph.D. Candidate, Atlanta, USA. [2]Emory University, Ph.D. Student, Atlanta, USA. [3]Emory University, Assistant Professor and Director, Atlanta, USA. [4]Emory University, Faculty, Atlanta, USA. [5]University of North Carolina at Chapel Hill, Assistant Professor, Chapel Hill, USA. [6]University of North Carolina at Chapel Hill, Lead Biostatistician, Chapel Hill, USA. [7]University of North Carolina at Chapel Hill, Professor, Chapel Hill, USA. [8]Emory University, MPH Student, Atlanta, USA. [9]Emory University, Research Assistant Professor, Atlanta, USA. [10]Emory University, Assistant Professor, Atlanta, USA. [11]Emory University, Senior Program Associate, Atlanta, USA. [12]Columbia University, Professor, New York, USA. [13]Emory University, Professor, Atlanta, USA

Abstract
Background: Phthalates are emerging endocrine disrupting chemicals that can negatively impact health, including adverse child development. The underlying biological mechanisms supporting these health associations are largely unknown, especially how prenatal exposure impacts the newborn metabolome at birth, hindering the development of strategies to improve maternal-child health. In this prospective analysis, we leveraged high-resolution metabolomics (HRM) to characterize the impact of prenatal phthalate exposures on the newborn metabolome.

Methods: Data collected from 278 mother-newborn pairs were available from the Atlanta African American Maternal-Child Cohort between 2014-2017. Eight phthalate metabolites were measured in maternal urine samples collected between 8- and 14-weeks gestation. Newborn dried-blood spot samples were collected at delivery and analyzed with an untargeted HRM workflow. Generalized linear models were run for each metabolite, adjusted for subject-specific and biological covariates, including creatinine.

Results: All phthalate metabolites were present across the maternal urine samples, with mono-ethylphthalate having consistently higher levels and more variability than other phthalate metabolites. After HRM, 7,228 prenatal metabolic features were detected. Pathway enrichment analyses revealed that each urinary phthalate metabolite was associated with multiple overlapping metabolic pathways, many of which are involved in inflammation and oxidative stress, including vitamin B5 - CoA biosynthesis, arachidonic acid metabolism, and cytochrome P450. Three metabolites (i.e., Coniferyl Aldehyde, 17-hydroxyprogesterone, and isoleucyl-glutamate) were confirmed using authentic chemical standards. These metabolites are linked to amino acid formation and tryptophan metabolism.

Conclusions: Our results suggest that phthalate exposure is linked to metabolic changes in the newborn metabolome, including inflammatory processes and oxidative stress.

531 – Aromatic compound metabolism and oxidative stress as potential mediators between exposure to ambient air pollution and hypertensive disorders of pregnancy--evidence from an MWAS study on Atlanta African American Maternal-Child Cohort

Haoran Cheng[1], Anne L. Dunlop[2], Zhenjiang Li[3], Douglas I. Walker[4], Jeremy A. Sarnat[4], Youran Tan[3], Dana Barr[5], P. Barry Ryan[5], Carmen Marsit[5], Dean P. Jones[5], Donghai Liang[6]
[1]Emory University, Master student, Atlanta, USA. [2]Emory University, Associate Research Professor, Atlanta, USA. [3]Emory University, PhD student, Atlanta, USA. [4]Emory University, Associate Professor, Atlanta, USA. [5]Emory University, Professor, Atlanta, USA. [6]Emory University, Assistant Professor, Atlanta, USA

Abstract

Background: Hypertensive disorders of pregnancies (HDP) are the most common medical disorder of pregnancy and a leading cause of maternal and infant morbidity and mortality. Previous studies have established an association between air pollution exposure and the occurrence of HDP, but the underlying biological mechanism is unclear, impeding targeted intention strategies.

Methods: 329 individuals from the Atlanta African American Maternal-Child cohort were included in the study (68 with HDP including 42 gestational hypertension and 26 preeclampsia). We employed liquid chromatography-high-resolution mass spectrometry to conduct metabolomics profiling on serum samples collected between 8-14 weeks of gestation and developed a spatiotemporally resolved model to estimate residential exposure to three common ambient air pollutants ($PM_{2.5}$, NO_2, and O_3) during four critical exposure windows (1yr prior to conception, 1st trimester, 1m and 1w prior to blood draw). We investigated overlapping features and pathways using a Meet-in-the-Middle Approach (MITM) and High-Dimensional Mediation Analysis (HDMA).

Results: 13,980 and 11,106 metabolic features were extracted from HILIC and C18 chromatography columns. 22 metabolites and 58 pathways involved in oxidative stress and systemic inflammation are significantly associated with air pollutant exposures during critical exposure windows and HDP, including phenylalanine, indole, benzoate, LysoPC, degradation of the aromatic compound, and propionate metabolism. Biliverdin and porphyrin metabolism pathways are also associated with exposures and outcomes, indicative of potential hepatic impairment due to aromatic compound metabolism.

Conclusions: Our findings suggest a potentially critical role of various aromatic compounds, indicative of oxidative stress, in the pathophysiology underlying the association between air pollution and HDP.

532 – Preliminary Investigation of the Use of Citizen Science and Real-Time Exposure Feedback on Indoor Air Quality in Tribal Homes

Darrah Sleeth[1], Muhammad Arashid[2], Claudia Charles[2], Steve Babits[3], Dean Goggles[4], Rod Handy[5], Scott Collingwood[6]
[1]University of Utah, Associate Professor, Salt Lake City, USA. [2]University of Utah, Research Assistant, Salt Lake City, USA. [3]Northern Arapaho Natural Resource Office, Environmental Scientist, Ethete, USA. [4]Northern Arapaho Natural Resource Office, Director, Ethete, USA. [5]University of Utah, Professor, Salt Lake City, USA. [6]University of Utah, Research Associate Professor, Salt Lake City, USA

Abstract

Native Americans continue to suffer environmental injustices and adverse health outcomes at a disproportional rate. One concern is poor indoor air quality (IAQ), which is critical since Americans spend ~90% of their time indoors. A partnership between the University of Utah and the Northern Arapaho tribe was developed to look at these issues on the Wind River Reservation, specifically PM2.5 and radon in member homes. The use of citizen science and self-deployment of exposure assessment tools was investigated, with the goal of reaching rural members. Participants were sent a kit of simple IAQ monitoring equipment for PM2.5 and radon. Detailed deployment instructions were developed, reviewed and pilot-tested by university investigators, a local community advisory board, and tribal researchers. Participants were then visited to evaluate how well these were deployed. As part of that kit, a PM2.5 monitor was designed and built that could display feedback to participants on their exposure levels in real-time. The goal was to investigate whether real-time exposure feedback resulted in reduced exposures. However, substantial challenges arose with self-deployment of the equipment. Significant reduction in PM2.5 concentration was not observed over a 2+ month deployment of monitors with real-time feedback compared to monitors without feedback. Based on these results, ongoing work includes: (1) utilizing researchers to install equipment and (2) modifications to the PM2.5 monitors to provide additional participant feedback at lower concentrations. This preliminary data analysis has provided important lessons learned for moving such research forward and increasing our understanding of IAQ in tribal homes.

534 – Exposure to perfluorooctanoic acid and effects on reproductive hormones and pubertal onset in a longitudinal study of young girls

Susan M Pinney, PHD, FACE[1], Cecily Shimpf Fassler[2], Gayle C Windham[3], Robert L Herrick[4], Changchun Xie[1], Lawrence Kushi[5], Frank M Biro[6]
[1]University of Cincinnati, Professor, Cincinnati, USA. [2]University of Cincinnati, Post doc, Cincinnati, USA. [3]California Dept. of Public Health, Professor, Richmond, USA. [4]Yolo County Health and Human Services, Epidemiologist, West Sacramento, USA. [5]Kaiser Permanente Division of Research, Director of Scientific Policy, Oakland, USA. [6]Cincinnati Children's Hospital Medical Center, Professor, Cincinnati, USA

Abstract
Background: Perfluoroalkyl substances (PFAS) cause changes in mammary gland development in rodents. A few human studies report delay in pubertal events with increasing perfluorooctanoic acid (PFOA) exposure but none have examined reproductive hormone levels at thelarche.

Methods: Girls recruited at 6-8 years underwent clinical examinations annually or semi-annually with sequential Tanner staging. PFAS concentrations were measured in the first serum sample of 704 girls; estradiol (E2), estrone (E1), testosterone (T) and dihydroandrostendione sufate (DHEAS) were measured in serum of 304 Cincinnati girls at time points around thelarche. Relationships between age at thelarche, pubarche and menarche and PFAS, body mass index (BMI), and reproductive hormones were analyzed using linear regression, survival and structural equation models.

Results: Median PFOA serum concentrations in Cincinnati (N=353, 7.3 ng/mL) and the San Francisco Bay Area (N=351, 5.8 ng/mL) were higher than the US population. In multivariable Cox proportional hazard models (adjusted for race, BMI), increasing serum PFOA was associated with a delay in pubarche (HR 0.83, 95% CI 0.70, 0.99) and menarche (HR 0.04, 95% CI 0.01, 0.25). PFNA, PFDeA and Me-PFOSA-AcOH were associated with later thelarche and Me-PFOSA-AcOH also with later pubarche. PFOA was inversely associated with DHEA-S ($p<0.01$), E1 ($p= 0.04$) and T ($p=0.03$) concentrations at six months prior to thelarche.

Conclusions: PFAS likely delay pubertal onset through the intervening effects on BMI and reproductive hormones. The decreases in DHEA-S and E1 associated with PFOA represent biological biomarkers of effect consistent with the delay in onset of puberty.

535 – Association between Prenatal Exposures to Per- and Polyfluoroalkyl Substances (PFAS) and Early Language Development in the ECHO Cohort

Patricia Cintora[1], Adaeze Wosu[2], Lesliam Quirós-Alcalá[3], Megan Woodbury[4,1], Sarah Geiger[5,6], Xiaotian Gao[7], Rachel Morello-Frosch[8], Anne Dunlop[9], Theresa Bastain[10], Anne Starling[11], Dana Dabelea[12], Carlos Camargo[13], Pi-I Lin[14], Rachel Kelly[15], Assiamira Ferrara[16], Lisa Croen[16], Tomas O'Connor[17], Susan Schantz[18,19]
[1]Beckman Institute for Advanced Science and Technology, University of Illinois Urbana-Champaign, Postdoc Research Associate, Urbana, USA. [2]Department of Epidemiology, Johns Hopkins Bloomberg School of Public Health, Research Scientist, Baltimore, USA. [3]Department of Environmental Health and Engineering, Johns Hopkins Bloomberg School of Public Health, Assistant Professor, Baltimore, USA. [4]College of Engineering, Northeastern University, Postdoc Research Associate, Boston, USA. [5]Beckman Institute for Advanced Science and Technology, University of Illinois Urbana-Champaign, Assistant Professor, Urbana, USA. [6]Department of Kinesiology and Community Health, 4University of Illinois at Urbana-Champaign,, Assistant Professor, Urbana, USA. [7]Department of Kinesiology and Community Health, 4University of Illinois at Urbana-Champaign, Assistant Professor, Urbana, USA. [8]University of California Berkeley, Professor, Berkeley, USA. [9]Department of Gynecology and Obstetrics, Emory University School of Medicine, Associate Research Professor, Atlanta, USA. [10]Department of Population and Public Health Sciences, Keck School of Medicine, Associate Professor, Los Angeles, USA. [11]Department of Epidemiology, Gillings School of Global Public Health, University of North Carolina, Assistant Professor, Chapel Hill, USA. [12]Lifecourse Epidemiology of Adiposity and Diabetes (LEAD) Center, Colorado School of Public Health, Professor, Director, Aurora, USA. [13]Department of Emergency Medicine, Massachusetts General Hospital, Harvard Medical School, Professor, Boston, USA. [14]Department of Population Medicine, Harvard Medical School and Harvard Pilgrim Health Care Institute, Health Scientist, Boston, USA. [15]14Channing Division of Network Medicine, Brigham and Women's Hospital and Harvard Medical School, Assistant Professor, Boston, USA. [16]Kaiser Permanent Northern California Division of Research, Senior Research Scientist, Oakland, USA. [17]Department of Psychiatry, University of Rochester, Professor, Rochester, USA. [18]Department of Comparative Biosciences, University of Illinois at Urbana-Champaign,, Professor Emerita, Urbana, USA. [19]Beckman Institute for Advanced Science and Technology, University of Illinois Urbana-Champaign, Professor Emerita, Urbana, USA

Abstract
Background/Aim: The relationship between prenatal exposure to per- and polyfluoroalkyl substances (PFAS) and child neurodevelopment remains inconclusive, and few studies have investigated the outcome of language development, in particular. This

study draws on data from five U.S.-based Environmental Influences on Child Health Outcomes (ECHO) cohorts to investigate associations between prenatal PFAS exposure and language development in early childhood.

Methods: We quantified concentrations of seven PFAS in maternal plasma or serum during pregnancy. We assessed language development using the communication scale from the Ages and Stages Questionnaire (ASQ) (n= 979) in children ages 1.5-5 years and the NIH Toolbox Picture Vocabulary Test (PVT) (n=299) in children ages 3-5 years. We define delayed language development as ASQ communication z-score <1.5. Associations between each prenatal PFAS and ASQ communication delay were examined in multilevel logistic regression models, while PVT continuous t-scores were examined in linear mixed models. Models were multivariable adjusted.

Results: Approximately 5.1% of children were below the 1.5 SD cut off for delay on the ASQ communication scale. Most associations between log2-transformed individual PFAS and PVT scores were positive though not statistically significant e [PFOA: β=0.38,95%CI: -1.04,1.80; PFOS: 0.91 (-0.49, 2.31); PFNA: 0.78 (-0.49,2.05); PFHxS:0.38 (-0.88,1.64); PFUNDA:1.12 (-0.06,2.41)], except for PFDA: 1.43 (0.11,2.73). We observed inverse, non-significant associations with NMFOSSA [-0.49 (-1.65,0.67)]. We did not observe any statistically significant relationships between PFAS and ASQ communication scores.

Conclusions: Prenatal concentrations of individual PFAS were not associated with PVT or ASQ scores. Future analyses will evaluate PFAS as a mixture.

536 – Association of Phthalates and DINCH Metabolite Exposure Levels with BMI in Children, Teenagers, and Adults: Human Biomonitoring CROME Study 2020-2021

Catherine Gabriel[1,2], Nafsika Papaioannou[3,4], Achilleas Karakoltzidis[3,4], Spyros Karakitsios[1,2], Denis Sarigiannis[5,6,7]
[1]HERACLES Research Center – CIRI, Aristotle University of Thessaloniki, Greece, Senior Researcher, Thessaloniki, Greece. [2]Environmental Engineering Laboratory, Department of Chemical Engineering, Aristotle University of Thessaloniki, Greece, Senior Researcher, Thessaloniki, Greece. [3]HERACLES Research Center – CIRI, Aristotle University of Thessaloniki, Greece, Pre-doctoral Researcher, Thessaloniki, Greece. [4]Environmental Engineering Laboratory, Department of Chemical Engineering, Aristotle University of Thessaloniki, Greece, Pre-doctoral Researcher, Thessaloniki, Greece. [5]Environmental Engineering Laboratory, Department of Chemical Engineering, Aristotle University of Thessaloniki, Greece, Professor, Thessaloniki, Greece. [6]HERACLES Research Center – CIRI, Aristotle University of Thessaloniki, Greece, Director, Thessaloniki, Greece. [7]Environmental Health Engineering, School for Advanced Study IUSS, Pavia, Italy, Professor, Thessaloniki, Greece

Abstract
Phthalates and DINCH substitute are synthetic organic chemicals and are widely used in household products (high molecular weight phthalates) and personal care products (low molecular weight phthalates). Phthalates were reported to have adverse effects in the human body with their obesogenic effect to be on high interest. Overweight and obesity is a major public health concern in childhood and adulthood related to puberty growth and metabolic disorders.
This study is investigating phthalates and DINCH metabolite exposure levels and their associations with BMI in three different age groups, children's, teenagers, and adults.

In the human biomonitoring CROME study, hold between 2020-2021, a total of 560 participants were recruited, 250 adults, 161 children, and 150 teenagers. The CROME study was initiated as a parent-children cohort investigating the levels of environmental pollutants and biochemical markers of exposure. All participants provided urine samples for exposure biomarker analysis and sociodemographic and dietary data. In total, the levels in human biofluids for the following exposure biomarkers have been quantified: MEP, MBzP, MiBP, MnBP, MCHP, MnPeP, MEHP, 5OH-MEHP, 5oxo-MEHP, 5cx-MEPP, MnOP, OH-MiNP, cx-MiNP, OH-MiDP, cx-MiDP, OH-MINCH and cx-MINCH.

The results for most of the metabolites show to be in higher concentrations than those reported by Koch et al 2017. Even phthalates substitute Hexamoll® DINCH® metabolites are shown to be slightly at higher levels than those presented in the literature 5 years before. The highest prevalence of overweight/obesity was reported in adults (20.04%) followed by teenagers (13.40%), and the lowest percentage to reported in children (12.42%).

537 – POSTER: Developing Chemical Signatures for 5 Categories of Household Products Using Non-Targeted Analysis

Zachary Stanfield[1], Kristin Favela[2], Alice Yau[3], Katherine Phillips[4], Antony Williams[5], Kristin Isaacs[4], John Wambaugh[4]
[1]US EPA, Postdoctoral Research Scientist, RTP, USA. [2]Southwest Research Institute, Analytical Chemist, San Antonio, USA. [3]Southwest

Research Institute, Principal Scientist, San Antonio, USA. [4]US EPA, Research Physical Scientist, RTP, USA. [5]US EPA, Cheminformatician, RTP, USA

Abstract
Consumer products are a major source of chemical exposure and therefore potential risk. It is important to understand what chemicals are typically present in different types of products not only for risk evaluation but also to assess new products for similarity with existing ones and identify any uncommon chemical ingredients. Non-targeted analysis (NTA) using two-dimensional gas chromatography time-of-flight mass spectrometry (GCxGC-TOFMS) was applied to 118 unique consumer products from 5 categories. A total of 489 probable chemical structures were tentatively identified using the NIST 17 spectral library (109 confirmed via chemical standards) across all products. Chemicals were characterized by functional use (in products) and chemical structural class. Fabric products contained the most total and unique chemicals (230 and 118, respectively) identifiable by GCxGC-TOFMS while kitchen products contained the least (63 and 30, respectively). Use of duplicated samples and repeats of the same individual products allowed for a within-product category similarity assessment, which showed highest variability in personal care products and lowest in kitchen products. Chemical ingredient signatures (including estimated sample concentration ranges) for each product type were obtained via a clustering analysis and a regularized classification model (97.52% training accuracy). This study provides a baseline set of chemical ingredients (that is, representative mixtures) across common types of consumer products, which will help in evaluating new and existing products. Separating constituent chemicals into typical and atypical might inform exposure assessment, in vitro bioactivity screening, and ultimately the risk related to using such products. This abstract does not reflect U.S. EPA policy.

538 – POSTER: Assessment of Fire Investigator Ensembles Against Chemical Vapor, Airborne Particle Infiltration, and Thermal Burden

Mark Gaskill[1], Bryan Ormond[2]
[1]NCSU, Research Assistant, Raleigh, USA. [2]NCSU, Assistant Professor, Raleigh, USA

Abstract
The evolution of Personal Protective Equipment (PPE) used in the fire service has historically focused on thermal protection and the thermal burden of firefighters. While these advancements in firefighter PPE have been impactful, the notion of protecting firefighters has now grown to include protection from fireground contaminants that have become largely associated with cancer in the fire service. While firefighter PPE and procedures have grown to mitigate these hazards, other occupations in the fire service have not, particularly fire investigators. To determine the effectiveness of fire investigator PPE, a selection of ensembles was tested against chemical vapor, aerosolized particulate, and thermal burden. The selection of ensembles used in testing included representative PPE commonly used by fire investigators as well as prototype garments selected for their suitability for use in fire investigations. Chemical vapor testing was conducted in accordance with the National Fire Protection Association's (NFPA) Man-In-Simulant Test protocol for determining the protection factor of a complete ensemble against the infiltration of chemical vapor. Aerosolized particulate testing was carried out utilizing a prototype manikin with integrated real-time aerosol detection to determine the localized infiltration of airborne particulate into the ensemble. Assessments of thermal burden were carried out using a thermal sweating manikin and measurements from ASTM F 2370 and ASTM F 1291. This evaluation of ensembles was conducted to provide recommendations to fire investigators for suitable PPE for use during fire investigations as well as performance requirements to be made to NFPA 921 Guide for Fire and Explosion Investigations.

539 – Exposure to Organophosphate Ester Flame Retardants and Plasticizers (OPEs) During Pregnancy and Autism-Related Outcomes

Jennifer Ames[1], Juanran Feng[2], Lyndsay Avalos[3], Emily Barrett[4], Theresa Bastain[5], Deborah Bennett[6], Jessie Buckley[7], Courtney Carignan[8], Patricia Cintora[9], Assiamira Ferrara[3], Akhgar Ghassabian[10], Ghassan Hamra[11], Monique Hedderson[3], Julie Herbstman[12], Ixel Hernandez[13], Margaret Karagas[14], Catherine Karr[15], Jordan Kuiper[16], Donghai Liang[17], Kristen Lyall[18], Cynthia McEvoy[19], Rachel Morello-Frosch[20], John Meeker[21], Thomas O'Connor[22], Jiwon Oh[23], Alicia Peterson[24], Lesliam Quiros-Alcala[11], Sheela Sathyanarayana[15,25], Rebecca Schmidt[26], Anne Starling[27], Tracey Woodruff[28], Heather Volk[7], Yeyi Zhu[29], Lisa Croen[3]
[1]Kaiser Permanente Northern California, Research Scientist I, Oakland, USA. [2]Kaiser Permanente Northern California, Data Consultant, Oakland, USA. [3]Kaiser Permanente Northern California, Research Scientist III, Oakland, USA. [4]Rutgers University, Assoc Professor, Piscataway, USA. [5]University of Southern California, Assoc Professor, Los Angeles, USA. [6]University of California, Davis, Professor, Davis, USA. [7]Johns Hopkins, Assoc Professor, Baltimore, USA. [8]Michigan State University, Assistant Professor, East Lansing, USA. [9]University of Illinois Urbana-Champaign, Postdoctoral Fellow, Urbana, USA. [10]New York University, Assoc Professor, New York, USA. [11]Johns Hopkins, Assistant Professor, Baltimore, USA. [12]Columbia University, Professor, New York, USA. [13]University of Southern California, Doctoral Student, Los Angeles, USA. [14]Dartmouth University, Professor, Hanover, USA. [15]University of Washington, Professor, Seattle, USA. [16]Johns Hopkins, Assistant Scientist,

Baltimore, USA. [17]Emory University, Assistant Professor, Atlanta, USA. [18]Drexel University, Assoc Professor, Philadelphia, USA. [19]Oregon Health & Science University, Professor, Portland, USA. [20]University of California, Berkeley, Professor, Berkeley, USA. [21]University of Michigan, Professor, Ann Arbor, USA. [22]University of Rochester, Professor, Rochester, USA. [23]University of California, Davis, Postdoctoral Fellow, Davis, USA. [24]Kaiser Permanente Northern California, Postdoctoral Fellow, Oakland, USA. [25]Seattle Children's Research Institute, Professor, Seattle, USA. [26]University of California, Davis, Assoc Professor, Davis, USA. [27]University of North Carolina, Assistant Professor, Chapel Hill, USA. [28]University of California, San Francisco, Professor, San Francisco, USA. [29]Kaiser Permanente Northern California, Research Scientist II, Oakland, USA

Abstract

Objective: We investigated whether urinary concentrations of organophosphate esters (OPEs) during pregnancy were associated with child autism-related outcomes in a large and diverse US sample.

Material and Methods: We included 3,246 mother-child pairs born between 2006-2020 from 15 cohorts in the NIH Environmental influences on Child Health Outcomes (ECHO) Program. Nine OPEs were measured in spot urine samples collected during mid- to late-pregnancy. We assessed child autism-related traits via a) parent report on the Social Responsiveness Scale (SRS) and b) clinical autism diagnosis from parent report and medical records. We examined relations of OPEs with SRS T-scores and autism diagnosis in multivariable linear and logistic regression models, respectively, using generalized estimating equations to account for clustering by cohort. OPEs were adjusted for urinary dilution and modeled continuously (log2-transformed), categorically (high [> median], low [≤ median], non-detect), or as detect/non-detect depending on the proportion of values above the limit of detection.

Results: The mean SRS T-score was 47.7 (SD=8.5) and 94 children had an autism diagnosis. Bis(1-chloro-2-propyl) phosphate (BCPP) and bis(butoxyethyl) phosphate (BBOEP) were associated with modest increases in autistic traits (BCPP adj-β for high vs. non-detect: 1.09, 95% CI: 0.18, 1.99; BBOEP adj-β for high vs. non-detect: 1.08, 95% CI: 0.69, 1.47). These analytes were also associated with greater odds of autism diagnosis (BCPP adj-OR: 1.25, 95% CI:1.03, 1.51; BBOEP adj-OR 1.33 [1.03, 1.73]).

Conclusion: Pregnancy exposures to OPEs, specifically BCPP and BBOEP, may be associated with slightly greater likelihood of traits and diagnosis of autism in childhood.

540 – Measuring in Uncharted Locations: Field-Calibrations of Low-Cost Sensors far from Reference Instruments

Kirsten Koehler[1], Neta Friedberg[2], Yi-Ting Lin[2], Misti Zamora[3], Colby Buehler[4], Lei Hao[5], Drew Gentner[6], Abhirup Datta[7]
[1]Johns Hopkins University, Associate Professor, Baltimore, USA. [2]Johns Hopkins University, Student, Baltimore, USA. [3]University of Connecticut Health, Assistant Professor, Farmington, USA. [4]Yale University, Student, New Haven, USA. [5]Johns Hopkins University, Research Specialist, Baltimore, USA. [6]Yale University, Associate Professor, New Haven, USA. [7]Johns Hopkins University, Assistant Professor, Baltimore, USA

Abstract

Calibrating networks of low-cost sensors can be resource intensive. Sensors often require unit-specific calibrations and can have strong dependencies on temperature, humidity, and interfering gases. Due to the limited range of conditions that can be reproduced in the laboratory, collocating sensors with reference grade instrumentation in the field is the preferred method to calibrate sensors. However, co-locating all sensors at a site with reference measurements can pose a logistic challenge. We evaluated a method in which we identified time periods where the atmosphere was well-mixed over an urban area to select training periods for fitting unit-specific calibration models using reference data from a non-collocated site. This method was evaluated in a low-cost sensor network operating in Baltimore, MD, USA. Sensors for NO_2 and CO_2 (Alphasense) were co-located at sites measuring reference-grade concentrations operated by the Maryland Department of the Environment for NO_2 or the National Institutes of Standards and Technology for CO_2 (two reference sites per gas). Random Forest models were constructed using time periods with the smallest relative differences between the concentrations at the two reference sites for training. We found minimal information loss when calibrating the sensor at one site with reference data from the other site located about 11 km away compared to calibration using the collocated reference data (<5% change in R^2 for NO_2; 6-33% change for CO_2). Identified training periods were used to calibrate 40+ other sensors in the network and maps of variability in concentrations were produced. Relationships between concentrations and sources are explored.

541 – Metabolomic Signatures as Potential Mediators in the Relationship between Ambient Fine Particulate Matter and Pre- and Early-term Births in the Atlanta African American Maternal-Child Cohort

Zhenjiang Li[1], Anne L. Dunlop[2], Jeremy A. Sarnat[3], Anke Huels[4], Stephanie M. Eick[4], Peng Gao[5], Youran Tan[1], Haoran Cheng[6], Anna Knight[7],

Dana Boyd Barr[8], Alicia K. Smith[3], Carmen Marsit[9], Dean P. Jones[9], Donghai Liang[4]
[1]Emory University, PhD student, Atlanta, USA. [2]Emory University, Associate Research Professor, Atlanta, USA. [3]Emory University, Associate Professor, Atlanta, USA. [4]Emory University, Assistant Professor, Atlanta, USA. [5]University of Pittsburgh, Assistant Professor, Atlanta, USA. [6]Emory University, Master student, Atlanta, USA. [7]Emory University, Instructor, Atlanta, USA. [8]Emory Unversity, Professor, Atlanta, USA. [9]Emory University, Professor, Atlanta, USA

Abstract
Background: African American (AA) are disproportionately exposed to higher levels of ambient fine particular matter ($PM_{2.5}$) and have higher rates of pre- (PTB, <37 weeks) and early-term birth (ETB, 37-38 weeks). We aimed to investigate whether specific metabolic signatures mediate the association between prenatal $PM_{2.5}$ exposure and the risk of PTB and ETB.

Methods: We included 329 pregnant AA people between 18-40 years of age without chronic medical conditions in Atlanta, GA from 2014-2018. Residential exposure to ambient PM2.5 was estimated using spatiotemporally-resolved air pollution model for 1 year before conception, the first trimester, and for 1 week and 1 month before the enrollment blood draw (6-17 weeks' gestation). Metabolic signatures in serum samples collected at enrollment were evaluated using an untargeted high-resolution metabolomics workflow. The mediating metabolic signals and biological pathways were investigated by the Meet-in-the-Middle (MITM) approach and high-dimensional mediation analysis (HDMA).

Results: Fifty-eight features, including adenosine and methyladenosine, were associated with $PM_{2.5}$. Eight features mediated the adverse effects of $PM_{2.5}$ on early birth, with one feature (mz: 202.0862, rt: 227.2s detected in a polar chromatography column with positive electrospray ionization) consistently found using both the MITM and HDMA. The pathway analysis identified six biological pathways, including vitamin B9 (folate) metabolism, and glycine and serine metabolism, indicating the potential mechanisms underlying $PM_{2.5}$ toxicity on early birth.

Conclusions: Metabolic signatures detected in early pregnancy, which play important roles in the urea cycle, serine and cysteine metabolism and folate metabolism, mediated the adverse effects of $PM_{2.5}$ on PTB or ETB.

542 – Regression Calibration of 15 Particle Components in the United States from 2000 to 2019

Bryan N. Vu[1], Heresh Amini[2], Yijing Feng[3], Joel Schwartz[4]
[1]Harvard University, Research Fellow, Boston, USA. [2]Icahn School of Medicine at Mount Sinai, Associate Professor, New York, USA. [3]Harvard University, Student, Boston, USA. [4]Harvard University, Professor, Boston, USA

Abstract
Estimates for air pollutants including $PM_{2.5}$ and its components derived from models utilizing remote sensing and land use variables provide fine spatiotemporal data that are critical for epidemiological research; however, these estimates may contain bias and measurement error. We aim to calibrate annual predictions of 15 particle components from 2000 – 2019 in the United States in four census regions in urban and nonurban areas to reduce measurement error and improve the estimates for epidemiological studies. Annual predictions of 15 particle components including organic carbon, elemental carbon, nitrate, and sulfate derived from an ensemble of machine learning models were first divided into urban and nonurban areas and further stratified into four census regions and calibrated with meteorological variables including temperature, relative humidity and planetary boundary layer height, land use categorizations, slope and aspect ratio of monitor location, population density, inverse weight of the probability of having a monitor, and distance to steel plants. Our regression calibration models achieved an R^2 between 0.32 and 0.94 with models for potassium and nickel performing the worst and models for sulfate and ammonium performance the best. Of the total 60 models, 55 achieved an R^2 greater or equal to 0.5 and 30 achieved an R^2 greater or equal to 0.75. Our results indicate that certain components are more susceptible to measurement error and predictions from our models are improved and more generalizable for epidemiologic studies.

543 – Raising Standards to Lower Diesel Emissions: Biomonitoring Evidence Reflects Success of Emissions Reductions

Kristin Knox[1], Robin Dodson[2], Ruthann Rudel[3], Claudia Polsky[4], Megan Schwarzman[5]
[1]Silent Spring Institute, Staff Scientist, Newton, USA. [2]Silent Spring Institute, Research Scientist, Newton, USA. [3]Silent Spring Institute, Director of Research, Newton, USA. [4]University of California, Berkeley, Clinical Professor of Law, Berkeley, USA. [5]University of California, Berkeley, Associate Project Scientist and Continuing Lecturer, Berkeley, USA

Abstract
Fine particulate air pollution is increasingly driving the global burden of disease, and diesel-powered vehicles are substantial contributors. California has diverged from U.S. federal policy, reducing mobile-source diesel emissions using the state's authority under the Clean Air Act to regulate air pollution more stringently than the federal government. By analyzing EPA National Emissions Inventory data from 1990-2014, we showed that policies enacted by the state during that time reduced emissions to the extent that by 2014 California was producing less than half the diesel emissions as would be expected had the state followed the same trajectory as the rest of the country. We then sought to determine whether NHANES biomonitoring data reflected Californians' lower exposures to diesel exhaust. Previous studies have identified fuel combustion, especially diesel emissions, as a dominant source of ambient polycyclic aromatic hydrocarbon (PAH) concentrations and have found that both ambient air and urinary levels of many PAHs are declining over time; they conclude that restricting motor vehicle emissions, especially from diesel engines, reduces PAH exposures. Similarly, we found lower levels of PAHs, including metabolites of fluorene, naphthalene, phenanthrene, and pyrene, in the urine of Californians compared with the rest of the U.S. population. Biomonitoring data of this kind can advance public policy discussions by demonstrating exposure reductions associated with rules controlling air pollution.

544 – POSTER: Personal Exposure to Particle-Bound PAHs in an Esophageal Cancer Pilot Study in Western Kenya

Ryan Chartier[1], Daniel Pope[2], James Watson[3], Diana Menya[4], Elisa Puzzolo[5], Iva Cukic[6], Edna Sang[7]
[1]RTI International, Research Chemist, RTP, USA. [2]University of Liverpool, Professor, Liverpool, United Kingdom. [3]University of Liverpool, PhD Candidate, Liverpool, United Kingdom. [4]Moi University, Professor, Eldoret, Kenya. [5]University of Liverpool, Senior Research Fellow, Liverpool, United Kingdom. [6]Manchester Metropolitan University, Research Fellow, Manchester, United Kingdom. [7]Moi University, Researcher, Eldoret, Kenya

Abstract
Western Kenya experiences elevated rates of esophageal squamous cell cancer (ESCC) even in population groups lacking common risk factors (smoking and alcohol). Cooking and heating using solid biomass fuel is common in this area and the emissions associated with combustion of these fuel sources may contribute to cancer risk. To investigate the potential relationship between HAP exposure and cancer risk, and to collect data to support the future generation of new hypotheses, we collected direct $PM_{2.5}$ samples from a subset of participants in an ESCC study in Western Kenya.

We collected 24-hour personal and kitchen $PM_{2.5}$ exposures from people diagnosed with ESCC (cases, N=15) and controls (N=23) using the RTI MicroPEM. The $PM_{2.5}$ samples were extracted and analyzed for 23 polycyclic aromatic hydrocarbons (PAHs) via gas-chromatograph/mass spectrometry. The use of LPG for cooking significantly reduced personal exposures compared to wood burning for both $PM_{2.5}$ (54.2 vs. 156.1 μg/m^3, p=0.027) and total PAHs (25.5 vs 121.4 ng/m^3, p=0.003). Smokers had 60% higher $PM_{2.5}$ exposures and 115% higher PAH exposures than non-smokers, but neither increase was significant.

Overall, no significant differences in exposures were observed between cases and controls. PAH concentrations were converted to benzo(a)pyrene equivalents and the mean calculated inhalation incremental lifetime cancer risk for all participants was 9^{-5}, similar to the high risk threshold (>10^{-4}), even without accounting for the additional burden of dermal and ingestion exposures. These data indicate that solid fuel combustion may significantly contribute to increased cancer risk in Western Kenya.

545 – Soil and Dust Metal Concentrations in Rural Indiana

Edem Sededji[1], Salma Akhter[1], Sa Liu[2], Jessica Huber[3], Aaron B Bowman[4], T. Eric Evans[5], Sherrie Steiner[6], Ellen Wells[7]
[1]Purdue University West Lafayette, PhD Student, School of Health Sciences, West Lafayette, USA. [2]Purdue University West Lafayette, Assistant Professor, School of Health Sciences, West Lafayette, USA. [3]Purdue University West Lafayette, Professor, Department of Speech, Language, and Hearing Sciences, West Lafayette, USA. [4]Purdue University West Lafayette, Professor, School Head, School of Health Sciences, West Lafayette, USA. [5]Blackford County Concerned Citizens (BCCC) group, President,Blackford County Concerned Citizens (BCCC) group, Hartford City, USA. [6]Purdue University Fort Wayne, Associate Professor, Department of Anthropology and Sociology, Fort Wayne, USA. [7]Purdue University Fort Wayne, Associate Professor, Department of Public Health, School of Health Sciences, West Lafayette, USA

Abstract
Chronic metal exposures may harm the nervous system. However, information about soil and dust metal contamination on a local level, especially rural areas, is limited. This analysis determined metal concentrations in soil and dust from Hartford City, Indiana, USA, due to

community concerns about industrial pollution. Additional samples from other rural areas of Indiana were also collected. Participants collected soil and dust samples from their own homes using a sampling kit; these were analyzed using x-ray fluorescence. N=36 kits from Hartford City, IN and N=9 kits from other rural locations have been analyzed to date. Mean soil Pb=48.8 ppm (standard deviation (SD)=44.5) from Hartford City and 18.2 ppm (SD=12.6) from the other locations. Pb dust concentrations were 79.1 ppm (SD=108.3) from Hartford City and 67.6 ppm (SD=101.7) from the other locations. N=3 (8%) of dust samples from Hartford City were higher than the US Environmental Protection Agency's Pb dust guideline (300 ppm). In Hartford City, mean soil As=13.2 ppm (SD=11.7) and 19.5 ppm (SD=19.8) in dust. These mean values are higher than the current Indiana state guideline for soil As (10 ppm). Mean soil Cd, and Cr were below guideline values in both soil and dust. Pb, As, and Cr dust concentrations were higher than soil concentrations. Soil Cd concentrations were higher than dust concentrations. Additional analyses are ongoing. Current results suggest Pb in dust and As in both soil and dust may be of concern in some parts of rural Indiana.

546 – Addressing Uncertainties in Oral Bioavailability of Chemical Contaminants

Shenghong Wang[1], Li Li[2]
[1]University of Nevada, Reno, Graduate student, Reno, USA. [2]University of Nevada, Reno, Assistant Professor, Reno, USA

Abstract
Understanding the oral bioavailability of chemical contaminants is essential for evaluating their exposures, toxicokinetics, health effects, and risks to humans. Most existing high-throughput physiologically based toxicokinetic (PBTK) modeling efforts rely on a constant gut absorption rate extrapolated from in vitro transmembrane permeability assays and ignore chemical loss due to first-pass metabolism. Such a simplifying approach may introduce uncertainty for certain chemicals. For instance, it does not account for the differences in conditions between the human intestine and transmembrane permeability assays, e.g., the aqueous boundary layer. Additionally, it assumes steady-state mass transfer, which may not hold for hydrophobic chemicals that require an extremely long time to reach the steady state, often exceeding the residence time of digesta in the gastrointestinal tract. This presentation introduces our improvement of the traditional PBTK modeling approach by accounting for non-steady-state gastrointestinal absorption of chemicals and pre- and post-absorption biotransformation. We evaluated our model using human oral absorption efficiency and time-dependent plasma concentration data, demonstrating its reliability and practicality. Inputs for the model include partition coefficients, molecular weight, and intrinsic hepatic clearance, all widely available or estimable using various quantitative structure-activity/property relationships, empowering this model as a practical tool for high-throughput screening of the multitude of chemicals in the market, or even before production. We also predict oral bioavailability for a wide range of chemicals of emerging concerns. The application of this new model improves our understanding of how chemical properties impact oral bioavailability and contributes to the development of New Approach Methods.

547 – Assessing Personal Air Exposure to VOCs in Hispanic Women Working as Domestic Cleaners in South Texas, 2022-2023

Inkyu Han[1], David Gimeno Ruiz de Porras[2], Elaine Symanski[3], Kelly Oyer-Peterson[4], Ana Pineda Reyes[5], Edward Brooks[6], George Delclos[7], Masoud Afshar[8], Kristina Whitworth[9]
[1]Temple University, Associate Professor, Philadelphia, USA. [2]UTHealth, Professor, San Antonio, USA. [3]Baylor College of Medicine, Professor, Houston, USA. [4]UTHealth, Doctoral Student, San Antonio, USA. [5]UTHealth, MPH Student, San Antonio, USA. [6]UTHealth San Antonio, Professor, San Antonio, USA. [7]UTHealth, Professor, Houston, USA. [8]UTHealth, Research Engineer, Houston, USA. [9]Baylor College of Medicine, Associate Professor, Houston, USA

Abstract
Domestic cleaners are essential in the US workforce, providing professional cleaning services. However, their work exposes them to health hazards associated with toxic chemicals, particularly volatile organic compounds (VOCs) present in cleaning and disinfection products. Further, domestic cleaners may also experience VOC exposures because of the neighborhoods they live in, as well as indoors in their own homes. Exposure to VOCs has been linked to respiratory and other health effects. Self-employed cleaners, who work independently rather than for a company, are especially vulnerable due to limited access to personal protective equipment, training, and other workplace protections. Partnering with a grassroots organization in San Antonio, Texas, we conducted a personal air exposure assessment of 30+ VOCs among 48, primarily Hispanic, female self-employed domestic cleaners. Participants wore a passive badge for three consecutive days during their activities in all locations, while a second badge was simultaneously worn only when participants were inside their homes. Preliminary findings showed that alkanes, terpenes, aromatics, and chlorinated VOCs comprised 49%, 38%, 10%, and 3% of the total personal air exposure to VOCs, respectively. Their own indoor home exposure contributed to ~70% of VOC exposure. Our next step is to identify work-related, environmental, and individual-level factors that affect exposure to VOCs and to evaluate

associations of total personal VOC exposures with biomarkers of inflammation and oxidative stress measured in urine, blood, and exhaled breath samples. Study outcomes should provide valuable insights for developing targeted interventions to promote safer working conditions and home environments for these vulnerable workers.

548 – Lessons Learned Along the Road to Hazardous Waste Clean-up

Susan M Pinney[1], Marilyn V Howarth[2], Kathleen M Vandiver[3], Adrian Wood[4], Bob Adams[5], Sharon Cook-Vargus[5], Lisa Crawford[6], Graham Mitchell[6], Martha K Stevenson[7], Suzanne Sullivan[7], Kayla McGill[8], Kent Meloy[9], Joyce Bentle[10]
[1]University of Cincinnati, Professor, Cincinnati, USA. [2]University of Pennslvania, Associate Professor, Philadelphia, USA. [3]MIT, Director, Cambridge, USA. [4]University of Pennsylvania, Program Coordinator, Philadelphia, USA. [5]University of Pennsylvania, Community Partner, Philadelphia, USA. [6]University of Cincinnati, Community Partner, Cincinnati, USA. [7]MIT, Community Partner, Cambridge, USA. [8]University of Cincinnati, Instructional Designer, Cincinnati, USA. [9]University of Cincinnati, Producer, Cincinnati, USA. [10]University of Cincinnati, Director, Cincinnati, USA

Abstract
Background: Hazardous Waste clean-up can be contentious. Information gradients and poor communication among stakeholders contribute to confusion, delays, and mistrust. There are very few educational resources for community organizations, governmental agencies and/or commercial entities to improve multi-directional communication while addressing an environmental hazard and navigating the cleanup process. Employing first-hand experiences and lessons learned, we provide a tool for groups beginning the process that supports the evolution of productive relationships between government, regulators and community members.

Methods: In a three-site collaboration among Community Engagement Cores we held a series of discussions distilling unique circumstances and commonalities among sites. We collected media of all types and developed three video modules to help other sites and stakeholders facing the often difficult and lengthy process of achieving hazardous waste clean-up. The "lessons learned" for these modules come from the Fernald OH uranium processing facility site, the Ambler PA asbestos manufacturing site, and the Wilmington MA Olin Chemical Superfund Site, and were developed by a multi-disciplinary team of community stakeholders, professionals in film development and academic partners.

Results: The content addresses: assembling a group including community members, regulators, elected officials and representatives of the commercial site entity; encouraging self-education; conducting an evaluation; productive protesting; speaking with the media; working towards agreement on big concepts; considering cost and feasibility. Samples from the video modules will be shown.

Impact Statement: The well-articulated lessons are presented in an engaging, interactive format and directed towards all stakeholders, providing insights about public health, community planning, and resource development.

549 – POSTER: Harnessing the power of metabolomics for the development of targeted intervention strategies

Blake Rushing[1,2], Timothy Fennell[3], Wimal Pathmasiri[1,2], Susan Sumner[1,2]
[1]University of North Carolina at Chapel Hill, Durham, USA. [2]Nutrition Research Institute, Kannapolis, USA. [3]Discovery Sciences, RTI International, Research Triangle Park, USA

Abstract
Untargeted metabolomics has the potential to revolutionize intervention strategies aimed at improving health and wellness by identifying aberrant metabolic targets associated with internal and external life-long exposures. We have developed a "Nutritional Pharmacology/Toxicology" concept that utilizes high-information metabolomics workflows to design nutritional interventions aimed at mitigating adverse effects or enhancing positive effects caused by exposures. Nutrients and other dietary compounds modulate similar biochemical processes, and by targeting specific metabolic pathways, they can counteract the negative effects of exposures or enhance their beneficial effects. This approach has the potential for precision nutrition, where nutrients and diets can be used to target gene-environment interactions and subpopulation characteristics. By deriving "nutrient cocktails," this approach presents an opportunity to modify the effects of exposures for more beneficial outcomes in public health. Nutritional Pharmacology/Toxicology, based on untargeted metabolomics data from human subject studies and experimental model systems, offers a promising approach to developing intervention strategies aimed at improving health and wellness, and can be applied to a wide variety of research areas. Ultimately, this approach could have a significant impact on public health by reducing the burden of disease caused by environmental exposures or by enhancing the beneficial effects of exposures such as prescription drugs or phytochemicals found in the diet.

551 – Associations between prenatal exposure to a mixture of phthalates, DINCH, and heavy metals and adverse birth outcomes in the EXHES Greece Cohort

Nafsika Papaioannou[1,2], Anthoula Chatzimpaloglou[3,4], Catherine Gabriel[5,6], Marios Maroulis[7,8], Maria Perikli[7,8], Spyros Karakitsios[5,9], Dimosthenis Sarigiannis[10,11,12]
[1]Environmental Engineering Laboratory, Department of Chemical Engineering, Aristotle University of Thessaloniki, PhD Candidate, Thessaloniki, Greece. [2]HERACLES Research Center on the Exposome and Health, Center for Interdisciplinary Research and Innovation, Researcher, Thessaloniki, Greece. [3]Environmental Engineering Laboratory, Department of Chemical Engineering, Aristotle University of Thessaloniki, Postdoctoral Researcher, Thessaloniki, Greece. [4]HERACLES Research Center on the Exposome and Health, Center for Interdisciplinary Research and Innovation, Postdoctoral, Thessaloniki, Greece. [5]Environmental Engineering Laboratory, Department of Chemical Engineering, Aristotle University of Thessaloniki, Senior Researcher, Thessaloniki, Greece. [6]HERACLES, Senior Researcher, Thessaloniki, Greece. [7]Environmental Engineering Laboratory, Department of Chemical Engineering, Aristotle University of Thessaloniki, Researcher, Thessaloniki, Greece. [8]HERACLES, Researcher, Thessaloniki, Greece. [9]HERACLES, Senior Reseacrher, Thessaloniki, Greece. [10]Environmental Engineering Laboratory, Department of Chemical Engineering, Aristotle University of Thessaloniki, Professor, Thessaloniki, Greece. [11]HERACLES, Director, Thessaloniki, Greece. [12]National Hellenic Research Foundation, Director and President of the Board, Athens, Greece

Abstract
Prenatal co-exposure to phthalates, DINCH, and metals are a societal concern due to their endocrine-disrupting properties. We aimed to investigate the associations of prenatal plasticizers and heavy metals co-exposure with birth outcomes in the EXHES study (n=60 pairs of mother-neonate from Greece). Pregnant women in the third trimester were recruited between 2020 to 2023 from Athens, Thessaloniki, Kavala, Crete, Naoussa, Xanthi, Serres and Chalkidiki. 14 urinary phthalates (MEP, MBzP, MiBP, MnBP, MCHP, MnPeP, MEHP, 5OH-MEHP, 5oxo-MEHP, 5cx-MEHP, MnOP, OH-MiNP, cx-MiNP, and OH-MiDP) and 2 DINCH metabolites (OH-MINCH and cx-MINCH) were quantified using an online SPE LC-MS/MS method. The analytical method was optimized and validated in the ICI/EQUAS scheme organized by the HBM4EU initiative. We quantified 11 toxic and essential metals (As, Cd, Co, Cu, Mn, Ni, Pb, Se, V, Zn, and Hg) in urine samples using an ICP-MS. The results were expressed as μg/g creatinine's. The birth outcomes were measured by field nurse/physician or self-reported in questionnaires. We applied elastic net regression to identify important predictors and explored the chemical mixture's joint effects using BKMR. MEP was the most abundant phthalate metabolite (95th percentile 95.7), while As was the most abundant heavy metal (95th percentile 51.5). In addition, our study revealed that weight at birth was strongly associated with Hg and Mn. In conclusion, this was the first study to measure the prenatal co-exposure to 16 metabolites of plasticizers and 11 meals to assess the effects on birth outcomes while covering north, central and south Greece.

552 – POSTER: Assessing the Impact of Historical Redlining and Modern-Day Segregation on Housing Quality: An Atlanta Case Study

Salvatore Milletich[1], Andres Manrique[1], Tamara Spikes[2], Anuj Nanavati[3], Sonia Karsan[4], Eric Coker[5], Christine Ekenga[6]
[1]University of Florida, Graduate Assistant, Gainesville, USA. [2]Emory University, PhD Student, Atlanta, USA. [3]Emory University, MPH student, Atlanta, USA. [4]Emory University, BA student, Atlanta, USA. [5]British Columbia Centre for Disease Control, Senior Scientist, Vancouver, Canada. [6]Emory University, Assistant Professor, Atlanta, USA

Abstract
Background: The United States federal government developed the Home Owner's Loan Corporation (HOLC) in 1930's which systemically withheld financial services based on the racial makeup of the individual's neighborhood. HOLC scored neighborhoods from A to D, with D areas marked in red (i.e., redlined) and deemed as places where property values were most likely to go down. Atlanta, like many urban cities in the United States, was subject to severe historical redlining, leading to a lack of investment in communities of color. Redlining has garnered significant attention in environmental health research. However, few studies have assessed associations between redlining and modern-day housing quality.

Methods: An ecological study was conducted to assess the relationship between historical redlining, modern-day segregation, and housing code violation complaints. Community-reported housing code violation complaints, reported by residents to the City of Atlanta from 2015-2019, were characterized as hazardous or non-hazardous code violations.

Results: Point pattern analysis and spatial clustering techniques revealed that total and hazardous complaints were highly concentrated in central and southern Atlanta aligning with C and D-graded areas, with hazardous complaints exclusively in D areas. A Bayesian multilevel Poisson regression revealed that modern-day segregation and vacant properties were significant predictors for total and hazardous complaints. Redlining was significant for total complaints, indicating that modern-day neighborhood-level issues drive the most hazardous complaints.

Conclusions: Historical redlining was associated with community-reported housing code violation complaints, while modern-day segregation was associated with the most hazardous complaint reports. Community-reported complaints can provide insight into housing quality.

553 – POSTER: Dermal Permeation of Polyfluoroalkyl Carboxylic Acids using Flow-Through Diffusion of Porcine Skin

Andrew Hall[1], Bryan Ormond[2], Ronald Baynes[3]
[1]North Carolina State Univerisity, Graduate Student, Raleigh, USA. [2]North Carolina State Univerisity, Assistant Professor, Raleigh, USA. [3]North Carolina State Univerisity, Professor, Raleigh, USA

Abstract
Per- and Polyfluoroalkyl Substances (PFAS) are found in a variety of places including cosmetics, rain jackets, dust, and water. All these sources provide dermal exposure to PFAS. In addition, firefighters are exposed to aqueous film-forming foams, liquid repellents in turnout gear, and moisture barriers, which likely contribute to the elevated levels of PFAS found in their blood serum. PFAS have been identified as possibly carcinogenic and as immunosuppressants. Since firefighters are dermally exposed to these compounds, there is a need to determine the skin permeabilities of different PFAS.

Using in-vitro flow-through diffusion, skin permeabilities were determined for 14C perfluorooctanoic acid (PFOA), perfluorohexanoic acid (PFHxA), and perfluorobutanoic acid (PFBA) using porcine skin. Preliminary tests were also done with 13C-labeled compounds to determine permeabilities through synthetic membranes due to the unavailability of 14C-labeled PFAS compounds. Tests were conducted over 8 hours with either acetone or synthetic sweat as the vehicle. PFBA was found to have greater permeability than PFHxA, likely due to having a smaller molecular weight. The dosing vehicle did not appear to impact permeability rates but impacted the disposition through the skin model. While PFAS compounds have a low permeability rate through skin membranes, PFAS stay in the skin, acting as a reservoir.

554 – Geospatial Analysis to Assess Linkage between Commercial Sources of Lead (Pb) and Childhood Blood Pb Levels

Nichole Kulikowski[1], Rogelio Tornero-Velez[2], Archana Lamichhane[2], Haesoo Kim[3,4], Zahra Gohari[3,4], Alexander Hall[5], Nicholas Spalt[6,1], Stephanie DeFlorio-Barker[7]
[1]Environmental Protection Agency, Physical Scientist, Washington DC, USA. [2]Environmental Protection Agency, Physical Scientist, Durham, NC, USA. [3]Environmental Protection Agency, Research Fellow, Durham, NC, USA. [4]Oak Ridge Institute for Science and Education, Research Fellow, Durham, NC, USA. [5]Environmental Protection Agency, Geographer, Cincinnati, OH, USA. [6]Environmental Protection Agency, Analyst, Washington DC, USA. [7]Environmental Protection Agency, Epidemiologist, Durham, NC, USA

Abstract
Lead (Pb) exposures in the US have declined significantly due to phaseout of Pb in gasoline, residential paint, and plumbing. However, Pb is still in use in some industries. We sought to identify commercial sources of Pb in the US and conduct geospatial analyses to identify associations between industrial sources and children's blood Pb level (BLL) exceedance (≥5 µg/dL) by census tract in the state of Michigan in 2018.

Peer-reviewed literature was used to identify relevant industrial sources of Pb. Pb-emitting industry sites in Michigan were geolocated by SIC code using Dun & Bradstreet, Federal Aviation Agency, and Mine Safety and Health Administration data. Aggregated children's BLL exceedance rates for each census tract were ranked within the state of Michigan. The top 20% of census tracts with the highest children's BLL exceedance were selected for geospatial analyses among Pb-site locations. Data extracted from the Toxic Release Inventory, CDC's Social Vulnerability Index, and US Census were linked by census tract. Geospatial analyses were performed using R and ArcGIS.

Unique SIC codes (n = 31) were identified and, in 2018, 11 were found in Michigan. This includes, but not limited to, shooting ranges, ship building facilities, and general aviation. We plan to investigate the top 20% of Michigan census tracts (n = 482) to determine whether there are associations.

The views expressed in this abstract are those of the authors and do not necessarily represent the views or policies of the U.S. Environmental Protection Agency.

555 – POSTER: Mixture Risk Assessment of PFAs and Phthalates in Teenagers-CROME Study 2020-2021

Denis Sarigiannis[1,2,3], Catherine Gabriel[4,5], Nafsika Papaioannou[6,7], Achilleas Karakoltzidis[6,7], Spyros Karakitsios[4,5]
[1]Environmental Engineering Laboratory, Department of Chemical Engineering, Aristotle University of Thessaloniki, Greece, Professor, Thessaloniki, Greece. [2]HERACLES Research Center – CIRI, Aristotle University of Thessaloniki, Greece, Director, Thessaloniki, Greece. [3]Environmental Health Engineering, School for Advanced Study IUSS, Pavia, Italy, Professor, Pavia, Italy. [4]HERACLES Research Center – CIRI, Aristotle University of Thessaloniki, Greece, Senior Researcher, Thessaloniki, Greece. [5]Environmental Engineering Laboratory, Department of Chemical Engineering, Aristotle University of Thessaloniki, Greece, Senior Researcher, Thessaloniki, Greece. [6]HERACLES Research Center – CIRI, Aristotle University of Thessaloniki, Greece, Pre-doctoral Researcher, Thessaloniki, Greece. [7]Environmental Engineering Laboratory, Department of Chemical Engineering, Aristotle University of Thessaloniki, Greece, Pre-doctoral Researcher, Thessaloniki, Greece

Abstract
Humans are exposed to thousands of chemicals through the environment, food, and daily-used products. Real-life mixtures are too complicated, and the risk to human health is very high. Per- and poly-fluoroalkyl substances (PFAS) and phthalates are synthetic organic chemicals with a high possibility of causing adverse effects on humans and the environment. It was reported to have endocrine-disrupting properties especially in teenagers.

Human biomonitoring studies and measurements of different groups of chemicals are valuable tools in human health risk assessment. In addition, network analysis may provide knowledge for real-life mixtures through chemical exposure patterns.
In this study, we have applied network analyses to the HBM dataset of CROME study in teenagers between 12-18 years old. We aimed to explore the correlations between the different variables, such as the exposure levels of individual PFAs compounds, individual phthalates metabolites and sums of group phthalates metabolites. In our models, we have used different determinants such as age, sex, and smoking. In addition, we have used linear regression analysis for continuous variables and logistic regression analysis for binary variables.

Participants were (mean ± SD) 14.4 (1.8) years of age and reported a BMI of 21.8 (3.8). Serum perfluorooctanoate (PFOA), perfluorooctane sulfonic acid (PFOS) and urinary monoethyl phthalate (MEP), mono-isobutyl phthalate (MiBP), mono-n-butyl phthalate (MnBP), found to be on high exposure levels and further analysis and correlations was performed focused on them.
In conclusion, network or principal component analyses allowed us to deeply exploit mixtures among PFAs and phthalates and aid in risk assessment.

556 – Oxidative Potential of the Particulate Matter Emitted from Common Household Sources

P. S.Ganesh Subramanian[1], Zhuying Dai[1], Vishal Verma[2]
[1]University of Illinois, Urbana-Champaign, Graduate Research Assistant, Urbana, USA. [2]University of Illinois, Urbana-Champaign, Assistant Professor, Urbana, USA

Abstract
Oxidative potential (OP) is a metric, which quantifies the ability of particulate matter (PM) to induce oxidative stress in humans and cause adverse health effects. Although people spend over 80% of their time indoors, OP of the PM emitted from indoor sources are largely unknown. We collected PM emitted by commonly used household sources (unscented-candles, scented-candles, incenses, cigarettes, ultrasonic-humidifiers, and essential-oil-diffusers) in a clean-controlled-environmental-chamber and measured their chemical composition and PM emission rates (PER). Three different assays; dithiothreitol (OP_{DTT}), glutathione (OP_{GSH}), and hydroxyl radical generation (OP_{OH}) were used to quantify OP.

Preliminary results showed that PER was highest for cigarettes (~1800 µg/min), followed by incenses (300-500 µg/min), and lower than 100 µg/min for the other sources. The PER from unscented-candles was substantially higher (~50x) in presence of a wind-draft. Over 75% of the total PM-mass comprised of organic-matter for cigarette and incense emissions, while over 50% consisted of elemental-carbon for unscented-candles in the presence of wind-draft.

Unscented-candle emissions in presence of a wind-draft showed the highest intrinsic OP_{DTT} and OP_{OH}, while OP_{GSH} was highest for cigarette emissions. This higher intrinsic OP_{DTT} and OP_{OH} could be attributed to the high elemental-carbon content of PM emitted from candles during wind-draft. Ultrasonic-humidifier and essential-oil-diffuser emissions had the lowest intrinsic OP_{DTT} and OP_{OH}, but their OP_{GSH} were comparable with the other sources. Experiments are underway for further OP and chemical composition measurements. We intend to use this dataset to develop an OP exposure assessment model to evaluate health risks from these sources.

557 – POSTER: INVESTIGATION OF METABOLOMIC SIGNATURES ASSOCIATED WITH METABOLIC SYNDROME IN TRIBUTYLTIN-EXPOSED ADIPOSE TISSUE

Dayna R Schultz[1,2], Thanasis Papageorgiou[3,2], Nafsika Papaioannou[1,2], Catherine Gabriel[1,2], Jan Kucera[4], Spyros Karakitsios[1,2], Dimosthensis Sarigiannis[2,1,5,6]

[1]Environmental Engineering Laboratory, Department of Chemical Engineering, Aristotle University of Thessaloniki, Thessaloniki, Greece. [2]HERACLES Research Center – CIRI, Aristotle University of Thessaloniki, Thessaloniki, Greece. [3]Environmental, Thessaloniki, Greece. [4]Research Centre for Toxic Compounds in the Environment (RECETOX), Faculty of Science, Masaryk University, Brno, Czech Republic. [5]Environmental Health Engineering, School for Advanced Study IUSS, Pavia, Italy. [6]National Hellenic Research Foundation, Athens, Greece

Abstract

Endocrine-disrupting chemicals (EDCs) are compounds that are known to disrupt a host of physiological processes in biota. Metabolic syndrome, which is a cluster of seemingly vague medical conditions frequently occurring together that have been linked to various health outcomes, are increasingly thought to be attributed, at least in part, to exposure to EDCs. More specifically, some EDCs may impact the normal development and function of adipose tissue, which was once considered an inert tissue. However, mounting evidence suggests it is a complex and metabolically active organ with considerable influence on the regulation of metabolism and energy homeostasis. To better understand the effects of EDCs on metabolic health outcomes, adipose tissue was exposed to a known lipogenic substance, tributyltin, to map the metabolomic signature associated with metabolite dysregulation. Differentiated Simpson-Golabi-Behmel syndrome (SGBS) adipocyte cells were exposed to 25 nM tributyltin (TBT) treatment. Samples were analysed using Reversed Phase (RP) and Hydrophilic Interaction (HILIC) Liquid Chromatography in positive and negative ionisation modes. Data preprocessing, batch correction, and metabolite identification, and statistical analyses were performed in R using xcms, MetaboAnalystR, and xMSannotator packages. Sample analysis resulted in the detection of more than 2000 features, with hundreds significantly dysregulated between treatments and control conditions. Of these, many were associated with the lipids family. Topology-based pathway analysis determined the most affected pathways were associated with metabolic syndrome. Future studies will further investigate changes in the adipocyte metabolome that may indicate susceptibility or early warning signs specific to health outcomes such as metabolic syndrome.

558 – Using mobile monitoring, urban infrastructure, and demographics to investigate local air pollution exposure misclassification in national land use regression models

Sarah Chambliss[1], Mark Campmier[2], Joshua Apte[3], Corwin Zigler[4]

[1]The University of Texas at Austin, Postdoctoral Fellow, Austin, USA. [2]University of California, Berkeley, Graduate Researcher, Berkeley, USA. [3]University of California, Berkeley, Associate Professor, Berkeley, USA. [4]The University of Texas at Austin, Associate Professor, Austin, USA

Abstract

National-scale linear regression modeling has been used to predict NO_2 and UFP at a spatial resolution intended to capture highly localized intraurban variation. However, some features of national models may lead to mischaracterization of both hyperlocal peaks and neighborhood-scale gradients that could impact studies focused on differences within a single urban area. We compare national model predictions against highly resolved (0.01 km^2) observations obtained via mobile modeling in the San Francisco Bay Area. Using a flexible machine learning technique, Bayesian Additive Regression Trees, we investigate how nonlinear relationships between urban sources and local pollution levels contribute to differences between observations and predictions within this geographic subdomain. Of particular relevance for Environmental Justice analyses, we also find that the degree of exposure misclassification varies with the racial/ethnic composition of the examined neighborhoods. Our findings inform the application and interpretation of nationally-modeled data for single-city studies and efforts to adjust or supplement exposure metrics for pollutants with high intraurban variability.

559 – Development of a Highly Sensitive Analytical Method to Detect 1,4-dioxane and Co-occurring Contaminants in Drinking Water and Blood

Sheng Liu[1], Elizabeth Lin[1], Nicole Deziel[2], Krystal Pollitt[2]

[1]Yale University, New Haven, USA. [2]Yale University, Associate Professor, New Haven, USA

Abstract

1,4-dioxane is a synthetic industrial chemical that has been widely used in commercial products. The USEPA classified 1,4-dioxane as a probable human carcinogen based on evidence from experimental animal studies. Approximately 7% of drinking water samples tested across the US exceed the federal cancer risk guideline level of 0.35 µg/L. Despite this significant environmental contamination, no federally enforceable drinking water standard has been established in large part due to the lack of understanding of 1,4-dioxane exposure and its health impacts to humans. While this chemical can be measured in drinking water, use of this environmental sampling matrix is subject to misclassification of exposure. While biological monitoring provides a more direct means of assessing exposure and has the advantage of aggregating across all sources and routes, low concentrations in blood present measurement challenges. We present an analytical protocol to simultaneously detect 1,4-dioxane and co-occurring contaminants (1,1-dichloroethane; 1,1,1-trichloroethane; trichloroethylene) in blood and water, providing sensitive detection to screen of low levels in the ng/L range. This method extracts samples by headspace solid-phase microextraction (carboxen/polydimethylsiloxane) and uses gas chromatography-high resolution mass spectrometry for analysis. We applied our novel method to paired drinking water and blood samples from healthy adults residing in Superfund sites on Long Island, NY with known groundwater 1,4-dioxane contamination. Our findings enable the association between environmental concentrations of 1,4-DX (and its co-occurring contaminants) and biological markers of exposure to be investigated, addressing important scientific questions in addition to urgent community questions about 1,4-DX exposure and human health.

561 – POSTER: Development of a Metabolomic Risk Score for Exposure to Traffic-Related Air Pollution: A Multi-Cohort Study

Susan Hoffman[1], Andrea Lane[2], Audrey Gaskins[3], Stefanie Ebelt[4], Timur Tug[5,6], Dean Jones[7], Donghai Liang[3], Anke Huels[3]
[1]Emory University, PhD Candidate, Atlanta, USA. [2]Duke University, Assistant Professor of the Practice, Durham, USA. [3]Emory University, Assistant Professor, Atlanta, USA. [4]Emory University, Associate Professor, Atlanta, USA. [5]Emory University, Visiting Scholar, Atlanta, USA. [6]TU Dortmund University, PhD Candidate, Dortmund, Germany. [7]Emory University, Professor, Atlanta, USA

Abstract

Background: This research aimed to develop and evaluate metabolomic risk score approaches reflecting the biological response to traffic-related air pollution (TRAP) exposure.

Methods: A metabolomics risk score can assess an individual's exposure based on their metabolite profile. A simulation study compared two metabolomic risk score methodologies: thresholding, where a p-value cutoff is used to select metabolites, and elastic net regression, which uses penalized regression to select metabolites. The methods performance was compared to assess 1) ability to correctly select metabolites associated with daily TRAP (particulate matter, black carbon, and nitrogen dioxide), and 2) ability of the risk score to predict daily TRAP exposure. Power calculations and false discovery rates (FDR) were calculated for each approach. This risk score methodology was applied to two real cohorts, the Center for Health Discovery and Wellbeing (TRAPHIC, n = 180) and Environment and Reproductive Health (EARTH, n = 200) studies.

Results: In simulations, elastic net regression consistently presented inflated FDR for both high and low effect sizes and across all three sample sizes (n = 200, 500, 1,000). Power to detect correct metabolites exceeded 0.8 when the sample size was 1,000. In the real data application assessing associations of metabolomics risk scores and TRAP, associations were largely null. Black Carbon was positively associated with the risk score in TRAPHIC data.

Conclusions: These results demonstrate that certain methodologies to construct metabolomics risk scores are statistically robust and valid; however, large sample sizes are required to apply this method successfully.

564 – Developing a Conceptual Scientific Workflow to Estimate Chemical Exposures

Daniel Vallero[1], Elaine Cohen Hubal[2]
[1]Center for Computational Toxicology & Exposure, U.S. Environmental Protection Agency, Senior Research Physical Scientist, RTP, NC, USA.
[2]Center for Public Health and Environmental Assessment, U.S. Environmental Protection Agency, Physical Scientist, Research Triangle Park, NC, USA

Abstract

Exposure scientists have developed models that use available data with other mechanistic information to understand how chemicals in products move through the natural and built environment. Each new problem incrementally advances data, models, and understanding. To efficiently address problems of increasing complexity and enable greater insights, more nimble modeling methodologies are required to assess important chemical exposure scenarios and pathways across the chemical/product lifecycle. A scientific workflow is a modeling methodology that is designed to explicitly describe and execute a series of data manipulation and computational steps to provide outputs tailored to decision-making contexts. For chemical assessments, scientific workflows are useful tools for efficient and transparent analyses, especially for considering impacts in complex systems. We began this effort with a test case using a data-rich chemical, 1,4-dioxane with the aim of building a conceptual workflow that could be successfully developed for both data-rich and data-poor compounds. To this end, the lessons learned from the 1,4-dioxane workflow was then tested against flame retardants and other data poor chemicals in consumer products. Ways to advance the workflows for efficient evaluation of a variety of chemicals and a range of decision contexts will be discussed.

565 – Method development for standardized microfiber reference materials

Kara Wiggin[1], Angela Aherrera[2], Ana Rule[3]
[1]Scripps Institution of Oceanography, UC San Diego, PhD Candidate, San Diego, USA. [2]Department of Environmental Health and Engineering, Johns Hopkins Bloomberg School of Public Health, Postdoctoral Scholar, Baltimore, USA. [3]Department of Environmental Health and Engineering, Johns Hopkins Bloomberg School of Public Health, Assistant Professor, Baltimore, USA

Abstract
Introduction: The potential health effects of airborne microfibers (MF), the most abundant type of microplastic in the environment are unknown. However, there are no standardized MF reference materials for toxicity testing utilizing the appropriate material, shape, and size, as current approaches rely almost exclusively on spherical polystyrene beads.

Objective: To develop a reproducible method of generating environmentally relevant MF samples for toxicity testing.

Methods: Microfibers were collected inside a garment factory using Inhalable samplers to assess real-world MF exposures. Filters were mounted on glass slides for imaging. Microfibers from eight different commonly used fabrics were generated in the lab through bead beating in a Qiagen tissue lyser at different time/frequency conditions. Post bead beating, 100µL was filtered and mounted on glass slides for imaging. One-hundred fibers from three replicates of each fabric type were measured for length and width on FIJI.

Results: The range of the MF length/width measured in the garment factory was 25-12745/7-129µm while the range of lab-generated fibers was 10-2551/7-106µm. Median size of MF from the garment factory were 28µm in width and 244µm in length while the average median sizes for lab-generated MF from the different fabrics (including polyester, rayon, and cotton) were 39µm in width and 190µm in length.

Conclusions: Our laboratory method can generate microfibers of consistent size distributions that are comparable to MF size dimensions seen in real-world exposure scenarios. The ability to generate standardized MF reference materials is vital for reproducible and environmentally relevant toxicity testing.

566 – Prenatal Exposure to Low-Level Air Pollution and Placental Inflammation: Investigating Effect Modification by Infant Sex

Emily Craig[1], Yan Lin Ph.D.[2], Sally Thurston Ph.D.[3], Yihui Ge[1], Philip Hopke Ph.D.[3], Kristin Scheible M.D.[4], Catherine Yount[5], Tom O'Connor P.h. D[6], David Rich P.h.D.[7], Junfeng (Jim) Zhang Ph.D.[8]
[1]Nicholas School for the Environment, Duke University, Ph.D. Student, Durham, USA. [2]Nicholas School for the Environment, Duke University, Post Doctoral Researcher, Durham, USA. [3]University of Rochester Medical Center, Professor of Biostatistics, Environmental Medicine, Rochester, USA. [4]University of Rochester Medical Center, neonatologist, Rochester, USA. [5]University of Rochester Medical Center, Ph.D. Student, Rochester, USA. [6]University of Rochester Medical Center, Professor - Department of Obstetrics and Gynecology, Rochester, USA. [7]University of Rochester Medical Center, Professor Department of Public Health Sciences, Department of Medicine, Department of Environmental Medicine, Rochester, USA. [8]Nicholas School for the Environment, Duke University, Professor of Global and Environmental Health, Durham, USA

Abstract
This study examined the association between gestational exposure to low-level air pollution and markers of placental inflammation and whether this association was modified by infant sex. We studied 313 pregnant women living in Rochester, NY, and estimated their trimester-specific exposure to PM2.5 and NO2 with high-resolution spatiotemporal models, and to PAHs with urinary 1-hydroxy-pyrene [1-OHP] levels. Generalized additive regression models were used to examine the trimester- and sex-specific associations between air pollution exposures and placental inflammation markers (protein-normalized IL-6 and TNF-α), adjusting for BMI, age, and smoking.

One interquartile range increase in 1-OHP in the first trimester (2.58 pg/µl), and entire pregnancy (2.70 pg/µl) were associated with 25.0%

(95:CI 0.56%, 55.5%) and 43.1% (95:CI 19.0%, 72.1%) increases in TNF-a, respectively. The increase in TNF-a levels associated with entire pregnancy PAH exposure was larger for male infants (43.5%, 95% CI: 0.56%, 92.7%) than female infants (-10.8%, 95% CI: -16.3%, 46.9%). Conversely, each IQR increase in PM2.5 (1.41 μg/m3) in the first trimester was associated with a -23.1% decrease (95%CI: -37.1%, 5.0%) in placental IL-6. Furthermore, an IQR increase in IL-6 in the third trimester was associated with a 25.6% increase (95%CI: 5.18%, 50.0%) in IL-6 among male infants and a -10.3% decrease (95%CI -24.3%, 6.4%) among female infants.

Air pollution exposure during pregnancy was associated with alterations in placental inflammation markers. The associations were modified by infant sex, with greater increases in TNF-a and IL6 among male infants associated with air pollution exposure.

567 – POSTER: Identifying the Relationship Between Carbon Dioxide Concentration, Occupancy Data, and Infection Risk of Airborne Microorganisms

Yoonhee Jung[1], Amanda Wilson[2]
[1]The University of Arizona, PhD Student, Tucson, USA. [2]The University of Arizona, Professor, Tucson, USA

Abstract
Introduction: The COVID-19 pandemic highlighted the importance of ventilation in built environments to reduce respiratory disease spread. CO_2 is a useful indicator of ventilation and human occupancy and can be further developed to inform microbial risk assessments. The study objectives are to (1) quantify the relationship between CO_2 concentration and occupancy, and (2) investigate the relationship between CO_2 and infection risk in published literature. Methods: A sensor study was conducted across different environments: lunchtime in an office breakroom, a lobby of an early learning center, a weekend in a household, and a family gathering. Awair Elements sensors were used for monitoring CO_2. Occupancy and foot traffic data were collected via direct observation and AI workplace sensors (VS121, Xiamen Milesight IoT). A systematic literature review was conducted to extract data from Embase and PubMed on relationships between CO_2 and infection risk in built environments. Results: The median CO_2 concentrations ranged from 919 ppm to 1434 ppm across environments, with concentrations exceeding the recommendation of the American Society of Heating, Refrigeration, and Air Conditioning Engineers: 1,000 ppm (schools) and 800 ppm (offices). The literature search resulted in 218 papers, with 64% published since the pandemic. Discussion/Conclusion: This study is ongoing; our next steps include completing article screening and continuing sensor data collection. Our preliminary data indicate that ventilation in a variety of built environments may be insufficient and that more research is increasingly being conducted on this topic in response to the COVID-19 pandemic.

568 – Assessing Drinking Water Quality in Private Wells

Mary H. Ward[1], Darrin A. Thompson[2], David C. Wheeler[3], Dana W. Kolpin[4], Rena R. Jones[5], Jonathan Hofmann[6], Laura Beane Freeman[7], Paul Bradley[8]
[1]National Cancer Institute, Division of Cancer Epidemiology and Genetics, Senior Investigator, Rockville, USA. [2]University of Iowa Center for Health Effects of Environmental Contaminants, Associate Director, Iowa City, USA. [3]Viriginia Commonwealth University, Associate Professor, Richmond VA, USA. [4]U.S. Geological Survey, Research Hydrologist, Iowa City, USA. [5]National Cancer Institute, Investigator, Rockville MD, USA. [6]National Cancer Institute, Senior Investigator, Rockville, USA. [7]National Cancer Institute, Senior Investigator, Rockville MD, USA. [8]US Geological Survey, Research Ecologist/Hydrologist, Columbia SC, USA

Abstract
The U.S. Safe Drinking Water Act provides health-based regulations to protect public water supplies (serving ≥25 persons) but private supplies (mostly wells) are exempt. Consequently, ~44 million people in the United States rely on unregulated drinking water. We developed random forest models to predict nitrate concentrations in private wells for the Agricultural Health Study (AHS), a cohort of pesticide applicators and their spouses in Iowa and North Carolina. At enrollment (1993-1997), about 70% used private wells for their drinking water. In a 2018-2019 tap water study, we measured nitrate and >400 inorganic and organic chemicals in tapwater samples from 47 AHS participants using private wells in hydrologically vulnerable aquifers in northeast Iowa. The most important explanatory variable in our models was well depth; other important variables included distance to sinkholes, slope length <1 km of the well, soil characteristics, distance to animal feeding operations, and agricultural land area <1 km. Thirteen percent of AHS Iowa participants' wells had predicted nitrate >10 mg/L NO3-N (regulatory limit). In the tapwater study, 32% of samples were >10 mg/L NO3-N. Pesticides and their transformation products were the most frequently detected organics (median detections: 8; interquartile range 1-10); herbicides were the major class of pesticide detected. Few participants were using water treatment systems that would remove these contaminants. Our results indicate the feasibility and the complexity of estimating nitrate concentrations in private wells. We demonstrate the need for comprehensive assessment of water quality and water treatment especially for wells located in vulnerable aquifers.

569 – POSTER: Comparative Analysis of Nephelometric and Gravimetric Techniques: Calibration and Agreement of PurpleAir, AirQo, and E-Sampler Devices in South Central Uganda

Sophia Le[1], Lauren Dudley[2], William Checkley[3], Kirsten Koehler[4], Laura Nicolaou[5]
[1]Johns Hopkins University, PhD Student, Baltimore, USA. [2]Johns Hopkins University, Student, Baltimore, USA. [3]Johns Hopkins University, Professor, Baltimore, USA. [4]Johns Hopkins University, Associate Professor, Baltimore, USA. [5]Johns Hopkins University, Assistant Professor, Baltimore, USA

Abstract
Air pollution disproportionately affects public health in low- and middle-income countries, especially in Africa, where reference-grade regulatory monitoring networks are scarce. Kampala, the capital of Uganda, has an established network of low-cost air quality monitoring devices (AirQo). Our research expands on this by deploying a complementary PurpleAir sensor network in rural South Central Uganda, Masaka City, and Kampala. Both AirQo and PurpleAir utilize dual Plantower sensors for high-resolution $PM_{2.5}$ measurements. Integrating these networks could enhance monitoring capabilities by capitalizing on the devices' complementary strengths.

To further improve monitoring, we installed a high-quality E-Sampler monitor that acquires both gravimetric and nephelometric measurements. Our study examines the agreement between these three sensing technologies, which is vital for understanding their efficacy in air quality monitoring. Preliminary Bland-Altman plot analyses comparing PurpleAir and AirQo devices indicate minimal bias in both Masaka and Kampala, with mean biases of -0.25 $\mu g/m^3$ and -0.16 $\mu g/m^3$, respectively. In Kampala, the plot illustrates that observed variance increases as the magnitude of the measurements increase, suggesting potential heteroscedasticity between the two devices. These findings underscore sensor performance discrepancies under diverse conditions, warranting further examination. Collocated PurpleAir and E-Sampler devices demonstrate robust agreement, suggesting that incorporating E-Sampler's gravimetric data could enhance PurpleAir's accuracy and overall monitoring efficacy.

Our investigation of device agreement elucidates strengths, limitations, and opportunities for refinement and optimization. Through collaboration with the AirQo team, our research bolsters air pollution monitoring capabilities in resource-constrained environments while concurrently fostering the development of environmental health literacy across the nation.

570 – Untargeted Metabolomics Study Reveals the Impact of Prenatal Exposure to EDCs on Metabolic Disorders

Thanasis Papageorgiou[1,2], Nafsika Papaioannou[1,2], Catherine Gabriel[1,2], Lorenzo Fabbri[3,4,5], Marta Gallego[3,4,5], Sarah Lyon-Caen[6], Remy Slama[7], Claire Philippat[7], Berit Granum[8], Catherine Thomsen[9], Léa Maitre[3,4,5], Maribel Casas[3,4,5], Martine Vrijheid[3,4,5], Spyros Karakitsios[1,2], Dimosthenis Sarigiannis[1,2,10,11]
[1]Environmental Engineering Laboratory, Department of Chemical Engineering, Aristotle University of Thessaloniki, Thessaloniki, Greece. [2]HERACLES Research Center on the Exposome and Health, Center for Interdisciplinary Research and Innovation, Balkan Center, Bldg. B, 10th km Thessaloniki-Thermi Road, Thessaloniki, Greece. [3]ISGlobal, Barcelona, Spain. [4]Universitat Pompeu Fabra (UPF), Barcelona, Spain. [5]Consorcio de Investigacion Biomedica en Red de Epidemiologia y Salud Publica (CIBERESP), Madrid, Spain. [6]Team of Environmental Epidemiology Applied to Reproduction and Respiratory Health, Institute for Advanced Biosciences (IAB), Grenoble Alpes University, Inserm, CNRS, 38700, Grenoble, France. [7]University Grenoble Alpes, Inserm U 1209, CNRS UMR 5309, Team of Environmental Epidemiology applied to Reproduction and Respiratory Health, Institute for Advanced Biosciences (IAB), 38000, Grenoble, France. [8]Department of Chemical Toxicology, Norwegian Institute of Public Health, Oslo, Norway. [9]Department of Environmental Immunology, Norwegian Institute of Public Health, Oslo, Norway. [10]Environmental Health Engineering, Institute of Advanced Study, Palazzo del Broletto - Piazza Della Vittoria 15, 27100, Pavia, Italy. [11]Director and President of the Board, National Hellenic Research Foundation, 48 Vasileos Constantinou Ave., Athens, Greece

Abstract
Several endocrine-disrupting chemicals (EDCs), such as phthalates, synthetic phenols and organophosphate pesticides, are found in a wide range of everyday products and are released into the environment due to human activity. The study aimed to evaluate the exposure of 308 pregnant women from three different European countries (Spain, France, and Norway) to these environmental contaminants and link them to metabolic disorders, such as diabetes mellitus, obesity and metabolic syndrome. Plasma samples from pregnant women were analyzed using Reversed Phase (RP) liquid chromatography and Hydrophilic Interaction Liquid Chromatography (HILIC) in both positive and negative ionization modes, to increase the coverage of the detected metabolites. Data pre-processing, batch effects correction, and metabolites annotation were performed using R's packages XCMS, MetaboAnalyst and xMSannotator. Samples analysis resulted in more than 1500 features. Most of these metabolites belonged to the family of lipids, specifically the Glycerophospholipids, Fatty Acyls and Sterol Lipids. Among these metabolites were metabolites of environmental pollutants, such as benzoic acid, and phthalate's metabolites. One of the most statistically significant detected pathways was Sphingolipid metabolism, which has been connected to the development and pathogenesis of insulin resistance and metabolic syndrome. Moreover, pathways

such as Glucose homeostasis and NAD biosynthesis were also dysregulated. Impairment to these pathways has been associated with diabetes pathogenesis. In conclusion, this study revealed potential biomarkers and perturbed biological pathways associated with exposure to environmental pollutants and explored the consequences on human health.

571 – Research to Action: Using Data Dashboards to Communicate Environmental Exposures

Muskaan Khemani[1], Mariangelí Echevarría-Ramos[2], Amruta Nori-Sarma[3], Gregory Wellenius[4], Jonathan I. Levy[4], Itai Kloog[5], Poornima Prabhakaran[6], Petter Ljungman[7], Kevin J. Lane[3]
[1]Boston University, Research Assistant, Boston, USA. [2]Mystic River Watershed Association, Climate Resilience Manager, Boston, USA. [3]Boston University, Assistant Professor, Boston, USA. [4]Boston University, Professor, Boston, USA. [5]Ben-Gurion University, Professor, Be'er Sheva, Israel. [6]Public Health Foundation India, Director, Center for Environmental Health, New Delhi, India. [7]Karolinska Institute, Senior Researcher, Solna, Sweden

Abstract
Introduction: Sharing of environmental and climate exposure data is vital for actionable research to support decision making processes. Interactive maps and data dashboards can support communication efforts, but often these maps are promulgated without community input or feedback, limiting their broader utility.

Methods: Across multiple studies, we developed local and global environmental exposure tools for research communication with stakeholders, building on partnerships with newspapers, government, and community organizations. In parallel, we have developed survey instruments and collected qualitative data regarding the ease of communication, practicality of maps, and presentation format to assess the utility of these data dashboard projects.

Results: Our environment and climate exposure tool for India focused on PM2.5 annually at the state and ward levels, as well as other climate and adaptation related variables (i.e. normalized difference vegetation index) and population density. Qualitative data indicated that data available at different spatial scales (city, ward, towns, villages) and additional data on land use cover, green space, and socioeconomic status would make the dashboard more useful for community needs. In another example study in the Mystic River Watershed of Massachusetts, our interactive dashboards included flood risk, presence of industrial facilities, and heat exposure. Data gathering for this work is ongoing during local community events, and has indicated the importance of spatial resolution, and interpretability of multi-layer data presentations.

Conclusions: Community action can be informed by fit-for-purpose data dashboards and story maps that are designed specifically to address community needs.

572 – Facilitating, supporting, and funding scientific efforts to advance environmental justice in the United States

Elise Elliott[1], Anna Rosofsky[2], Ellen Mantus[3]
[1]Health Effects Institute, Staff Scientist, Boston, USA. [2]Health Effects Institute, Senior Scientist, Boston, USA. [3]Health Effects Institute, Director of Science, Boston, USA

Abstract
Historically marginalized communities in the United States often experience disproportionate exposures to environmental contaminants and health effects. These communities often face substantial cumulative impacts - the totality of exposures, including air pollutants and other environmental and social stressors, from multiple sources. Although recent descriptive research and federal and state actions have focused a spotlight on these exposure and health inequities, knowledge and funding gaps exist regarding solutions-oriented research and cumulative impact assessment. To address these gaps, the Health Effects Institute (HEI) has launched a new environmental justice program to facilitate, support, and fund scientific research, special projects, and research translation to advance environmental justice for historically marginalized communities in the United States. Here, we will discuss the new program and lessons learned from HEI's multi-sector environmental justice workshop that informed these efforts. HEI plans to fund different research tracks and is rethinking traditional funding models. The new environmental justice program aims to identify best practices for funding environmental exposure and health research and community-driven science. Through this new program, HEI will serve as a convener for multiple sectors, support community partnerships to find solutions to address environmental inequities, and partner with a broad array of stakeholders across academia, community groups, non-governmental organizations, and policymaking to advance environmental justice.

573 – POSTER: Development of a field quantification method for tetrachloroethylene (PCE) in exhaled breath using proton transfer reaction time-of-flight mass spectrometry (PTR-TOF-MS)

Jung Hyun Lee[1], Marwan Alajlouni[1], Alaina K. Bryant[1], Brandon E. Boor[2], Antonious Tasoglou[3,4], Sa Liu[2]
[1]Purdue University, student, West Lafayette, USA. [2]Purdue University, Professor, West Lafayette, USA. [3]RJ Lee Group, Inc., Monroeville, USA. [4]Montrose Environmental, Emerging Technology Project Manager, Denver, USA

Abstract
This study aimed to evaluate the feasibility of a novel on-site method for quantifying exposure to tetrachloroethylene (PCE), a volatile organic compound (VOC) known to cause adverse health effects, by analyzing exhaled breath using proton transfer reaction time-of-flight mass spectrometry (PTR-TOF-MS), a high-resolution real-time VOC analyzer. Exhaled breath samples were collected from residents living in a Midwestern community with multiple PCE contamination sites. We compared the results from PTR-TOF-MS with those from gas chromatography-mass spectrometry (GC-MS) and explored the effects of sampling time and day (morning vs. afternoon, weekday vs. weekend), the sampling mode (online direct breath sampling with PTR-TOF-MS vs. offline analysis of bag samples with PTR-TOF-MS), as well as using moving averages and raw signals to quantify low peaks. The results of the PTR-TOF-MS and GC-MS analysis were correlated with PTR-TOF-MS being more sensitive than GC-MS. PCE concentrations were higher in samples collected in the morning than those in the afternoon (0.11 vs. 0.05 ppb, $p = 0.015$), and higher in those collected during the weekend than those on weekdays (0.11 vs. 0.06 ppb, $p = 0.025$). The offline PTR-TOF-MS analysis and noise-removal technics resulted in more stable peaks and an average 9.25% increase in the signal compared to the online mode. Using the method developed, PCE was detected in 70 of 105 samples collected from the participants with an average concentration of 0.59 ppb (range: 0.11 to 5.49). Our findings highlight the potential of PTR-TOF-MS analysis of exhaled breath in assessing VOC exposures in community settings.

574 – Challenges in Data Sharing for New Jersey Biomonitoring Studies

Elisabeth Cook[1], Emily Lax[2,3], Colleen Donohoe[4,5], Chang Ho Yu[6], Eric Bind[7], Zhihua (Tina) Fan[8]
[1]New Jersey Department of Health, Research Scientist, Ewing, USA. [2]New Jersey Department of Health, CDC-PHAP Associate, Ewing, USA. [3]Centers for Disease Control and Prevention, PHAP Associate, Ewing, USA. [4]New Jersey Department of Health, APHL-CDC Fellow, Ewing, USA. [5]APHL, Fellow, Ewing, USA. [6]New Jersey Department of Health, Program Manager, Data Management and Analysis, Ewing, USA. [7]New Jersey Department of Health, Program Manager, Biomonitoring, Ewing, USA. [8]New Jersey Department of Health, Director, Environmental and Chemical Laboratory Services, Ewing, USA

Abstract
NJDOH's Biomonitoring program (NJB) conducts various types of biomonitoring projects including a population biomonitoring surveillance study for 120+ environmental contaminants in NJ, prenatal screening for toxic metals, and investigational studies. NJB shares findings with participants and the public to increase public awareness of environmental exposure and ultimately improve public health. Data sharing challenges include making data available to interested stakeholders, maximizing the data's meaningfulness, and overcoming limitations of national and state portals.

NJB collaborated with the NJ Environmental Public Health Tracking (NJEPHT) team to utilize the data sharing and tracking portal to develop effective sharing approaches, including generated data's relevance to NJEPHT for each biomonitoring study, what data should be shared to address public's interest, and useful parameters and variables for public inquiries. Harmonizing NJB data with NJEPHT considers concentration unit, chemical name vs. symbol, treatment of non-detect data for emerging contaminants including lead and PFAS, and providing metadata and full datasets. Data visualization and modernization is underway by optimizing data presentation format (table vs. graphs) and use of relevant descriptive statistics (geometric mean vs. median). These efforts will make NJB data accessible, available, and informative as a baseline for future NJ-based biomonitoring studies. Prenatal screening data are better shared by storytelling through our portals. Traditional data sharing methods, such as mail and calls, are still utilized to share data generated from biomonitoring projects with different study designs. The number of website visits and public feedback will be evaluated to assess the effectiveness of data-sharing methodology.

575 – Community Health Workers/Promotores Increasing Environmental Health Community Research Capacity

Venezia Ramirez[1], Sandra Serrano[2,3], Leticia Ortiz[3], Jill Johnston[4]
[1]University of Southern California, Research Coordinator, Los Angeles, USA. [2]University of Southern California, Project Assistant, Los Angeles,

USA. [3]Esperanza Community Housing, Promotora, Los Angeles, USA. [4]University of Southern California, Associate Professor, Los Angeles, USA

Abstract
Promotores (community health workers) are an innovative method for research institutions to foster knowledge spaces outside of academia, translate results to policy change to advance health equity, and leverage community expertise to create solutions by the community, for the community. In collaboration with Promotores from Esperanza Community Housing, the USC Environmental Justice Research Lab co-created the "Breast health and the Environment among Latinas in Los Angeles" (BELLA) study to investigate impacts of urban oil drilling on women's breast cancer risk in Los Angeles with the target population focused on Latina residents living in South Los Angeles.

This community engaged participatory research (CBPR) study advanced community education and research capacity, evaluated impacts of oil drilling operations on mammographic density, quantified exposure to oil-related mixtures, and contributed new insight on the intersection between environmental justice, oil-extraction chemicals, and breast cancer in the community.

We will describe best practices and lessons learned from engaged Promotores as environmental health researchers to assess the environmental impact on breast health, apply scientific approaches for gathering health data, and incorporate community knowledge to ensure the results are reported in a comprehensive manner to community members. This process facilitates a multidirectional communication structure that encourages interaction between research institutions to share research findings, and community members to share their stories to increase researchers' understanding of community concerns. We share ways in which we are building the foundation for research institutions to organize mindful research initiatives in solidarity with frontline communities that are working towards achieving environmental justice.

576 – Investigating PFAS Levels in Paired Silicone Wristband and Blood Serum Samples

Taylor Hoxie[1], Ankita Bhattacharya[2], Rachel Bauer[2], Ying Guo[2], Sharon Zhang[3], Nick Herkert[3], Chris Higgins[4], Courtney Carignan[5], Heather Stapleton[6]
[1]Duke University, PhD Candidate, Durham, USA. [2]Michigan State University, PhD Student, East Lansing, USA. [3]Duke University, Research Project Manager, Durham, USA. [4]Colorado School of Mines, Professor, Golden, USA. [5]Michigan State University, Professor, East Lansing, USA. [6]Duke University, Professor, Durham, USA

Abstract
Exposure to per- and polyfluoroalkyl substances (PFAS) occurs via contaminated drinking water and food; however, individuals are also exposed to PFAS in the indoor environment. This study sought to evaluate PFAS exposure from the ambient indoor environment by utilizing silicone wristbands as personal passive samplers.

Paired wristband and human serum samples (n=87) were collected from participants residing in a midwestern community that had PFAS impacted drinking water. Each participant wore a wristband for 7 days to assess personal exposure to PFAS. Individuals in this study ranged from 7 to 84 years and the majority were white (84%) and female (66%).

Serum samples were analyzed for 35 PFAS via LC-MS/MS and the most frequently detected analytes were PFHxS, n-PFOS, br-PFOS, PFHpS and PFOA. Wristbands were analyzed for 43 total PFAS using both LC-MS/MS and GC-HRMS. The most frequently detected analytes on the wristbands were 6:2 diPAP (LC-MS/MS) as well as 6:2 FTOH and MeFOSE (GC-HRMS). Analyses revealed a sex-dependent difference in exposure to 6:2 diPAP and age-dependent differences in exposure to 6:2 FTOH and MeFOSE.

The current data indicate that PFAS are commonly detected on wristbands, and that exposure to PFAS precursors in the indoor environment (e.g. diPAPs) are greatest. To our knowledge, no other studies have been published which quantify volatile PFAS precursors on silicone wristbands via GC-HRMS. This information can be useful for understanding PFAS exposures from indoor environment sources (e.g. paint, furniture, carpeting, etc).

577 – Exploring mechanisms to better utilize low-cost sensor exposure data in community-engaged research

Sara Duncan[1], Nick Lundy[2]
[1]Western Carolina University, Assistant Professor, Cullowhee, USA. [2]Western Carolina University, Undergraduate student, Cullowhee, USA

Abstract
In the last decade, the $PM_{2.5}$ measurement space has seen an influx in low-cost air quality monitors. However, the accuracy of these monitors is not as high as more expensive, but highly validated instruments regularly used for exposure assessment. While the absolute value of measurements from low-cost monitors may not be trustworthy, perhaps the relative values between baseline and high exposure provide a better metric particularly for use in community-engaged research.

To test the usefulness of low-cost air quality instruments in detecting sources, two high quality PDR-1500s and four Airbeam3 monitors were co-located at several locations. At a busy campus bus stop, during periods without direct sources nearby, both monitor types recorded similar baseline values (~ 4 μg/m³ for PDRs and ~ 7 μg/m³ for Airbeams). During periods of nearby sources such as shuttle bus congestion and nearby cigarette smoking, both monitor types detected elevated concentrations. However, concentrations recorded by the Airbeams consistently were biased high.

While these results confirm prior research regarding performance of low-cost monitors, these monitors are still useful for community-engaged exposure research as they are inexpensive, relatively easy to operate, and provide immediate feedback for volunteer users. Upon collection of low-cost sensor data, rather than conducting analysis with actual concentrations, values can be tagged with binary markers (e.g., high vs low), or can be further differentiated into additional bins such as those used for the EPA's Air Quality Index. This presentation will provide insight into a method to better employ low-cost monitors for community-engaged research.

578 – POSTER: Addressing Environmental Justice Concerns in a Working-Class Midwestern Community Related to Tetrachloroethylene (PCE) Contamination

Alaina Bryant[1], Kaleigh Karageorge[1], Jung Hyun Lee[1], Seohyung Kim[1], Sa Liu[2]
[1]Purdue University, Graduate Student, West Lafayette, IN, USA. [2]Purdue University, Assistant Professor, West Lafayette, IN, USA

Abstract
As part of the EPA Environmental Justice (EJ) Video Challenge for Students, our team researched a possible EJ case in a working-class community in Indiana. Historical mismanagement of tetrachloroethylene (PCE) has led to groundwater contamination and the risk of vapor intrusion into homes. PCE is a known neurotoxicant and possible human carcinogen, and residents have raised concerns that lower socioeconomic families living in the Superfund site are at a greater risk of exposure. These families often face everyday stresses which prevent them from moving and impact their participation in local projects.
In Phase I, we used publicly available data from the EPA, studied literature on EJ and PCE, and conducted community interviews to identify concerns and potential EJ issues in the community. We found that the community has a higher poverty rate than the state average at 19.6% which increased to 26.9% as one moved closer to the Superfund site. The cancer incidence rate for this region was 540.9 new cancer cases per 100,000 people, the highest reported in the state. These findings among others suggest that lower socioeconomic residents are at an increased risk of exposure to PCE.

In Phase II, we collaborated with community partners to address the EJ issues identified in Phase I by developing a three-action strategy: conducting intercept interviews to identify additional barriers to participation, developing a targeted multimedia campaign on vapor intrusion, and hosting free community events to engage specific populations. These actions are ongoing and will be conducted throughout Summer and Fall 2023.

579 – LC-MS/MS Characterization and Automated Data Analysis of Biological Transformation Products of Per- and Polyfluoroalkyl Substances (PFAS)

Sheng Liu[1], David Dukes[2], Jeremy Koelmel[1], Carrie Mcdonough[3], Krystal Pollitt[4]
[1]Yale University, New Haven, USA. [2]Stony Brook University, Stony Brook, USA. [3]Carnegie Mellon University, Assistant Professor, Pittsburgh, USA. [4]Yale University, Associate Professor, New Haven, USA

Abstract
Per- and polyfluoroalkyl substances (PFAS) are persistent and widely used synthetic chemicals. They have been linked to a wide range of adverse human health effects, including delayed metacognition and behavioral problems in children, impaired kidney function and kidney cancer, testicular cancer, obesity, and lipid dysregulation. Assessment of PFAS exposure has focused on the parent PFAS compounds and transformation products from naturally occurring microbial pathways in waste waters and landfills. Our understanding, however, of mammalian PFAS transformation is limited. This study aims to identify biological PFAS transformation products and generate mass spectral libraries to facilitate automated search and identification of PFAS transformation products from raw data files. Biological

transformation of 27 PFAS of various chemical classes, including alcohols, sulfonamides, carboxylic acids, ethers, and esters, etc., were evaluated using mouse liver S9 fractions with an in vitro assay. PFAS standards were incubated with mouse liver S9 fractions at 37 °C with gentle shaking for 60 minutes. Liquid chromatography-tandem mass spectrometry (LC-MS/MS) was used to analyze the supernatant after incubation and identify potential PFAS transformation products. Four major PFAS transformation pathways were proposed, including glucuronidation; sulfation; dealkylation and oxidation. A total of 13 transformation products were identified and their MS/MS spectra were obtained. Class-specific fragment peaks were annotated and incorporated into FluoroMatch 3.0 to enable future identifications of products from the same chemical series. Time-course experiments of varying incubation times showed different formation rates for different transformation products. Non-targeted analysis further revealed additional molecular features that were associated with PFAS transformation.

580 – POSTER: Firefighters' Exposure to PFAS from their Turnout Gear during its Lifespan of Use

Nur-Us-Shafa Mazumder[1], Andrew Hall[1], R. Bryan Ormond[2]
[1]North Carolina State University, Graduate Research Assistant, Raleigh, USA. [2]North Carolina State University, Assistant Professor, Raleigh, USA

Abstract
Firefighting occupation is now a "Group-1" carcinogen, meaning firefighting is a known carcinogen to humans. Recent focuses are on Per- and polyfluoroalkyl substances (PFAS) in addition to polycyclic aromatic hydrocarbons, phenols, and phthalates. Firefighters' exposure to PFAS from the turnout gear itself has become an important research focus. Turnout composites of PFAS-based durable water repellent (DWR) finish subjected to cycles of ultra-violet light, thermal exposures, and laundering to realistically age the materials. All the aged samples showed some level of yarn breakage on the surface in digital microscopic imaging, which suggests that fiber along with DWR finishes are coming off during the ageing process. In ToF-SIMS analysis, the aged samples showed a significantly lower number of CF3+ion detection compared to the new fabrics. In XPS analysis, fluorine detection from two different PFAS-finished Fabrics in new conditions were 44 and 50 atomic percentages (At%), which eventually reduces to 16 and 14.7 At% respectively after ageing. All the surface analysis techniques suggest that PFAS chemistries are subjected to break and wash-off during the ageing process. PFOA and PFHxA were detected in QqQ-MS-MS analysis after the high-pressure methanol solvent extraction. The results suggest that a significant amount of PFAS is coming-off during the lifespan of the use of the turnout gear and could be a potential exposure source to PFAS for firefighters.

581 – Ambient Ammonia Patterns in the Gelderse Vallei: A Design and First Results of an Air Quality Network

Serigne Lô[1], Gerard Hoek[2], Wietske Dohmen[3], Dick Heederik[4], Roel Vermeulen[4]
[1]Utrecht University, PhD-student, Utrecht, Netherlands. [2]Utrecht University, Associate Professor, Utrecht, Netherlands. [3]Utrecht University, Assistant Professor, Utrecht, Netherlands. [4]Utrecht University, Professor, Utrecht, Netherlands

Abstract
High ammonia deposition exerts stress on biodiversity and is currently a major topic in the Netherlands. Monitoring spatial and temporal patterns of ambient ammonia concentrations is an important action in limiting biodiversity loss. There are currently few sites, beyond nature reserve sites, where ammonia is measured routinely. For this reason, we set up an air quality monitoring network in the framework of the Regiodeal Foodvalley project. The study area (Gelderse Vallei, the Netherlands) includes livestock dense areas, but also nature reserves as the Utrechtse Heuvelrug and the Veluwe. We selected 73 sites covering the area that varied in modeled concentration of ammonia (RIVM estimated Grootschalige Concentratie Nederland at 1*1 km) and distance to intensive livestock farms. Sites were located both in nature reserves and living environments, the latter being back yards of volunteering local residents. Ammonia samples were taken monthly from March 2022 with Gradko NH3 diffusion tubes in collaboration with the volunteers. First results of the averaged ammonia measurements in 2022 showed considerable spatial variability, up a factor 8 difference between sites. Highest concentrations were found in areas with a high density of livestock farms (maximum = 17 farms /km2). Averaged concentrations correlated well with the number of farms within a 1500m circle ($r = 0.77$). Concentrations peaked in March at all sites, including those in nature reserves, and were on average 2.5 times higher than in other months ($p<0.01$). March 2022 was a month with minimal rainfall, following the start of the season for manure application.

583 – POSTER: Preliminary Analysis of Chemical and Non-Chemical Stressors Using PRAPARE to Assess Birth Outcomes

Jacob Donovan[1], Nicolle Tulve[2], Kent Thomas[2]
[1]ORAU, National Student Services Contractor, Research Triangle Park, USA. [2]EPA, Research Physical Scientist, Research Triangle Park, USA

Abstract
Children and adults are exposed to a variety of chemical and non-chemical stressors from all facets of daily life. These stressors, alone and together, can lead to changes in health and well-being. Our research approach seeks to understand these complex relationships and their impact on health outcomes. The National Children's Study Vanguard Study database contains a large amount of environmental, survey, and outcome data. Data were collected from more than 5,000 mothers in the pilot studies which ran from 2009-2014. Secondary analysis of the available datasets provides a unique opportunity to explore approaches for combining quantitative and qualitative data. One approach is to use PRAPARE (Protocol for Responding to and Assessing Patients' Assets, Risks, and Experiences), a national standardized tool for categorizing non-chemical stressors into social determinants of health (SDoH) clusters, to generate scores. Cluster scores were divided into three separate groups to describe SDoH attributes for each participant: low (0-2 SDoH), medium (2-4 SDoH), and high (5-9 SDoH). Maternal biomarker measurements for chemicals such as BPA, phthalates, PAHs, and pesticides were organized into tertiles. PRAPARE scores and chemical tertiles were combined and used to assess associations between chemical and non-chemical stressors and child birth weight. Lower birth weight has been associated with various chemicals including PAHs, phthalates, and pesticides as well as non-chemical stressors. This poster will highlight preliminary findings.

584 – Cadmium Exposure in Adults across Europe: Combining Environmental and Harmonized HBM Data from the HBM4EU Aligned Studies Survey

Janja Snoj Tratnik[1], David Kocman[1], Milena Horvat[1], Anna-Maria Andersson[2], Anders Juul[2], Eva Jacobsen[3], Kristin Ólafsdóttir[3], Jana Klanova[4], Lenka Andryskova[4], Beata Janasik[5], Wojciech Wasowicz[5], Natasa Janev Holcer[6,7], Sónia Namorado[8], Inês Coelho[8], Loïc Rambaud[9], Margaux Riou[9], An Van Nieuwenhuyse[10], Brice Appenzeller[11], Marike Kolossa-Gehring[12], Till Weber[12], Marta Esteban-López[13], Argelia Castaño[13], Liese Gilles[14], Laura Rodriguez Martin[14], Greet Schoeters[14], Ovnair Sepai[15], Eva Govarts[14]
[1]Jožef Stefan Institute, Ljubljana, Slovenia. [2]Copenhagen University Hospital-Rigshospitalet, Copenhagen, Denmark. [3]University of Iceland, Reykjavik, Iceland. [4]Masaryk University, Brno, Czech Republic. [5]Nofer Institute of Occupational Medicine, Lodz, Poland. [6]Croatian Institute of Public Health, Zagreb, Croatia. [7]Faculty of Medicine University of Rijeka, Rijeka, Croatia. [8]National Institute of Health Dr. Ricardo Jorge, Lisbon, Portugal. [9]Santé publique France, Saint Maurice, France. [10]Laboratoire National de Santé, Dudelange, Luxembourg. [11]Luxembourg Institute of Health, Strassen, Luxembourg. [12]German Environment Agency, Berlin, Germany. [13]Instituto de Salud Carlos III, Madrid, Spain. [14]Flemish Institute for Technological Research, Mol, Belgium. [15]UK Health Security Agency, Harwell, United Kingdom

Abstract
Alignment of national or regional HBM studies under the European Human Biomonitoring Initiative (HBM4EU) enabled a refined assessment of cadmium (Cd) exposure among European population of adults and identification of main sources of exposure. Cadmium was measured in urine of 20-39 years old adults from nine European countries (Iceland, Denmark, Poland, Czech Republic, Croatia, Portugal, Germany, France, Luxembourg) sampled in the years 2014-2021 (n=2510). The measurements in urine were quality assured by the HBM4EU quality assurance/quality control scheme, and study participants' questionnaire data were post-harmonized. Spatially resolved external data, namely Cd concentrations in soil, agricultural areas, phosphate fertilizer application, traffic density and point source Cd release were collected for the respective statistical territorial unit (NUTS). The age-dependent alert values for Cd in urine were exceeded by 16% of the study participants. Association analysis showed an important contribution of vegetarian diet to the overall exposure, with 35% higher levels in vegetarians as opposed to non-vegetarians. Increase in Cd levels due to smoking was 25%. Using NUTS2-level external data, positive associations between HBM data and percentage of cropland and use of Cd-containing mineral phosphate fertilizer were revealed, which indicates a significant contribution of mineral phosphate fertilizers to human Cd exposure through diet. The findings of the study support the recommendation by EFSA to reduce Cd exposure as also the estimated mean dietary exposure of adults in the EU is close or slightly exceeding the tolerable weekly intake. It also indicates that regulations are not protecting the population sufficiently

585 – Blood metal levels by e-cigarette use and device characteristics in a Maryland cohort

Anna Tillery[1], Angela Aherrera[2], Rui Chen[1], Joyce Lin[1], Mina Tehrani[2], Donia Moustafa[3], Kathrin Schilling[4], Olgica Balac[4], Jana Mihalic[5], Ana Navas-Acien[4], Ana Rule[5]
[1]Johns Hopkins University, PhD Candidate, Baltimore, USA. [2]Johns Hopkins University, Postdoctoral Fellow, Baltimore, USA. [3]US Environmental Protection Agency, Washington, DC, USA. [4]Columbia University, New York City, USA. [5]Johns Hopkins University, Baltimore, USA

Abstract

Introduction: Electronic cigarettes rapidly evolved from "Mod" devices mostly used by smokers as a risk reduction strategy to "Pod" devices with disposable cartridges. E-cigarette heating components and the generated aerosol are rich in metals. Differences between Mod and Pod devices may impact metal exposure. Inhalation of metals found in e-cigarette aerosols can lead to adverse health outcomes such as asthma and lung cancer.

Objective: To compare metal levels in blood of e-cigarette users and nonusers and by e-cigarette use behaviors and device type.

Methods: We recruited 75 participants: 18 Mod, 26 Pod, and 31 non-users. Sociodemographics, use behaviors, and device characteristics were collected by survey. Blood metals were analyzed by inductively-coupled plasma mass spectrometry (ICP-MS). Analysis was characterized by primary device used. Chi-squared, ANOVA, and regression analysis were used to assess relationships between variables and groups.

Results: Pod users had higher Cr (1.3 times, $p=0.01$), Pb (1.4 times, $p=0.06$), and Se (1.1 times, $p=0.003$) in blood compared to nonusers. Mod users had higher Cr (1.3 times, $p=0.007$) and Ni (1.5 times, $p=0.007$) in blood compared to nonusers.

Conclusions: E-cigarette users had higher blood Cr, Pb, Se, and Ni levels than non-users. E-cigarette heating components and differences in power between devices could explain differences in metal exposures. This study found that device type and behaviors determine exposure to certain metals and could indicate differential health risks for e-cigarette users. This information could help inform potential interventions and strategies.

586 – POSTER: The Role of Next-Generation Sequencing in Wastewater-Based Epidemiology

Adam Gushgari[1], Himeshi Samarasinghe[2], Kavya Boyina[1], John Cosgrove[1]
[1]Eurofins Pandemic Prevention Services, Sacramento, USA. [2]Eurofins Pandemic Prevention Services, Principal Scientist, Sacramento, USA
Abstract
Wastewater-based epidemiology (WBE) is an approach where detection and quantification of biological and chemical health indicators in wastewater are used as proxies for estimating disease and health burdens in the community. Pathogen detection and quantitation in wastewater are often achieved via molecular diagnostic tools incorporating polymerase chain reaction (PCR). Over time, next-generation sequencing (NGS) of wastewater samples has become increasingly popular due to NGS data providing deeper and more comprehensive insights into the wastewater microbiome and community health. We conducted a systematic review of the literature to identify current global application of NGS in wastewater surveillance. NGS was successfully used to detect a wide array of pathogens and other health indicators in wastewater. NGS data often detected novel SARS-CoV-2 variants in wastewater ahead of molecular methods. Despite the benefits, application of NGS in wastewater surveillance has been limited, likely due to higher cost, requirements for advanced data analytics, and relative unfamiliarity with the techniques. For the field of WBE to successfully expand beyond the SARS-CoV-2 pandemic, the power and benefits of NGS methodologies should be integrated alongside the traditionally applied molecular tools. We identified limited representation of developing nations among WBE projects that incorporated NGS, although the few publications from developing nations highlighted the utility of NGS in public health crises response when individual clinical surveillance was not feasible. We identify the significance of allocating resources to develop NGS in WBE, especially in economically-disadvantaged settings, to increase global health equity and emergency preparedness.

587 – POSTER: Characterizing the organizational landscape for environmental justice and community health promotion in Austin, Texas, and Chicago, Illinois

Caroline Youssef[1], Hue Mai[2], Deborah Salvo[2], Sharmilee Nyenhuis[3]
[1]University of Illinois, Chicago, USA. [2]University of Texas Austin, Austin, USA. [3]University of Chicago, Chicago, USA

Abstract
Background: Austin and Chicago have well-documented are home to a high number of low-income, minoritized populations with high vulnerability to climate impacts. Local organizations focused on environmental justice and community health promotion can help address the social determinants of health responsible for spatial health inequities.

Objective: To systematically identify and characterize environmental justice and community health agencies in Austin and Chicago.

Methods: We developed an organizational mapping protocol to search for local organizations with focus on environmental justice and/or community health. Using search terms about organizations, climate/environmental justice, health, and setting (Austin/Chicago) we ran searches using Google Scholar, and screened the first 100 results per search (sufficient to reach saturation) to confirm eligibility. Data from eligible organizations was abstracted and thematically coded.

Results: We identified 20 eligible organizations in Austin and 19 in Chicago. Organizations were coded into 11 non-mutually exclusive themes: Land and Wildlife Conservation (n=7), Urban Green Space (n=2), Green Building and Energy Efficiency (n=10), Air Quality and Pollution (n=4), Water and Waste Management Advocacy (n=6), Community Development (n=11), Environmental Law and Policy Advocacy (n=11), Environmental Education (n=9), Climate Change (n=9), Environment and Health (n=5), and Transportation Equity (n=2).

Conclusion: Comprehensively assessing the presence and focus of these types of agencies in places of interest helps diagnose unmet needs and identify potential research-community partnerships. Future research will use network analysis to identify linkages across organizations in both cities, to advance our understanding of existing and/or lacking community supports for climate resilience in high-need settings.

588 – The Impact of Phthalate Exposure on the Gut Microbiota of Pregnant African American People in Atlanta, Georgia

Natalie Shen[1], Anne Dunlop[2], Dana Boyd[3], Elizabeth Holzhausen[4], Bridget Chalifour[4], Yijuan Hu[5], Ziyin Yang[6], Parinya Panuwet[7], Grace Lee[8], Peter Ryan[9], Elizabeth Corwin[10], Tanya Alderete[11], Donghai Liang[7]
[1]Emory University, Graduate Student, Atlanta, USA. [2]Emory University, Associate Research Professor, Atlanta, USA. [3]Emory University, Professor, Atlanta, USA. [4]University of Colorado Boulder, Postdoctoral Associate, Boulder, USA. [5]Emory University, Associate Professor, Atlanta, USA. [6]Emory University, Doctoral Student, Atlanta, USA. [7]Emory University, Assistant Professor, Atlanta, USA. [8]Emory University, Senior Public Health Program Associate, Atlanta, USA. [9]Emory University, Research Professor, Atlanta, USA. [10]Columbia University, Research Professor, New York City, USA. [11]University of Colorado Boulder, Assistant Professor, Boulder, USA

Abstract
Background: Phthalates, present in household and personal care products, may disrupt hormones and harm fetal and reproductive health. These compounds interact with the gut microbiota, potentially increasing risk of adverse pregnancy outcomes. However, few studies have investigated the relationship between phthalate exposure and gut microbiota in pregnant African Americans (AAs), an understudied population unequally burdened with phthalate exposure and preterm birth.

Methods: We used the Atlanta AA Maternal-Child cohort data between 2016-2020 (N = 193). Maternal rectal swabs and urine samples were collected at 8-14 weeks and 24-30 weeks gestation for gut microbiota profiling via 16S rRNA gene sequencing and phthalate exposure, which was measured by urinary metabolite concentration and dichotomized into high and low exposure levels using the median threshold. We compared relative abundance of bacterial genera and alpha-diversity between high and low phthalate exposure groups. We used univariate Spearman correlation and multivariable linear regression to examine the associations between phthalate metabolite concentrations and gut bacterial abundance.

Results: Those with higher concentrations of MBzP had greater gut alpha-diversity. Higher concentrations of specific phthalates negatively associated with the relative abundance of certain beneficial gut bacterial taxa (MiBP with Verrucomicrobiota, MEHP with Bifidobacterium) and positively associated with the relative abundance of some pathogenic gut taxa (MBP and MEHP with Proteobacteria, MEP with Peptostreptococcus).

Conclusion: Our findings provide evidence that maternal gestational phthalate exposure may associate with altered gut microbiota alpha-diversity. Moreover, higher phthalate exposure associates with increased abundance of some pathogenic bacterial taxa and decreased abundance of some beneficial bacterial taxa.

589 – Experiences with Low-cost Optical Particle Sensors for Indoor Exposure Assessment in Residential Field Studies

Mingyu Wang[1], Brent Stephens[2], Mohammad Heidarinejad[3], Insung Kang[4], Aditya Singh[1], Ruben Rodriguez[5], David Chang[6], Gina Solomon[7], Israel Rubinstein[8], Zane Elfessi[9], Kaveeta Jagota[10], Nancy Karpen[11]
[1]Illinois Institute of Technology, Doctoral Candidate, Chicago, USA. [2]Illinois Institute of Technology, Professor and Department Chair, Chicago, USA. [3]Illinois Institute of Technology, Assistant Professor, Chicago, USA. [4]Illinois Institute of Technology, Senior Research Associate, Chicago, USA. [5]Central California Environmental Justice Network, Community Organizer & Air Monitoring Specialist, Coalinga, USA. [6]Public Health Institute, Project Coordinator, Oakland, USA. [7]Public Health Institute, Program Director, Oakland, USA. [8]Jesse Brown Veterans

Affairs Medical Center, Associate Chief of Staff, Chicago, USA. [9]Jesse Brown Veterans Affairs Medical Center, Emergency Medicine Pharmacist, Chicago, USA. [10]Jesse Brown Veterans Affairs Medical Center, Senior Research Coordinator, Chicago, USA. [11]Jesse Brown Veterans Affairs Medical Center, Registered Nurse, Chicago, USA

Abstract

Indoor particulate matter (PM) is of increasing concern for human health. A wide variety of low-cost optical particle sensors are now commercially available for use. While a large body of recent work has evaluated the performance of low-cost particle sensors in laboratory and outdoor field settings, less is known about the overall utility and performance of low-cost particle sensors for long-term exposure assessment in field studies of indoor air. This work summarizes recent experiences with PurpleAir PA-II monitors deployed in two residential field studies of indoor air interventions. In the first study, PA-II monitors were deployed indoors in 30 homes of agricultural workers in California's San Joaquin Valley, as well as at several nearby outdoor locations, to evaluate the impacts of a custom evaporative cooler filtration solution for mitigating exposure to infiltrating wildfire smoke. In the second study, PA-II monitors were deployed inside the homes of U.S. military veterans with chronic obstructive pulmonary disease (COPD) (25 households to date, recruitment ongoing) to evaluate the impacts of portable HEPA air cleaners on indoor PM concentrations and COPD outcomes (again with a smaller number of monitors deployed at nearby outdoor locations). Our experiences with these deployments provide insights into: 1) intra-model variability (assessed via periodic co-location measurements of over 50 sensors), 2) different challenges with data collection and recovery from SD and WiFi versions of PA-II monitors in the field, 3) the utility of such monitors for evaluating indoor PM concentrations and exposures at baseline and during/after interventions.

590 – Household Air Pollution Exposures During Cooking with Traditional and LPG Stoves During Pregnancy in Rural Bangladesh

Jonathan Thornburg[1], Sajia Islam[2], S.M. Rokonuzzaman[2], Michelle McCombs[1], Sk Masum Billah[3], Camille Raynes-Greenow[3]
[1]RTI International, Research Triangle Park, USA. [2]icddr,b, Dhaka, Bangladesh. [3]University of Sydney, Sydney, Australia

Abstract

Household air pollution is a major health risk in South Asia. The CHANge trial aims to assess the impact of liquefied petroleum gas (LPG) for cooking to reduce household air pollution (HAP) exposure on perinatal morbidity and mortality compared to usual cooking practices in Bangladesh. One objective of the trial compared reduction in personal exposure to HAP in the LPG fuel arm compared to the traditional cooking arm. MicroPEM monitors measured $PM_{2.5}$, black carbon (BC), and brown carbon (BrC) exposures on 268 pregnant women, equally divided between the two arms. Valid data from women that completed one assessment each trimester were included in the analysis. Mean $PM_{2.5}$ exposure in the LPG arm was 21.2 µg/m3 (19.8%) lower than the control arm. Decreases in BC of 15.6 µg/m3 and in BrC of 5.5 µg/m3 accounted for 99% of the $PM_{2.5}$ reduction. Ambient $PM_{2.5}$ and BC, measured with collocated PurpleAir sensors and filter samples, that infiltrated indoors contributed to a high baseline personal exposure. HAP was the source of BrC exposure as the mean ambient concentration was near the detection limit. Personal $PM_{2.5}$ and BC varied seasonally and was driven by the ambient levels. The first and third trimester sample collection periods had the highest levels and corresponded to the "dry" season when rice parboiling and brick production were at their peaks. Second trimester samples were collected during the monsoon season. The reduction in $PM_{2.5}$, BC, and BrC in the LPG arm and our understanding of covariates supports the pending intention-to-treat analysis.

591 – POSTER: The Use of Low-Cost Sensors for Worker Exposure Assessment: Testing the Impact of Various Particulate Matter Compositions on Sensor Performance

Samuel Stowe[1], Vilcassim Ruzmyn[2]
[1]The University of Alabama at Birmingham, PhD Student Researcher, Birmingham, USA. [2]The University of Alabama at Birmingham, Assistant Professor, Birmingham, USA

Abstract

Today's low-cost Particulate Matter (PM) sensors have several advantages over the current industry standard devices that would make them optimal for use in occupational settings, however, previous studies have demonstrated limitations to these sensors. Low-cost PM sensors use light scattering techniques similar to those used in current industry standard devices, however, their capacity to measure concentrations over 100 µg/m3 and the impact of particle composition on sensor performance needs further evaluation. Therefore, the goal of this study was to test two generations of a low-cost PM sensor device - the 'Airbeam' - against a Thermo Scientific Personal Data RAM 1500's (industry standard device) and investigate their ability to measure PM of varying compositions. During laboratory calibrations both generations of Airbeam were placed into an airtight chamber with one PDR 1500 containing a pre-weighed filter and for field calibrations, the two devices were run simultaneously in a selected occupational environment. During these calibrations, they were exposed to three common types of aerosols: biomass smoke, exhaust from a leaf blower, and construction dust. Our results showed a linear relationship and high R2 between the Airbeams' and the PDR 1500's measurements at around 150 μg/m3 and below. However, the Airbeam's measurements plateaued at around 300 μg/m3. Particle composition also impacted the calibration coefficients. In conclusion, our findings demonstrate while the Airbeams have the capacity to be successfully used to measure occupational exposures to PM, varying chemical compositions can impact their accuracy and should be taken into consideration when used in research.

592 – Development of a Novel Limit Of Detection (LoD) "Lite" Approach For Wastewater Assay Validation with a Focus on Public Health Needs in Developing Nations

Himeshi Samarasinghe[1], Maftuna Gafurova[2], Ellen Cavanaugh[3], Adam Gushgari[4], Kavya Boyina[4], Rochelle Holm[5], Petros Chigwechokha[6], John Cosgrove[4]
[1]Eurofins Pandemic Prevention Services, Principal Scientist, Sacramento, USA. [2]Eurofins Wastewater, Laboratory Director, Sacramento, USA. [3]Eurofins Wastewater, Sacramento, USA. [4]Eurofins Pandemic Prevention Services, Sacramento, USA. [5]University of Louisville, Louisville, USA. [6]Malawi University of Science and Technology, Mikolongwe, Malawi

Abstract
Wastewater-based epidemiology (WBE) has emerged as a valuable approach for community-wide surveillance of biological and chemical agents of public health significance. Wastewater surveillance requires rapid diagnostic assay development and validation to swiftly respond to novel pathogens and emerging variants of concern. Quantitative polymerase-chain-reaction (qPCR)-based diagnostic tools necessitate robust validation and optimization for accurate and sensitive target detection in the complex wastewater matrix, one component of which is the establishment of the Limit of Detection (LoD) of the assay. The standard practice for establishing LoD often requires testing of the serially diluted target in replicates of 20 or more across multiple technicians and days with repetition. This time-consuming and resource-intensive process is not feasible in economically-disadvantaged populations, especially in times of public health crises. In this project, we have partnered with University of Louisville and Malawi University of Science and Technology to evaluate the performance of a more targeted LoD approach, termed "LoD Lite". The Eurofins Wastewater laboratory in Sacramento, California will establish the LoD of a novel qPCR cholera diagnostic assay in wastewater following standard protocols while the laboratory in Malawi will establish "LoD Lite" with fewer replicates and repetitions. Result comparison between the two methods will be used to evaluate and optimize the "Lite" approach. The "Lite" approach can considerably lower resource and time requirements for LoD establishment, allowing labs in developing nations such as Malawi, to swiftly onboard novel diagnostic assays in response to public health crises more effectively.

593 – Soil Lead is Elevated in High Poverty Areas of a USA Suburb

Ben K. Greenfield[1], Jannatul Ferdous[2,3], Jarrod N. Koski[4,5], Eloho Unufe[2], Geoffrey S. Siemering[6], Thressa Willett[7]
[1]Public Health Program, Muskie School of Public Service, University of Southern Maine, Associate Professor, Portland, ME, USA. [2]Environmental Sciences Department, Southern Illinois University Edwardsville, Graduate Assistant, Edwardsville, IL, USA. [3]Jacobs, Environmental Consultant, Chicago, IL, USA. [4]Environmental Sciences Department, Southern Illinois University Edwardsville, Undergraduate Research Assistant, Edwardsville, IL, USA. [5]LightStream Consulting Group, CTO, Edwardsville, IL, USA. [6]Soil Science Department, University of Wisconsin-Madison, Contaminant Transport and Bioavailability Specialist/ Project Manager, Madison, WI, USA. [7]Public Health Program, Muskie School of Public Service, University of Southern Maine, Graduate Assistant, Portland, ME, USA

Abstract
Lead is an industrial pollutant and a neurotoxin, and neurodevelopmental hazards are greater for low-income children. Few studies have directly examined the spatial relationship between environmental lead concentrations and sociodemographic indicators of community vulnerability. We spatially compared roadside surface soil lead concentrations to industrial sources and community poverty levels in the Metro East region of St. Louis (Illinois, USA). The study location includes residential and agricultural areas and an industrial suburb with a history of lead processing and smelting. Sampling included randomly chosen sites, stratified by census tract poverty, and fenceline sites, adjacent to industrial lead sources. In linear regression, soil lead was moderately and significantly ($R^2 = 0.45$, $p < 0.001$, $n = 115$) higher in sites closer to the Mississippi river and to industrial lead source locations. Poverty was also higher closer to the Mississippi river ($R^2 = 0.36$, $p < 0.001$), and near sites with elevated soil Pb ($R^2 = 0.29$, $p < 0.001$). These results indicate that residents of low-income communities in a typical USA suburb face the added environmental injustice of increased lead in their local soils.

594 – POSTER: Personal Exposome Trackers for Precision Occupational Health

Xinyue Zhang[1], Sawson Taheri[2], Michael Snyder[3]
[1]Stanford University, Research Scientist, Palo Alto, USA. [2]Stanford University, Research Engineer, Stanford, USA. [3]Stanford University, Chair and Professor, Stanford, USA

Abstract
Exposures to hazardous substances or conditions in the workplace have a significant impact on one's health, both in the short-term and long term. Occupational exposures come from a wide range of sources, including physical, chemical, biological and ergonomic factors. However, current monitoring practices often focus on a limited set of exposures, which cannot fully capture the complex exposures workers may encounter. Moreover, fixed-location monitors fail to accurately reflect occupational exposures at the personal level. Here, we propose a wearable device that captures the diverse airborne exposome and allows users to get a comprehensive view of their personal exposures at the workplace. The wearable device, termed Personal Exposome Tracker (PET), is an active sampling device that collects all traditional environmental parameters, such as temperature, humidity, PM2.5, PM10 and volatile organic compound (VOC), at the personal level. PET is compact and suitable for workers to carry around while performing day to day functions. More importantly, PET contains a compartment to collect chemical and biological exposomes, respectively. Paired with next generation sequencing (NGS) and high-resolution mass spectrometry (MS), PET provides the most detailed view of the occupational airborne exposome. The onboard Global Positioning System (GPS) sensor provides an additional layer of exposure information, especially for outdoor work environments. Therefore, PET is a powerful exposometer that provides insightful information to aid precision occupational health.

597 – POSTER: Arsenic and Depression: An Analysis of NHANES Data (2013-2020)

Lauren Giacobbe[1], Maria Argos[2], Mary Turyk[3]
[1]University of Illinois Chicago, PhD Student, Chicago, USA. [2]University of Illinois Chicago, Associate Professor, Chicago, USA. [3]University of Illinois Chicago, Professor, Chicago, USA

Abstract
Arsenic is a naturally occurring heavy metal in the air, soil, and water; it is also abundant in food sources such as rice and fish. However, known outcomes of repeated exposure include skin abnormalities, cardiovascular disease, cancer, and neurological damage. Depression has also been tentatively linked with heavy metal exposure. We looked for any meaningful association between total arsenic exposure and depression outcomes in 3,333 participants in the National Health and Nutrition Examination Survey data (2013-2020) in a complete-case analysis. Creatinine-adjusted total urinary arsenic (μg/L) was set at a 50 μg/L threshold; likely depression was established via the Patient Health Questionnaire. Covariates included marriage, education, income, age, race/ethnicity, BMI, hypertension, nutrition, acuity, and stress. Results were stratified by sex a priori. There were doubled odds of depressive outcomes for arsenic exposure >50 μg/L and was found to be higher for women (AOR=3.77, 95%CI=1.18-12.01) as compared to men (AOR=1.18, 95%CI=0.44–3.16). Modeling by a lower threshold of 10 μg/L (AOR=1.37, 95%CI=1.01–1.85) produced more conservative outcomes. Including acuity and stress indicators did not produce any meaningful outcomes; however, effect sizes did appear to change more for women. Adjusting for nutrition modified the arsenic-depression association when stratified by male (AOR=2.81, 95%CI=1.07-7.36) and female (AOR=2.61, 95%CI=0.72-9.50) sex. In conclusion, we found elevated odds of depressive outcomes given higher arsenic exposure, especially for women. Future investigations should incorporate better-defined depression outcomes and covariates and a longitudinal design should be considered.

599 – Pediatric Exposure to Environmental Manganese and Adolescent Postural Balance and in the Marietta CARES Cohort

Danielle McBride[1], Heidi Sucharew[2], Amit Bhattacharya[3], Erin Haynes[4]
[1]University of Kentucky, Post Doc, Lexington, USA. [2]University of Cincinnati, Research Professor, Cincinnati, USA. [3]University of Cincinnati, Director, Cincinnati, USA. [4]University of Kentucky, Chair and Professor, Lexington, USA

Abstract
Background: Manganese (Mn) is both an essential micronutrient to pediatric neurodevelopment and an irreplaceable raw input in the global steel industry. Epidemiological studies demonstrate a dose-response relationship between Mn exposure and neuromotor function ranging from subclinical effects to a clinically defined syndrome.

Methods: The objective of this study was to investigate the association between Mn exposure in childhood years 7-9 and adolescent years 13-18, as measured in blood, hair and toenails, and neuromotor function, as measured by postural balance, in participants from the longitudinal Marietta Communities Actively Researching Exposures Study (CARES) cohort.

Results: The CARES participants who completed biological assessments and adolescent balance testing (n=121) were 97.5% white, 54% female with a mean age of 16.1 years (range 13-18). Geometric mean (GM) biomarker concentrations for hair Mn were 408 ng/g in childhood and 189 ng/g in adolescence. GM for blood Mn was 9.8 μg/L in childhood and 10 μg/L in adolescence. For each ng/g increase childhood ln hair Mn, adolescent ln sway area increased 0.14 cm2 (95% CI: 0.43-0.25) and 0.05 cm (95% CI: 0.005-0.10) respectively under test condition foam open. For each μg/L increase adolescent ln blood Mn, adolescent ln sway area increased 0.45 cm2 (95% CI: 0.06-0.83) and 0.44 cm2 (95% CI: 0.04-0.85) under test conditions eyes open and closed, respectively.

Conclusion: This study supports childhood and adolescence as critical Mn exposure windows and postural balance as a sensitive neuromotor test to assess the impact of pediatric Mn exposure on neuromotor function.

600 – Assessing indoor air quality interventions in schools impacted by air pollution sources

Nancy Carmona[1], Aline Maybank[2], Edmund Seto[3], Jeffry Shirai[4], B.J. Cummings[5], Lisa Hayward[5], Elena Austin[6]
[1]University of Washington, PhD Candidate, Seattle, USA. [2]University of Washington, Undergraduate Research Assistant, Seattle, USA. [3]University of Washington, Professor, Seattle, USA. [4]University of Washington, Research Scientist, Seattle, USA. [5]University of Washington, Community Engagement Manager, Seattle, USA. [6]University of Washington, Assistant Professor, Seattle, USA

Abstract
Exposure science has the potential to improve indoor air quality (IAQ) in school settings. The Healthy Air, Healthy Schools community-engaged study was established to deploy air quality sensors and identify sources of ultrafine particles (UFPs) in classrooms in urban and rural settings in Washington State. The study also examined the removal efficiency of portable air cleaners (PACs) with HEPA filters in classrooms and found that after the HEPA units were deployed, UFPs in classrooms dropped to about one-tenth of outdoor levels. To advance the promotion of PAC use in schools, we conducted semi-structured interviews on the motivators and barriers to acceptance and use of PACs with school district administration and operations personnel, parents, and environmental health officials in Washington State. A total of 13 participants were recruited via email. We summarized recommendations for improving school IAQ, highlighting air quality testing. Interviewees reported that sensor technology can be used to 1) better understand the impact of PACs indoors and 2) evaluate whether PACs are an effective solution in school settings. Overall, schools want to understand the scale of IAQ concerns, regardless of their PAC use. We found that schools want a data-driven approach to determine whether there is a benefit from the PACs. Large school districts report using instruments such as the TSI 7545 monitor to investigate air quality concerns. This study uniquely focused on the need for sensor technology to highlight the importance of IAQ monitoring in schools impacted by poor indoor air quality.

601 – Differential Impacts of Wildfire Smoke Exposure on Allergy, Cardiovascular and Respiratory Health

Jennifer Stowell[1], Ian Sue Wing[2], Yasmin Romitti[3], Gregory Wellenius[1], Patrick Kinney[4]
[1]Boston University School of Public Health, Postdoc, Boston, USA. [2]Boston University, Professor, Boston, USA. [3]Boston University, Student, Boston, USA. [4]Boston University School of Public Health, Professor, Boston, USA

Abstract
While much is known about the impacts of wildfire-specific PM2.5 on cardiorespiratory health, current literature often focuses on regional exposure, specific fire episodes, and shorter time windows. We sought to quantify the impact of wildfire PM2.5 on emergency department visits (ED) for respiratory, cardiovascular, and allergy outcomes across multiple years in California (2012-2019) and identify potential effect modification due to age, sex, and race. Using a case-crossover logistic regression approach and lag0-7 wildfire PM2.5 per 10μg/m3, we identified increased risk of ED for total respiratory (OR: 1.031 [1.027,1.035]), asthma (OR: 1.082 [1.074,1.090]), allergies (OR: 1.080 [1.069,1.090]), and COPD (OR: 1.032, [1.020,1.044]). We did not observe significant increases in risk for total and specific cardiovascular outcomes. Stratifying on age, we identified elevated risk of respiratory ED in individuals 1-84 years, allergy ED visits in children 1-19 years and adults 20-84 years, asthma ED in children 1-19 years and adults over 20, pneumonia ED visits in children 1-4 years, and COPD ED visits in adults 20-84 years. We did not observe increased risk for bronchitis or acute upper respiratory infections among any age group. When stratifying on sex, we observed increased risk of an ED visit for asthma among females compared to males. Stratification on race demonstrated increased risk of allergy, respiratory, and asthma ED for all groups excluding American Indian/Alaska Natives. COPD was elevated in whites only. Additional analyses will analyze hospitalizations and explore effect modification by percentiles of temperature, extreme heat, other diagnosis categories, and comparisons with ambient PM2.5.

602 – Community knowledge, exposure science, and scientific uncertainty: The Michigan PBB Oral History Project

Erin Lebow-Skelley[1], Brittany Fremion[2], Martha Quinn[3], Jane-Ann Crowley[4,5,6], Melanie Pearson[7], Amy Schulz[8]
[1]Emory University Rollins School of Public Health, Assistant Director of Programs, Atlanta, USA. [2]Central Michigan University, Professor, Mt. Pleasant, USA. [3]University of Michigan, Research Area Specialist Lead, Ann Arbor, USA. [4]Pine River Superfund Citizen Task Force, Member, St. Louis, USA. [5]PBB Leadership Team, Member, Alma, USA. [6]PBB Citizens Advisory Board, Member, Alma, USA. [7]Emory University Rollins School of Public Health, Director of Research Projects, Atlanta, USA. [8]University of Michigan, Professor, Ann Arbor, USA

Abstract
In 1973, a flame retardant (polybrominated biphenyl or PBB) was accidentally shipped and mixed into livestock feed in Michigan, exposing farm animals across the state and entering the human food supply. After nine months, a chemist turned farmer discovered the mix-up and the state quarantined more than 500 farms. Researchers estimate that nearly everyone living in the state at the time, some 8 million people, were exposed to PBB. However, uncertainty about the human health risks resulting from the exposure delayed swift health action and guidance. Instead, citizens documented their experiences, mobilized, and demanded a response to their concerns through protests, litigation, and voluntary participation in pilot health studies. Community residents impacted by the PBB contamination wanted to share these stories and provide lessons for future events. And so, the Michigan PBB Oral History Project conducted 68 interviews with community members to document and learn from this community's experience. Using a grounded theory qualitative approach, we systematically analyzed 31 interview transcripts to identify patterns and themes. We identified themes reflecting community knowledge, skills, and leadership in the midst of uncertainty and inaction and developed a conceptual model showing how community capacity is mobilized by a large-scale environmental health threat. Oral histories provide lessons from the PBB disaster relevant to similar ongoing and persistent environmental contaminations, and demonstrate the importance of incorporating community knowledge into exposure science. Our presentation demonstrates the significance of qualitative methods for social and environmental health research and action.

603 – POSTER: Changes in Children's Time Spent Indoors and Outdoors: A Comparative Analysis of Pre- and Post-March 2020 Using Data from the American Time Use Survey

Jenna Honan[1], Paloma Beamer[1], Jacqueline Larson[1], Olusola Ogunseye[1], Helena Solo-Gabriele[2], Alesia Ferguson[3]
[1]Department of Community, Environment and Policy, Mel and Enid Zuckerman College of Public Health, University of Arizona, Tucson, AZ, USA. [2]Department of Chemical, Environmental, and Materials Engineering, Miami, FL, USA. [3]Department of Built Environment, North Carolina Agricultural and Technical State University, Greensboro, NC, USA

Abstract
This study examines the amount of time children (age $<$ 18) in the United States spend indoors versus outdoors using data from the American Time Use Survey (ATUS). Data are available from January 2003 through September 2021. A comparative analysis of data collected before versus after March 2020 was conducted to explore potential changes in children's time use patterns. The results reveal that pre-March 2020, children spent a geometric mean of 23.9 minutes/day (geoSD=2.9, range=(0, 680)) outdoors, while after March 2020, a timeframe notably marked by varying degrees of shelter-in-place orders around the nation, this decreased to 20.4 minutes/day (geoSD=3.0, range=(0, 1170)). Although this is a difference of only about 3 minutes/day, this change is statistically significant ($p<0.001$). There was greater variability among post-March 2020 participants, indicating that while the average time outdoors decreased, some substantially increased their time outdoors. No differences were observed by sex, although log-time spent outdoors decreased with age ($\beta=-0.05$, $p<0.001$). The results challenge the commonly-used estimate from the U.S.EPA in 1989 of Americans spending on average 90% of time indoors, instead suggesting a new average of about 98.5%. These data provide relevant parameters for use in exposure models that consider differences in indoor vs. outdoor exposures. The findings underscore the need for policymakers, educators, and parents to promote a healthy balance between indoor and outdoor time, especially during times of pandemic-related disruptions. Further research is warranted to explore the long-term effects of these changes on children's health.

604 – Labeling Liabilities: Effectively Evaluating Risk From Low-Risk Labeling Claims

Jordan Shore, Allison Person
RHP Risk Management, Sr. Associate Health Scientist, Chicago, USA

Abstract
In the early 1990s, the U.S. EPA began the Design for the Environment (DfE) Safer Product Labeling program, which has since been

renamed the Safer Choice Program, as a voluntary program to encourage industry to consider human and environmental health effects of chemicals and technologies. Along with Safer Choice labels, additional language such as "earth friendly" and "non-toxic" designed to inform consumers of a lower risk product has invited criticism and litigation over the validity of the labeling claims.

Such lawsuits have posed the question of 'how can a product be labeled "non-toxic" if the product contains ingredients that can potentially harm humans, animals, or the environment?'. While certain ingredients that meet the Safer Choice criteria in cleaners, paints, and soaps, may still have the potential to harm humans or animals under certain conditions, the answer lies in one of the fundamental principles of toxicology: the dose makes the poison.

Understanding the "dose" of a chemical that makes its way into the environment is a non-trivial feat that entails a comprehensive characterization of a source-receptor relationship. While the in-use product concentration (e.g. cleaner inside the container) may have the potential to cause harm, these do not accurately represent probable exposure scenarios for receptors in the environment or in users. Additionally, Safer Choice labeling is based on using the safest possible ingredients to make a high-quality product but does not mean the product is inert. This poster will explore such a relationship and discuss regulatory guidance regarding ecological risk assessment.

605 – POSTER: Quantitative Structure Use Relationships: Technical Summit Highlights

Peter Egeghy[1], Andreas Ahrens[2], Jon Arnot[3], Richard A. Becker[4], Mark Bonnell[5], Sean Collins[6], Paul DeLeo[4], Michelle Embry[7], Todd Gouin[8], Kristin Isaacs[1], Elke Jensen[9], Rosemary Zaleski[10]
[1]Office of Research and Development, U.S. Environmental Protection Agency, Durham, NC, USA. [2]formerly European Chemicals Agency, Helsinki, Finland. [3]ARC Arnot Research and Consulting Inc., Toronto, Ontario, Canada. [4]American Chemistry Council, Arlington, VA, USA. [5]Environment and Climate Change Canada, Gatineau, Quebec, Canada. [6]Health Canada, Ottawa, Ontario, Canada. [7]Health and Environmental Sciences Institute, Washington, D.C., USA. [8]TG Environmental Research, Sharnbrook, United Kingdom. [9]The Dow Chemical Company, Midland, MI, USA. [10]Lumina Consulting, L.L.C., Hillsborough, NJ, USA

Abstract
The Quantitative Structure Use Relationship (QSUR) Summit, held on November 2–4, 2022, focused on advancing the development, refinement, and use of QSURs to support chemical substance prioritization, risk assessment, and mitigation. QSURs utilize chemical structures to predict the function of a chemical within a formulated product or an industrial process. This presumed function can then be employed to develop chemical use categories or other information necessary to refine exposure assessments, such as estimated weight fractions within a given formulated product or releases into the environment from chemical or product manufacturing and use. The invited expert meeting was attended by 38 scientists from Canada, Finland, France, the UK, and the USA, representing government, industry, and academia, with expertise in exposure science, chemical engineering, risk assessment, formulation chemistry, and machine learning. Workshop discussions emphasized the importance of collection and sharing of data and quantification of estimates. Key challenges were identified, and collaborative approaches to address those challenges were put forward. These included ideas to aggregate information while protecting proprietary product composition and other confidential business information and potential opportunities for sectors to provide frame formulations for products with industrial uses. Discussions led to proposals for a wide range of applications beyond exposure and risk modeling, including sustainable formulation discovery. In addition, discussions continue on the development, execution, and dissemination of case studies tied to explicit problem formulation illustrating how QSURs can supply or derive information on chemical functions, concentrations, and exposures.

606 – The Impact of Climate Change on Pesticide Exposure Biomonitoring

Carly Hyland[1,2], Cynthia Curl[2]
[1]UC Berkeley, Assistant Professor, Berkeley, USA. [2]Boise State University, Boise, USA

Abstract
Introduction: Climate change is increasingly impacting exposure and vulnerability to pesticides, which may require changes in typical study methods and sampling strategies.

Methods: We examine climate change-induced challenges we faced in two recent studies employing urinary pesticide biomonitoring in a cohort of pregnant people and a farmworker population, and discuss implications for designing future studies.

Findings: We conducted urinary pesticide biomonitoring studies in 2021 and 2022 that were impacted by extreme heat events and altered weather patterns in Idaho. These events may have directly and indirectly changed participants' exposure and vulnerability to the pesticides we monitored, such as by altering the stability of the metabolites in the urine samples, or increasing participants' pesticide concentrations due to a decreased physiological ability to metabolize environmental chemicals during heat stress. Further, some farmworker participants perceived that more pesticides were being used to "make up time" after the delayed agricultural season.

Discussion: In addition to its potential impacts on exposure and vulnerability to pesticides, climate change directly altered our sampling strategies in these studies. Suggestions for future work include building additional time for data collection to respond to unanticipated events (e.g., agricultural seasons starting earlier or later than normal due to weather changes), considering the time of day of sample collection (e.g., whether to collect samples during the hottest part of the day), and implementing additional strategies to keep urine samples cool during collection and transport to the laboratory. Studies are needed that examine the impact of extreme heat on pesticide exposure.

607 – Cannabis Products and Trends in a Novel Cohort: The VapeScan Study

Katlyn McGraw[1], Elizabeth Oelsner[2], Siyue Gao[3], Nancy Lolacono[4], Dona Sangapalaarachchi[4], Vesna Ilievski[4], Justin Liu[5], Daichi Shimbo[2], Ana Navas-Acien[2], Tiffany Sanchez[2]
[1]Columbia University, Postdoctoral Fellow, New York, USA. [2]Columbia University, PI, New York, USA. [3]Columbia University, Data Analyst, New York, USA. [4]Columbia University, Researcher, New York, USA. [5]Columbia University, Reseracher, New York, USA

Abstract
Cannabis products represent a potentially important source of environmental contaminants, including metals and pesticides. We characterized cannabis use in a cohort of young adults based in the New York City area. VapeScan is an ongoing prospective study of subclinical cardiopulmonary disease among vape users and non-vaping/non-smoking controls. Participants undergo 3 visits 9 months apart. We designed and implemented a cannabis questionnaire at visit 2 to characterize cannabis products and their methods of use beyond cannabis use status. As of March 2023, VapeScan has enrolled 201 participants. Participants mean (SD) age was 27.6 (7.91) years, 54% were female, 49% self-selected white race. At baseline, 20 (10%) participants were current cannabis only users, 39 (19.4%) were e-cigarette only users, 20 (10%) were dual cigarette and e-cigarette users, 48 (23.9%) used all three, and 74 (36.8%) were non-users. Of the 89 participants who have completed visit two, 56 (62.9%) reported using any cannabis products. Of participants using cannabis containing tetrahydrocannabinol (THC) (n=51), 31 (35.6%) reported consuming it via smoking, 42 (28.7%) via eating/drinking, and 22 (25.3%) via vaping. Of the 30 (34.5%) participants who reported using cannabidiols (CBD), 8 (47.1%) reported consuming it via eating/drinking, 5 (29.4%) via topical application and 2 (11.8%) via vaping. Findings suggest cannabis use, including THC and CBD, is prevalent among vapers in the New York area and may represent an underrecognized source of environmental contaminants. Future research will characterize the relation between environmental contaminants in cannabis products from the participants and their matched biospecimens.

608 – Applying EAS-E Suite to support chemical decision making

Jon Arnot[1,2], Alessandro Sangion[3], Trevor Brown[4], James Armitage[5], Liisa Toose[3], Lauren Hughes[3], Li Li[6]
[1]ARC Arnot Research and Consulting, Toronto, Toronto, Canada. [2]University of Toronto, Toronto, Canada. [3]ARC Arnot Research and Consulting, Toronto, Canada. [4]ARC Arnot Research and Consulting, Halifax, Canada. [5]AES Armitage Environmental Sciences, Ottawa, Canada. [6]University of Nevada Reno, Reno, USA

Abstract
Thousands of new and existing chemicals legally require chemical assessment and the necessary information is limited or non-existent for most chemicals. The Exposure And Safety Estimation (EAS-E) Suite was developed to integrate advances in exposure science in a user friendly, freely accessible on-line platform to support state of the science decision making and guide chemical safety and sustainability. EAS-E Suite includes numerous datasets of curated measured chemical information (e.g., physical-chemical properties, in vitro and in vivo toxicokinetics) and various models and tools to address data gaps and corroborate measured data. For example, EAS-E Suite includes quantitative structure-activity relationships (QSARs) for predicting chemical property information (e.g., partitioning, half-lives). Multi-media mass-balance models for simulating chemical emission rates throughout their life-cycle, indoor and outdoor fate and distribution, bioaccumulation, persistence, toxicokinetics, and exposure to a wide range of ecological receptors (plants, invertebrates, fish, birds, and mammals) and humans are included. The tools can be applied individually or holistically. For example, the PROduction-To-EXposure High Throughput (PROTEX-HT) model simulates aggregate exposures to representative ecological receptors and humans from chemical structure, functional use category and production volume. EAS-E Suite requires only a chemical CAS, name, or SMILES notation for the application of dozens of models and tools for exposure and risk assessment. Users can easily replace the

default parameters initially provided by the platform with preferred values. Case study applications of EAS-E Suite are presented to highlight its utility in several different decision contexts and on-going work to expand and test EAS-E Suite is summarized.

609 – Prenatal and Early Childhood Exposure to Organic Pollutants Through Hair Analysis

Linda Macheka[1], Paul Palazzi[2], Alba Iglesias-Gonzalez[1], Brice BMR Appenzeller[3], Florence Zeman[4,5]
[1]Human Biomonitoring Unit, Department of Precision Health, Luxembourg Institute of Health, Postdoctoral Fellow, Luxembourg, Luxembourg. [2]Human Biomonitoring Unit, Department of Precision Health, Luxembourg Institute of Health, Research Engineer, Luxembourg, Luxembourg. [3]Human Biomonitoring Unit, Department of Precision Health, Luxembourg Institute of Health, Unit Head, Luxembourg, Luxembourg. [4]Unité Toxicologie Expérimentale et Modélisation Institut National de l'Environnement Industriel et des Risques, Principal Investigator, Verneuil-en-Halatte, France. [5]PériTox, UMR_I 01, CURS, Université de Picardie Jules Verne, Chemin du Thil, Principal Investigator, Amiens, France

Abstract
Introduction: Exposure to endocrine disruptors during pregnancy may affect children from birth, early childhood, through to adulthood. Due to behaviour patterns and smaller body sizes, internal exposure is often amplified in younger children, particularly during critical growth periods. This study assessed exposure during pregnancy and early childhood exposure using 152 biomarkers of persistent pollutants, pesticides and their metabolites.

Methods: Exposure biomarkers from 32 chemical families were determined in 46 hair samples collected from pregnant women, and in hair collected at 3.5 years in corresponding children. Concentrations levels were compared and regressions were used to investigate any associations between maternal and child exposure to the same compounds.

Results and Discussion: Ninety endocrine disruptors were detected in mother-child pairs, with 18 of these having detection frequencies of > 95 % in both groups. Seven compounds showed significantly positive correlations between mother-child pairs. Among these, TCPy, Fipronil, Fipronil Sulfone and Fenhexamid influenced birth weight and length, and these continued to persist in the children at 3.5 years. The highest median concentrations of 120 pg/mg and 43 pg/mg for both mothers and children, respectively, were for Bisphenol A, suggesting similar exposure sources.

Conclusions: While there was a notable difference in the level of exposure between mother and child pairs for some compounds, children are more likely to suffer greater impacts compared to adults. Hair analysis for the biomonitoring of endocrine disruptors proves to be a valuable tool to assess these exposure levels and identify potential exposure risks.

610 – POSTER: Determination of Mn in blood with electrochemical point-of-care sensor: a validation study

Zhizhen Wu[1], Elena Boselli[2], Erin Haynes[3], Ian Papautsky[4]
[1]University of Illinois Chicago, Postdoctoral Fellow, Chicago, USA. [2]University of Illinois Chicago, PhD student, Chicago, USA. [3]University of Kentucky, Professor, Lexington, USA. [4]University of Illinois Chicago, Professor, Chicago, USA

Abstract
Manganese (Mn) is a trace element essential for numerous enzyme-related physiological processes. However, excessive Mn can induce neurotoxicity, and elevated Mn in children are associated with developmental delays, including reduced memory function, attention deficits, and poor academic performance. Blood Mn levels provide a reliable indicator of Mn exposure. The normal Mn range in blood is 5-18 μg/l, and values greater than twice the upper limit correlate with overt disease outcomes. The gold standard for Mn determination in blood is inductively coupled plasma mass spectrometry (ICP-MS), which is associated with high costs and long turnaround times. Moreover, collecting a relatively large blood volume by venipuncture can be problematic in pediatric subjects. In this study, we demonstrated and validated a low-cost electrochemical point-of-care sensor for reliable Mn monitoring. A cohort of children aged 7-16 years in Southside Chicago, IL was used to validate the sensor. The developed system requires only a few drops of blood (~150 μl), and thus is especially suited for pediatric subjects. The sensor system exhibited a linear range of 5-500 μg/l and detection limit of 0.5 μg/l. We compared the results of blood sample analysis (n=58) with those of ICP-MS from the NY state certified laboratory and achieved an accuracy of 86.3%. The study showed a slightly elevated Mn level (11.2 μg/l) compared to the nationwide average (10.7 μg/l). The study validates the Mn sensor system, and highlights the potential of the point-of-care electrochemical sensors for accurate, low-cost, and accessible monitoring of Mn.

611 – POSTER: Investigation of metal concentrations in toenails as biomarkers for chronic exposure to welding fumes

Chang Geun Lee[1], Jung Hyun Lee[1], Sa Liu[2], Ulrike Dydak[3], Jae Hong Park[4]
[1]Purdue University, PhD student, West Lafayette, USA. [2]Purdue University, Assistant Professor, West Lafayette, USA. [3]Purdue University, Professor, West Lafayette, USA. [4]Purdue University, Associate Professor, West Lafayette, USA

Abstract
Chronic overexposure to metals in various occupational settings, including welding, brazing, and soldering, has been linked to a wide range of diseases. For example, manganese (Mn) can cause neurotoxic effects and hexavalent chromium (Cr) compounds are well-known carcinogens. To better understand dose-response of chronic exposure to metals, biomarkers for chronic exposure are crucial. In a previous study, we found that toenail Mn concentration was an excellent biomarker for chronic exposure to Mn. However, the use of concentrations of other metals in toenails as biomarkers to estimate exposure to corresponding metals has not been explored. In this study, we collected welding fumes and toenail clippings from 14 welders and 13 non-welders (controls) and measured metal concentrations to investigate the relationship between metal concentrations in welding fumes and toenails for Mn, iron (Fe), zinc (Zn), Cr, and copper (Cu). Additionally, we calculated the area under curves (AUCs) of the receiver operating characteristics (ROC) curves to assess the ability of toenail metal levels to distinguish between welders and controls. In the results, Mn, Fe, and Cr concentrations in welding fumes and toenails were significantly higher in welders than non-welders. Mn concentrations in toenails had the most capability to distinguish between welders and non-welders (AUC = 0.91), followed by Fe (0.73), Cr (0.58), Zn (0.54), and Cu (0.45). These findings highlight the potential of toenail metal concentrations as a biomarker for chronic exposure to metals.

612 – Assessing the Effectiveness of Firefighter Protective Clothing at Mitigating Exposures to Fireground Vapors and Particulate Hazards

R. Bryan Ormond[1], Mark Gaskill[2], Andrew Hall[2], Chandler Probert[2]
[1]NC State University, Assistant Professor, Raleigh, USA. [2]NC State University, Doctoral Student, Raleigh, USA

Abstract
Over the last decade, the fire service has become increasingly aware of their daily exposure to hazardous substances resulting in the occupation being classified as a Group 1 known human carcinogen in 2022. Structural firefighters, wildland firefighters, and fire investigators are routinely exposed to toxic byproducts of combustion in the gases, vapors, smoke, and soot released from burning structures, vehicles, or forests. Numerous studies have characterized the chemical composition of smoke during fire propagation and suppression showing that firefighters are exposed to significant amounts of volatile organic chemicals, polycyclic aromatic hydrocarbons, phthalates, phenols, flame retardants, and per- and polyfluoroalkyl substances (PFAS). Contact with these contaminants typically occurs in spite of the firefighters wearing their personal protective equipment (PPE) properly, indicating that the current firefighter PPE is not effective at mitigating fireground exposures. This presentation will discuss laboratory methods that can be used to assess the vapor and particulate protective performance of firefighter and fire investigator PPE, how they relate to reducing firefighter exposures, and how they may be extended out of the laboratory and into the field. In vitro flow-through diffusion experiments to measure dermal absorption of fireground contaminants, including select PAHs and PFAS, provide an understanding of the significance of the dermal route of exposure. System-level evaluations of vapor and particulate protection of various firefighter ensembles were conducted through the Man-In-Simulant-Test (MIST) and the Fluorescent Aerosol Screening Test (FAST), respectively. Results of these evaluations indicate substantial limitations in current protective ensembles, specifically for exposure to fireground vapors.

613 – Initiating and Sustaining Community Engaged Research in Rural Space

Danielle Koonce[1,2], Sherri White-Williamson[3], Courtney Woods[4], Rebecca Witter[5], Shea Tuberty[6], Carol Babyak[7]
[1]University of Maryland, PhD Candidate, College Park, USA. [2]EJCAN, Board Member, Clinton, USA. [3]EJCAN, Founder & Board Chair, Clinton, USA. [4]UNC Chapel Hill, Associate Professor, Chapel Hil, USA. [5]Appalachian State University, Assistant Professor, Boone, USA. [6]Appalachian State University, Professor, Boone, USA. [7]Appalachian State University, Associate Professor, Boone, USA

Abstract
Sampson County, NC, a largely rural county with a high proportion of Black and Latinx residents, hosts several environmental hazards, including industrial swine and poultry production facilities, municipal solid waste and construction waste landfills and wood pellet facilities. In 2020, the first nonprofit, environmental justice organization in the county,Environmental Justice Community Action Network (EJCAN) was formed. There have been numerous challenges to establishing trust within a community that has long been ignored and

forgotten. Furthermore, challenges of obtaining resources to educate and empower community members are ever present. EJCAN has worked hard to develop strong, ethically sound and mutually respectful collaborations with university partners, research institutions, and local church and community organizations to build an empowering community-based, research and educational network. Through these partnerships, EJCAN aims to center local communities' concerns and expertise in our research to understand environmental exposures and impacts. This presentation will discuss the lessons we have learned, the positive outcomes that community-engaged research has created, and possible next steps for expanding our collaborations across the county and beyond.

615 – POSTER: Leveraging Spatial Measurements of PM, VOCs, and Metals to Quantify Community Exposures and Risk: Results from the Hazardous Air Pollution and Monitoring Assessment Project (HAP-MAP)

Peter DeCarlo[1], Ellis Robinson[2], Mina Tehrani[3], Andrea Chiger[4], Carolyn Gigot[4], Roger Sheu[3], Megan Claflin[5], Ed Fortner[6], Manjula Canagaratna[5], Conner Daube[7], Benjamin Werden[8], Robert Roscoli[5], Jordan Krechmer[5], Harald Stark[5], Scott Herndon[5], Scott Van Bramer[9], Ana Rule[10], Kirsten Koehler[1], Tara Yacovitch[5], Thomas Burke[11], Keeve Nachman[1]
[1]Johns Hopkins University, Associate Professor, Baltimore, USA. [2]Johns Hopkins University, Assistant Research Engineer, Baltimore, USA. [3]Johns Hopkins University, Postdoctoral Scientist, Baltimore, USA. [4]Johns Hopkins University, PhD Student, Baltimore, USA. [5]Aerodyne Research Inc., Principal Scientist, Billerica, USA. [6]Aerodyne Research Inc., Senior Scientist, Billerica, USA. [7]Aerodyne Research Inc., Research Associate, Billerica, USA. [8]Aerodyne Research Inc., Postdoctoral Scientist, Billerica, USA. [9]Widener University, Professor, Widener, USA. [10]Johns Hopkins University, Assistant Professor, Baltimore, USA. [11]Johns Hopkins University, Professor, Baltimore, USA

Abstract
Fenceline communities frequently bear the brunt of environmental pollution, and current regulatory monitoring approaches fail to generate the data needed to document these disproportionate exposures. We addressed this critical data gap by combining low-cost sensor technology with state-of-the-art analytical instrumentation deployed on a mobile van and at a fixed site to characterize spatial concentration gradients for hazardous air pollutants (I.e. particulate matter, metals, and VOCs) in a highly industrialized urban corridor in southeastern Pennsylvania. We quantified the pollutant burden contributed by petrochemical and other industrial activity by comparing spatial concentrations of air pollutants measured from the mobile laboratory in the study area to the fixed site measurements (approximating "urban background"). Low-cost PM sensors were also deployed at several locations throughout the study area and were hosted by community groups involved in the study. Gradients in hazardous air pollutants were observed with higher levels, and consequently higher exposure risk, found near facility fencelines. Formaldehyde measurements provide an example to illustrate gradients of exposure across the study area. Mobile measurements performed in fenceline areas averaged 1ppb higher than the urban background levels observed at the fixed site. The observed differences in concentration and therefore exposure for a range of hazardous air pollutants were then placed in a cumulative risk framework to elucidate the spatial variability of risks at the neighborhood scale. This presentation will give an overview of the HAP-MAP study design and results.

616 – CHALLENGES FACED WHEN ANALYZING ARCHIVED WHOLE BLOOD FOR TRACE ELEMENTS: RESCUING CLOTTED BLOOD SAMPLES

Patrick Parsons[1,2], Christopher Palmer[3,4], Kayla Mehigan[3], Jeffrey Wickliffe[5], Maureen Lichtveld[6]
[1]Wadsworth Center, NY State Dept of Health, Director, Div Environmemntal Health Sciences, Albany, USA. [2]Dept of Environmental Health Sciences, University at Albany, Professor, Albany, USA. [3]Wadsworth Center, NY State Dept of Health, Research Scientist, Albany, USA. [4]Dept of Environmental Health Sciences, University at Albany, Assistant Professor, Albany, USA. [5]Dept of Environmental Health Sciences, The University of Alabama at Birmingham, Professor, Birmingham, USA. [6]School of Public Health, University of Pittsburgh, Professor, Pittsburgh, USA

Abstract
Analyzing whole blood for trace elements requires a suitable anticoagulant, either EDTA (preferred) or heparin. Blood may be stored at ambient temperature for a several days, refrigerated for several weeks, or frozen for long term storage. Clotted blood is a criterion for specimen rejection based on clinical laboratory standards. Clots are sometimes inevitable when blood is collected under less-than-ideal conditions. In research studies, there may be no option to collect new blood samples and clots present an analytical challenge. In one such study, 898 blood specimens were archived at −70 °C, and then shipped on dry ice to our laboratory for trace element analysis. When thawed, almost all blood samples had large clots making it impossible to remove a representative aliquot. A pilot study was conducted using sonication to break up clots. We compared the effects of sonication on blood-based reference materials (RMs) and study samples to ensure repeatability and accuracy. Results show clotted blood samples were successfully homogenized using indirect sonication and analyzed successfully for six trace elements: Pb, Cd, Hg, Se, Mn, and Sn. Repeatability of sonicated blood was similar to that for un-sonicated blood for these six elements. Small but statistically significant differences were evident for Hg, and Mn in the RMs. There was an increased risk of over-sonication that degraded the blood matrix, so care was needed to optimize indirect sonication

parameters. A total of 893 clotted blood samples were rescued for trace element analysis using indirect sonication, with just 5 samples lost to over-sonication.

617 – The Respiratory Impacts of Primary and Secondhand Exposure to Electronic Cigarette Aerosols

Ruzmyn Vilcassim[1], Samuel Stowe[2], Rachel Majumder[3]
[1]The University of Alabama at Birmingham, Assistant Professor, Birmingham, Alabama, USA. [2]The University of Alabama at Birmingham, PhD Candidate, Birmingham, Alabama, USA. [3]The University of Alabama at Birmingham, Student, Birmingham, Alabama, USA

Abstract
Currently around 13-16 million people in the United States use electronic cigarettes, while 5.3 million middle and high schoolers reported ever using e-cigarettes in 2021. E-cigarette aerosols may contain toxic compounds that can adversely affect lung health. However, despite this increase in use, and exposure to potential toxicants, the health effects of primary and secondhand exposure to e-cigarette aerosols are still not fully understood. Therefore, the main objective of this study is to assess the adverse respiratory impacts associated with primary and secondhand exposures to e-cigarette aerosols in young adults, with a primary focus on lung function and respiratory symptoms. We aim to recruit thirty participants (10 primary exposed (vapers), 10 secondhand exposed, and 10 non-smoking/non-vaping controls), aged 19 to 35, and examine their lung function (Forced Expiratory Volume in the first second [FEV1], Forced Vital Capacity [FVC] and airway reactance using Force Oscillation Technique (FOT)) and symptoms at three different time periods across 6 months. Preliminary results from 18 participants have demonstrated that vapers and secondhand exposed individuals had a lower FEV1/FVC ratio as compared to controls, which indicates possible airway obstruction. Those who vaped also experienced significantly higher mean respiratory symptom scores compared to secondhand exposed individuals and control participants ($p<0.05$). Our results indicate that primary vapers may be at increased risk of experiencing adverse respiratory effects compared to non-smoking/non-vaping individuals. Respiratory effects for secondhand exposed individuals were mixed compared to controls, and thus, needs further investigation.

618 – Advancing Exposome Research for ADRD to Enable Precision Environmental Health Approaches to Treatment and Prevention

Richard Kwok
NIA, Program Director, Bethesda, USA

Abstract
The complexity and heterogeneity of Alzheimer's Disease and Related Dementias (ADRD) are a consequence of a dynamic interaction between genes and the environment over the lifespan and a growing body of epidemiologic, genomic and mechanistic research pointing to a significant contribution of various environmental factors (i.e. the exposome) on ADRD risk and resilience.

The National Institute on Aging (NIA) has embarked on a series of initiatives towards a Precision Environmental Health approach to ADRD risk reduction and disease prevention and to advance our understanding of ADRD etiology through the next-generation of Environmental Health research occurring at the intersection of G (epigenome/genome) perturbations x E (environmental exposures) x D (data-omics i.e. exposome, epigenome, genome, proteome, metabolome). Additionally, NIA supports the development of novel artificial intelligence / machine learning (AI/ML) approaches to identify patterns and relationships across large, complex, and disparate datasets to draw meaningful insights into the etiology of complex diseases such as ADRD.

This talk will highlight infrastructure enhancements for exposome research on ADRD including new research opportunities for understanding the mechanisms by which environmental toxicants influence the molecular trajectories of brain aging and ADRD and contribute to health disparities. Opportunities for unique collaborations between the environmental and ADRD communities will be discussed in addition to research perspectives between the communities and challenges in advancing exposome research to enable a precision environmental health approach to ADRD treatment and prevention.

619 – Subsewershed Surveillance in Rural Central Appalachia Reveals Pathogen Signal Is Influenced by Inflow and Infiltration

Amanda Darling, Madeline Deck, Gabriel Maldonado, Clayton Markham, Benjamin Davis, Amy Pruden, Alasdair Cohen
Virginia Tech, PhD Candidate, Blacksburg, USA

Abstract
Despite the advantages of wastewater based surveillance (WBS) for monitoring community health, in the U.S. it has been used primarily in densely populated communities with relatively well-maintained and funded sewer infrastructure. Consequently, there is a lack of knowledge with regard to how WBS may be used in rural communities, and the potential of wastewater based epidemiology to better understand health outcomes and disparities. In this study, we initiated a year-long monthly wastewater monitoring campaign in a small, rural sewershed in Southwest Virginia with known infrastructural challenges, including sewer main leaks and groundwater and surface water infiltration and inflow (I&I). In addition to collecting wastewater treatment plant (WWTP) influent, we sampled 12 sewershed nodes at each branch line to the sewer conveyance system to assess pathogen signal, and potential signal loss due to I&I and sewer system design characteristics. ddPCR was performed for RSV, Influenza A, Influenza B, SARS-CoV-2 and five antibiotic resistance genes (sulI, tetA, intI1, vanA, and blaCTX-M). Due to a 21% lower detection rate of pathogen signal at the influent as compared to downstream of a large residential facility, preliminary results point to the risk of underestimating pathogen circulation when sampling only WWTP influent in a system with persistent I&I.

620 – Breast Cancer Relevant Occupational Exposures Facing Immigrant Women

Kristin Knox[1], Jennifer Ohayon[2], Ruthann Rudel[3], Rachel Morello-Frosch[4]
[1]Silent Spring Institute, Staff Scientist, Newton, USA. [2]Silent Spring Institute, Research Scientist, Newton, USA. [3]Silent Spring Institute, Director of Research, Newton, USA. [4]University of California, Berkeley, Professor, Berkeley, USA

Abstract
First-generation immigrants comprise roughly 15% of the U.S. population, and past research indicates that rates of breast cancer rates rise for many immigrant groups after relocating to the U.S. Given this, we investigated how occupational exposures may contribute to an increase in breast cancer risk for U.S. immigrants. We analyzed data from the American Community Survey Public Use Microdata Sample (PUMS) to characterize which occupations are most prevalent for immigrant women ages 18-65. We then combined these results with data on occupational chemical exposures from the Women's Occupation and Risk from Chemicals (WORC) tool, which identifies occupations with chemical exposures likely associated with increased breast cancer risk. Across the country, immigrant women most commonly work as maids, nurses, cashiers, janitors, and care aides. Among the 20 most prevalent occupations, maids, janitors and nurses face the highest number of probable chemical exposures, while postsecondary teachers, home health aides, and nursing assistants face the highest combination of probable plus possible chemical exposures. Chemical exposures of interest include those found in fragrances, cleaning and maintenance product ingredients, phthalates, antimicrobials, and alkylphenols. A number of chemicals in these classes are considered to be mammary gland carcinogens, mammary gland developmental toxicants, and/or endocrine disruptors. By identifying key occupations where immigrant women are employed, and which ones may be most hazardous from an exposure perspective, this research provides a basis for future research and action aimed at protecting immigrant health.

621 – POSTER: A Cohort Study of Residential Proximity to Unconventional Oil and Gas Development and Birth Defects in Ohio

Casey Gaughan[1], Keli Sorrentino[1], Zeyan Liew[1], Nicholaus Johnson[1], Cassandra Clark[1], Mario Soriano, Jr[2,3], Julie Plano[1], Desiree Plata[4], James Saiers[2], Nicole Deziel[5]
[1]Yale School of Public Health, New Haven, USA. [2]Yale School of the Environment, New Haven, USA. [3]Princeton University, Princeton, USA. [4]Massachusetts Institute of Technology, Boston, USA. [5]Yale School of Public Health, Associate Professor, New Haven, USA

Abstract
Background: Chemicals used or emitted from unconventional oil and gas development (UOGD) include reproductive/developmental toxicants.

Methods: We conducted a registry-based cohort study of 965,237 live births in Ohio from 2010-2017. We assigned UOGD exposure based on maternal residential proximity at birth to active UOGD wells and a new water-specific proximity metric incorporating groundwater flow patterns. 4,653 individuals with structural birth defects were identified from state birth records and a birth defect surveillance program. We estimated odds ratios (ORs) and 95% confidence intervals (CIs) for all defects combined and specific birth defect types (neural tube, congenital heart, musculoskeletal, oral clefts, hypospadias) using binary exposure metrics (any versus no UOG wells within 10 km and any versus no upgradient UOG wells within 10 km), controlling for multiple confounders.

Results: The odds of a birth defect in children born to mothers within 10 km of UOGD were 1.13 times higher than those born to unexposed mothers (95%CI: 0.98-1.30). Odds were greater for neural tube defects (OR: 1.57; 95%CI: 1.12-2.19), congenital heart defects (OR: 1.25; 95%CI:

0.94-1.67), and limb reduction defects (OR: 1.99; 1.18-3.35). Odds for hypospadias were inversely related to UOGD exposure (OR: 0.62; 95%CI: 0.43-0.91). Odds of any structural defect were greater in magnitude but less precise with the water-specific metric (OR: 1.30; 95%CI: 0.85-1.97), in rural areas (OR: 1.21, 95%CI: 0.77-1.91), and areas with high social vulnerability (OR: 1.26, 95%CI: 0.99-1.60).

Conclusions: We observed elevated, but not always statistically significant, ORs across multiple birth defects in relation to UOGD exposure.

622 – Bidirectional Effects of Exposures and Aging: Characterizing the Aging Exposome

Ram Gouripeddi[1,2,3], Caden Steward[3], Julio Facelli[1,3]
[1]Center of Excellence for Exposure Health Informatics, University of Utah, Salt Lake City, USA. [2]Clinical and Translational Science Institute, University of Utah, Salt Lake City, USA. [3]Department of Biomedical Informatics, University of Utah, Salt Lake City, USA

Abstract
Effects of a lifetime of environmental and lifestyle exposures on aging or age-associated diseases are not well understood. Exposures not only directly affect biological pathways but also have mutagenic and epigenetic mechanistic influences on disease and well-being. Aging leads to reduced reserve capacity to compensate for the effects of exposures due to changes in biomolecular and cellular kinetics and dynamics. While general effects are well documented, details on how temporal patterns of exposures may affect systemic and targeted aging are not well understood. Characterizing differential, additive effects of continuous low-level and intense sporadic multi-agent exposures require advanced big data and artificial intelligence methods.

Quantifying bidirectional effects of environment and aging requires time series of data from all contributing exposures, which can span endogenous processes within the body, biological responses of adaptation to environment, and socio-behavioral factors. Gaps in measured data may need to be filled with computationally modeled data. Essentially, the challenge in generating aging exposome is the absence of readily available records for individuals over the course of their life. Instead, these would need to be assimilated from historic person reported data (e.g. residential location, durations, behaviors) along with publically available data. This could lead to potential gaps and uncertainties that would need inform on how the exposomic records can be used for aging research.

We present a pragmatic approach to generation of longitudinal exposomic and aging records as required for different study archetypes. Such records can then be used to understand the bidirectional effects of exposures and aging.

623 – PFAS Contaminated Groundwater and Prevalence of Chronic Health Conditions among Medicare Beneficiaries

Ana Rappold
US Environmental Protection Agency, Branch Chief, Chapel Hill, USA

Abstract
Per- and polyfluoroalkyl substances (PFAS) are a class of persistent organic pollutants with emerging environmental and regulatory concerns. The California Water Board's Groundwater Ambient Monitoring and Assessment (GAMA) water quality database was used to estimate PFAS contamination within California's groundwater resources but non-systematic testing for PFAS may have introduced a selection bias. For each ZIP code population centroid, we calculated an inverse distance squared weighted estimate of PFAS groundwater contamination for all tested wells within 20 km and then Windsorized to the 95th percentile to reduce the influence of outliers. The Centers for Medicare and Medicaid Services' (CMS) Chronic Conditions Database was used to calculate the prevalence of 27 chronic health conditions among US Medicare beneficiaries aged 65 and older by ZIP code of residence. We restricted our analyses to California counties that derived 25 percent or more of their drinking water from groundwater. A Poisson regression model was used to estimate associations between PFAS concentration in groundwater and prevalence of chronic conditions adjusting for a PFAS-testing propensity score, demographic characteristics, and the prevalence of lung cancer as a negative control. Preliminary results suggest an association between PFAS contaminated groundwater and several chronic health conditions including hypertension, hypothyroidism, glaucoma, and hyperlipidemia. In addition to strengthening our understanding of the chronic health impacts of PFAS, this project demonstrates the utility of synthesizing data across multiple disciplines to explore these emerging contaminants within the broader context of environmental justice. This abstract does not represent EPA policy.

624 – Mapping indoor air pollution hotspots in London: evidence for for policy update and behaviour adjustment

Hanbin Zhang[1], Chuyan Zhang[2], study team COPE[2], Benjamin Barratt[2]
[1]Imperial College London, Early Career Research Fellow, London, United Kingdom. [2]Imperial College London, London, United Kingdom

Abstract
People spend 90% of their time indoors. Home environment is fairly well understood, but exposures in other indoor environments, are more heterogeneous. A very large database of personal exposure to air pollution collected via wearable sensors, including more than 16,000 person-days of GPS and air pollution exposure data at one-minute time resolution from 125 individuals with chronic obstructive pulmonary disease (COPD), provides a unique opportunity to investigate this.
We first excluded data points of being at home and at office, as well as high-speed transition points. Then, we used DBSCAN clustering method to classify GPS data points and determined number of clusters and number of points in each cluster. After that, we ran a bunk imputation program to translate geographic coordinates into human-readable addresses via Application Programming Interface (API) calls into OpenStreetMap. We used the average geographic coordinate of GPS points in each cluster as a spatially representative point for each cluster.

The indoor locations most frequently visited by COPD participants were hospitals, bars, supermarkets, schools, apartments and places of worship. Highest exposure of PM2.5 and CO were found in café/restaurant/pub at 24.1 μg/m3 and 0.2 ppb, respectively. The highest exposure of NO2 was found in the transport microenvironment at 17.0 ppb. Recreations had the lowest concentrations.

We suggest communities to lobby together with scientists to regulatory bodies about routine monitoring and implementing pollution guideline values in hospitality sector and healthcare service centres; before that, participants can consider adjusting their behaviour according to our microenvironment pollution ranking.

625 – Strengthening Wastewater Surveillance: Improving Core Principles, Standardization Procedures, and Data Analysis Techniques

Adam Gushgari
Eurofins Pandemic Prevention Services, Wastewater Surveillance Technical Expert, Phoenix, USA

Abstract
Wastewater surveillance has emerged as a powerful tool in the fight against the SARS-CoV-2 pandemic, with its applications expanding to other environmental matrices. To maximize the value of wastewater-based epidemiology (WBE) data, improvements to core principles, standardization procedures, and data analysis techniques are crucial. This presentation discusses methods to enhance the efficacy of wastewater surveillance data, including improving analytical capabilities, standardizing laboratory processes, implementing regulatory guidance, refining data modeling and estimation procedures, and promoting data transparency. The importance of serving the needs of disadvantaged communities, including those in economically challenged and developing nations, is also emphasized as a core principle of WBE to promote universal health equity and advance understanding of communicable disease prevalence. Prioritizing these principles can strengthen WBE data and increase global adoption of this cost-effective health surveillance technique while securing continued public support and buy-in for future initiatives.

626 – Comparison of concentrations of per-and polyfluoroalkyl substances (PFASs) in maternal serum and cord blood of twins

YeEun Seo[1], Na-Youn Park[2], Hye-Ri Jeon[2], Hyo-Bang Moon[3], Jeongim Park[4], Younglim Kho[5]
[1]Department of Food Technology & Service, Eulji University, Master's Degree, Seongnam, Korea, Republic of. [2]Department of Health, Environment & Safety, Eulji University, Doctor Degree, Seongnam, Korea, Republic of. [3]Department of Marine Science and Convergence Engineering, Hanyang University, Professor, Ansan, Korea, Republic of. [4]Department of Environmental Health Sciences, Soonchunhyang University, Professor, Cheonan, Korea, Republic of. [5]Department of Health, Environment & Safety, Eulji University, Professor, Seongnam, Korea, Republic of

Abstract
Per-and polyfluoroalkyl substances (PFAS) are classified as highly fluorinated aliphatic substances which have 1 or more carbon atoms where the hydrogens have been replaced by fluorine atoms. PFASs tend to accumulate because of the long half-life in vivo. So, the

environmental presence and potential human toxicity have increased social and scientific interest. In this study, the meternal serum and cord blood of twins were analyzed to compare PFAS concentrations. The sample preparation was performed by solid-phase-extraction. Then, quantitative analysis was performed by UPLC-MS/MS. As a result of analyzing the concentration of PFAS in maternal serum, PFNA, PFDA, PFUnDA, and L-PFOS were detected in all samples, with the highest concentrations detected in the order of L_PFOS, PFNA, PFUnDA, and PFDA. In addition, as a result of analyzing PFAS concentrations in cord blood of twins, PFNA, PFDA, PFnDA, and L-PFOS were detected more than 90%, and with the highest concentrations detected in the order of PFOA, L-PFHxS, L-PFOS, PFNA, br-PFOS and PFDA. However, alternative PFASs were detected on average less than 10% in maternal serum and cord blood of twins. In conclusion, various PFAS material were detected in maternal serum and cord blood of twins, it is highly likely that they affected fetal health.

This work was supported by Korea Environment Industry & Technology Institute(KEITI) through Technology Development Project for Environmental Diseases Prevention and Management Program(or Project), funded by Korea Ministry of Environment(MOE) (2022003310004).

627 – Exploring the external exposome in a US-based preconception cohort study: Pregnancy Study Online (PRESTO)

Elizabeth Lin[1], Emily Johnson[2], Kayley DeLay[3], Mary Willis[4], Martha Koenig[5], Elizabeth Hatch[6], Amelia Wesselink[7], Lauren Wise[6], Krystal Pollitt[8]
[1]Yale University, Phd Candidate, New Haven, USA. [2]Yale University, Phd student, New Haven, USA. [3]Yale University, Graduate student, New Haven, USA. [4]Boston University, Assistant Professor, Boston, USA. [5]Boston University, Research Fellow, Boston, USA. [6]Boston University, Professor, Boston, USA. [7]Boston University, Research Assistant Professor, Boston, USA. [8]Yale University, Associate Professor, New Haven, USA

Abstract
Environmental contaminants can impact reproductive health, including fertility. Wearable passive samplers can overcome challenges of traditional exposure assessment methods (i.e., active sampler with pumps) by comprehensively evaluating personal chemical exposure (the external exposome). This study explored the feasibility of assessing environmental exposures of couples trying to conceive using a wristband air sampler. As part of Pregnancy Study Online (PRESTO), an internet-based preconception cohort study, Fresh Air wristbands were mailed to 139 female participants (aged 21-45) between June and November 2021. Participants were from 39 states in the US and were all trying to convince. For a subset of these participants (n=24), wristbands were also given to their male partners, for within-household analyses. Participants wore wristbands continuously for five days before being returned for chemical analysis by thermal desorption gas chromatography high resolution mass spectrometry. A snapshot of participants' external exposome revealed exposures to a diverse range of phthalates, pesticides, and flame retardants. Targeted analysis provided quantitative personal exposure concentration of 43 compounds across 8 chemical groups. Our suspect screening analysis identified 491 high confidence chemicals, including 107 endocrine disruptors. Exposure patterns were explored by residential location, income level, race/ethnicity, occupation, and lifestyle factors. Self-reported questionnaires confirmed unique detected exposures from food products, pharmaceuticals, and smoking activities (cigarette, marijuana). This study sheds light on the environmental exposures of females and males across the US, demonstrating the feasibility of using non-invasive mail-based exposure assessment tools to characterize preconception exposures in future studies.

628 – Assessing Impacts of Highly-Spatially Resolved Environmental Exposures Using Electronic Health Records of Understudied Populations

Flannery Black-Ingersoll[1], Kate Burrows[2], Stephanie Grady[1], Katharine Teigen[1], Rachel Oblath[3], Amruta Nori-Sarma[1]
[1]Boston University School of Public Health, Boston, USA. [2]Brown, Providence, USA. [3]Boston University School of Medicine, Boston, USA

Abstract
Growing evidence demonstrates adverse mental health effects of exposure to extreme heat. However, many vulnerable populations, including those who are uninsured or unhoused, may be left out of conventional epidemiological data sources such as insurance claims data. In order to evaluate the effects of extreme heat in these understudied and high-risk populations, we collaborated with healthcare providers to develop techniques for cleaning and geocoding data from a city-level emergency psychiatric services system. This study focused on the greater Boston metropolitan area and draws on electronic health record data (2005-2021) on home, mobile, and emergency room encounters with the Boston Emergency Services Team (BEST). Substantial data cleaning was implemented to assign address/location at time of encounter. Geocoding was conducted using a custom, locally stored geocoder to protect identifiable patient data. Geospatial environmental exposures were then assigned to patients using highly spatially and temporally resolved data on hourly ambient temperature (1x1km), air pollution, and residential greenness. We constructed methods that leverage a unique set of electronic health record data for environmental exposure assessment and developed a cohort of previously understudied populations

for further environmental epidemiologic research. These methods demonstrate the capacity, through collaborations with service providers like BEST, to study highly vulnerable populations in other cities and locations across the country and the world.

629 – Indoor monitoring campaign to assess long-term fine particulate matter for epidemiological studies

Sun-Young Kim[1], Kyumin Kim[2], Seokmin Ji[2], Miyoun Shin[1], Soyoung Han[3], Seung-Hyun Cho[4], Hee-Young Yoon[5], Jin Woo Song[6], Korea ILD study group Korea ILD study group[7]
[1]National Cancer Center of Korea, Goyang, Korea, Republic of. [2]Korea University, Seoul, Korea, Republic of. [3]Yonsei University, Seoul, Korea, Republic of. [4]Research Triangle Institute, RTP, USA. [5]Soonchunhyang University Hospital, Seoul, Korea, Republic of. [6]Asan Medical Center, Seoul, Korea, Republic of. [7]Korea ILD study group, Korea, Korea, Republic of

Abstract
Background: While there was accumulating evidence of the association between long-term exposure to fine particles (PM2.5) and health endpoints, one of the key limitations was suggested as reliance on outdoor PM2.5. As many studies reported the difference between outdoor and indoor air quality, the difference could affect subsequent health effect analyses. The impact could be larger for patients with chronic diseases who spend much time indoors. However, most epidemiological studies of indoor PM2.5 adopted short-term measurements limited in time and space. We aimed to assess long-term exposure to indoor PM2.5 at homes of Idiopathic Pulmonary Disease (IPF) patients in South Korea for epidemiological application.

Methods: We performed a longitudinal indoor air pollution monitoring campaign for 124 IPF patients from 15 hospitals from July 2021 through August 2023. For each patient, we collected real-time measurements of PM2.5 and integrated filter samples to analyze black carbon and 33 inorganic elements for five days including both weekdays and weekends in each of the four seasons using RTI MicroPEM. Based on PM2.5 measurements corrected by gravimetric measurements, we computed hourly averages to investigate temporal patterns and four-season averages to assess long-term exposure.

Results: Household indoor PM2.5 concentrations were higher in daytime than nighttime, higher on Friday and weekends than the other weekdays, and the lowest in summer. Four-season average concentration was 13.83 ug/m3 (standard deviation=10.77) with some variation by household characteristics.

Conclusions: Our exposure assessment approach will help assess the association of long-term indoor PM2.5 and prognosis of IPF patients.

630 – Community Engagement in the National Academies PFAS Guidance Study

Elizabeth Boyle
National Academies of Sciences, Engineering, and Medicine, Senior Program Officer, Washington, DC, USA

Abstract
People in communities exposed to PFAS have questions about how much PFAS exposure they have and the appropriate clinical follow-up they should receive to protect their health. The Agency for Toxic Substances Disease Registry (ATSDR) published a PFAS clinical guidance, but the guidance did not provide specific recommendations on testing for PFAS, how to interpret the results, or what clinical follow-up based on PFAS exposure might look like. The lack of specific advice frustrated many people living in or treating patients from PFAS-exposed communities. In response to the community concerns, the ATSDR and the National Institute of Environmental Health Sciences (NIEHS) requested the National Academies convene an ad hoc committee to develop a report that would develop principles and recommendations for testing for PFAS exposure and clinical care to help ATSDR update its clinical guidance. In conducting this study, the National Academies aimed to include community perspectives throughout the study process. The community engagement plans involved appointing a group of over 40 community liaisons to provide perspectives to the staff and the committee throughout the study's data collection phase. The community liaisons suggested speakers for meetings, provided input on discussion topics, spoke at meetings, and self-assembled to provide a letter to the committee with items they wished the report would address. This presentation will detail the process used to engage communities in this National Academies study and the efforts to maintain the independence and

objectivity of the National Academies.

631 – POSTER: Comparison of costs and results of mobile monitoring designs for environmental epidemiology

Sun-Young Kim[1,2], Magali Blanco[2], Amanda Gassett[2], Lianne Sheppard[2]
[1]National Cancer Center of South Korea, Goyang, Korea, Republic of. [2]University of Washington, Seattle, USA

Abstract
Background: There has been emerging interest in mobile monitoring (MM) to investigate the association between air pollution and human health. Recent studies applied MM measurements to develop prediction models that assess individual-level long-term air pollution concentrations. MM provides increased flexibility over stationary designs, and myriad possible designs could be proposed to monitor a geographic area. Cost is an important feature in selecting a design. This study aims to compare costs to prediction model performance for different designs, with implications for environmental epidemiology.

Methods: We based our analysis on MM data of ultrafine particle measurements collected from the Adult Changes in Thought Air Pollution study, conducted in 2019-2020 in Seattle, WA. By assuming that costs such as equipment and investigator oversight are stable, we focused on the incremental cost of staff effort in different MM designs and its relationship with model performance. We simulated different designs by selecting subsets of limited seasons, days of week, hours of day, and visits from the complete dataset. We fit prediction models for each subset and calculated cross-validation (CV) statistics.

Results: Cost decreased by half when monitoring included limited seasons or weekday business hours only instead of four seasons and all days, while CV R2 also reduced to ~.73 and ~.39, respectively from 0.8. When the number of visits was reduced to 25%, the cost decreased to one third with a reduced R2 of ~.74.

Conclusion: Our study provides practical guidance in mobile monitoring designs to assess the health effect of long-term air pollution.

632 – POSTER: A Geospatial Analysis of Demographic Characteristics and Environmental Burdens in Communities Near Superfund Sites, Long Island, New York

Fintan Mooney[1,2], Jill Kelly[1], Joshua Warren[1,2], Nicole Deziel[1]
[1]Yale School of Public Health, New Haven, USA. [2]Yale Superfund Research Center, New Haven, USA

Abstract
Background: Nassau and Suffolk Counties of Long Island, New York, have over 30 federal Superfund sites. Superfund sites may contribute to increased exposures and health risks in nearby communities. We assessed proximity to Superfund sites on Long Island in relation to community demographics and presence of other environmental burdens.

Methods: Data were obtained for 2022 census blocks from the U.S. Environmental Protection Agency's Environmental Justice Screening and Mapping Tool (EJScreen) and included proximity to Superfund sites, presence of other environmental hazards, and sociodemographic variables from the U.S. Census American Community Survey. Bivariable statistical tests were performed overall and by county to evaluate demographic factors and presence of additional environmental hazards in relation to proximity to Superfund sites.

Results: Preliminary findings show that areas closer to Superfund sites (above median proximity) have a statistically significantly higher proportion of racially minoritized groups, lower English proficiency, and a lower proportion of high school graduates; the magnitude of associations were higher in Suffolk versus Nassau Counties. Moreover, communities near Superfund sites have statistically significantly higher proportions of old housing units; proximity to hazardous waste facilities; and ambient fine particulate matter concentrations.

Conclusion: Our findings suggest that socially vulnerable populations in Long Island are at increased risk for adverse environmental exposures associated with Superfund sites. The health risks of living near Superfund sites may be compounded by proximity to other environmental hazards. This study underscores the need for a more comprehensive exposure assessment of at-risk populations living near Superfund sites on Long Island.

633 – POSTER: Changes in Living Behavior in Japan Following the Spread of COVID-19 Infections

Satoshi Nakai[1], Rika Miura[2], Lujia Shi[2], Junki Hosoda[3], Jun Kohzaki[3]
[1]Yokohama National University, Professor, Yokohama, Japan. [2]Yokohama National University, Student, Yokohama, Japan. [3]National Institute of Technology and Evaluation, Researcher, Tokyo, Japan

Abstract
Exposure to chemicals from consumer products are largely dependent to consumer behaviors, such as human activity pattern and how to use consumer products. Pandemic of COVID-19 must have changed some consumer behaviors. We collected the data of lifestyle patterns at home from the web questionnaire surveys in 2015 and 2021. Responses were obtained from over 2,000 adults nationwide. The items compared were the time, place, and frequency of sleeping, eating, bathing, getting ready, working, studying, hobbies, laundry, cooking, and cleaning (in each room). No significant changes were observed in terms of "time" for activities traditionally performed at home, such as sleeping and eating. On the other hand, the percentage of respondents who work at home on weekdays increased from 23% in the previous survey to 39%, and the average time required increased from 2.2 hours in the previous survey to 8.1 hours in this survey. The percentage of respondents who eat lunch at home increased from 76% in the previous survey to 87%, and the average time required for lunch slightly decreased from 27 minutes in the previous survey to 22 minutes in the current survey.

Experiments were also conducted to measure ethanol exposure concentrations when infection control products were used. It was shown that the exposure concentration was higher with higher usage volume, but that it could be removed by the usage method (e.g., wiping after spraying) and that ventilation could reduce the exposure of the product user by up to 75%.

634 – POSTER: Use of dried blood spot protein adducts for biomonitoring pesticide exposures in agricultural workers

Rothboury Doung, Eric Huang, Rebecca Richter, Michael Paulsen, Clement Furlong, Michael MacCoss, Christopher Simpson, Judit Marsillach, Ashley Phillips
University of Washington, Seattle, USA

Abstract
Chemical agents are widely used in both developed and developing countries to control pests and weeds, leading to enhanced crop yields and reduced crop losses worldwide. Consequently, agricultural workers are highly susceptible to chemical-related injuries, making it crucial to improve their safety and well-being by enhancing the biomonitoring of workers in the agricultural industry to pesticides such as organophosphorus (OP) insecticides. The conventional biomonitoring methods have several limitations, including the requirement of pre-exposure baseline activity levels, reduced accuracy at lower levels of inhibition, and a demanding sample collection and shipping process. To overcome these limitations, our study aims to evaluate the use of mass spectrometry for identifying OP-modified blood proteins as a biomarker of OP exposure and assess the use of dried blood spots (DBS) as a less invasive biospecimen collection process. Our long-term goal is to use DBS protein adductomics for exposomic analysis and for establishing exposure-adverse outcome relationships. For this project, we chose to monitor butyrylcholinesterase (BChE), a common biomarker of OP exposures, and albumin, known to be modified by certain environmental exposures, using plasma and DBS samples collected from the same agricultural worker prior to and at the end of the spray season. Using immunoprecipitation coupled to LC-MS/MS, our results showed that exposures can be identified in both plasma and DBS, even at low level exposures, which traditional biomonitoring methods could miss. Additionally, the use of DBS adductomics improves the biospecimen collection process and reduces the need for baseline sample from each worker.

635 – Categorization of Skilled Nursing Facilities based on Population Vulnerability and Fine Particulate Air Quality Monitor Deserts in the Mountain West

Madison "Dewie" Roth[1], Savannah D'Evelyn[2], Jihoon Jung[3], Luke Montrose[4]
[1]Warner College of Natural Resources, Colorado State University, Fort Collins, USA. [2]Pacific Northwest Agricultural Safety and Health Center, University of Washington, Seattle, USA. [3]Department of City and Regional Planning, University of North Carolina, Chapel Hill, USA. [4]Department of Environmental and Radiological Health Sciences, Colorado State University, Fort Collins, USA

Abstract
Wildfire smoke activity is increasing in the western US and smoke exposures are impacting downwind communities. Wildfire smoke is a complex mixture of gasses and particles. However, ambient concentrations of fine particulate matter (PM2.5) are commonly used as a proxy measurement for the mixture because it is a dominant constituent in smoke and it has known deleterious impacts on the lower

airways. The purpose of this project was to highlight PM2.5 air quality monitor coverage across the Mountain West region in relation to vulnerable elderly populations using a vulnerability index. Our hypothesis for this project was that some SNFs in the Mountain West fall in "monitor deserts" which we designate as areas with limited access to accurate air quality data and among those facilities, some may be more vulnerable than others. Using ArcGIS software and a vulnerability index, we identified monitor coverage based on EPA guidelines and categorized SNFs into facilities receiving precise air quality readings and facilities located in monitor deserts as well as the most vulnerable facilities based on our vulnerability index. We found that the average distance from facilities to air quality monitors was 15.83mi when only considering regulatory monitors whereas this changed to 14.07mi when we included average distance from both regulatory and low cost monitors. Our findings indicate where high risk communities, ranging in vulnerability, are that could benefit from more precise air quality data to better inform decisions about their interaction with unsafe ambient air including wildfire smoke events.

637 – Metabolomic perturbations related to later breast cancer incidence in a population living near a uranium refinery

Susan M Pinney[1], Cecily Shimpf Fassler[2], Jeanette Buckholz[3], Elizabeth Shaughnessy[4], Changchun Xie[4]
[1]University of Cincinnati, Professor, CINCINNATI, USA. [2]University of Cincinnati, Post doc, Cincinnati, USA. [3]University of Cincinnati, Research Associate, Cincinnati, USA. [4]University of Cincinnati, Professor, Cincinnati, USA

Abstract
Background: Ionizing radiation is recognized as a risk factor for breast cancer, but recent studies of rodents and mammary cell lines have noted that the heavy metal uranium exhibits estrogenic activity.

Hypotheses: We hypothesize that certain metabolic perturbations exist in women who later develop BC and in response to cumulative environmental U exposure.

Methods: We estimated uranium exposure for 4832 women enrolled in the Fernald Community Cohort using algorithms developed by the CDC Fernald Dosimetry Reconstruction project. Using 324 female breast cancer cases diagnosed during 32 years of follow-up (1990–2022) we conducted proportional hazards analyses to determine the association between exposure and breast cancer incidence. We then conducted a pilot exposomic metabolomics study in 50 cases and 50 controls, using serum obtained 1-5 years prior to diagnosis, to explore the biological basis of the association between uranium and breast cancer using an a priori selected a list of hormone-related steroids for targeted analyses.

Results: The mean uranium particulate exposure (ug/m3-years) for women in the three exposure groups (low, medium, high) was 0.086, 0.354, 1.596, and for cases was 0.35 vs. 0.19 for controls. In comparisons of 54 cases and 46 controls in the targeted metabolomics study, in plasma collected prior to diagnosis, we found statistically significantly higher levels of corticosterone, epiallogpregnanolong sulfate, and a progesterone metabolite and lower levels of dehydroepiandrosterone sulfate (DHEAS) and thyroxine. When comparing those with uranium exposure ≤ the median to those >median, we found statistically significant higher levels of cortisone.

638 – Mapping Cumulative Exposure to Hazardous Air Pollutants

Lauren Padilla, Dan Peters, Libby Mohr
Environmental Defense Fund, Boston, USA

Abstract
The people living in neighborhoods along petrochemical industry corridors have long understood that they suffer disproportionately from the health harms of breathing toxic emissions. However, the sources, specific substances and total quantities of these emissions are difficult to trace, especially when multiple chemical plants, often clustered together, emit toxins with similar health effects that add up to a cumulative adverse outcome.

Monitoring for air toxics is sparse even in areas with the highest cumulative exposure, and most stations don't measure the full array of compounds contributing to risk. Thus, insight into the spatial distribution of health risk and associated sources is limited to modeled estimates of exposure concentrations based on industry's self-reported estimates of emissions.
The aim of this work was to address key challenges of working with modeled exposure estimates (large size, hard to access, bias, incomplete toxicity and location information) to map the communities that are most burdened by cumulative toxic air pollution from the petrochemical industry and unpack the chemicals and facilities driving the exposure risk in those places. Using EPA's Risk Screening Environmental Indicators dataset, we created a user-friendly, national, high resolution, interactive risk map and characterized bias and

uncertainty. With invaluable user feedback from community groups who have expressed a need for this information for public hearings, we were able to share the map more broadly with additional contextual information while maintaining an intuitive user experience.

639 – Personal, Indoor and Regional Particle Exposure Variations During a Major Wildfire Smoke Episode in the Pacific Northwest and Implications for Exposure Reduction

Lilian Liu[1], Ching-Hsuan Huang[1], Elena Austin[2], Edmund Seto[3]
[1]Department of Environmental & Occupational Health Sciences, University of Washington, Research Assistant, Seattle, WA, USA.
[2]Department of Environmental & Occupational Health Sciences, University of Washington, Assistant Professor, Seattle, WA, USA.
[3]Department of Environmental & Occupational Health Sciences, University of Washington, Professor, Seattle, WA, USA

Abstract
Wildfire events have become increasingly prevalent in the Pacific Northwest in recent years, affecting regional $PM_{2.5}$ air pollution concentrations. However, regional air pollution may not fully represent indoor or personal exposures due to varying levels of particle infiltration and filtration, and time activity between outdoor and indoor microenvironments. To better assess these variations, our research group collected time-resolved particle measurements during a major wildfire event in the Seattle, Washington region that reflect personal, household indoor, and household outdoor air pollution concentrations. We compared these to hourly fixed site government measurements during the same smoke event that reflect regional $PM_{2.5}$ concentrations. Findings indicated higher outdoor than indoor or personal particle concentrations as well as a significant correlation between personal and indoor concentrations. Results showed around two-fold higher regional $PM_{2.5}$ concentration compared to the personal and indoor concentrations. While there were differences in outdoor sensor concentrations by site, overall, these differences were relatively similar to the regional $PM_{2.5}$ concentrations. These findings support public health guidance to shelter indoors during wildfire smoke events. We discuss these findings in the context of recent evidence on the effectiveness of portable HEPA air cleaners and box-fan filter strategies in both household and community/ congregate settings.

640 – POSTER: Comparative assessment of personal exposure to environmental contaminants using silicone wristbands, Fresh Air wristbands, and biological monitoring

Elizabeth Lin[1], Nicholas Herkert[2], Duncan Hay[3], Ellen Cooper[4], Heather Stapleton[5], Krystal Pollitt[6]
[1]Yale University, PhD Candidate, New Haven, USA. [2]Duke University, Postdoctoral Associate, Durham, USA. [3]Duke University, Research Tech, Durham, USA. [4]Duke University, Research Scientist, Durham, USA. [5]Duke University, Professor, Durham, USA. [6]Yale University, Associate Professor, New Haven, USA

Abstract
Wearable passive samplers are increasingly being adopted as a non-invasive approach for assessing personal exposure to complex mixtures of environmental contaminants. Various silicone-based sampler designs have emerged that capture different exposure pathways. Silicone wristbands are a loop of silicone rubber that is in direct contact with the air and an individual's skin, providing a combined estimate of inhalation and dermal exposures. Fresh Air wristbands mount a silicone bar in a perforated case to exclusively evaluate inhalation exposures. A comparative assessment of chemical exposure profiles detected by these two wearable passive samplers was conducted for 30 adults residing in the US, contrasting sampler designs as well as analytical and data processing workflows. Silicone and Fresh Air wristbands were simultaneously worn by participants for five days and then returned for chemical analysis together with urine samples collected on days 2, 4, and 6. Using targeted analysis, the same panel of phthalates, organophosphates, and polycyclic aromatic hydrocarbons (PAHs) were detected by both wristband designs. Suspect screening analysis of silicone wristbands revealed 1330 compounds, predominantly intermediate- and semi-volatile organic compounds, including fatty acyls and lipids (cholesterol, squalene). Fresh Air wristbands identified 2261 chemicals, mostly volatile and intermediate-volatile organic compounds. Both wristband designs were positively correlated with paired pool urine samples for organophosphate flame retardants. Findings of this study improve our understanding of the performance characteristics of different silicone-based wearable passive samplers, informing future use in population studies for personal exposure assessment of a diverse panel of environmental chemicals.

643 – Development of a PFAS Continuing Medical Education (CME) Module for Clinicians

Nicholas Newman[1], Frank M Biro[2], Susan M Pinney[3]

[1]Cincinnati Children's Hospital Medical Center, Associate Professor, Cincinnati, USA. [2]Cincinnati CHildren's Hospital Medical Center, Professor, Cincinnati, USA. [3]University of Cincinnati, Professor, Cincinnati, USA

Abstract
The public now has become more aware of potential PFAS exposure through recent research publications. The State of Ohio is requiring water systems to report PFAS levels, posted on an Ohio.gov website and reported in the annual report to water consumers. Moreover, the Ohio River Valley has a long history of PFAS exposure through contamination of the Ohio River, the primary source of drinking water.

Following phone calls from area physicians to the Community Engagement Core (CEC) Leader Newman, Center for Environmental Genetics (CEG) Director Pinney and CEG member Biro regarding concerns about patients being exposed to PFAS, the CEC created a PFAS Continuing Medical Education (CME) module "Perfluorochemicals ("Forever Chemicals") and Human Health Effects: An update for Clinicians". The 45-minute recorded lecture incorporates CEG scientists' research and addresses several topics relevant to clinicians including a) sources of PFAS chemicals in your patients, b) the extent of PFAS exposure in the past & currently, c) identifying patients whose health may be impacted by PFAS exposure, d) risks and health effects of PFAS exposure, and e) ways to counsel patients exposed to PFAS. The module is posted on the University of Cincinnati Continuing Medical Education website and is available free of charge to physicians including awarding CME units. To our knowledge, this is one of the first CME modules focusing on the health effects of an environmental exposure.

644 – POSTER: Using mobile monitoring methods to characterize disparities in outdoor exposure to traffic related air pollutants at schools

Elena Austin[1], Rajni Oshan[2], Timothy Larson[3], Jeffry Shirai[2], Timothy Gould[2], David Hardie[2], Edmund Seto[4]
[1]University of Washington, Assistant Professor, Seattle, USA. [2]University of Washington, Seattle, USA. [3]University of Washington, Professor Emeritus, Seattle, USA. [4]University of Washington, Professor, Seattle, USA

Abstract
There is a significant disparity in the siting of schools, with schools serving higher proportions of Black and Hispanic students or students from lower socio-economic backgrounds more likely to be located near sources of air pollution. Understanding the impact of these disparities on exposure to traffic related air pollution (TRAP) is challenging. Mobile monitoring (MM) campaigns have been used to identify spatial distribution of air pollution sources and to develop annual average exposures to air pollutants at fine spatial scales. This presentation describes the design considerations of a MM campaign intended to capture variations in exposure to TRAP at schools located within the Seattle Metropolitan area. We applied MM methods to characterize annual exposures to TRAP at 14 area Schools. A location-allocation algorithm was used to select schools spanning the spatial variation in distance to major sources of air pollution, including flight paths, highways and point sources. A MM platform equipped with fast response instruments, collected TRAP measurements and size fractioned ultrafine particle (UFP) concentrations. Data was collected using a time balanced sampling scheme, from Fall 2022 through Summer 2023, and each location received 28 visits of duration 4 minutes per visits. We report significant differences in exposure to UFP and other TRAP pollutants as well as in size distribution of particles. This approach may provide a novel method to quantify disparities in exposure and could aid policy makers in identifying and mitigating schools with a higher burden of exposure to outdoor pollutants.

645 – Mitigating Risk and Hazards: A Culturally Appropriate Educational Intervention for Crop Farmworkers on Occupational Dust Exposure and Heat-Related Illnesses.

Andres Manrique[1], Salvatore Milletich[2], Noe Bautista[3], Tracy Irani[4], Tara Sabo-Attwood[4], Eric Coker[5]
[1]University of Florida, Graduate Assistant, Gainesville, USA. [2]University of Florida, MPH student, Gainesville, USA. [3]Redlands Christian Migrant Association, Health Advocate, Mulberry, USA. [4]University of Florida, Professor & Department Chair, Gainesville, USA. [5]British Columbia Centre for Disease Control, Senior Scientist, British Columbia, Canada

Abstract
The agricultural industry is one of the most hazardous occupations, with workers being exposed to biological, chemical, meteorological, and ergonomic hazards. Occupational studies have identified exposure to particulate matter (PM) as an emerging concern among crop farmworkers which is concerning based on the association between exposure to particulates and respiratory health outcomes. Furthermore, heat-related illnesses (HRIs) among crop farmworkers are a present issue with growing concerns over increasing global temperatures. Despite concerns, crop farmworkers have a limited understanding of workplace hazards and mitigation strategies due to a lack of educational material and trainings. To address this, we developed culturally appropriate educational interventions on

occupational dust exposure (ODE) and HRIs using a placebo-attention study design. A pre- and post-test was developed to examine the knowledge, perceived severity, self-efficacy, and behavioral intent of Florida's crop farmworkers, with a focus on migrant and H2A workers. We acquired information on training history, self-reported HRIs and respiratory symptoms, and demographics. Initial results suggest that H2A workers are younger and have a higher level of education but receive fewer safety trainings when compared to the domestic workforce. The prevalence rate of self-reported moderate HRIs symptoms and chronic phlegm was 50% and 5%, respectively. There was an increase in knowledge, self-efficacy, and behavior intent for participants in the ODE intervention. Ongoing work includes evaluating an additional 100 participants. Our first-time evaluation suggests the greatest knowledge gain will be seen in those receiving the occupational dust exposure intervention as this is a novel topic for crop farmworkers.

646 – PFAS Exchange – An Online Resource Center to Educate and Empower PFAS-Impacted Communities

Laurel Schaider[1], Phil Brown[2], Courtney Carignan[3], Andrea Amico[4], Dana Colihan[5], Alan Ducatman[6], Elizabeth Friedman[7], Kimberly Garrett[8], Alexandra Goho[9], Erik Haugsjaa[10], Aaron Maruzzo[11], Kira Mok[12], Cheryl Osimo[13,14]
[1]Silent Spring Institute, Senior Scientist, Newton, MA, USA. [2]Northeastern University, Professor, Boston, MA, USA. [3]Michigan State University, Assistant Professor, East Lansing, MI, USA. [4]Testing for Pease, Co-Founder, Portsmouth, NH, USA. [5]Slingshot, Co-Executive Director, Portland, ME, USA. [6]Alan Ducatman LLC, Self-employed, Morgantown, WV, USA. [7]Children's Mercy Hospital, Medical Director of Environmental Health, Kansas City, MO, USA. [8]Northeastern University, Postdoctoral Research Associate, Boston, MA, USA. [9]Silent Spring Institute, Director of Communications, Newton, MA, USA. [10]Silent Spring Institute, Software Engineer, Newton, MA, USA. [11]Silent Spring Institute, Research Assistant, Newton, MA, USA. [12]Northeastern University, Undergraduate, Boston, MA, USA. [13]Massachusetts Breast Cancer Coalition, Executive Director, Barnstable, MA, USA. [14]Silent Spring Institute, Outreach Coordinator, Newton, MA, USA

Abstract
Per- and polyfluoroalkyl substances (PFAS) are a class of highly persistent, bioaccumulative, and toxic compounds with ubiquitous exposures among the general population. PFAS have contaminated drinking water supplies globally, including hundreds of communities across the United States. For residents of PFAS-impacted communities, barriers to action include challenges interpreting blood and water test results, rapidly evolving information about PFAS, and a lack of environmental health training among clinicians to address community member concerns.

The PFAS Exchange (pfas-exchange.org) is an innovative online resource center that educates and empowers PFAS-impacted communities. Features include:

- What's My Exposure Tool – an interactive tool that allows users to enter blood and water test results and immediately obtain a context-rich report, including comparisons to relevant standards and benchmarks, with personalized graphs and text headlines, along with information to help users take steps to reduce their exposure.
- PFAS Sites and Community Resources Map – a GIS-based map with contamination sites (known and presumptive), state-level information about community groups, regulations, and health and exposure studies, developed in collaboration with the PFAS Project Lab.
- Resources Library – up-to-date and essential information for PFAS-impacted communities, including information about accessing blood tests, PFAS health effects, and drinking water guidelines and treatment. In partnership with clinicians, we developed a Resources for Clinicians page, which features a Continuing Medical Education (CME) course and medical screening guidance documents.

The PFAS-Exchange is a product of the federally-funded PFAS-REACH Study, which is led by a researcher-community partnership of environmental health researchers, social scientists, and community activists.

647 – Use of Personal Care Products and Asthma Morbidity Among Urban Children

Meleah Boyle[1], Magdalena Fandiño-Del-Rio[2], Rachelle Koehl[3], Nadia Hansel[3], Elizabeth Matsui[4], Meredith McCormack[3], Lesliam Quirós-Alcalá[2]
[1]University of Maryland School of Public Health, College Park, USA. [2]Johns Hopkins University Bloomberg School of Public Health, Baltimore, USA. [3]Johns Hopkins University School of Medicine, Baltimore, USA. [4]University of Texas at Austin Dell Medical School, Austin, USA

Abstract
Background: Over 4 million U.S. children suffer from asthma and emerging evidence indicates that chemicals in personal care products (PCP) may play a role in asthma development and control, but research remains sparse.

Methods: We used generalized linear models to examine associations between PCP use and asthma symptoms in the prior two weeks (maximal symptom days, cough without a cold, and exercise-related symptoms) among 110 urban children 8-17 years with asthma. Product use included individual products and the number of products used in the prior week by type of application (i.e., aerosol, wash-off, leave-on hair products, leave-on face/body products). Models were adjusted for age, sex, race, insurance status, season, and having a caregiver who is a smoker.

Results: Participants were majority Black (87%), male (56%), and 8-11 years (66%). In the prior week, participants used an average of 7 PCPs (range: 3-23). Use of aerosol products, shampoo, hair sheen, and nail polish were associated with increased odds of maximal symptom days [number of aerosol products (aOR:1.36; CI: 1.17, 1.59), hairspray (aOR:1.63; CI:1.14, 2.33), perfume (aOR:1.40; CI:1.11, 1.77); liquid shampoo (aOR:1.34; CI:1.05, 1.73), hair sheen (aOR:1.41; CI:1.00, 2.00), and nail polish (aOR:2.42; CI:1.72, 3.41)]. Hair products were positively associated with exercise-related symptoms [hair coconut oil (aOR:1.40; CI:1.01, 1.95), hair gel (aOR:1.85; CI:1.26, 2.70), hair oil (aOR:1.17; CI:0.88, 1.55) and hair sheen (aOR:2.10; CI:1.44, 3.06)].

Conclusion: Our findings indicate that some PCPs may contribute to asthma morbidity, but larger studies are needed to confirm our findings and identify chemicals of concern.

648 – POSTER: Intercity and Intracity variation of indoor-outdoor NO_2 and VOCs in urban homes in multiple cities in India

Rajdeep Singh[1], Prince Vijay[1], Vinayak Sahota[2], Shreya Dubey[1], Sonali Borse[2], Harish Phuleria[3]
[1]Indian Institute of Technology Bombay, PhD Scholar, Mumbai, India. [2]Indian Institute of Technology Bombay, Project Assistant, Mumbai, India. [3]Indian Institute of Technology Bombay, Associate Professor., Mumbai, India

Abstract
Past studies have shown that exposure to traffic-related air pollutants such as nitrogen dioxide (NO_2) and volatile organic compounds (VOCs) are associated with various adverse health risks, particularly in urban areas. This study aims to understand the indoor-outdoor NO_2 and VOCs exposures in multiple cities across India. This study is a part of the APEAL (Longitudinal Effects of Air Pollution Exposure on Adolescents Lungs) study, a cohort study in four cities in India. This pilot study was conducted across four cities in India - Delhi, Mumbai, Bangalore, and Mysore. with 7-day integrated indoor and outdoor NO2 and VOCs measurements in 7-9 homes in each city using passive sampling tubes during the summer of 2022. Delhi shows the maximum outdoor NO_2 (60±8.5 μg/m^3), which is 41%, 40%, and 66% higher than Mumbai, Bangalore, and Mysore, respectively. Similar trend was observed for total VOC, with Delhi (98.1±30.7 μg/m^3) having the highest outdoor levels. Indoor concentrations were higher for both pollutants but statistically significant for NO_2 ($p<0.05$). Indoor/outdoor NO_2 ratios were highest for Mumbai and lowest for Delhi, whereas for VOCs, the ratios were similar. Our study highlights that outdoor air pollution is primarily driven by traffic, however, the indoor sources such as gasstove cooking and incense/oil lamp burning for religious purposes can significantly affect indoor exposures, which must be accounted for estimating true exposures. Further analysis is underway to understand the relationship of indoor/outdoor exposures with home characteristics and traffic variables such as distance to the nearest road.

649 – Air pollution monitoring and health outcomes in LatAm: a snapshot of the current situation

Tailine C. dos Santos[1], Sebastian Diez[2], Josefina Urquiza[3], Colleen Marciel F. Rosales[4,5]
[1]Universidade de São Paulo, Post doctorate, Sao Paulo, Brazil. [2]Universidad del Desarrollo - UDD, Santiago, Chile. [3]Universidad Tecnológica Nacional - UTN, Córdoba, Argentina. [4]OpenAQ, Washington D.C., USA. [5]University of California, Davis, CA, USA

Abstract
Air pollution has a wide range of detrimental effects on public health, spanning from respiratory and cardiovascular issues to metabolic and neurological disorders. As such, characterizing atmospheric pollutants is of paramount importance, particularly in regions where environmental injustice is a pervasive issue. This study concentrates on Latin America, a region where numerous countries face heightened vulnerability due to factors such as limited access to healthcare and inadequate availability of information on air quality and health records. The scarcity of data significantly hampers the development of effective public policies, as establishing links between causes and consequences remains fraught with uncertainties. To address this issue, our research aims to investigate the relationship between the current state of air quality information and its corresponding health outcomes in the Latin American region. To achieve this objective, we have utilized the latest report on air quality data availability from OpenAQ, along with health impact metrics from the Global Burden of Disease study. This initial analysis reveals that countries without air quality monitoring (41% of the countries in the region) exhibit an average death rate approximately 20% higher than countries with monitoring in place and roughly 35% higher than

countries with fully open data. However, it is crucial to note that these findings do not imply causality, but rather offer a snapshot of the current situation. In order to develop strategies that positively impact the overall health of the region, it is essential to take other relevant factors into consideration.

650 – Assessing the Exposure and Health Impacts of Prescribed Burning in California

Michael Jerrett[1], Rachel Connelly[2], Claire Bekker[3], Jason Su[4], Diane Garcia-Gonzales[5], Miriam Marlier[6]
[1]UCLA, Professor, Los Angeles, USA. [2]UCLA, Researcher, Los Angeles, USA. [3]UCLA, PhD Student, Los Angeles, USA. [4]UC Berkeley, Professional Researcher, Berkeley, USA. [5]UCLA, Research Scientist, Los Angeles, USA. [6]UCLA, Assistant Professor, Los Angeles, USA

Abstract
Background: Prescribed burning is a method to reduce the fuel load in forests by setting controlled fires at strategic locations to minimize the future possibility of larger uncontrolled wildfires igniting or spreading. Little is known about the exposure and health effects of prescribed burning compared to those of wildfires.

Objective: To estimate the excess mortality due to prescribed burning in California for 2018.

Methods: We fit the Stochastic Time-Inverted Lagrangian Transport (STILT) particle dispersion receptor-based model for fire and non-fire days in 2018 (79 days in total) to reflect various meteorological conditions. These were used to develop atmospheric sensitivity "footprints" that provide a flexible means of assessing various scenarios based on changes in the emissions from prescribed burns. We assembled data from CAL FIRE on the locations of prescribed burns throughout the state in 2018. These were spatially joined with the Fire Inventory from NCAR (FINN)v2.5 database of fire emissions at the 3 km scale to isolate emissions associated with prescribed burning. This methodology enabled us to predict emission changes at county receptor locations and to calculate county-wide mortality estimates.

Results: Prescribed burning contributed minimally to ambient levels of PM2.5, with concentrations increasing maximally by about 0.5 ug/m3. Using ZIP-code level mortality data, we estimated 45 excess deaths from prescribed burning in 2018 and about 11,500 excess deaths from wildfire-generated PM2.5.

Conclusion: Our results demonstrate relatively small effects from prescribed burning in 2018 compared the deaths attributable to wildfire smoke. Researchers need to evaluate the efficacy of prescribed burning.

651 – Assessment of Anxiety and Ultrafine Particle Exposure Among Adolescents

Ashley Turner[1], Patrick Ryan[2,3], Sherrill Ingram[4], Cole Brokamp[1], Seung-Hyun Cho[5], Jeffrey Strawn[3,6], Chris Wolfe[5]
[1]Cincinnati Children's Hospital Medical Center, Postdoctoral Researcher, Cincinnati, USA. [2]Cincinnati Children's Hospital Medical Center, Professor, Cincinnati, USA. [3]University of Cincinnati, Professor, Cincinnati, USA. [4]Cincinnati Children's Hospital Medical Center, Clinical Research Coordinator III, Cincinnati, USA. [5]RTI International, Principal Exposure Research Scientist, Research Triangle Park, USA. [6]Cincinnati Children's Hospital Medical Center, Associate Professor, Cincinnati, USA

Abstract
Background: Acute exposure to ambient PM2.5 has been linked to the exacerbation of mental health and anxiety in children. However, the effects of personal and sub-daily exposures to air pollutants, including ultrafine particles, on symptoms of anxiety is unknown.

Methods: We sampled personal UFP exposure using a high-resolution monitor (Partector, Naneos) while collecting brief anxiety symptom surveys for one week among a sample of adolescents with a clinical anxiety disorder. Symptoms recorded by the surveys included generalized anxiety symptoms as well as panic and social anxiety symptoms. Logistic univariate mixed effects models were used to determine the relationship between daily median UFP exposure and anxiety symptom scores.

Results: Of the 13 participants analyzed, median weekly personal UFP exposures ranged from 242 particles/cc to 5,113 particles/cc per participant. On average, participants reported elevated anxiety symptom scores in the evening compared to the morning. Mixed effects models revealed participants with greater daily median UFP exposure were more likely to report greater fear of going to school in the morning (Odds Ratio [OR]: 6.04, CI: 0.66,16.25), and being restless (OR: 1.58, CI: 0.05, 3.40) or feeling afraid that something bad might happen (OR: 1.78, CI: 0.14, 4.00) throughout the day, though our results were not significant.

Conclusion: In our panel study of children with anxiety, personal exposure to UFP is associated with elevated anxiety symptoms. Reducing exposure to air pollution may be a method to improve symptoms of anxiety among adolescents.

652 – Centering the Stories Behind the Numbers: Mixed Methods Research for Advancing Understanding of Environmental Harm

Rebecca Witter[1], Danielle Koonce[2,3], Sherri White-Williamson[3], Courtney Woods[4], Carol Babyak[5], Shea Tuberty[5], Ben Pluska[6], Bekah Nielsen[6], Dana Powell[1]

[1]Appalachian State University, Associate Professor, Boone, USA. [2]University of Maryland, PhD Candidate, College Park, USA. [3]Environmental Justice Community Action Network, Clinton, USA. [4]University of North Carolina, Associate Professor, Chapel Hill, USA. [5]Appalachian State University, Professor, Boone, USA. [6]Appalachian State University, Undergraduate Researcher, Boone, USA

Abstract

Quantitative measures of water and air pollutants and related health effects provide important indicators for assessing the well-being of communities impacted by environmental injustice. Yet, the extent of environmental harm cannot be fully understood without the perspectives of those who experience it directly. In rural eastern North Carolina (USA), Sampson County residents live with the largest landfill in the state, concentrated animal feeding operations for the industrialized production of hogs and poultry, energy facilities for the industrialized production of "biogas" and "biomass", and intensifying hurricanes and flooding. In 2021 we established a transdisciplinary, community-driven water quality testing initiative in response to residents' concerns about contamination. In addition to assessing levels of toxicity in ground and surface water, our research examines people's experiences with exposure. We spend time with, listen to, and converse with people living in high-risk areas; we participate in and document state regulatory and community-based meetings; and we conduct semi-structured interviews. This mixed methods approach enables us to examine variation in perceptions of environmental harm, chart the historical legacies that undergird current conditions, and identify community-informed methods for enacting justice. When employed in isolation of social science approaches, quantitative measures can abstract, even erase, people's lived experiences, resulting in limited understanding of multiple and accumulating harms. Thus, we argue against the tendency in STEM to reduce stories collected through qualitative research to "nuance". A convergence approach to environmental exposure research advances understanding of harm by better accounting for the ways impacted people experience and explain it.

653 – Risk Assessment of Di(2-Ethylhexyl) Phthalate (DEHP) and Bisphenol A (BPA) Exposure to Neonates Undergoing Surgery for Congenital Heart Defects

Amanda Alvarado[1], Kimberly Valle[1], Valerie Martinez[1], J. William Gaynor[2], Nancy B. Burnham[3], Susan C. Nicolson[2], Thomas L. Spray[2], Richard F. Ittenbach[4], Frederick M. Henretig[2], Erin E. Zullo[5], J. Laurenson Ward[6], William W. Russell[7], David C. Bellinger[8], Hannah Lu[9], Adriana Espinosa[10], Asa Bradman[11]

[1]University of California, Merced, M.S., Merced, USA. [2]Children's Hospital of Philadelphia, M.D., Philadelphia, USA. [3]Children's Hospital of Philadelphia, C.R.N.P., Philadelphia, USA. [4]Cincinnati Children's Hospital Medical Center, Ph.D., Cincinnati, USA. [5]Children's Hospital of Philadelphia, B.S.N, Philadelphia, USA. [6]Children's Hospital of Philadelphia, B.A., Philadelphia, USA. [7]Children's Hospital of Philadelphia, B.S., Philadelphia, USA. [8]Boston Children's Hospital, Ph.D., Boston, USA. [9]University of California, Berkeley, B.A., Berkeley, USA. [10]University of California, Merced, Merced, USA. [11]University of California, Merced, Ph.D., Merced, USA

Abstract

Background: Di(2-Ethylhexyl) phthalate (DEHP) and bisphenol A (BPA) are anthropogenic chemicals used to manufacture plastics. DEHP and BPA are endocrine disruptors and have been linked to poor airway functioning, preterm birth, delayed mental and motor development, and cancer risk. Humans are commonly exposed to DEHP and BPA from consumer products or when they leach into food and drinks from food-contact materials. Exposures in health-care environments occur when these chemicals leach into IV fluids or directly into blood or other body fluids in contact with medical equipment. Newborns undergoing surgery for congenital heart defects (CHD) experience high exposures to DEHP and BPA and are particularly vulnerable to chemical exposures due to their young age.

Objective: To perform a risk assessment evaluating BPA and DEHP exposures in 18 infants who underwent heart defect surgery.
Methods: Reverse-dosimetry was used to estimate the infant dose based on postoperative urinary BPA and DEHP metabolite concentrations. Exposure-doses were compared to U.S. EPA chronic reference doses (RfD) to calculate a hazard quotient (HQ).
Results: Operations were performed in 18 infants (mean age, 5 ± 4 [SD] days). The median BPA HQ was 0.28 (SD =0.2), with a maximum of 0.8. The median DEHP HQ was 51.2 (SD=86.8), with a maximum of 353.4.

Significance: We observed elevated exposures to DEHP in neonates undergoing cardiac surgery that may have long-term impacts on child well-being. Exposures to DEHP and BPA are likely to remain elevated as long as the infants are supported by medical interventions, which may last weeks.

654 – POSTER: Ethylene Oxide at the Industrial Fenceline in Southeastern Louisiana: Spatially-Resolved Concentrations and Plume Measurements from Fast Time-Resolution Mobile Monitoring During HAP-MAP

Ellis Robinson[1], Amira Yassine[2], Shivang Agarwal[2], Mina Tehrani[3], Andrea Chiger[2], Carolyn Gigot[2], Jonathan Bent[4], Kai Skog[5], Megan Claflin[6], Conner Daube[7], Elizabeth Lunny[8], Rob Roscioli[6], Scott Herndon[6], Ana Rule[9], Kirsten Koehler[10], Thomas Burke[11], Keeve Nachman[10], Peter DeCarlo[10]

[1]Johns Hopkins University, Assistant Research Engineer, Baltimore, USA. [2]Johns Hopkins University, PhD Student, Baltimore, USA. [3]Johns Hopkins University, Postdoctoral Scientist, Baltimore, USA. [4]Picarro Inc., Senior Technical Application Lead, Santa Clara, USA. [5]Picarro Inc., Research Scientist, Santa Clara, USA. [6]Aerodyne Research Inc., Principal Scientist, Billerica, USA. [7]Aerodyne Research Inc., Research Associate, Billerica, USA. [8]Aerodyne Research Inc., Senior Scientist, Billerica, USA. [9]Johns Hopkins University, Assistant Professor, Baltimore, USA. [10]Johns Hopkins University, Associate Professor, Baltimore, USA. [11]Johns Hopkins University, Professor Emeritus, Baltimore, USA

Abstract

Ethylene oxide (EtO) is a reactive volatile organic compound (VOC) with applications as both a petrochemical feedstock and a biocide used for surface disinfection in medical sterilization and food safety. EtO's useful reactivity in industrial applications also makes it hazardous to human health. U.S. EPA classifies EtO as a carcinogen, with a cancer inhalation unit risk (IUR) value much higher than other common hazardous VOCs. Because of the associated risk, there are significant exposure concerns around facilities using or producing EtO. Despite this risk, however, there have been few ambient measurements of EtO, which typically are stationary and have long time-integration. Here we present in-situ, fast (1 hz) time-resolution measurements of EtO in the industrial corridor of Louisiana between Baton Rouge and New Orleans, where we used two different real-time EtO instruments aboard mobile platforms. EtO represents the majority of hazard in this domain, as evaluated by the sum of facility-level RSEI scores (Risk Screening Environmental Indicator, U.S. EPA). We report EtO concentrations at fine (sub 1-km) spatial scales across an area spanning roughly 65 km. Mean EtO concentration across all (n=1,227) 200m grid cells in our domain was sub-100 ppt. A small subset (n=27) of near-source grid cells had concentrations above 100 ppt, with a maximum of 745 ppt. We also present comparisons of high concentration (>5 ppb) EtO plumes that were measured simultaneously from the two EtO instruments.

655 – Community-Driven Investigations of Fugitive Emissions from a Local Coal Terminal in Curtis Bay, South Baltimore, Maryland

Lauren Deanes[1], Matthew Aubourg[2], Kristoffer Spicer[3], Caila Wagar[4], Gregory Sawtell[5,6], Meleny Thomas[7], Shashawnda Campbell[8], Christopher Heaney[9]

[1]Johns Hopkins University, PhD Candidate, Baltimore, USA. [2]Johns Hopkins University, Research Assistant, Baltimore, USA. [3]Johns Hopkins University, Research Program Manager, Baltimore, USA. [4]Johns Hopkins University, JD/MPH Candidate, Baltimore, USA. [5]South Baltimore Community Land Trust, Zero Waste Just Transition Director, Baltimore, USA. [6]Community of Curtis Bay Association, Board Member, Baltimore, USA. [7]South Baltimore Community Land Trust, Executive Director - Development without Displacement, Baltimore, USA. [8]South Baltimore Community Land Trust, Environmental Justice Coordinator, Baltimore, USA. [9]Johns Hopkins University, Associate Professor, Baltimore, USA

Abstract

A major coal terminal is one of 70 regulated sources of air pollution in Curtis Bay, South Baltimore, Maryland. Consequently, local community organizations and residents have long-standing concerns about the adverse impacts of industrial air pollution on their local environment and health, and an explosion at the terminal (December 2021) intensified these concerns. In partnership with local community organizations and businesses, we established an air quality monitoring network throughout Curtis Bay in May 2022. Collectively, these monitors measure particulate matter, black carbon, various gases, and meteorological data. We supplemented these air quality data with visual observation logs of coal terminal activity patterns (e.g., train movement, "dust suppression" sprinklers, etc.), coded from trail camera time-lapse videos. We performed t-tests of the difference of log10-transformed mean air pollutant concentrations during time periods of downwind vs. non-downwind conditions (relative to the terminal) and when coal terminal activities were vs. were not observed. Preliminary results showed that concentrations of all measured particulate matter size fractions were, on average, higher when the wind was blowing downwind from the coal terminal towards nearby air monitors, as opposed to when the wind was blowing from other directions. Additionally, when the terminal's "dust suppression" system was engaged, concentrations of larger particles, expected to be those from coal, tend to be higher than when the system was not engaged. This combination of camera and air monitoring data provides a useful tool for estimating fugitive dust events in the community and attributing them to activity patterns at the terminal.

656 – Assessment of indoor concentrations of termiticides and their risk to humans

Kazushi Noro[1], Kasumi Yamaguchi[2], Takashi Amagai[3], Yuichi Miyake[4]
[1]University of Shizuoka, Assistant Professor, Shizuoka, Japan. [2]University of Shizuoka, Student, Shizuoka, Japan. [3]University of Shizuoka, Professor, Shizuoka, Japan. [4]Yokohama National University, Associate Professor, Yokohama, Japan

Abstract
Chlorpyrifos has been used as a termiticide and was banned in 2003 because it is one of the causative agents of sick building syndrome. Currently, neonicotinoids, pyrethroids, phenyl pyrazoles, and their enhancers are used as alternative termiticides. However, there is concern about health risks to people living in rooms sprayed with the alternative termiticides because these termiticides are neurotoxic, carcinogenic, and genotoxic. Data on indoor concentrations of the alternative termiticides is lacking despite their importance for human health risk assessment. Here we show that 18 alternative termiticides were detected in the range of <67 μg g-1 in house dust samples and <18 ng m-3 in indoor air samples. The concentrations of bifenthrin, permethrin, and fipronil were relatively high in house dust samples. The concentrations of S-421, cyproconazole, and fipronil were relatively high in house air samples. Indoor air was the major route of 1-day exposure to humans, rather than house dust. Risk assessment by margin of exposure showed the highest risk for fipronil. Relatively high concentrations of fipronil were detected in the one-year-old house because fipronil could be sprayed on the house as a termiticide prior to occupancy. These results suggest that analysis of indoor air in homes within a few years of termiticide application is important for assessing the human health risks of termiticides.

657 – Exploring the Size Detection Limit: Characterizing Microplastics in Household Indoor Air

Lauren Ead[1], Kathryn Harris-Howard[2], Matthew Adams[3]
[1]University of Toronto Mississauga, Department of Geography, Geomatics, and the Environment, MSc Candiate, Mississauga, Canada.
[2]University of Toronto Mississauga, Office of the Vice-Principal, Research, Senior Research Associate, Imaging, Mississauga, Canada.
[3]University of Toronto Mississauga, Department of Geography, Geomatics, and the Environment, Associate Professor, Mississauga, Canada

Abstract
Microplastics (MPs) prevail in all environments and can enter the body through various exposure routes. Of these, inhalation is considered the most significant, with indoor air contributing to the bulk of MP uptake. Respirable MPs, specifically those ≤1μm, warrant particular concern as particles of this size can deposit deep into the lungs and may circulate systemically. Nevertheless, current techniques cannot reliably identify MPs of this size, indicating a need to optimize detection methods. Selective fluorescent dyes (e.g., Nile red) offer a solution, as they enable the detection of fluorophores as small as 0.1μm and are common in MP research. This study is two-phase, consisting of passive and active sampling campaigns, and aims to investigate the prevalence of respirable MPs in Canadian households via fluorescent tagging and spectroscopic techniques. In phase 1, particle fallout was captured for one-month (n=10), treated with Nile red, and examined via confocal microscopy at various excitation wavelengths. Images were acquired at an incremental depth and analyzed to define particle abundance, morphology, and size. Raman spectroscopy will be used to characterize particles (≥0.5μm) and explore whether fluorescent patterns exhibited at specific wavelengths offer insights into particle composition. In phase 2, a subset of homes (n=5) will be sampled for one-week using a low-flow, cyclone-based sampler and then analyzed via Raman to characterize the physicochemical properties of coarse and fine particulates. Hybrid systems, like Atomic Force Microscopy-Raman, may be explored to investigate MPs ≤0.5μm. This study aims to provide comprehensive insights into respirable MPs to enhance exposure assessment.

658 – Developing Metadata Specifications for Representing Exposure Health Data

Ram Gouripeddi[1,2,3], Katherine Sward[1,2,3,4], Julio Facelli[1,2,3], Mollie Cummins[1,2,4]
[1]Center of Excellence for Exposure Health Informatics, University of Utah, Salt Lake City, USA. [2]Clinical and Translational Science Institute, Salt Lake City, USA. [3]Department of Biomedical Informatics, University of Utah, Salt Lake City, USA. [4]College of Nursing, University of Utah, Salt Lake City, USA

Abstract
Representing exposure health studies and the exposome often require the use of semantic standards in order to provide a shared representation, support performance of secondary analysis and achieve reproducibility of the data and results. In order to specify appropriate representations for use in research studies, we categorized all exposomic data in six major categories: sensor, computationally modeled, clinical, person-generated and aggregate streams. For each of these we developed metadata specifications

in or order to harmonize and standardize these data. For example, the sensor metadata specifications consists of instrument, deployment and output domains. In addition these specifications also describes semantic values each facet of the data including vocabularies such as (1) Children's Health Exposure Analysis Resource ontology, (2) Chemical Entities of Biological Interest Ontology, (3) Environment Ontology, (4) Exposure ontology, and (5) Unified Medical Language System Metathesaurus. All these specifications are logically stored in a graph-based metadata repository for public consumption. In addition, data belonging to each of these streams are represented as 'events' that have spatial and temporal dimensions allowing for reasoning across different root objects. For example, this approach allows for assessing the relationship between environmental and health condition occurrences. We are evaluating these metadata and events for their representativeness and completeness of all domains and will present our initial results of the same. We envision when completely developed the specifications will help exposure health researchers to select appropriate metadata representation standards for representing their data, and use within informatics platforms for harmonizing results arising from different data streams.

660 – POSTER: Temporal and Spatial Variability of Ultrafine and PM2.5 Personal Exposure Patterns

Seung-Hyun Cho[1], Christopher Wolfe[1], Ryan Chartier[1], Sherrill Ingram[2], Michelle McCombs[1], Ashley Turner[2], Patrick Ryan[2]
[1]Research Triangle Institute, Research Triangle Park, USA. [2]Cincinnati Children's Hospital Medical Center, Cincinnati, USA

Abstract
Evidence from toxicology studies suggest that ultrafine particles (UFP) play a significant role in overall particulate matter (PM) toxicity. However, this has not been validated in epidemiological settings mainly because accurate, directly measured exposure data are not readily available. Our on-going study includes 7-day personal exposure assessment of 30 asthmatic adolescents (Cincinnati, OH) using UFP and PM2.5 sensors carried in a small backpack alongside a GPS logger.

Sampling took place between May and November 2021 when most students were attending virtual classes and, on average, 70% of the measurements were collected from home. The mean UFP hourly average was highest at home (7857 particles/cc), followed by transit (5110) and school (3572). UFP exposure was highest in the evening (18-24pm; 11220 #/cc), followed by night (midnight-6 am; 8987), and lowest in the morning (6am-noon; 3337). PM2.5 showed a similar spatial pattern as UFP with the mean hourly average at home of 15.4 ug/m3 (from 1.2 to 122.2 between individuals) (vs. transit [14.3] and school [8.7]).

The 1-minute correlation between UFP and PM2.5 varied between individuals with Spearman correlation coefficients ranging from 0.06 to 0.78, with mean of 0.47. These correlations varied by microenvironment with the highest correlation at home (0.42), followed by -0.39 (other) and 0.09 (transit), Further examination of mobile app-based activity data will improve our understanding of activities that lead to peak UFP and PM2.5 exposures. This novel research approach will provide insights into distinct and joint effects of varying size fractions of PM on lung function changes.

661 – Investigating the impact of household air pollution on COVID-19 in the Household Air Pollution Intervention Network (HAPIN) Trial

Lindsay J. Underhill[1], Dana B. Barr[2], Jennifer L. Peel[3], Maggie L. Clark[3], Thomas Clasen[2], Ajay Pillarisetti[4], Yunyun Chen[2], Bonnie N. Young[3], Kalpana Balakrishnan[5], Michael Johnson[6], William Checkley[7], Kyle Steenland[2]
[1]Washington University School of Medicine, St. Louis, MO, USA. [2]Rollins School of Public Health, Emory University, Atlanta, GA, USA. [3]Colorado State University, Fort Collins, CO, USA. [4]University of California, Berkeley, CA, USA. [5]Sri Ramachandra Institute for Higher Education and Research, Chennai, India. [6]Berkeley Air Monitoring Group, Berkeley, CA, USA. [7]Johns Hopkins School of Medicine, Baltimore, MA, USA

Abstract
Objective: While substantial evidence exists to support an association between COVID-19 and ambient air pollution exposure, no study to date has examined the relationship between COVID-19 and household air pollution (HAP) from solid fuel cookstove combustion. Additionally, since existing studies have been observational, causality assessments have been limited. To address these gaps, the current study investigates the impact of HAP exposure on COVID-19 infection in the Household Air Pollution Intervention Network (HAPIN) Trial, a randomized controlled trial of a liquefied petroleum gas (LPG) stove intervention in Guatemala, India, Peru, and Rwanda.

Methods: In the HAPIN trial, 3200 households were randomized (1:1) to an LPG stove intervention or control group. Personal HAP exposure and dried blood spot (DBS) sampling was conducted in pregnant women (<35 years), their resulting infants (<1 year), and other adult women (>40 years). In a subset of 1576 households, immunoglobin G (IgG) SARS-CoV2-specific antibodies were measured in DBS.

Results: Preliminary results demonstrate a SARS-CoV-2 antibody prevalence of 7.1% in pregnant women (n=338), 8.1% in children (n=1038), and 6.8% in other women participants (n=118), with an overall prevalence of 8.0%. SARS-CoV-2 antibody prevalence was 8.8% in controls (n=742) and 6.7% in intervention participants (n=752).

Conclusion: This study addresses a critical gap in the literature regarding HAP exposure and COVID-19. Overall, COVID-19 infection prevalence was higher than expected and varied by participant and intervention groups. Ongoing intention-to-treat and exposure-response analyses may provide further insight into the potential impact of the LPG stove intervention and HAP exposure on COVID-19 infections.

662 – Assessment of particulate air pollution and toxicity due to festive firework in Mumbai, India

Nidhi Prajapati[1], Sonali Borse[2], Atique Barudgar[3], Vinayak Sahota[2], Shreya Dubey[4], Harish Phuleria[5]
[1]Indian Institute of Technology Bombay, Research Intern, Mumbai, India. [2]Indian Institute of Technology Bombay, Project Assistant, Mumbai, India. [3]Indian Institute of Technology Bombay, Junior Research Fellow, Mumbai, India. [4]Indian Institute of Technology Bombay, PhD Scholar, Mumbai, India. [5]Indian Institute of Technology Bombay, Associate Professor, Mumbai, India

Abstract
Extensive use of fireworks during festivals and events, deteriorates air quality and potentially causes adverse health impacts. The study presents a comprehensive approach to studying the changes in aerosol's optical and chemical properties, along with their oxidative potential (OP), as a result of the use of the fireworks during the Diwali festival period in India. $PM_{2.5}$ was measured at ten urban residential sites across Mumbai, India during Diwali, a festival of lights, in October 2022, for 12 hours during the day and night for 5 days, each. Water-extracted PM samples were subjected to two acellular assays (OP_{AA} and OP_{DTT}), coupled with an evaluation of several PM2.5 chemical components. Mean Diwali night-time $PM_{2.5}$ increased by >200% and >300% over daytime and pre-Diwali night-time levels, respectively, exceeding national regulatory standard (60µg/m3) by 6-7 folds. Black carbon (BC) levels ranged from 2.9 to 65.9µg/m3, whereas, OP_{DTT} and OP_{AA} levels were highest in Santacruz (12.2 nmol min-1m-3) and Borivali (5.5 nmol min-1m-3) neighborhoods. OP_{DTT} and OP_{AA} varied significantly, with the highest toxicity obtained during Diwali night ($p<0.05$), likely driven by the redox-active metals, key ingredients of the fireworks for the color and glitter. $PM_{2.5}$, BC, and OP spike on Diwali night, strongly implicate the toxicity induced by particles from fireworks. Furthermore, the weak correlation between $PM_{2.5}$ and OP suggests that PM mass concentration alone may not be sufficient to safeguard public health. This study emphasizes the importance of effective regulation to address short-term exposure to $PM_{2.5}$ from such episodic events.

663 – Estimating PM_{10} Exposures Using Moderate-Cost PM Sensors: Gravimetric Validation and Spatial Variability in Two US Fenceline Communities as part of the HAP-MAP Study

Mina Tehrani[1], Shivang Agarwal[2], Ana Rule[3], Kirsten Koehler[4], Peter DeCarlo[4]
[1]Johns Hopkins University Department of Environmental Health and Engineering, Postdoctoral Scientist, Baltimore, USA. [2]Johns Hopkins University Department of Environmental Health and Engineering, PhD Student, Baltimore, USA. [3]Johns Hopkins University Department of Environmental Health and Engineering, Assistant Professor, Baltimore, USA. [4]Johns Hopkins University Department of Environmental Health and Engineering, Associate Professor, Baltimore, USA

Abstract
Particulate matter (PM) exposure is associated with many adverse, often particle-size dependent, health effects. While focus has traditionally been on $PM_{2.5}$, PM_{10} toxicity and spatial variability are increasingly recognized, and make accurate measurements at fine spatial scales necessary to obtain reliable exposure and risk estimates. Advances in technology have resulted in the proliferation of low-cost PM sensors able to provide valuable $PM_{2.5}$ data; however, most of these technologies suffer from poor accuracy in PM_{10} measurement.

In this study, gravimetric validation of novel moderate-cost PM sensors (Quant-AQ MODULAIR-PM), which combine nephelometry with optical particle counting (OPC), was conducted by comparison with Micro-Orifice Uniform Deposit Impactor (MOUDI) measurements. Spatial variability and emission sources of PM_{10} in two urban-industrial areas in the US, Chester, PA and an industrial corridor of Louisiana, were investigated.

Strong linear associations between gravimetric MOUDI impactor measurements and MODULAIR-PM sensors were observed for $PM_{2.5}$ (slope: 1.08, χ^2: 51) as well as PM_{10} (1.17, 7.9), suggesting the MODULAIR-PM is accurate in measurements of both PM size fractions. In Chester, PA, measurements from 5 sensors demonstrated higher spatial variability for PM_{10} than for $PM_{2.5}$. In industrial areas of Louisiana,

PM_{10} 24-hour average concentrations routinely exceeded the National Ambient Air Quality Standard limit of 150 µg m^{-3}. Wind direction and speed analysis provides an indication of likely sources of the PM_{10}. These measurements suggest that additional monitoring of PM_{10} is warranted to minimize exposures and protect health.

664 – Wastewater-based Epidemiology to Identify Susceptibility to Neurodegenerative Diseases within Populations through Regional Analysis of Environmental Toxin Occurrence

Melanie Engstrom Newell[1], Ayesha Babbrah[2], Anumitha Aravindan[2], Alexandria Ellershaw[2], Sharvaree Kulkarni[3], Raj Rathnam[2], Erin Driver[4], Devin Bowes[5], Katherine Henke[5], Rolf Halden[6]
[1]Arizona State University, Graduate Research Assistant, Tempe, USA. [2]Arizona State University, Undergraduate Student Worker, Tempe, USA. [3]Arizona State University, High School Intern, Tempe, USA. [4]Arizona State University, Research Scientist, Tempe, USA. [5]Arizona State University, PostDoc, Tempe, USA. [6]Arizona State University, Tenured Professor, Tempe, USA

Abstract
Human exposure to environmental toxins has been associated with neurodegenerative diseases (NDDs). The pesticide paraquat has been associated with Alzheimer's Disease (AD) and Parkinson's Disease (PD), the herbicide glyphosate with PD, and the mycotoxin β-N-methylamino-L-alanine (BMAA) with amyotrophic lateral sclerosis (ALS). Current diagnostic tools for identifying NDDs at the individual level are associated with significant delays (18 months for AD, 6 months for PD, 3 years for ALS), highlighting a need for alternative strategies. Wastewater-based epidemiology (WBE) aims to quantify drivers of disease at the population level by measuring markers indicative of human health/exposure in wastewater and was utilized to identify susceptibility to NDDs within communities. Samples of untreated, raw wastewater (n = 484) were collected between December 2021 to November 2022 from wastewater treatment plants in regions with high and low prevalence rates of NDDs within the United States. Identification and quantification of paraquat, AMPA, and BMAA were optimized using liquid chromatography tandem mass spectrometry and subsequently mapped with published NDD prevalence rates. Resultant maps and associations allowed us to test the hypothesis that wastewater concentrations of paraquat, AMPA, and BMAA are significantly higher in highly prevalent regions for NDDs as compared to wastewater from low NDD prevalent regions (Spearman's rho >/= 0.60). This work expands the use of WBE to quantify environmental toxins serving as markers of NDDs to create a non-invasive and inclusive screening tool for identifying populations in need of preventative measures to improve the lives of future patients and their caregivers.

665 – POSTER: Non-targeted Analysis using Human Liver Microsomes to Explore Metabolites of Isothiazolinone Preservatives

Hye-Ri Jeon[1], Sunghoon Yeo[2], Ye Eun Seo[3], Youhglim Kho[1]
[1]Department of Health, Environment & Safety, Eulji University, Student, Seongnam-si, Korea, Republic of. [2]SCIEX Korea, application specialist, Seoul, Korea, Republic of. [3]Department of Food Technology & Service, Eulji University, Student, Seongnam-si, Korea, Republic of

Abstract
Isothiazolinone preservatives are widely used as preservatives in various household, personal care, and consumer products. However, these isothiazolinone preservatives can have adverse effects on human health, such as skin irritation, allergies, and inhalation toxicity. Therefore, it is necessary to conduct a risk assessment through biomonitoring to evaluate the potential harmful effects on human health. However, appropriate biomarkers for estimating human exposure have not been identified, and there is a lack of relevant research on these compounds. Therefore, in this study, isothiazolinones were exposed to human liver microsome (HLM), and the metabolites were investigated through non-targeted analysis. The target compounds were Methylchloroisothiazolinone/ methylisothiazolinone (CMIT/MIT), Benzisothiazolinone (BIT), Octylisothiazolinone (OIT), and Dichlorooctylisothiazolinone (DCOIT), and non-targeted analysis was performed using LC-Q-ToF by suspect-screening analysis. In this study, we identified primary metabolites of the target compounds using HLM. The obtained metabolites were characterized by comparing their spectra and chemical structures. Malonamic acid derivatives were observed in all compounds except BIT. This provided information on the metabolic products of isothiazolinone preservatives. Based on these results, we can identify biomarkers of isothiazolinone preservatives for human biomonitoring and developing analytical methods.

Acknowledgement: This work was supported by Korea Environment Industry & Technology Institute (KEITI) through Technology Development Project for Safety Management of Household Chemical Products Program (or Project), funded by Korea Ministry of Environment (MOE) (2022002970005).

666 – The Informatics of the Autoimmune Exposome

Ram Gouripeddi[1,2,3], Sejal Mistry[3,1], Mollie Cummins[1,2,4,3], Naomi Schlesinger[5], Katherine Sward[1,2,3,4], Julio Facelli[1,2,3]
[1]Center of Excellence for Exposure Health Informatics, University of Utah, Salt Lake City, USA. [2]Clinical and Translational Science Institute, University of Utah, Salt Lake City, USA. [3]Department of Biomedical Informatics, University of Utah, Salt Lake City, USA. [4]College of Nursing, University of Utah, Salt Lake City, USA. [5]Division of Rheumatology, Department of Internal Medicine, University of Utah, Salt Lake City, USA

Abstract
About 9% of the world's population is affected by autoimmune diseases, and it now understand that that the exposome plays a significantly larger role than genetics in their pathophysiology. An individual's lifetime exposure to external and internal factors, including food, common products, the air, and even infections. Recent work from our group has shown the role of such exposures in the development of type 1 diabetes mellitus. Exposures not only directly affect biological pathways causing the release of autoantigens and formation of neoantigens, but also have mutagenic and epigenetic mechanistic influences on disease and well-being. Interplay among exposures, pathogens, and host immune systems can be complex. Exposures and pathogens can directly, and in combination alter the state of the host to a diseased state. Pathogens can alter exposure interactions with a host immune system, and exposures can alter interactions between pathogen and host immune system. Finally, the altered state of the host immune system due to exposures or pathogens can increase the host's susceptibility to pathogens and exposures, respectively. In other words, data representing immunome, the microbiome and the exposome need to account for not only for sequential but also for concurrent occurrences. In this presentation, we lay out the informatics and temporal machine learning implications for understanding the intersection of the exposome – microbiome – immunome, and discuss its implications for autoimmune disease research including type 1 diabetes, rheumatoid arthritis and gout.

667 – Distinguishing pollutant sources in indoor and outdoor settings using low-cost black carbon (BC), particulate matter (PM) and gaseous sensors

Tian Xia[1], Jeff Gearhart[2], Stuart Batterman[3]
[1]University of Michigan, Postdoc, Ann Arbor, USA. [2]The Ecology Center, Research Director, Ann Arbor, USA. [3]University of Michigan, Professor, Ann Arbor, USA

Abstract
Low-cost sensors for particulate matter (PM) and NO_2 are increasingly deployed to supplement the regulatory air quality monitoring networks operated by states and the US EPA. These sensors provide consistent and very localized measurements, but most do not provide the speciation or other measurements needed or helpful to identify pollutant sources. Further, the performance of low-cost sensors can vary significantly from federal reference methods (FRM) used by EPA, and has not been well documented across a broad set of environmental conditions. Our objectives in this study were to compare performance of selected low-cost black carbon (BC) and PM sensors and use their data to distinguish pollutant sources in indoor and outdoor settings in Detroit, Michigan at moderately polluted urban/industrial areas. Six sensor systems that measure PM, NO_2, black carbon (BC), and other pollutants, namely PurpleAir (PA), Clarity Node-S (CN), TSI BlueSky Air Quality Monitor 8145 (BS), Met One Aerocet-531S (AC), DSTech ObservAir (OA), and microAeth MA200 (MA), were examined. BS, OA and MA were collocated with FRM measurements for a month. Then OA, PA and CN were collocated for outdoor road-side measurements, and AC, OA and MA were collocated for indoor air quality (IAQ) applications, specifically to examine the penetration of outdoor pollutants into school classrooms, and the ability to assess changes in IAQ due to filter interventions. The results showed acceptable BC and $PM_{2.5}$ agreements between sensors after calibration (r^2 ranged 0.52 – 0.85), while PM_{10} shows poorer correlation. Advantages, disadvantages, and precautions for using each sensor were discussed.

668 – Application of Statistical Methods for Mixtures: Examples from the Powering Research through Innovative Methods for mixtures in Epidemiology (PRIME) program

Bonnie Joubert[1], Mary Turyk[2], Chris Gennings[3], Huan Yun Chen[2], Marie Lynn Miranda[4], Thomas Webster[5], Katherine Ensor[6], David Dunson[7], Brent Coull[8], Marianthi-Anna Kioumourtzoglou[9]
[1]National Institute of Environmental Health Sciences, Durham, USA. [2]University of Illinois Chicago, Chicago, USA. [3]Icahn School of Medicine at Mount Sinai, New York, USA. [4]University of Notre Dame, South Bend, USA. [5]Boston University School of Public Health, Boston, USA. [6]Rice University, Houston, USA. [7]Duke University, Durham, USA. [8]Harvard T.H. Chan School of Public Health, Boston, USA. [9]Columbia University Mailman School of Public Health, New York, USA

Abstract

Human exposure to a complex and changing mixture of environmental chemicals can present analytical challenges, but a variety of statistical methods for mixtures now exist. This includes over 37 methods from the NIEHS Powering Research through Innovative Methods for mixtures in Epidemiology (PRIME) program and other approaches. However, the application of mixtures methods to complex exposomics data has not been extensively tested. We leveraged the well-known NHANES telomere dataset (n=1003, 18 chemicals) and a small publicly accessible exposomics dataset from the NIEHS Human Health Exposure Analysis Resource (HHEAR, n=667, >50 chemicals) to apply statistical methods from PRIME and other mixtures methods. We selected methods to assess overall effect estimation and/or toxic agent identification (variable selection), with R packages in a public repository. We excluded methods designed for time-varying exposures/outcomes, causal estimation, pattern identification, prediction, or assessment of interactions, leaving the following: Bayesian Multiple Index Model (BMIM), Repeated holdout Weighted Quantile Sum regression (RH-WQS), Total Explained Variation (TEV), Bayesian variable selection for understanding mixtures in environmental exposures (BSSVM), and Computational improvements for Bayesian multivariate regression models based on latent meshed Gaussian processes (GriPS). We considered the unique design, context of application, and value added from each method, and how diverse methods could be considered systematically. Follow-up work may address research questions such as pattern identification or assessment of interactions, consider concepts of time, causal estimation, and application in larger complex datasets. Overall, our assessment contributes to a consensus understanding of how mixtures of complex exposures impact human health.

669 – POSTER: Prenatal exposure to ambient pesticides and gestational diabetes mellitus risk in a population-based study in Arizona

Kimberly Parra[1], Leslie Farland[2], Robin Harris[2], Melissa Furlong[2]
[1]University of Arizona, Graduate Research Associate, Tucson, USA. [2]University of Arizona, Professor, Tucson, USA

Abstract
Introduction: Organophosphates (OP), pyrethroids (PYR), and carbamates (CAR) are agricultural pesticides with glucotoxic properties. There are limited studies examining effects of these pesticides during pregnancy on gestational diabetes mellitus (GDM). We assessed associations between prenatal exposures to individual OP/PYR/CAR pesticides and GDM.

Methods: From the AzPEARS (Arizona Prenatal Environment And Reproductive outcomes Study) birth certificate cohort of all Arizona births, we included individuals aged 14-49 years, having a singleton, live-birth from 2014-2020 (n=476,852). Maternal residential addresses from birth certificates were linked to the state's Pesticide Use Registry. Pesticide exposure was dichotomized as ever exposed within 500 meters of the home residence for each trimester and pesticide, and GDM was derived from birth certificates. Adjusted odds ratios (aORs) and 95% confidence intervals (CIs) were estimated using multivariable logistic regression in single pollutant models, adjusting for maternal age, race/ethnicity, education, conception year, and birth season.

Results: The prevalence of GDM was 7.8%, consistent with nationwide estimates. We observed positive associations with GDM for several pyrethroids during early pregnancy, including bifenthrin, cyfluthrin, esfenvalerate, and permethrin, ranging from aOR=1.15 to 2.39. Any OP exposure during the third trimester increased risk by 10% (95%CI: 1.00, 1.21), with positive associations for the specific OPs bensulide, chlorpyrifos, tribuphos, and ethephon. We found no significant associations with exposures to CAR.

Conclusion: Prenatal exposures to some PYRs early in pregnancy and OPs in the third trimester were associated with higher odds of GDM. Pesticide exposures during pregnancy may be associated with adverse maternal health outcomes.

670 – POSTER: Development of toolkits for wildfire smoke response in schools and childcare facilities using low-cost sensors

Orly Stampfer[1,2], Julie Fox[3], Ryan W. Allen[4], Stephanie Farquhar[5], Christopher Zuidema[6], Edmund Seto[7], Catherine J. Karr[5,8]
[1]University of Washington, Doctoral student, Seattle, USA. [2]Washington State Department of Health, Indoor Air Quality Epidemiologist, Olympia, USA. [3]Washington State Department of Health, Air Quality Supervisor, Olympia, USA. [4]Simon Fraser University, Professor, Burnaby, Canada. [5]University of Washington, Professor, Seattle, USA. [6]University of Washington, Clinical Assistant Professor, Seattle, USA. [7]University of Washington, Associate Professor, Seattle, USA. [8]Northwest Pediatric Environmental Health Specialty Unit, Director, Seattle, USA

Abstract
During wildfire smoke, staff in schools and childcare facilities need data to mitigate health risks to children. The Washington State Department of Health (DOH) partnered with Local Health Jurisdiction (LHJ) staff to develop protocols based on data collected at 4 childcare facilities with stationary outdoor, stationary indoor, and handheld low-cost PM2.5 sensors, capturing a wildfire smoke period at

two of the facilities. UW researchers developed toolkits using PM2.5 data collected with low-cost sensors and co-located gravimetric samplers at four schools during wildfire smoke. UW used this data to assess sensor accuracy, and to characterize within-school variation in PM2.5. UW used random sampling of the data to simulate "grab sampling" using a handheld sensor. To address feasibility and practical use of the toolkits, UW conducted interviews with 15 staff from schools, a childcare facility, LHJs, school districts, and air quality agencies. UW found that sensor averages were 1.0-2.1x higher than gravimetric concentrations. Using sensor data corrected with the gravimetric data, the greatest within-school difference in concentration means during school hours during wildfire smoke was 57 μg/m3. 14 of 15 interviewees were interested in using low-cost sensors to support decision-making despite potential data quality issues. We will present how sensor data and qualitative data supported toolkit development in these two projects, and discuss recommendations for future use.

672 – Feedback from Minnesota Families on Report-back Methods for their Child's Biomonitoring Results

Jessica Nelson[1], Suad Salad[2], Iftu Hunte[2], Sheila Amenumey[3], Carin Huset[3]
[1]Minnesota Department of Health, Epidemiologist and Program Director, St. Paul, USA. [2]Minneapolis Public Schools, Minneapolis, USA. [3]Minnesota Department of Health, St. Paul, USA

Abstract
As the use of biomonitoring expands in public health practice, an essential question is how best to share individual results with participants in an accessible, informative, and culturally appropriate way.

The Minnesota Department of Health's Healthy Kids Minnesota program is assessing environmental exposures in preschool-age children across the state. Recruitment is conducted by partners at Early Childhood Screening Programs in school districts, local public health, and tribal nations. Families who agree to participate complete an interview and help their child give a urine sample. Urine samples are analyzed for metals, pesticides, environmental phenols, phthalates, flame retardants, and air pollution chemicals. Results are provided to families in three mailings. Families from the first program cycle (n=453) received the first mailing of their child's results in July 2022.

To assess the effectiveness of our report-back approach and learn about ways it could be improved, MDH worked with recruitment partner Minneapolis Public Schools (MPS) to follow up with families about their experience receiving the mailing. In August 2022, MPS multicultural staff conducted structured telephone interviews with 37 families (58% of those contacted). Families who spoke languages other than English were prioritized for contact.

Results of semi-quantitative and qualitative analyses showed that, despite best practice plain language and language translation, our current approach is not working for many families. There were differences in families' understanding of their child's results based on preferred language and caregiver's educational attainment. We are using these findings to improve accessibility of upcoming results return materials.

673 – POSTER: Understanding the Relationship Between Environmental Exposures and the Risk of Pediatric Obesity using Unsupervised Machine Learning

Catherine Petersen[1,2], Sejal Mistry[3,2], Ramkiran Gouripeddi[4,5], Julio Facelli[1,6]
[1]University of Utah, Undergraduate Research Assistant, Salt Lake City, USA. [2]Center of Excellence for Exposure Health Informatics, Salt Lake City, USA. [3]University of Utah, Salt Lake City, USA. [4]University of Utah, Assistant Professor, Salt Lake City, USA. [5]Center of Excellence for Exposure Health Informatics, Assistant Professor, Salt Lake City, USA. [6]Center of Excellence for Exposure Health Informatics, Distinguished Professor, Salt Lake City, USA

Abstract
Exposures in prenatal and early life lead to development of disease phenotypic conditions. While general effects are well documented, details on how temporal patterns of exposures affect health are not well understood. Characterizing differential and additive effects of continuous low-level and intense sporadic multi-agent exposures require advanced big data and artificial intelligence methods. The challenge to generate and utilize imperfect exposure records spans three themes using unsupervised machine learning methods to generate exposome-health associations following multi-omics data integration. The objective of this research was to determine the role of transitions from prenatal to postnatal exposure profiles in pediatric obesity risk.

In order to generate multi-omics exposure records, we will utilize our existing methods with the Exposure Health Informatics Ecosystem, to generate events that are semantically coherent and available with spatiotemporal attributes at the highest available granularity.

Next, we will use longitudinal unsupervised K-means clustering (kShape) with shape respecting distances on these exposure trajectories. The clustering performance will be evaluated using the Silhouette score and the Calinski-Harabasz score. We will evaluate prenatal and postnatal exposure trajectories until the onset of specific phenotypic conditions. We will use principal component analysis to understand contributions of specific exposures and combinations of exposures in relationship to Chemical Entities of Biological Interest. Potential limitations are in the quality of the trajectories which could be addressed by other methods such as shapelet transformations.

674 – Urine Mercury Biomonitoring: Development of State Exposure Guidance, Intervention and Testing Protocols, and Case Examples

Jessica Nelson[1], Erin Batdorff[2], Michael Xiong[3], John Gilkeson[3], Lisa Strong[4], Sheila Amenumey[4]
[1]Minnesota Department of Health, Epidemiologist and Program Director, St. Paul, USA. [2]Tucson VA Medical Center, Tucson, USA. [3]Minnesota Pollution Control Agency, St. Paul, USA. [4]Minnesota Department of Health, St. Paul, USA

Abstract
The Minnesota Department of Health Biomonitoring Program, along with an inter-disciplinary team of partners, uses urine mercury testing to identify individuals exposed to mercury through the use of skin lightening products and offer interventions to reduce exposure. The presence of high levels of mercury in skin lightening products is a growing public health concern. These products, while illegal, are widely available locally and online. Because mercury is not listed as an ingredient, individuals are unaware they are at risk for exposure. Work is needed on many fronts to reduce this harmful practice, which has deep roots in colorism and the preference for lighter skin.

Through our experience with the Minnesota Family Environmental Exposure Tracking study, follow-up clinical screening projects, and involvement in other exposure cases, we have refined our approach to using urine mercury biomonitoring as a key strategy in this effort. Working jointly with the MDH Public Health Laboratory, Minnesota Poison Control, Minnesota Pollution Control Agency, and local public health agencies, we have developed: 1) exposure and health-based guidelines for interpreting urine mercury results to inform provider and public health responses, and 2) protocols for conducting home visits to assess mercury contamination and offer additional urine testing.

Using these tools, we have successfully responded to a number of elevated urine mercury cases resulting from the use of skin lightening products, working with our partners to share information with exposed individuals, remove contaminated products from homes, and conduct follow-up home visits as needed to be sure exposure has been reduced.

675 – POSTER: Characterization of Adult Mouth Surface Area and Bite Volume

Jacqueline Larson[1,2], Jenna Honan[1], Olusola Ogunseye[1], Helena Solo-Gabriele[3], Alesia Ferguson[4], Paloma Beamer[1]
[1]Mel and Enid Zuckerman College of Public Health, University of Arizona, Tucson, USA. [2]National Institutes of Health, Bethesda, USA. [3]University of Miami, Coral Gables, USA. [4]North Carolina Agricultural and Technical State University, Greensboro, USA

Abstract
A key parameter required by many non-dietary ingestion exposure and risk models for which there is limited data is the contact surface area during mouthing events. A total of 20 participants were recruited through flyers placed around a university campus and on social media. Participants took part in a survey to obtain socio-demographic information and in a series of trials that included putting on lipstick and placing their neutral lips on a piece of paper with a scale, reapplying the lipstick and kissing a piece of paper with pursed lips, placing lips around a straw, and biting into an apple. The surface area was calculated from the prints using ImageJ software. The neutral lip print measurements allowed us to reach an estimated total mouth surface area of the adult participants (mean 9.7 cm2 ± 2.2 cm2). The straws were cut along the length, unfolded, and secured to a paper before measuring the surface area of the lips. These straw measurements were compared to the full lip measurements to determine the percentage of the lips (11% on average) came into contact with the straw's surface. The volume of the apple bite was calculated using water displacement methods. Male participants had a 6.2% larger surface area (0.6 cm2) than female participants. In the volume displacement the male participants had a volume (19.3 cm3) nearly double the measurement of the females (10.8 cm3).

676 – POSTER: Exploring the chemical exposome among children and adults from India, South Africa, China, Switzerland and the US using wearable passive samplers

Elizabeth Lin[1], Jeremy Koelmel[2], Paul Stelben[3], Emily Johnson[4], Jonathan Chevrier[5], Song Tang[6], Maryam Aniya Khalili[7], Shelly Miller[8], Zeyan Liew[9], Wan-Ling Tseng[9], Maria Viviana Gonzalez Serrano[10], Dusan Licina[11], Nicholas Herkert[12], Heather Stapleton[13], Elizabeth Hatch[14], Amelia Wesselink[15], Lauren Wise[14], Krystal Pollitt[16]
[1]Yale University, PhD Candidate, New Haven, USA. [2]Yale University, Postdoctoral Associate, New Haven, USA. [3]Yale University, Student Researcher, New Haven, USA. [4]Yale University, PhD Student, New Haven, USA. [5]McGill University, Associate Professor, Montreal, Canada. [6]Chinese Center For Disease Control And Prevention, Associate Professor, Beijing, China. [7]University of Colorado Boulder, PhD Candidate, Boulder, USA. [8]University of Colorado Boulder, Professor, Boulder, USA. [9]Yale University, Assistant Professor, New Haven, USA. [10]École polytechnique fédérale de Lausanne, PhD Candidate, Lausanne, Switzerland. [11]École polytechnique fédérale de Lausanne, Assistant Professor, Lausanne, Switzerland. [12]Duke University, Postdoctoral Associate, Durham, USA. [13]Duke University, Professor, Durham, USA. [14]Boston University, Professor, Boston, USA. [15]Boston University, Research Assistant Professor, Boston, USA. [16]Yale University, Associate Professor, New Haven, USA

Abstract
We are exposed to a broad range of airborne environmental contaminants, including volatile organic compounds, polycyclic aromatic hydrocarbons (PAHs), pesticides, and phthalates, which can lead to an increased risk of adverse health outcomes. Wearable passive samplers are promising exposomic tools for understanding the compositional diversity of these complex exposure mixtures at a population scale. In this study, we used the Fresh Air wristband to assess personal exposures to environmental contaminants of children (<18 years) in rural areas of South Africa (n=151) and urban areas of the US (n=30) as well as adults in rural areas of India (n=5) and urban areas of China, Switzerland, and the US (n=378). Exposures were compared across geographic regions and age groups. Using targeted analysis, 68 contaminants spanning 16 chemical groups, including phthalates, pesticides, PAHs, and flame retardants, were identified. Non-targeted analysis of the Fresh Air wristbands further revealed personal exposure to elevated levels of distinct pesticides, such as dichlorvos, DDT, and DEET, respectively detected in participants from urban centers of China, rural South Africa, and the US. Cumene, a natural component of coal tar and petroleum, was found in nearly all cohorts. US children were exposed to tris(1-Chloro-2-propyl) phosphate (TCPP) at least 27 times more than fluorene, while South African children's exposure to TCPP was only 3 times higher than fluorene. Findings provide insights into the chemical exposome across the life course in different geographies, providing a valuable resource for assessing the associated impact on health and risk assessment.

678 – POSTER: Utilizing a Novel Method to Explore the Impact of an Active Green Wall on Airborne Biological and Chemical Contaminants

Phoebe Mankiewicz[1], Elizabeth Lin[2], Chandrima Bhattacharya[3], Elizabeth Hénaff[4], Krystal J. Godri Pollitt[5], Anna Dyson[6]
[1]Yale CEA, PhD Candidate, New Haven, USA. [2]Yale School of Public Health, PhD Candidate, New Haven, USA. [3]Weill Cornell Medicine, PhD Student, NYC, USA. [4]NYU Tandon School of Engineering, Assistant Professor, NYC, USA. [5]Yale School of Public Health, Associate Professor, New Haven, USA. [6]Yale CEA, Hines Professor, New Haven, USA

Abstract
Urban inhabitants spend upwards of 90% of their time indoors where relationships between airborne chemical contaminants and microbiome dynamics have emerged as significant contributors to poor indoor environmental quality (IEQ) and human health outcomes. Active green infrastructure designed to improve IEQ by pulling indoor air through plants and root-associated microbial communities have been studied for over 50 years. Current hypotheses posit that leaf photosynthesis alongside microbial metabolisms could simultaneously reduce human exposure to indoor CO_2 and airborne chemicals, as well as diversify human microbiomes, each of which have been linked to measurable health benefits. Although laboratory results are often promising, few field studies have been done, and many metrics, such as the use of total volatile organic compounds, are problematic due to both airborne chemical dynamics and potential interactions with biological compounds associated with plant-based ecosystems. In addition, indoor microbiome studies often experience challenges in collecting sufficient biological material to reliably conduct DNA sequencing analysis. Here we present a pilot study utilizing a novel data collection protocol using two forms of passive samplers that can be subsequently processed using paired metagenomic and chemical analyses, designed to address these challenges. This process was deployed in a typical New York City office containing an active green wall to determine how the green wall may simultaneously influence microbial and chemical components of the airstream. The results suggest that this method could make significant contributions towards more completely understanding not only active green infrastructure impacts on IEQ, but indoor environmental exposures more broadly.

679 – Exposures to Pesticides in an Occupational Chilean Cohort

Melissa DeSantiago[1], Lesliam Quirós-Alcalá[2], Christian Lindh[3], Catterina Ferreccio[4], Sandra Cortés A[5]
[1]Johns Hopkins University, PhD Candidate, Baltimore, USA. [2]Johns Hopkins University, Assistant Professor, Baltimore, USA. [3]Lund University,

Associate Professor, Lund, Sweden. [4]Pontificia Universidad Católica de Chile, Professor, Santiago, Chile. [5]Pontificia Universidad Católica de Chile, Associate Professor, Santiago, Chile

Abstract
The intensification of agriculture in Chile has increased potential exposure to agrochemicals, including pesticides linked to a wide array of adverse health effects. The Maule region is one of Chile's leading agricultural regions with the 2ndlargest sales of pesticides in the country. However, biomonitoring efforts in Latin America, including Chile, are scant, limiting the evaluation of the potential role that pesticides may play in the etiology of chronic diseases. We conducted a study to assess exposure to pesticides commonly and historically used in Maule. Spot urine samples were collected from a subset of adults participating in the ongoing prospective Maule Cohort study (n=470) in Chile. Seventeen pesticide biomarkers, including organophosphates, pyrethroids, and phenoxyacetic acids were quantified in urine samples using LC-MS/MS. Detection frequencies for pesticide biomarkers in urine samples were between 32-100%. Biomarkers of chlorpyrifos (p50=1.79 mg/L), profenofos (p50=0.084 mg/L), 3-phenoxybenzoic acid (p50=0.78 mg/L), 2,4-dichlorophenoxyacetic acid (p50=0.013 mg/L), 2,4-Dichlorophenol (p50=2.38 mg/L), 4-hydroxy-pyrimethanil (p50=0.60 mg/L), and hydroxy tebuconazole (p50=0.31mg/L) were detected in >70% of the study population. Median TCPy concentrations, chlorpyrifos biomarker, were almost twice as high as adults reported in 2009-2010 from the U.S. general population (0.97 mg/L) prior to U.S. restrictions on chlorpyrifos. For the most frequently detected biomarkers, concentrations were generally higher in female versus male participants. Analyses are ongoing and will include further characterization and interrogation of patterns of exposure to pesticides among participants and evaluation of exposure determinants to identify modifiable risk factors in this vulnerable population.

680 – POSTER: Associations between Individual Exposure to Fine Particulate MatterElemental Constituent Mixtures and Blood Lipid Profiles: A PanelStudy in Chinese People Aged 60-69 Years

Jiao Nan, Tiantian Li, Jianlong Fang, Song Tang, Xiaoming Shi
National Institute of Environmental Health, Chinese Center for Disease Control and Prevention, Beijing, China

Abstract
Dyslipidemia may be a potential mechanism linking Fine Particulate Matter (PM2.5) to adverse cardiovascular outcomes. However, the associations between PM2.5 and blood lipid have been inconsistent and the joint effect of PM2.5 elemental constituent on lipid profiles and the role of each constituent are still unclear. We aims to investigate the overall associations of PM2.5 elemental constituent with lipid parameters, and assess the relative importance of each constituent for lipid parameters. A panel study that involved sixty-nine older individuals with five repeated measurements (237 blood samples in total) was performed. Each participant wore a PM2.5 monitor for 72 hours before each health examination, including a questionnaire survey, physical examination, and biological sample collection. Bayesian kernel machine regression (BKMR) model was implemented to estimate the joint effect of the seventeen elemental constituent on various lipid parameters. We found that significant increases in total cholesterol (TC), lowdensity lipoprotein cholesterol (LDL-C), and non-high-density lipoprotein cholesterol (non-HDL-C) levels in older adults when exposed to the mixture of PM2.5 elemental constituents. Copper and titanium had higher posterior inclusion probabilities than the other constituents ranging from 0.76 to 0.90 and 0.74 to 0.94. This study highlights the importance of identifying critical hazardous PM2.5 constituents that may cause adverse cardiovascular outcomes in the future.

681 – PFAS Associations with Thyroxine-Binding Globulin (TBG) and Thyroxine-Bound TBG

Mary Turyk[1], Robert Sargis[2], Meghan Cerpa[3], Saria Awadalla[4]
[1]University of Illinois at Chicago School of Public Health, Professor, Chicago, USA. [2]University of Illinois at Chicago, Associate Professor, Chicago, USA. [3]University of Illinois at Chicago School of Public Health, Doctoral Student, Chicago, USA. [4]University of Illinois at Chicago School of Public Health, Assistant Professor, Chicago, USA

Abstract
PFAS are persistent pollutants that circulate bound to proteins and can disrupt thyroid homeostasis. Most circulating thyroxine (T4) is carried in the blood by thyroxine-binding globulin (TBG), with a smaller portion more loosely bound to albumin and transthyretin, and a very small percentage free. We measured total and free T4 (TT4/FT4), TBG, %TBG-bound TT4 (TT4-TBG) and %albumin-bound (TT4-ALB) and MeFOSAA, EtFOSAA, PFHxS, PFNA, PFDA, PFUnDA, Sm-PFOS, n-PFOS, n-PFOA, PFHpS and FOSA in 468 participants from the Great Lakes Fish Consumer Study. Linear regression models, adjusting for age and BMI, were used to evaluate associations between PFAS and thyroid parameters. Sex-specific estimates were also evaluated due to the influence of sex steroid hormones on TBG synthesis in the liver. Associations of PFAS with TBG, TT4-TBG, and TT4-ALB were stronger in men than women. In men, MeFOSA, FOSA and PFOA were negatively associated with TBG; while FOSA, PFOA, PFDA and PFUnDA were negatively associated with TT4-TBG and positively associated

with TT4-ALB. In women, PFDA was negatively associated with TT4-TBG and positively associated with TT4-ALB, but PFAS were not associated with TBG. PFOA was negatively associated with FT4 in men. PFAS were not associated with TT4. These findings suggest that exposure to some PFAS may decrease concentrations of TBG and competitively bind to TBG, thereby disrupting the levels of bound and free thyroid hormones in the blood and potentially altering cellular availability and the half-life of thyroxine. Further work is needed to better understand the origins of these disruptions and their biological impact.

682 – Get the Lead Out: Electrochemical Sensor for Blood Lead Level Detection in Children

Elena Boselli[1], Zhizhen Wu[2], Erin Haynes[3], Ian Papautsky[4]
[1]University of Illinois Chicago, PhD student, Chicago, USA. [2]University of Illinois Chicago, Postdoctoral researcher, Chicago, USA. [3]University of Kentucky, Faculty, Lexington, USA. [4]University of Illinois Chicago, Faculty, Chicago, USA

Abstract
Lead (Pb) has long been recognized as systemic toxicant for humans. However, there is growing concern regarding the adverse health effects at levels of exposure once considered safe. Even at low levels of environmental Pb exposure, children are particularly vulnerable to neurophysiological, neurobehavioral, and cognitive impairments. To assess Pb in the human body, blood Pb level (PbB) is regarded as a reliable biomarker, with a safety limit of 5μg/dL in children. The onset of symptoms resulting from Pb overexposure often occurs at a later stage, leading to long-term adverse effects. Hence, it is of paramount importance to frequently monitor PbB. Gold standard techniques for PbB detection at trace levels rely on complex equipment, complicated sample preparation, and highly trained personnel, whereas electrochemical techniques offer a viable alternative for point-of-care (POC) PbB detection. This study seeks to validate an electrochemical system based on square wave stripping voltammetry (SWASV) for POC PbB detection. Advantages include simple sample preparation (6x dilution of whole blood), small sample volume of few blood drops (200μL), easy operation, and short time to results (1hr). To demonstrate the practical utility of our sensor system, we evaluated its performance in the analysis of blood samples collected from children in the Southside Chicago. Compared to the gold standard of inductively coupled plasma mass spectrometry (ICP-MS), our system achieved average accuracy around 77% (n=21). We envision our approach as a valuable tool for more frequent monitoring of PbB, especially in communities where access to laboratory testing is impractical or cost prohibitive.

683 – Physician Education on the Health Concerns of PFAS - Challenges and Successes in Engagement

Elizabeth Friedman[1,2,3], Sherri Homan[4], Nicholas Newman[5,6,7], Susan Buchanan[8,9]
[1]Mid-America Pediatric Environmental Health Specialty Unit (MAPEHSU), Kansas City, MO, USA. [2]Children's Mercy, Kansas City, Physician, Kansas City, USA. [3]University of Missouri, Kansas City Medical Center, Kansas City, USA. [4]Mid-America Pediatric Environmental Health Specialty Unit (MAPEHSU), Columbia, MO, USA. [5]University of Cincinnati Departments of Pediatrics and Environmental & Public Health Sciences, Cincinnati, OH, USA. [6]Cincinnati Children's Hospital Medical Center, Cincinnati, OH, USA. [7]Great Lakes Pediatric Environmental Health Specialty Unit, Cincinnati, OH, USA. [8]University of Illinois in Chicago, Chicago, USA. [9]Great Lakes Pediatric Environmental Health Specialty Unit, Chicago, USA

Abstract
Background: The Pediatric Environmental Health Specialty Units (PEHSUs) provide education and consultation regarding reproductive and children's health through integration of environmental health into clinical care and public health practice. Per- and polyfluoroalkyl substances (PFAS) are a family of over 9000 chemicals that are persistent and bioaccumulative, and there are concerns regarding long-term health effects. The PEHSUs have been charged with educating health professionals serving PFAS contaminated communities about the health effects of PFAS exposure and recommendations for clinical care for highly exposed patients.

Methods: The MAPEHSU partnered with several PFAS-REACH partners to develop one CME video. The video was posted on the Children's Mercy, Kansas City CME portal which required registration for viewers to obtain CME credit, as well as the Silent Spring Youtube channel.

The Great Lakes PEHSU developed two educational/promotional videos and a CME video which was available through the University of Cincinnati CME portal which required registration. All CME videos were made available free of charge.

Results: The Children's Mercy hosted video had 35 viewers; 22 received CE credits. The Silent Spring hosted video had 345 views. The University of Cincinnati hosted video had 39 viewers; all received CE credit. 36 viewers were recruited through the promotional videos which had 738 views in total.

Discussion (lessons learned): MAPEHSU received several complaints that accessing the video through the CME portal was cumbersome which resulted in Silent Spring's hosting of the video. The release of YouTube videos promoting a PFAS CME activity helped increase physician participation.

684 – Worrisome Water: Investigating Wells & Wellness in Rural Eastern NC

Courtney Woods[1], Carol Babyak[2], Shea Tuberty[3], Danielle Koonce[4,5], Sherri White Williamson[6], Rebecca Witter[2], Abby Hrabosky[7], Dana Powell[2]
[1]UNC Chapel Hill, Associate Professor, Chapel Hill, USA. [2]Appalachian State University, Associate Professor, Boone, USA. [3]Appalachian State University, Professor, Boone, USA. [4]University of Maryland, Graduate Student, College Park, USA. [5]Environmental Justice Community Action Network, Board Member, Clinton, USA. [6]Enviornmental Justice Community Action Network, Board Member, Clinton, USA. [7]Appalachian State University, Graduate Student, Boone, USA

Abstract
North Carolina ranks among the top states for the number of residents who rely on private wells for their drinking water supply. Yet, in rural communities across the state, a range of environmental and climate-related hazards pose a threat to well water quality. Our team, comprised of university researchers and environmental justice (EJ) organization leaders, established a free water quality testing campaign in Sampson County. The county is located in a flood-prone region of the state and hosts several industrial facilities, including for agricultural and biomass production and waste storage. In collaboration with local churches, the team has tested approximately 150 wells for fecal contaminants, heavy metals, nutrients and legacy and novel perfluoroalkyl substances (PFAS). Roughly 10% of the samples exceeded a maximum contaminant level for a heavy metal or nutrient, and many samples exceeded a secondary standard for iron, aluminum, or manganese. Approximately 30% of samples tested were close to or exceeded EPA's proposed maximum contaminant level (4 ppt) for PFOA or PFOS, with measurements as high as 1200 ppt. Few samples (~5%) tested positive for fecal coliform, with levels up to 146 MPN. In addition to these findings, this presentation will describe the co-occurrence of specific contaminants and proximity of the wells to known hazards.We will also discuss lessons learned from our community-engaged processes for collecting samples, reporting the findings back to residents and connecting residents to mitigation/treatment services.

685 – Measuring structural racism and its association with air pollution exposures

Christine Ekenga[1], Emily White[2]
[1]Emory University, Assistant Professor, Atlanta, USA. [2]Emory University, Research Assistant, Atlanta, USA

Abstract
Background: Hazardous air pollutants have been associated with a variety of adverse health outcomes among minoritized communities in the United States. Structural racism may play a role in elevated exposure to HAPs among these communities. The purpose of this study is to examine associations between county-level structural racism and exposure to air toxics in Georgia.

Methods: We obtained air toxics data from the Environmental Protection Agency's Air Toxics Screening Assessment and sociodemographic data from the American Community Survey. We utilized county data on education, employment, incarceration, economic status, political participation, home ownership, and residential segregation to create a composite county-level structural racism index. Relative Risks (RR) were estimated for associations between neighborhood characteristics and high (top 10%) exposure to air toxics. We explored effect modification by county-level structural racism. We also interviewed residents from 5 Georgia counties.

Results: Neighborhoods with the highest quartile of African American residents had elevated exposures to air toxics when compared to neighborhoods with the lowest quartiles of African American residents (RR = 2.08, 95% CI = 1.31-3.30). This association was attenuated in counties with high levels of structural racism. Qualitative interviews revealed air quality perceptions.

Conclusion: Our results suggest that structural racism is a potential pathway through which neighborhoods with high proportions of African Americans are exposed to air toxics. Study results may be utilized to inform both social policy and air pollution policy in Georgia.

686 – Applications in Wastewater-based Epidemiology

Erin Driver[1,2], Devin Bowes[3]

[1]Arizona State University, Tempe, USA. [2]University of Arizona, Tucson, USA. [3]Boston University, Boston, USA

Abstract
The use of wastewater-based epidemiology (WBE) has been proven to possess the ability to capture community-level human health, behavior, exposure, and activity through near real-time assessments. The field has rapidly evolved from almost exclusively measuring indicators of licit and illicit drug use across communities at an international-level, to now capturing SARS-CoV-2 at varying scales (building-, community-, city-level) in order to monitor and mitigate community spread. Over the past decade, some practitioners have proposed expanding WBE into novel realms of public health, however, few have reported on full-scale execution. The work presented herein is a tableau of novel and successful case studies conducted over the course of the past 5 years.

687 – Indoor Air Quality Findings from a Gas to Induction Pilot in New York City

Misbath Daouda[1], Annie Carforo[2], Heather Miller[3], Jennifer Ventrella[2,4], Yu Ann Tan[5], Michelle Feliciano[6], Jessica Tryner[7], Steven Chillrud[8], Roisin Commane[9,10], Andrew Hallward-Driemeier[10], Diana Hernández[11], Michael Johnson[3], Darby Jack[11]
[1]Mailman School of Public Health of Columbia University, Graduate Research Student, New York, USA. [2]WEACT for Environmental Justice, New York, USA. [3]Berkeley Air, Berkeley, USA. [4]10. Milano School of Policy, Management and Environment, The New School, New York, USA. [5]RMI, Carbon-Free Buildings, Boulder, USA. [6]Affordable Energy Association, New York, USA. [7]Access Sensors Technologies, Fort Collins, USA. [8]Lamont-Doherty Earth Observatory of Columbia University, Lamont Research Professor, Palisades, USA. [9]Lamont-Doherty Earth Observatory of Columbia University, Palisades, USA. [10]Department of Earth and Environmental Studies, Columbia University, New York, USA. [11]Mailman School of Public Health of Columbia University, New York, USA

Abstract
In the US, gas is the fuel predominantly used to provide residents with energy services such as cooking and heating. Yet, gas combustion generates several health-damaging pollutants, including nitrogen dioxide (NO_2) and carbon monoxide (CO), and gas stoves emit these pollutants directly into the home living area. The goal of this pilot was to evaluate the effect of a gas-to-induction transition on indoor air quality in low-income housing.

Twenty NYCHA households with gas stoves were randomized to an intervention (replacement of gas stove with induction stove) or a control arm. Between October 2021 and July 2022, three 168-hr long monitoring campaigns were conducted to assess baseline and end-line in 1) indoor air quality (NO_2, CO, and $PM_{2.5}$) using the Home Health Box (Access Sensor Technologies, Fort Collins, CO) and passive Ogawa NO_2 samplers and 2) stove usage using stove use monitors (Wellzion). Additionally, indoor air quality was further evaluated during controlled cooking tests (CCTs) carried out in both gas and induction homes with high-precision instruments for NO_2 (2BTech) and CO_2 and CH_4 (LI-COR).

Between baseline and end-line, 168-h average NO_2, CO, and $PM_{2.5}$ concentrations decreased in both study arms, likely due to seasonality factors. Adjusted models suggest that daily NO_2 concentrations were 35.7% (95% CI: -46.5, -22.7%) lower in induction homes compared to gas homes in the post-intervention period. During CCTs, the median background NO_2 concentration (18 ppb) in gas kitchens rose to 197 ppb. Comparatively, in induction kitchens, the median NO_2 concentration negligibly changed (from 11 to 14 ppb).

688 – Ambient air pollution and serum cardiometabolic proteome in young adults with obesity history

Jiawen Liao[1], Wu Chen[1], Chenyu Qiu[2], Jesse Goodrich[3], Elizabeth Costello[2], Carmen Jiawen Chen[4], Tanya L. Alderete[5], Lida Chatzi[2], Frank Gilliland[2], Zhanghua Chen[1]
[1]Department of Population and Public Health Sciences, Keck School of Medicine of the University of Southern California, Postdoc, Los Angeles, USA. [2]Department of Population and Public Health Sciences, Keck School of Medicine of the University of Southern California, Los Angeles, USA. [3]Department of Population and Public Health Sciences, Keck School of Medicine of the University of Southern California, Assistant Professor, Los Angeles, USA. [4]Department of Population and Public Health Sciences, Keck School of Medicine of the University of Southern California, PhD student, Los Angeles, USA. [5]Department of Integrative Physiology, University of Colorado Boulder, Assistant Professor, Boulder, USA

Abstract
Current evidence of air pollution exposure and blood proteomic markers is mostly based on a small number of markers (such as TNF-a and IL6) and have demonstrated mixed results. In this study, we aim to investigate proteomic biomarkers involved in cardiometabolic disease onset for their associations with air pollution exposure.

In a cohort of young adults with a history of obesity in childhood, serum levels of 384 cardiometabolic-related proteins were measured by Olink Cardiometabolic Explore panel at baseline (N = 132, age: 19.4 ± 1.3 years) and follow-up (N = 86, age: 24.0 ± 0.8 years). Exposure to air pollutants including $PM_{2.5}$, PM_{10}, NO_2 and O_3 were estimated at 1-month and 1-year average, respectively. Marginal linear models using generalized estimating equation were used to estimate the associations between air pollution exposure and the proteome, controlling for covariates. Over representative analysis (ORA) and Ingenuity Pathway Analysis (IPA) were conducted to assess enriched pathways, identify diseases and biofunctions, respectively.

We found that air pollution exposures were associated with the cardiometabolic proteome. 1-month NO_2,1-month $PM_{2.5}$ and 1-year PM_{10} air pollution were significantly associated with IL-1 receptor like-1, plasminogen activator and Amyloid-like protein 1 (FDR < 0.05). ORA using Gene Ontology found enriched pathways of collagen-containing extracellular matrix and insulin like growth factor binding (FDR < 10E-5). IPA identified significant activation of blood platelets, angiogenesis, and immune response of cells significantly enriched with air pollutant exposures (Z-score > 2, FDR < 10E-3).

Ambient air pollution exposure perturbated the cardiometabolic proteome related to cell signaling, immune responses, and inflammation.

689 – Changing process of community engagement for environmental health research in low- and medium-income countries – Example from Nigeria

Adetoun Mustapha
Nigerian Institute of Medical Research, Adjunct Research Fellow, Lagos, Nigeria

Abstract
The process for engaging communities to gain support for environmental health research is changing in low- and medium-income (LMIC) countries for various reasons that include increasing literacy levels, more access to information through the internet and local administration involvement in getting ethics approval for environmental health studies. This talk will examine the challenges and opportunities of engaging different communities for exposure science studies, using example of a study of air pollution and respiratory health in schoolchildren in Warri, Delta State, Nigeria in 2004 and another study of ambient air pollution levels and seasonal variation using low-cost sensors in Lagos State, Nigeria in 2023. The requirements to include adequate resources and time for community engagement and the need to understand the power structure, traditional rights and cultural sensitivities in different populations to secure social approvals for environmental health research will be discussed. The emerging community engagement model is enabling the inclusion of local knowledge in study designs, fostering community participation in decision-making and formulating public health interventions and facilitating translation of research findings into policy actions.

690 – POSTER: Suspect and non-target screening of chemicals in household cleaning products, and their toxicity assessment

Jeonghoon Park[1], Hyejin Yun[2], Chungsik Yoon[1], KYUNG-DUK ZOH[3]
[1]Seoul National University, PhD Candidate, Seoul, Korea, Republic of. [2]Seoul National University, MS student PhD Candidate, Seoul, Korea, Republic of. [3]Seoul National University, Professor, Seoul, Korea, Republic of

Abstract
In this study, 31 household spray-type cleaning products from 10 categories (bathroom/toilet, kitchen, glass, metal, air conditioner, automobile, carpet, multipurpose, floor, and bowling ball) were examined to identify the substances contained in these products and evaluate their resulting toxicities. After substance detection via LC-QTOF-MS, suspect and non-target screenings were performed using UNIFI software. A total of 48 substances were identified during suspect screening. Among them, surfactant (freq. 22/31(71.0%)), plasticizer (freq. 22/31(74.2%)), antioxidant (freq. 20/31(64.5%)) were detected with high frequencies. Among the compounds, tributyl citrate acetate was the most frequently detected (frequency 13/31, 41.9%). In addition, 51 substances were identified via non-target analysis. Among them, oleamide, a type of surfactant, showed a high frequency of detection (22/31(71.0%)). The toxicities of the 98 substances (suspect: 47 (except for confidence level 5) and non-target: 51) identified were evaluated using the European Chemicals Agency (ECHA) database and Toxtree. Among substances identified, 20 were classified as toxic. The suspect and non-target screening technique applied in this study proved to be a useful method that can be applied when there are many non-specific chemicals and their components in the consumer products are not well known.

693 – Measuring participant exposure to terpenes in a forest bathing clinical trial

Chaja M. Levy[1], Christopher D. Simpson[2], Anne M. Riederer[3], Deepak Bhandari[4], Lalith K. Silva[5], Benjamin C. Blount[6], Gregory N. Bratman[7]
[1]University of Washington, Research Assistant, Seattle, USA. [2]University of Washington, Professor, Seattle, USA. [3]University of Washington, Clinical Associate Professor, Seattle, USA. [4]CDC/National Center for Environmental Health, Chief, VOC Biomarkers Laboratory, Atlanta, USA. [5]CDC/National Center for Environmental Health, Research Chemist, Atlanta, USA. [6]CDC/National Center for Environmental Health, Chief, Tobacco and Volatiles Branch, Atlanta, USA. [7]University of Washington, Assistant Professor, Seattle, USA

Abstract
Forest bathing, and other time spent in forests, is associated with health benefits including reduced stress, anxiety and blood pressure. We are conducting a randomized crossover trial to investigate whether inhalation of terpenes and other volatile organic compounds (VOCs) present in forest air plays a role. Forty adult volunteers complete two 60-min seated forest bathing sessions in old growth forest near Mt. Rainier (WA, USA) wearing powered air purifying respirators (PAPRs). PAPRs contain VOC+particle filters in one session and sham (particle only) filters in the other, with sequence randomized for each participant and an 8-day minimum washout period between sessions. The trial is powered to detect the effect of forest VOC inhalation on the anticipated increase in heart rate variability, an indicator of parasympathetic nervous system activation. Field work concludes September 2023. We are also measuring α- and β-pinene and other forest terpenes in air and serum to characterize absorbed dose. Air samples are collected for 4+ hr during the session using thermal desorption tubes and low-flow pumps. Tubes are analyzed for 22 terpenes using GC/MS at Eurofins (Edmonton, Canada). Serum is collected immediately before and after the session and analyzed for α- and β-pinene, β-myrcene, D-3-carene, limonene, β-carophyllene, and α-humulene at CDC (Atlanta, USA). Preliminary results show α- and β-pinene and other terpenes at detectable levels in both air and serum. In this presentation, we share forest air and serum terpene concentrations and absorbed dose estimates from this first NIH-funded clinical trial of forest bathing and health.

694 – Using proton transfer reaction time-of-flight mass spectrometry for exhaled breath analysis in a community environmental health study

Sa Liu[1], Jung Hyun Lee[2], Marwan Alajlouni[2], Alaina Bryant[2]
[1]Purdue University, Assistant Prof., West Lafayette, USA. [2]Purdue University, Graduate Student, West Lafayette, USA

Abstract
Chlorinated volatile organic compounds (CVOC), such as tetrachloroethylene (PCE), trichloroethylene, and vinyl chloride, are ubiquitous, legacy contaminants, with widespread occupational and community exposures. Evidence of the associations between exposure and health effects of these compounds are primarily from occupational exposures and animal studies, leaving the health impact of community exposure inadequately characterized. One of the major challenges has been quantifying low levels of exposure among individuals living near contamination sites. We conducted a pilot cross-sectional study to evaluate the associations between CVOC exposure and neurobehavioral performance among children living near multiple contamination sites with CVOC being the primary contaminants. We used a cutting-edge assessment tool, proton transfer reaction time-of-flight mass spectrometer (PTR-TOF-MS) to analyze CVOC in exhaled breath, a novel method to characterize the current exposure. The number of years living in contaminated sites is to be developed as a surrogate of the cumulative exposure. Outcome was evaluated by the NIH Toolbox, the Wide Range Achievement Test Fifth Edition, and the Behavior Assessment System for Children Third Edition. Among seventy-three participants, PCE exposure as measured by PTR-MS ranged from 2.04 to 150 count per second (cps), with a mean of 17.3 and a standard deviation of 18.2 (cps). Children's cognitive function, problem-solving ability, adaptability, and academic performance were found to be negatively associated with the current PCE exposure. Data analysis is ongoing. Exhaled breath analysis with PTR-TOF-MS appeared to be a promising tool for field assessment in community environmental health field research.

695 – Houston Data to Action: Community Air Monitoring

P. Grace Tee Lewis[1], Bridgette Murray[2], Tracy Stephens[3], Rachel White Roy[4], Denae King[5]
[1]Environmental Defense Fund, Senior Health Scientist, Houston, USA. [2]Achieving Community Tasks Successfully, Executive Director, Houston, USA. [3]Achieving Community Tasks Successfully, Director of Research, Houston, USA. [4]Achieving Community Tasks Successfully, Epidemiologist, Houston, USA. [5]Texas Southern University, Bullard Center for Climate and Environmental Justice, Associate Director, Houston, USA

Abstract

Historical redlining policies and lack of zoning in Houston fostered clustering of chemical manufacturing, refining facilities, and other industrial sources in environmental justice (EJ) communities. These facilities emit not only criteria air pollutants but also highly reactive air toxics, some which are recognized as known carcinogens by the International Agency for Research on Cancer. In light of extensive and well documented history of non-compliance at chemical manufacturing facilities in Harris County, there is an urgent need to understand air pollutant exposures posing health risks to overburdened, disadvantaged EJ neighborhoods. State and federal regulatory air monitor networks are insufficient in meeting the needs of these communities as existing sensors are almost never located in their residential areas. As a result communities have been empowered to create their own air sensor networks to generate the data needed to demand action by government.

We will present a case study of community air monitoring network designed, deployed, and operated by Achieving Community Tasks Successfully (ACTS) in Pleasantville, Houston, Texas to illustrate how community science can be used for advocacy and to address residents' desires for better understanding of exposures to improve health. We will share the participatory action research approach used to develop their network, lessons learned, data considerations, how to make data actionable, challenges, solutions, and successes in undertaking this work.

697 – Trace element exposure and gestational diabetes mellitus: Evidence from regions in the USA and China

Qiurun Yu[1], Hongcheng Wei[1], Yuqing Zhang[2], Mingzhi Zhang[1], Yankai Xia[1]
[1]Nanjing Medical University, Nanjing, China. [2]Women's Hospital of Nanjing Medical University, Nanjing, China

Abstract
Background: Trace element is found to be associated with metabolic disorders but the associations for gestational diabetes mellitus (GDM) remain unclear. Furthermore, a few research has compared the effects across regions.

Objectives: To explore the associations between trace element exposure and GDM.

Methods: All participants were drawn from three surveys, 2005-2016 National Health and Nutrition Examination Survey (NHANES) in USA, and two prospective cohorts conducted in Nanjing and Suzhou, respectively. In both surveys, fasting plasma glucose and 2-h plasma glucose after oral glucose tolerance test were performed for GDM diagnosis, and concentrations of trace metals in urine were measured. The generalized linear regression model, weighted quantile sum (WQS) and Bayesian kernel machine regression (BKMR) were jointly conducted to investigate the association between trace element exposure and GDM.

Results: A total of 8501 participants from NHANES, 837 participants from Nanjing cohort, and 902 participants form Suzhou cohort were included for data analysis. In NHANES, we found exposure to hydrargyrum and arsenic were significantly associated with GDM. Combined the results from Nanjing and Suzhou cohorts, arsenic and molybdenum were risk factors for GDM, and cobalt exposure hag a certain protective effect on GDM. No interaction is found between the GDM-related key trace elements.

Conclusion: Exposure to trace elements can contribute to GDM, and arsenic is an essential risk factor both in the UAS and China. Further research is needed to validate this result and to explore the involved mechanisms to facilitate early identification and intervention of GDM-related risk factors.

698 – An Order of Operations for Multidisciplinary Community-Based Traditional Knowledge (TK) Research

Jonathan Dellinger
UW-Whitewater, Assistant Professor, Whitewater, USA

Abstract
This paper explores, from a non-indigenous academic perspective, best practices for developing culturally responsive research by informing empirical and biomedical research with traditional/indigenous forms of knowledge (TK). Federally funded research in collaboration with Native American tribes faces a need to interface with cultural competency and tribal self-determination. These goals are often paired with a desire to integrate "Traditional Knowledge" and "indigenous methods" in research. Such efforts can serve as positive forces to promote Native American self-determination in health research, if properly integrated with the scientific method. Thus, TK is an important starting point for defining the outcomes researchers seek to test empirically.

A key hazard of translating TK to biomedical and empirical research is the conflation of postmodern philosophy and critical theoretical methodology with science (e.g. post-positivist methodologies and experimentation). Due to postmodernism's association with advocacy, it is ill-suited to hypothesis testing. While both subjectivist and objectivist methodologies are indispensable for addressing issues of the human condition such as health disparities, these methodologies use different approaches to answer different questions and cannot be used interchangeably. I propose a methodological order of operations for multidisciplinary research in which TK informs hypothesis generation and testing, which in turn can influence policy decisions and advocacy. Establishing such an order of operations is essential to avoid cultural appropriation, reap the benefits of the empiric method, and ensure the integrity of Native American and academic partnerships.

701 – Inverse Associations between circulating PFAS and accelerated biological aging in a general statewide population-based sample

Kristen Malecki[1], Amy Schultz[2], Rachel Pomazal[3], Joseph Clark[4], Michal Engelman[5]
[1]University of Illinois at Chicago, Professor, Division Director, Chicago, USA. [2]University of Wisconsin Madison, Scientists, Madison, WI, USA. [3]University of Wisconsin Madison, Student, Madison, WI, USA. [4]University of Wisconsin Madison, Post Doc, Madison, WI, USA. [5]University of Wisconsin Madison, Professor, Madison. WI, USA

Abstract
Perflourinated compounds (PFAS) have been found in multiple environmental media and associated with numerous adverse outcomes including immune dysregulation, cancer and adverse birth outcomes. Despite known endocrine disrupting properties, recent research has found variable associations with aging related biomarkers and PFAS exposure among general population based samples. Using a sub-sample of 253 participants from the Survey of the Health of Wisconsin, a randomly selected state-wide representative sample, we examine associations between 38 different PFAS and whole-blood based DNA methylation analyzed on 850 K EPIC Chip Array. Several "biological aging clocks" derived from variable constellations of CpGs were examined including Pheno-Age determined on lab based indicators of aging (e.g. total cholesterol, glucose etc.), Grim Age modeled using pre-mature mortality, and a newer DunedinPACE age representing accelerated page of aging. Perfluoro-n-octanoic acid (PFOA), Perfluoro-n-octanesulfonic acid (PFOS), and Perfluoro-n-hexanesulfonic acid (PFHXS) were detected in more than 99% of samples. Adjusted and unadjusted linear regression models (covariates: age, sex, education, income, white blood cell count, survey year) were used to examine log transformed PFAS concentrations. Adjusted models found inverse associations between PFOA with Dunedin (-.04, p<.00) and Grimage (-1.0, p <.01) but not Pheno-Age; PFOS with Dunedin (-.04, p<.00) and PhenoAge (-1.1, p<.03) but not GrimAge; and, PFHXS with all three Dunedin (-.03, p<.00), PhenoAge (-.96, p<.0 and GrimAge (-.67, p<.02). Results are consistent with recent findings. Further examination is needed to understand the complex associations between PFAS exposures and specific differential DNA methylation changes beyond biological aging clocks.

702 – A multi-class method for the analysis of environmental chemicals in urine by high-performance liquid chromatography-tandem mass spectrometry

Kurunthachalam Kannan[1], Hongkai Zhu[2], Sridhar Chinthakindi[2]
[1]Wadsworth Center, Research Scientist, Albany, USA. [2]Wadsworth Center, Research Affiliate, Albany, USA

Abstract
Biomonitoring of human exposure to environmental chemicals has gained momentum in recent years. Biomonitoring methods often include analysis of a single class of chemicals with similar chemical properties. In this study, we describe a method that involves solid-phase extraction (SPE) coupled with liquid chromatography-tandem mass spectrometry (LC-MS/MS) and capable of measuring 121 environmental chemicals comprising plasticizers (PMs; n=45), environmental phenols (EPs; n=45), and pesticides (n=31) through a single extraction of urine. Urine samples were incubated with 20 µL of β-glucuronidase/arylsulfatase (4000 units/mL urine) (from Helix pomatia) buffered at pH 5.5 for 2 h at 37oC for optimal deconjugation conditions. We compared two extraction methods, namely liquid-liquid extraction and SPE, and the latter with ABS Elut NEXUS® cartridges was optimized to yield best extraction efficiencies. For increased resolution and chromatographic separation, two methods involving Ultra AQ C18® and BetasilTM C18® columns were used. The MS/MS analyses were performed under both negative and positive ionization modes. The optimized method yielded excellent intra- and inter-day variabilities (relative standard deviation: 0.40–11%) and satisfactory recoveries (80–120%) for >95% of the analytes. The limits of detection were ≤ 0.1 for 101 analytes and between 0.1 and 1.0 ng/mL for 18 analytes. The optimized SPE LC-MS/MS method was validated through the analysis of standard reference materials and proficiency test urine samples and further applied in the analysis of 21 real urine samples to demonstrate simultaneous determination of 121 environmental chemicals in urine samples.

1085 – DEFAULT VALUES AND READ-ACROSS PROPOSAL FOR DERMAL ABSORPTION BASED ON HUMAN SKIN IN VITRO STUDIES DATASET WITH PESTICIDES

Danilo Sarti[1], Jörg Wagner[2], Fabiana Palma[3], Heloisa Kalvan[3], Marcela Giachini[4], Daniele Lautenschalaeger[5], Vanessa Lupianhez[5], Janaina Pires[6], Marina Sales[7], Patricia Faria[8], Lisa Bertomeu[9], Marine Le Bras[10]
[1]Consultant, MAYNOOTH, Ireland. [2]Knoell Germany GmbH, Mannheim, Germany. [3]Instituto ProHuma de Estudos Cientificos, Jundiai, Brazil. [4]Corteva Agriscience,, Barueri, Brazil. [5]Bayer CropScience, São Paulo, Brazil. [6]Ouro Fino Quimica S.A, Ribeirão Preto, Brazil. [7]Syngenta Protecao de Cultivos, São Paulo, Brazil. [8]BASF S.A., São Paulo, Brazil. [9]Knoell France SAS, Lyon, France. [10]Knoell Consult Schweiz GmbH, Basel, Switzerland

Abstract
The skin is the main route of exposure to plant protection products for operators, workers, residents, and bystanders. Assessing dermal absorption is key for evaluating pesticide exposure risks. However, conducting in vitro dermal absorption studies using human skin is resource-intensive. Therefore, establishing realistic regulatory default values is essential for non-dietary risk assessment and ensuring safety. This project analysed 468 in vitro dermal absorption studies with human skin compiled by ProHuma. Combined with a prior database by Aggarwal et al. (2015), the dataset comprised 759 GLP-compliant studies covering 25 formulation types and 248 active substances at various concentrations. The assessment confirmed earlier findings that dermal absorption shows no correlation with molecular weight or log Pow and does not increase linearly with dilution. Based on the comprehensive ProHuma/ECPA dataset, dermal absorption default values are proposed: 2% for solid concentrates, 5% for liquid concentrates, 30% for solid dilutions, and 34% for liquid dilutions. These values, derived conservatively using the 95th percentile, better represent formulations and agricultural use cases, reducing the need for unnecessary studies. The project also suggests read-across between different formulation types. By proposing more realistic default values and read-across approaches, it streamlines decision-making and supports safety in agriculture.

1087 – Exploring the Effects of Climate Change on Fungal Contamination, Azole Resistance,and Mycotoxins in Primary Schools

Renata Cervantes[1,2], Pedro Pena[1,2], Bianca Gomes[1,3], Marta Dias[1,2], Carla Viegas[4,5]
[1]H&TRC—Health & Technology Research Center, ESTeSL—Escola Superior de Tecnologia e Saúde, Instituto Politécnico de Lisboa, 1990-096 Lisbon, Portugal, researcher, Lisbon, Portugal., Researcher, Lisboa, Portugal. [2]NOVA National School of Public Health, Public Health Research Centre, Comprehensive Health Research Center, CHRC, NOVA University Lisbon, Portugal, Researcher, Lisboa, Portugal. [3]CE3C—Center for Ecology, Evolution and Environmental Change, Faculdade de Ciências, Universidade de Lisboa, 1749-016 Lisbon, Portugal I, Researcher, Lisboa, Portugal. [4]H&TRC—Health & Technology Research Center, ESTeSL—Escola Superior de Tecnologia e Saúde, Instituto Politécnico de Lisboa, 1990-096 Lisbon, Portugal, researcher, Lisbon, Portugal., Full professor and Researcher, Lisboa, Portugal. [5]NOVA National School of Public Health, Public Health Research Centre, Comprehensive Health Research Center, CHRC, NOVA University Lisbon, Portugal, Full professor and Researcher, Lisboa, Portugal

Abstract
Climate change amplifies indoor fungal growth, driven by altered temperature and moisture, escalating mycotoxin production and resistant strains [1]. Children's extended indoor exposure, especially in schools, raises health concerns [2]. Fungi impact indoor air quality and pose risks, including mycotoxin-related health issues [3]. Our goal is to assess azole-resistant fungi and mycotoxin exposure in European schools, informing risk management. We'll characterize contamination and resistance, evaluate mycotoxins, study health impacts, and address climate-related factors. Aiming to create an Integrated Risk Assessment tool, we'll connect sources, exposures, and health outcomes, enabling effective contamination control and risk reduction, safeguarding children's well-being

1088 – Association between Air Pollution and Sleep Efficiency in 6 - 7 years old Children living in East Harlem

Terry Thompson[1], Cecilia Alcala[1], Malika Garg[1], Maida Galvez[2]
[1]Icahn School of Medicine at Mount Sinai, Assistant Professor, New York City, USA. [2]Icahn School of Medicine at Mount Sinai, Professor, New York City, USA

Abstract
Environmental toxicants, and exposures to pesticides may adversely impact children's growth and development. Children are particularly vulnerable to environmental exposures because of the sensitive nature of children's neurodevelopment. Yet, little is known about the relationship between air quality and sleep efficiency among children. To address this knowledge gap, this research team will work closely with residents of East Harlem, a predominantly low-income, Black and LatinX community to examine potential relationships

between indoor air quality and its effects on sleep patterns among Hispanic children, 6-7 years old living in East Harlem, NY. Specific research aims are: 1) Convene and engage with a Community Advisory Board (CAB) quarterly to elicit meaningful input on study design, implementation and dissemination of how air quality impacts child sleep. 2) Administer two study instruments, (questionnaire, diary) in Spanish and English to household caregivers to assess sleepiness and sleep quality patterns of child sleep behavior. 3) Quantify PM2.5 exposure using indoor air monitors, and children's 24-hour activities wearing continuous wrist-worn Actigraph watches over 7 days, to examine potential associations between air quality and sleep. We will apply Community Advisory Board (CAB) involvement and community-partnered approaches throughout all stages of this study to ensure the needs of the study population are met. The research team hopes that meaningful application of this approach will deepen community trust for future research projects.

1089 – SCIENTIFIC APPROACHES IN DEVELOPMENT OF PESTICIDES OCCUPATIONAL RISK ASSESSMENT IN BRAZIL

Fabiana Palma[1], Thiago Santos[2], Daniele Lautenschalaeger[3], Marcela Giachini[3], Jeane Fonseca[2]
[1]Instituto ProHuma, Jundiai, Brazil. [2]ANVISA, Brasilia, Brazil. [3]Instituto ProHuma, Sao Paulo, Brazil

Abstract
Over the last few years, Brazil has been discussing the implementation of Occupational Risk Assessment, a new legislation including a technical Guidance is being developed by the Brazilian Health Agency – ANVISA and will be implemented soon. In this scenario of technical discussion, the ProHuma Institute was created to support the data development, scientific projects and knowledge exchange.

Due to this intense scientific discussion, a five year-Technical Cooperation Agreement was signed in August 2020 between the Brazilian Health Agency - ANVISA and ProHuma with the objective of developing an occupational exposure database representative for the Brazilian operator, including the joint performance technical activities, aiming to subside the assessment of occupational risk, and residents and bystanders to pesticides in Brazil. The following projects are concluded or in progress: (i) Technical Publications – Data generation on the Brazilian Agricultural Scenario; (ii) Development of the Brazilian Occupational Exposure Database; (iii) Default values for area treated per day for both tractor and aerial application; (iv) Proposition of transfer coefficients (TC) for adoption in the assessment of re-entry workers exposure; (v) Technical events to joint a theoretical and experiential knowledge; (vi) Proposal of default values and read-across approach for dermal absorption based on in vitro studies on human skin; (vii) conducting local Exposure Studies.

This technical partnership enables a collaborative arrangement and increase of knowledge, supporting the implementation of the Brazilian Risk Assessment for Operators, Workers, Residents and Bystanders. The outcome is the enhancement in human health safety and overall society benefit.

1090 – Association between urinary haloacetic acid concentrations and liver injury among women: Results from the Tongji Reproductive and Environmental (TREE) study

Yanling Deng[1], Qiang Zeng[2]
[1]Emory University, postdoc, Atlanta, USA. [2]Huazhong University of Science and Technology, professor, Wuhan, China

Abstract
Experimental studies have shown that disinfection byproducts (DBPs) including haloacetic acids (HAAs) can cause liver toxicity, but evidence in humans is sparse. We aimed to explore the associations between HAA exposures and liver injury. We included 922 women between December 2018 and January 2020 from the Tongji Reproductive and Environmental (TREE) cohort study in Wuhan, China. Urinary HAA concentrations including trichloroacetic acid (TCAA) and dichloroacetic acid (DCAA) and serum indicators of liver function including alanine aminotransferase (ALT), aspartate aminotransferase (AST), and glutamine transaminase (GGT) were measured. Liver injury was defined as if any of serum indicator levels were above the 90th percentile. Multivariate logistic and linear regression models were fitted to assess the associations of urinary HAA concentrations with the risk of liver injury and liver function indicators. Urinary DCAA concentrations were associated with an elevated risk of liver injury [odds ratio (OR) for one-unit increase in ln-transformed DCAA concentrations: 1.82; 95% confidence interval (CI): 1.12, 2.96]. This association was only observed among non-drinkers (P for interaction=0.025). We also found that one-unit increase in ln-transformed DCAA concentrations was positively associated with ALT levels (percentage change: 9.81%; 95% CI: 0.82%, 19.59%) and negatively associated with AST/ALT (percentage change: -7.03%; 95% CI: -12.08%, -1.69%). In addition, urinary DCAA concentrations in relation to increased GGT levels was only observed among passive smokers (P for interaction=0.044). Our findings suggest that exposure to DCAA but not TCAA is associated with liver injury among women.

International Society of Exposure Science (ISES)
c/o Infinity Conference Group
1035 Sterling Road, Suite 202
Herndon VA 20170
USA

ISBN 978-1-7138-9644-9
9 781713 896449

CURRAN ASSOCIATES INC.
proceedings.com

49th International Computer Music Conference (ICMC 2024)

Sound in Motion

Seoul, South Korea
7 - 13 July 2024

International
Computer
Music
Association